Test Item File and Instructor's Resource Guide
COLLEGE ALGEBRA
FOURTH EDITION
Larson / Hostetler

David C. Falvo
The Pennsylvania State University
The Behrend College

HOUGHTON MIFFLIN COMPANY Boston   New York

Sponsoring Editor: Christine B. Hoag
Senior Associate Editor: Maureen Brooks
Managing Editor: Catherine B. Cantin
Senior Project Editor: Karen Carter
Associate Project Editor: Rachel D'Angelo Wimberly
Production Supervisor: Lisa Merrill
Art Supervisor: Gary Crespo
Marketing Manager: Charles Cavaliere

Copyright © 1997 by Houghton Mifflin Company. All rights reserved.

No part of this work may be reproduced or transmitted in any form or by any means, electronic or mechanical, including photocopying and recording, or any information storage or retrieval system without the prior written permission of Houghton Mifflin Company unless such copying is expressly permitted by federal copyright law. Address inquiries to College Permissions, Houghton Mifflin Company, 222 Berkeley Street, Boston, MA 02116.

Printed in the United States of America.

International Standard Book Number: 0-669-41755-6

123456789-VG 00 99 98 97 96

PREFACE

The *Test Item File and Instructor's Resource Guide for College Algebra, Fourth Edition*, is a supplement to the text by Roland E. Larson and Robert P. Hostetler.

We begin with a Survey of Assessment Techniques. This discusses a broad range of standard and alternative assessment strategies, including many examples and suggestions for grading the latter type.

Part 1 of the test item file is a bank of questions arranged by text section. To assist you in selecting questions and administering examinations, each question is followed by a two- or three-item code line. The first item of the code indicates the level of difficulty of the question: routine (1) or challenging (2). (Challenging questions require three or more steps to obtain the solution.) If the solution to the problem requires the use of a graphing utility, the next code is (T) for technology-required. The last item in the code is the answer (Ans).

Part 2 of the test item file includes a bank of chapter tests. For each chapter, there are three tests with multiple choice questions and two tests with open-ended questions. The tests are geared to 50-minute class periods, and the final exams are primarily intended as samples to give instructors ideas for questions.

Finally, the last part of this test item file includes the answers to the chapter tests and final exams. I have made every effort to see that the answers are correct. However, I would appreciate very much hearing about any errors or other suggestions for improvement.

Computerized versions of this test item file with User Manuals are available for the IBM-PC, IBM-compatible computers, and the Macintosh. The program has these advantages:

- Multiple-choice test questions can be converted to an open-ended format at the touch of a button.
- Each multiple-choice test question offers a choice of five possible responses, one of which is correct.
- The distracters, or incorrect answers, are designed to be common errors or to look like reasonable answers.
- Questions can be selected from the computerized version in a variety of ways to accommodate differing class needs, including previewing questions on screen, entering the question number after consulting this test item file, and selecting randomly by section, question type, or level of difficulty. In addition, common math notation symbols are available for editing or adding questions.

This Test Item File and Instructor's Resource Guide is the result of the efforts of Larson Texts, Inc.

David C. Falvo
The Pennsylvania State University
The Behrend College

CONTENTS

	Notes to the Instructor	vi
PART 1	**Test Items**	1
Chapter P	Prerequisites	1
Chapter 1	Equations and Inequalities	52
Chapter 2	Functions and Their Graphs	126
Chapter 3	Zeros of Polynomial Functions	200
Chapter 4	Rational Functions and Conics	244
Chapter 5	Exponential and Logarithmic Functions	275
Chapter 6	Systems of Equations and Inequalities	304
Chapter 7	Matrices and Determinants	348
Chapter 8	Sequences and Probability	391
PART 2	**Chapter Tests and Final Exams**	439
PART 3	**Answer Keys to Chapter Tests and Final Exams**	546

Notes to the Instructor

ASSESSMENT

The purpose of assessment in mathematics is to improve learning and teaching. A key component in contemporary education, assessment must be used to broaden and inform, not restrict, the educational process. This view is well stated in the NCTM *Curriculum and Evaluation Standards for School Mathematics*.

"Assessment must be more than testing; it must be a continuous, dynamic, and often informal process." (p. 203)

According to the Mathematical Sciences Education Board (MSEB), there is a need for mathematics assessments that accomplish several goals.
• Promote the development of mathematical power for all students
• Measure the full range of mathematical knowledge, skills, and processes specified by the national Council of Teachers of Mathematics (NCTM) curriculum standard
• Communicate to students and teachers what mathematics students already know, as well as the mathematics they have yet to learn

Formal Assessment Strategies

College Algebra, Fourth Edition, is a comprehensive teaching and learning package. The text and an array of text-tied supplements offer a variety of vehicles for progress assessment.

In the text, Chapter Tests, Cumulative Tests, and Focus on Concepts are designed to help measure retention and understanding. *Interactive College Algebra: A Self-Guided Study Companion on CD-ROM for Windows* offers pre-, post-, and self-tests with answers. The supplements package for *College Algebra*, Fourth Edition, offers substantial resources to help you prepare formal tests and quizzes. This *Test Item File and Instructor's Resource Guide* contains many test items with answers for each section of the text. These multiple-choice and open-ended test items are organized by text section and coded as routine or challenging and technology-required if appropriate. A bank of Chapter Tests (Forms A through E) and final exams with answer keys completes the *Guide*. The test questions are also available in a computerized format, offering the instructor the opportunity to create customized tests. The *Study and Solutions Guide* is particularly useful for student self-assessment.

Alternative Assessment Strategies

In addition to the formal assessment strategies, various forms of alternative assessment may help the instructor (and student) reveal the student's capabilities, as well as gather information to determine what students "know" (understand) as compared to what students "know how to do" in mathematics. Alternative assessment is a means for evaluating student progress using non-traditional assessment tools. The information gathered by alternative assessment can be used in various ways. Students can use it to appraise themselves and their mathematical abilities; instructor can use it to make informed decisions about the instruction of their students. Some formats for informal and ongoing assessment are: student-created problems (problem posing), journals and portfolios, projects or research assignments, and demonstration or performance assessment.

There are several advantages to using a variety of measures of achievement.
• One type of assessment cannot serve all informational needs.
• Receiving information from multiple sources leads to more informed decision making.
• Traditional paper-and-pencil tests may be incomplete measures of achievement.
• Using a variety of assessment methods enables the instructor to obtain a more equitable measure of a student's mathematical progress with less potential for biases than traditional measures.

The goals of alternative assessment are to:
• find out what the student already knows.
• evaluate the depth of the student's conceptual understanding and his or her ability to transfer this understanding to new and different situations.

- evaluate the student's ability to communicate his or her understanding mathematically, make mathematical connections, and reason mathematically.
- plan the mathematics instruction in order to achieve the objectives.
- report individual student progress and show growth in mathematical maturity.
- analyze the overall effectiveness of the mathematics instruction.

Scoring

There are several ways to score alternative assessment assignments. The most simplified scoring involves grading the assignment on a scale of 0–3. If the student gives a clear, coherent explanation with appropriate diagrams or graphs, a score of 3 is given; if the student's work shows understanding of the assignment but contains computational errors and/or minor flaws in the explanation, a score of 2 is given; if the work contains serious conceptual errors in addition to flagrant computational errors, a score of 1 is given; and if little or no attempt is made to complete the assignment, a score of 0 is given.

Formats

Problem Posing Students can be asked to create their own problems for tests, with answers. They can then exchange questions and check each other's answers. Asking students to create review sheets can also provide a means for you to assess student understanding of the essentials. Note that many of the in-text Group Activities encourage students to create problems or correct solutions.

Journals and Portfolios Students can keep mathematics portfolios of problems and write journals of attitudes, understandings, and applications of mathematics as part of an instructor's assessment strategy. Students can be asked to write explanations for others; they can identify and explain mathematical ideas; and they can design exercises to find out whether someone understands a mathematical relationship. Following are suggestions for Journal and Portfolio topics.

<u>Attitudes</u> What is the best experience the student has ever had in a math class? What did the student like best/least about math class this week? Of all the resources available to help students understand math (CD-ROM, videos, tutorial software, study guide, solutions guide, and so on), which did they find most useful and why? How would the student evaluate his or her own progress this week? Did any of the applications based on real data interest the student, and, if so, why?

<u>Understanding</u> Ask students to compare and contrast, and explain mathematical relationships as appropriate. Students could be asked to choose an exercise that has given them trouble, and describe the approaches they used to try to solve the problem. Writing can encourage more active learning.

<u>Applications</u> Students might be asked to create one of each basic type of problem, together with its solution, for a section. Compare and contrast problem-solving strategies used. Some instructors have successfully used concept maps: having students list all the concepts they consider important in a section and draw arrows to indicate the relationships among the ideas.

<u>Explanations</u> Have students explain to a friend, in writing, how to solve a particular problem. What advice would they give to help him or her avoid errors?

<u>Exercises</u> Ask students to keep all or some instructor-selected homework exercises in a notebook. Collect this Portfolio occasionally, and grade a subset of the problems. You may also choose to give a grade for general effort based on how many of all of the problems were tried.

<u>In-class activities</u> A summary of in-class problems can also be part of the portfolio, as well as review handouts from the instructor.

Projects Projects provide the student with the opportunity to express his or her mathematical ability in a format other than paper-and-pencil tests. Some sources for projects include the following.

<u>Reports</u> Students can be asked to research and write a report on the contributions of a mathematician. The Historical Notes in the text introduce several mathematicians and their work as well as relevant mathematical artifacts, and students can be asked to elaborate on this material.

<u>Chapter Projects</u> The multi-part Chapter Projects can be assigned to groups, pairs, or individual students. Students might be asked to present the findings of the project to the class.

Demonstration or Performance Assessment   Instructors can assess student comprehension by affording them the opportunity to explain mathematical concepts in their own words—using tables, graphs, diagrams, and examples—for the entire class or within a small-group setting. Students may reveal their level of comprehension, show their ability to transfer skills, demonstrate their computational skills, and so on.

An instructor might look for opportunities to further involve students in the learning process. For example, if technology (calculator, graphing calculator, or computer) is being used in conjunction with an overhead viewing device, consider asking a student to press the keys to help demonstrate simple ideas. This can encourage active class participation, while the instructor monitors the discussion.

As you begin to integrate alternative assessment into your course, it is important to start slowly to avoid becoming overwhelmed. Try one form of alternative assessment, then add others to your repertoire. As you select problems from the text or design your own, you might consider characteristics such as open-ended response, multiple correct answers, no stringent time limits, and opportunities for learning during the assessment process. Take your time, and enjoy what you are doing. A combination of formal and informal assessment can help you evaluate each student most effectively.

COOPERATIVE LEARNING

Benefits of Cooperative Learning

Cooperative learning, or small group instruction, is an effective tool for achieving many instructional goals. Group work can be done in pairs, or in groups of as many as six individuals. Often thought of as peer tutoring or teaching, group work can also be used for group investigations and activities. In the text, you will find Group Activities and Chapter Projects that are designed for use in a cooperative learning setting, although some may also be appropriate for individual assignments. In addition, the more challenging exercises, technology-required exercises, and Explorations are also good problems for group work.

When making small-group assignments, consider having each group in the class work on the same problem, then join in a class-wide discussion of the results. Alternatively, consider assigning different parts of a multi-part problem to each group, then have the groups share all results for analysis and interpretation by the entire class.

Group work can empower students and enhance self-esteem. When groups are formed at the beginning of the course, students often report that they enjoy getting to know members of the group well. Group members can learn to be responsible for each other, phoning other students to share notes or picking up copies of handouts for an absent classmate. If groups change during the term, students enjoy meeting new members.

Group work encourages students to focus on learning, instead of merely earning grades. Students often feel that they can ask questions or admit confusion to peers in groups. Students may listen better to a peer explanation since they tend to communicate more directly and effectively with each other.

Group work teaches the vital life skills of cooperation, respect for others, listening, and speaking that are used in committees and working groups in the community and on the job. The instructor can monitor in-class group work dynamics, adjusting the focus and mediating as necessary.

Group work can help meet diverse needs in a class. Students working with partners can share complementary skills. For example, a student who grasps the algebra but for whom English is a second language can work with a student whose English is good, but who needs help with the mathematics.

Managing the Cooperative Learning Classroom

Instructors must be aware of two possible characteristics of groups that may make them less effective: the group may have trouble staying on task and one student may dominate by doing the work while the others copy. To eliminate the first problem (if the group work occurs during class), circulate around the room, listen, and adjust the focus of each group as necessary. To help with the second problem, talk to the dominant students. Help them to realize that they learn by explaining to others as they formulate, organize, justify, and express their skills and knowledge. Give these high achievers recognition for helping others. Let students know that when anyone else in the group arrives at a new understanding, all members can be proud of the achievement.

It can help students to give them general guidelines to consider when working on problem solving in groups.
1. Agree on what you have to do.
2. Make a plan.
3. Listen to each other's ideas.
4. Praise each other's ideas and see if you can build on another's idea.
5. Ask for help when you need it; give help when asked. If you finish your part and others are still working, volunteer your help.
6. Finish the project together. Proofread your work.
7. Discuss what you did well together and what you could do differently next time.

MULTICULTURAL ISSUES

Students should be aware of the rich cultural heritage of mathematicians throughout the centuries. They also should be aware that mathematics is created and used by people like themselves. Throughout *College Algebra*, Fourth Edition, Historical Notes and real-life applications can help promote this awareness.

LANGUAGE ISSUES

Students for whom English is a second language, or who have limited English proficiency, may need help from the instructor and fellow students to express their interest and abilities in mathematics. As an instructor, consider the following suggestions.
- At the beginning of each class (or during the first class of the week), outline the material you plan to discuss.
- When you use a key math term for the first time, and when you use a key term that you feel should be emphasized, write it on the board.
- When you write a definition or example on the board, review sheet, or overhead transparency, label it.
- A student who is adept at the mathematics, but for whom English is a second language, can benefit from working with a student whose English skills are good, but who may need help with the mathematics.
- Use figures, graphs, and diagrams. Note that the text contains numerous figures, helpful side comments accompanying the solutions to examples, straightforward language, and key terms in boldface type, and that color is used to enhance readability.

FLEXIBILITY AND ORDER OF TOPICS

College Algebra, Fourth Edition, was designed to be flexible with respect to the order of coverage of core topics, adapting easily to a wide variety of course syllabi and teaching styles. The text begins with Prerequisites, a review chapter. All or part of this material may be covered or it can be omitted. Graphing is introduced in Chapter P with the discussion of the Cartesian plane and Chapter 1 with graphs of equations. The early introduction of graphs make more figures available to illustrate subsequent topics, giving students further insight and opportunities for greater understanding. Technology is integrated at appropriate points for easy access and for student awareness.

STUDY SKILLS

College Algebra, Fourth Edition, encourages the development of strong study skills. Study Tips are interspersed throughout the text offering specific help and insights to improve understanding, learning, and retention. Additionally, there are opportunities throughout the text for self-assessment and review in the Chapter and Cumulative Tests and throughout *Interactive College Algebra* in the pre-, post-, and self-tests.

The supplements package reinforces the text's emphasis on good study skills.
- The *Study and Solutions Guide* contains summaries of the highlights of each chapter.
- A multiple-tape set of Videotapes accompanies the text. Keyed to the text, these include full content coverage, animations, and study tips.
- Tutorial Software includes opportunities for additional exercises and diagnostics.

CHAPTER P
Prerequisites

❑ P.1 Real Numbers

1. Determine how many natural numbers there are in the set:
   $\{-3, -\frac{1}{2}, 2, 0.3535\ldots\}$.
   (a) 1   (b) 2   (c) 3
   (d) 4   (e) None of the numbers are natural numbers.

   1—Answer: a

2. Determine how many integers there are in the set:
   $\{-3, -\frac{1}{2}, 2, 0.3535\ldots\}$.
   (a) 1   (b) 2   (c) 3
   (d) 4   (e) None of the numbers are integers.

   1—Answer: b

3. Determine how many integers there are in the set:
   $\{5, -16, \frac{2}{3}, 0\}$.
   (a) 4   (b) 3   (c) 2
   (d) 1   (e) None of the numbers are integers.

   1—Answer: b

4. Determine how many rational numbers there are in the set:
   $\{-3, -\frac{1}{2}, 2, 0.3535\ldots\}$.
   (a) 4   (b) 3   (c) 2
   (d) 1   (e) None of the numbers are rational.

   1—Answer: a

5. Determine how many irrational numbers there are in the set:
   $\{-3, -\frac{1}{2}, 2, 0.3535\ldots\}$.
   (a) 4   (b) 3   (c) 2
   (d) 1   (e) None of the numbers are irrational.

   1—Answer: e

6. Use a calculator to find the decimal form of the rational number: $\frac{4}{9}$.
   (a) 0.4   (b) 0.44   (c) 0.43
   (d) $0.\overline{4}$ (e) None of these

   1—T—Answer: d

7. Use a calculator to find the decimal form of the rational number: $\frac{62}{495}$.

(a) 0.125 (b) $0.1\overline{25}$ (c) 0.13

(d) 0.1 (e) None of these

1—T—Answer: b

8. Use a calculator to find the decimal form of the rational number: $\frac{7}{12}$.

(a) 0.58 (b) 0.6 (c) $0.5\overline{83}$

(d) $0.58\overline{3}$ (e) None of these

1—T—Answer: d

9. Use a calculator to find the decimal form of the rational number: $\frac{10}{11}$.

1—T—Answer: $0.\overline{90}$

10. Use a calculator to find the decimal form of the rational number: $\frac{173}{330}$.

1—T—Answer: $0.5\overline{24}$

11. Plot the real numbers on the number line: $\left\{\frac{2}{3}, -4, 1, -\frac{3}{2}, 2\right\}$.

1—Answer:

12. Plot the real numbers on the number line: $\left\{-\frac{5}{3}, 3, -1, 4, -\frac{1}{4}\right\}$.

1—Answer:

13. Use inequality notation to describe: b is at least 5.

(a) $b > 5$ (b) $b \geq 5$ (c) $b < 5$

(d) $b \leq 5$ (e) None of these

1—Answer: b

14. Use inequality notation to describe: x is positive.

1—Answer: $x > 0$

15. Use inequality notation to describe: x is nonnegative.

(a) $x > 0$ (b) $x < 0$ (c) $x \geq 0$

(d) $x \leq 0$ (e) None of these

1—Answer: c

16. Use inequality notation to describe: y is no larger than 10.

    (a) $y \leq 10$ (b) $y < 10$ (c) $y \geq 10$

    (d) $y > 10$ (e) None of these

    1—Answer: a

17. Use inequality notation to describe: y is at least 5.

    (a) $y \leq 5$ (b) $y < 5$ (c) $y \geq 5$

    (d) $y > 5$ (e) None of these

    1—Answer: c

18. Describe the subset of real numbers represented by the inequality: $x < 3$.

    1—Answer: The set of real numbers that are less than 3.

19. Describe the subset of real numbers represented by the inequality: $y \geq 5$.

    1—Answer: The set of real numbers that are at least 5.

20. Use inequality notation to describe the set of real numbers that are less than 4 and at least -2.

    (a) $-2 < x < 4$ (b) $-2 < x \leq 4$ (c) $-2 \leq x < 4$

    (d) $-2 \leq x \leq 4$ (e) None of these

    1—Answer: c

21. Use inequality notation to describe the set of real numbers that are more than 5 and at most 10.

    (a) $5 \leq x \leq 10$ (b) $5 \leq x < 10$ (c) $5 < x < 10$

    (d) $5 < x \leq 10$ (e) None of these

    1—Answer: d

22. Use inequality notation to describe the set of real numbers that are at least -1 and at most 3.

    (a) $-1 \leq x \leq 3$ (b) $-1 < x < 3$ (c) $-1 \leq x < 3$

    (d) $-1 < x \leq 3$ (e) None of these

    1—Answer: a

23. Use a calculator to order the numbers from smallest to largest: $\left\{\frac{45}{99}, \frac{2}{5}, \frac{7}{5}, \frac{152}{333}, \frac{23}{5}\right\}$.

    2—T—Answer: $\frac{2}{5}, \frac{45}{99}, \frac{152}{333}, \frac{7}{5}, \frac{23}{5}$

24. Use a calculator to order the numbers from smallest to largest: $\left\{\frac{13}{2}, \frac{28}{5}, \frac{650}{99}, 6.56, 6.065\right\}$.

    2—T—Answer: $\frac{28}{5}, 6.065, \frac{13}{2}, 6.56, \frac{650}{99}$

25. Which of the following is a true statement?

    (a) $|-3| + |6 - 5| \leq -(-4)$ (b) $|14 + (-2)| - |-16| = 28$ (c) $|16 - (-4)| - |3 - 5| \geq 12$

    (d) Both a and b are true.   (e) Both a and c are true

    1—Answer: e

4 Chapter P Prerequisites

26. Which of the following is a true statement?
- (a) $|-3| - |4 - 6| = 1$
- (b) $|-5| + |-13| = -|-18|$
- (c) $|-6 - (-3)| \geq |-4|$
- (d) Both a and b are true.
- (e) Both b and c are true.

1—Answer: a

27. Which of the following is a true statement?
- (a) $\dfrac{-5}{|-5|} = 1$
- (b) $-|-5| + |-3 - 2| \geq -1$
- (c) $|-6 + 4| < |-3 - 8|$
- (d) Both a and c are true.
- (e) Both b and c are true.

1—Answer: e

28. Which of the following is a true statement?
- (a) $-3|-3| < |-3(-3)|$
- (b) $\dfrac{4}{|-12|} \leq \dfrac{|-12|}{-4}$
- (c) $|16| - |12| \geq |16 - 12|$
- (d) Both a and c are true.
- (e) Both b and c are true.

1—Answer: d

29. Which of the following is a true statement?
- (a) $4|-3| - 6 = -18$
- (b) $\dfrac{|-3|}{-3} \leq 0$
- (c) $|-5 - (-7)| = |-5| - |-7|$
- (d) Both a and b are true.
- (e) Both b and c are true.

1—Answer: b

30. Evaluate: $|-3 + 2|$.
- (a) 1
- (b) -1
- (c) 5
- (d) -5
- (e) None of these

1—Answer: a

31. Evaluate: $|-4| - |-2|$.
- (a) -2
- (b) 2
- (c) -8
- (d) 6
- (e) None of these

1—Answer: b

32. Evaluate: $-3|-6| + |-1|$.
- (a) -19
- (b) 19
- (c) 17
- (d) -17
- (e) None of these

1—Answer: d

33. Evaluate: $-|-5| - 5$.
- (a) 0
- (b) -10
- (c) 10
- (d) 25
- (e) None of these

1—Answer: b

34. Use absolute value notation to describe: The distance between x and 5 is at least 6.

   (a) $|x + 5| > 6$ (b) $|x - 5| > 6$ (c) $|x - 5| \geq 6$
   (d) $|6 - x| \geq 5$ (e) None of these

   2—Answer: c

35. Use absolute value notation to describe: 6 is at most 3 units from x.

   (a) $|x - 3| < 6$ (b) $|x - 6| \leq 3$ (c) $|x - 6| \geq 3$
   (d) $|x - 3| \geq 6$ (e) None of these

   2—Answer: b

36. Use absolute value notation to describe: The distance between x and 7 is greater than 2.

   (a) $|x - 7| \geq 2$ (b) $|x - 2| < 7$ (c) $|x - 7| > 2$
   (d) $|x - 2| > 7$ (e) None of these

   2—Answer: c

37. Use absolute value notation to describe: y is closer to 5 than y is to -6.

   2—Answer: $|y - 5| < |y + 6|$

38. Use absolute value notation to describe: The distance between x and 16 is no more than 5.

   (a) $|x - 5| \leq 16$ (b) $|x - 5| > 16$ (c) $|x - 16| \leq 5$
   (d) $|x - 16| > 5$ (e) None of these

   2—Answer: c

39. Find the distance between -43 and 16.

   (a) 27 (b) -27 (c) -59
   (d) 59 (e) None of these

   1—Answer: d

40. Find the distance between x and -42.

   (a) $x - 42$ (b) $x + 42$ (c) $|x - 42|$
   (d) $|x + 42|$ (e) None of these

   1—Answer: d

41. Find the distance between $-\frac{2}{3}$ and $-\frac{1}{2}$.

   (a) $-\frac{1}{3}$ (b) $\frac{1}{6}$ (c) $\frac{1}{2}$
   (d) $\frac{7}{6}$ (e) None of these

   1—Answer: b

42. Find the distance between x and -3.

(a) $|x - 3|$ (b) $|x + 3|$ (c) $x - 3$
(d) $x + 3$ (e) None of these

1—Answer: b

43. Determine the distance between a and b given $a > b$.

(a) $a - b$ (b) $b - a$ (c) $a + b$
(d) ab (e) None of these

1—Answer: a

44. The sales at a local store were projected to be $13,750 per week. The accuracy of this projection is considered good if the actual sales differ from the projected sales by no more than $1375.00. Determine from weekly sales listed below any weeks where the projection was not considered good.

(a) $12,370 (b) $14,980 (c) $15,025
(d) $12,475 (e) The projection was good in each of these cases.

2—Answer: a

45. The sales at a local store were projected to be $13,750 per week. The accuracy of this projection is considered good if the actual sales differ from the projected sales by no more than $1375. Determine from weekly sales listed below any weeks when the projection was not considered good.

(a) $13,150 (b) $14,120 (c) $12,350
(d) $12,750 (e) The projection was good in each of these cases.

2—Answer: c

46. The sales at a local store were projected to be $15,970 per week. The accuracy of this projection is considered good if the actual sales differ from the projected sales by no more than $1597. Determine from the weekly sales listed below any weeks when the projection was not considered good.

(a) $17,470 (b) $14,520 (c) $15,370
(d) $16,570 (e) The projection was good in each of these cases.

2—Answer: e

47. While traveling you enter the interstate near the 234 mile marker, then exit near the 130 mile marker. Determine the number of miles traveled on the interstate.

(a) 364 (b) 104 (c) 234
(d) 130 (e) None of these

1—Answer: b

48. While traveling you enter the interstate near the 125 mile marker, then exit near the 81 mile marker. Determine the number of miles traveled on the interstate.

(a) 125 (b) 81 (c) 44
(d) 206 (e) None of these

1—Answer: c

49. While traveling you enter the interstate near the 26 mile marker, then exit near the 180 mile marker. Determine the number of miles traveled on the interstate.

    (a) 154   (b) 206   (c) 180

    (d) 26   (e) None of these

    1—Answer: a

50. Identify the terms of the expression: $3x^2 - 6x + 1$.

    (a) $3, -6, 1$ (b) x (c) $3x^2, -6x, 1$

    (d) $3x^2, -6x$ (e) None of these

    1—Answer: c

51. Identify the terms of the expression: $6x^4 - 3x^2 + 16$.

    (a) $6, -3, 16$ (b) x (c) $6x^4, -3x^2, 16$

    (d) $6x^4, -3x^2$ (e) None of these

    1—Answer: c

52. Identify the terms of the expression: $12x^3 - 6x^2 + 2$.

    (a) $12, -6, 2$ (b) x (c) $12x^3, -6x^2, 2$

    (d) $12x^3, -6x^2$ (e) None of these

    1—Answer: c

53. Identify the terms of the expression: $2x^4 - 3x^3 + 2x + 1$.

    1—Answer: $2x^4, -3x^3, 2x, 1$

54. Identify the terms of the expression: $6x^3 - 2x^2 + 4x - 3$.

    1—Answer: $6x^3, -2x^2, 4x, -3$

55. Evaluate $3x^2 - 4x$ for $x = 2$.

    (a) -2 (b) 4   (c) 28

    (d) 2   (e) None of these

    1—Answer: b

56. Evaluate $2x^2 - 4x$ for $x = 3$.

    (a) 0   (b) -6 (c) 24

    (d) 6   (e) None of these

    1—Answer: d

57. Evaluate $4x^2 - 3x$ for $x = 2$.

    (a) 10   (b) 58   (c) 2

    (d) 0   (e) None of these

    1—Answer: a

58. Evaluate: $6x^2 - 2x$ for $x = 3$.

    1—Answer: 48

8 Chapter P Prerequisites

59. Evaluate: $2x^2 - 5x$ for $x = 4$.

   1—Answer: 12

60. Identify the property illustrated by $5(\frac{1}{5} + x) = 5(\frac{1}{5}) + 5(x)$.

   (a) Commutative (b) Distributive (c) Associative
   (d) Inverse (e) None of these

   1—Answer: b

61. Identify the property illustrated by $\left(\frac{\sqrt{7}}{2} - y\right) + \left(-\frac{\sqrt{7}}{2} + y\right) = 0$.

   (a) Inverse (b) Identity (c) Associative
   (d) Commutative (e) None of these

   1—Answer: a

62. Identify the property illustrated by $3[x + (-1)] = 3x + 3(-1)$.

   (a) Commutative (b) Associative (c) Distributive
   (d) Identity (e) Inverse

   1—Answer: c

63. Identify the property illustrated by $3 + (2 + 7) = (3 + 2) + 7$.

   1—Answer: Associative

64. Identify the property illustrated by $7(\frac{1}{7}) = 1$.

   1—Answer: Inverse

65. Perform the operations. Write fractional answers in reduced form.

   $\frac{1}{2} - \frac{1}{6} + \frac{3}{4}$

   (a) $\frac{17}{12}$ (b) $\frac{13}{12}$ (c) 0
   (d) Undefined (e) None of these

   1—Answer: b

66. Perform the operations. Write fractional answers in reduced form.

   $\frac{3}{8} - \frac{2}{3} + \frac{1}{4}$

   (a) $\frac{2}{9}$ (b) $\frac{31}{24}$ (c) $\frac{19}{24}$
   (d) $-\frac{1}{24}$ (e) None of these

   1—Answer: d

67. Perform the operations. Write fractional answers in reduced form.

   $\frac{2}{5} - \frac{1}{3} + \frac{7}{10}$

   (a) $\frac{43}{30}$ (b) $\frac{1}{30}$ (c) $\frac{23}{30}$
   (d) $\frac{2}{3}$ (e) None of these

   1—Answer: c

68. Perform the operations. Write fractional answers in reduced form.

    $\frac{3}{5} - \frac{1}{2} + \frac{3}{10}$

    1—Answer: $\frac{2}{5}$

69. Perform the operations. Write fractional answers in reduced form.

    $\frac{2}{5} - \frac{1}{3} + \frac{3}{10}$

    1—Answer: $\frac{11}{30}$

70. Use a calculator to evaluate the expression. Round your answer to two decimal places.

    $\frac{1.25 - 3.89}{4.2}$

    1—T—Answer: -0.63

71. Use a calculator to evaluate the expression. Round your answer to two decimal places.

    $\frac{3.84 - 2.51}{3.6}$

    1—T—Answer: 0.37

72. Use a calculator to evaluate the expression. Round your answer to two decimal places.

    $\frac{2.41(3.86 - 10.25)}{2.42}$

    1—T—Answer: -6.36

73. Use a calculator to evaluate the expression. Round your answer to two decimal places.

    $\frac{3.21(6.14 + 2.56)}{-2.5}$

    1—T—Answer: -11.17

❏ P.2 Exponents and Radicals

1. Write the expression as a repeated multiplication: $(2x)^3$.

    1—Answer: $(2x)(2x)(2x)$

2. Write the expression as a repeated multiplication: $(3y)^4$.

    1—Answer: $(3y)(3y)(3y)(3y)$

3. Write the expression as a repeated multiplication: $(-3a)^2$.

    1—Answer: $(-3a)(-3a)$

10   Chapter P   Prerequisites

4. Write the expression using exponential notation: $(2a)(2a)(2a)(2a)$.
   (a) $2a^4$
   (b) $8a^4$
   (c) $(2a)^4$
   (d) All of these
   (e) None of these

   2—Answer: c

5. Write the expression using exponential notation: $(-3x)(-3x)$.
   (a) $-3x^2$
   (b) $-(3x)^2$
   (c) $(-3x)^2$
   (d) $6x^2$
   (e) None of these

   2—Answer: c

6. Write the expression using exponential notation: $(5x)(5x)(5x)$.
   (a) $5x^3$
   (b) $15x^3$
   (c) $125x^3$
   (d) $15x$
   (e) None of these

   2—Answer: c

7. Evaluate: $(2^3 \cdot 3^2)^{-1}$.
   (a) -72
   (b) $\frac{1}{46,656}$
   (c) $-\frac{1}{36}$
   (d) $\frac{1}{72}$
   (e) None of these

   1—Answer: d

8. Evaluate: $(6^{-2})(3^0)(2^3)$.
   (a) $\frac{2}{9}$
   (b) -288
   (c) -216
   (d) $\frac{1}{6}$
   (e) None of these

   1—Answer: a

9. Evaluate: $(4)^{-2}(3)^0(-1)^2$.
   (a) 0
   (b) -8
   (c) -16
   (d) $\frac{1}{16}$
   (e) None of these

   1—Answer: d

10. Evaluate: $(2)^3(2)^{-3}$.
    (a) 1
    (b) 0
    (c) -36
    (d) -64
    (e) None of these

    1—Answer: a

11. Evaluate: $\dfrac{4(2)^{-1}}{(3)^{-2}(2)}$.
    (a) $\frac{2}{3}$
    (b) 0
    (c) 9
    (d) $\frac{4}{9}$
    (e) None of these

    1—Answer: c

Section P.2 Exponents and Radicals

12. Use a calculator to evaluate the expression. Round to three decimal places.

$$\frac{2^6}{3^4}$$

2—T—Answer: 0.790

13. Use a calculator to evaluate the expression. Round to three decimal places.

$$\frac{2^{-5}}{3^{-4}}$$

2—T—Answer: 2.531

14. Use a calculator to evaluate the expression. Round to three decimal places.

$$(8^{-2})(3^3)$$

2—T—Answer: 0.422

15. Use a calculator to evaluate the expression. Round to three decimal places.

$$(-4)^5(3^{-2})$$

2—T—Answer: -113.778

16. Evaluate $3x^2y^{-4}$ when $x = -1$ and $y = -2$.

1—Answer: $\frac{3}{16}$

17. Evaluate $2x^4 + 3x$ when $x = -3$.

(a) 1287 (b) 153 (c) -15
(d) 1215 (e) None of these

1—Answer: b

18. Evaluate $7(-x)^3$ for $x = 2$.

(a) -1 (b) -42 (c) -56
(d) 2744 (e) None of these

1—Answer: c

19. Evaluate $4x^{-2}$ for $x = 3$.

(a) 36 (b) -24 (c) $\frac{1}{144}$
(d) $\frac{4}{9}$ (e) None of these

1—Answer: d

20. Evaluate $3x^0 - x^{-2}$ for $x = 4$.

(a) -16 (b) $\frac{47}{16}$ (c) $\frac{15}{16}$
(d) -5 (e) None of these

1—Answer: b

12   Chapter P   Prerequisites

21. Simplify: $\left(\dfrac{x^{-3}y^2}{z}\right)^{-4}$.

(a) $\dfrac{z^4}{x^7y^6}$ (b) $\dfrac{y^2z^4}{x^7}$ (c) $\dfrac{x^{12}z^4}{y^8}$

(d) $\dfrac{z^4}{x^{12}y^8}$ (e) None of these

1—Answer: c

22. Simplify: $\left(\dfrac{3x^2y^3}{xw^{-2}}\right)^3$.

(a) $9w^6x^3y^9$ (b) $9w^{-8}x^8y^{27}$ (c) $27w^6x^3y^9$

(d) $3w^6x^3y^9$ (e) None of these

1—Answer: c

23. Simplify: $3x^2(2x)^3(5x^{-1})$.

(a) $30x^{-6}$ (b) $\tfrac{6}{5}x^6$ (c) $\tfrac{24}{5}x^4$

(d) $120x^4$ (e) None of these

1—Answer: d

24. Simplify: $(3x^2y^3z)^{-2}(xy^4)$.

1—Answer: $\dfrac{1}{9x^3y^2z^2}$

25. Simplify: $(-2x^2)^5(5x^3)^{-2}$.

1—Answer: $-\dfrac{32x^4}{25}$

26. Simplify: $\left(\dfrac{x^{-5}y^2}{z^2}\right)^{-3}$.

(a) $\dfrac{x^{-8}y^{-1}}{z^{-1}}$ (b) $\dfrac{x^{15}z^6}{y^6}$ (c) $\dfrac{z^5}{x^{-2}y^{-1}}$

(d) $\dfrac{x^{-15}y^6}{z^6}$ (e) None of these

1—Answer: b

27. Simplify: $x^2 \cdot x^3 \cdot x^4$.

(a) x^{24} (b) x^9 (c) $(x^3)^9 = x^{27}$

(d) x^{10} (e) None of these

1—Answer: b

28. Simplify: $[(A^2)^3]^2$.

(a) A^{12} (b) A^7 (c) A^{10}

(d) A^8 (e) None of these

1—Answer: a

Section P.2   Exponents and Radicals   13

29. Simplify: $(-x^2)^3(-x^3)^2$.
    - (a) x^{10}
    - (b) $-x^{10}$
    - (c) $-x^{12}$
    - (d) x^{12}
    - (e) None of these

    1—Answer: c

30. Simplify: $(-x^2)(-x)^3(-x)^4$.
    - (a) $-x^{24}$
    - (b) x^{24}
    - (c) $-x^9$
    - (d) x^9
    - (e) None of these

    1—Answer: d

31. Simplify: $(-3x^2)^2(-3x)^3$.

    1—Answer: $-243x^7$

32. Simplify: $(-3x^2)^3(-3x)^2$.

    1—Answer: $-243x^8$

33. Simplify and write the answer without negative exponents.
    $$(-2a^2b^3)(-3ab)^3$$
    1—Answer: $54a^5b^6$

34. Simplify and write the answer without negative exponents.
    $$\frac{-3y^{-2}}{(2y)^{-3}}$$
    1—Answer: $-24y$

35. Simplify and write the answer without negative exponents.
    $$(-2a^0b^{-2})(3b^{-1}a^{-2})^{-2}$$
    1—Answer: $-\frac{2}{9}a^4$

36. Simplify: $-(3x^2)^2(-3x^2)^3$.

    1—Answer: $243x^{10}$

37. Simplify: $(-3x^2)^3(3x^2)^2$.

    1—Answer: $-243x^{10}$

38. Use the rules of exponents to write without negative exponents.
    $$2x^{-2}(2x^2y)^0$$
    1—Answer: $\frac{2}{x^2}$

39. Use the rules of exponents to write without negative exponents.
    $$\left(\frac{3x^2}{y^{-2}}\right)^{-1}$$
    1—Answer: $\frac{1}{3x^2y^2}$

14   Chapter P   Prerequisites

40. Use the rules of exponents to write without negative exponents.

$$\left(\frac{b^2}{3a}\right)^{-2}$$

   1—Answer: $\dfrac{9a^2}{b^4}$

41. Use the rules of exponents to write without negative exponents.

$$\left(\frac{x^{-2}y^3}{4}\right)^{-2}$$

   1—Answer: $\dfrac{16x^4}{y^6}$

42. Evaluate: $\dfrac{1}{81^{-1/2}}$.

   (a) $\dfrac{1}{9}$
   (b) 9
   (c) $-\dfrac{1}{9}$
   (d) -9
   (e) None of these

   1—Answer: b

43. Evaluate: $\left(\dfrac{8}{27}\right)^{-2/3}$.

   1—Answer: $\dfrac{9}{4}$

44. Evaluate: $\left(\dfrac{1}{64}\right)^{-2/3}$.

   (a) -16
   (b) $\dfrac{1}{16}$
   (c) $\dfrac{1}{512}$
   (d) -512
   (e) None of these

   1—Answer: e

45. Evaluate: $\left(\dfrac{1}{64}\right)^{3/2}$.

   (a) $\dfrac{1}{512}$
   (b) -512
   (c) $\dfrac{1}{16}$
   (d) -16
   (e) None of these

   1—Answer: a

46. Evaluate: $\dfrac{1}{27^{-1/3}}$.

   (a) $\dfrac{1}{3}$
   (b) 3
   (c) $-\dfrac{1}{3}$
   (d) $-\dfrac{1}{9}$
   (e) None of these

   1—Answer: b

47. Use a calculator to approximate the number. Round to three decimal places.

    $\sqrt[3]{51}$

    1—T—Answer: 3.708

48. Use a calculator to approximate the number. Round to three decimal places.

    $\sqrt[3]{48}$

    1—T—Answer: 3.634

49. Use a calculator to approximate the number. Round to three decimal places.

    $\dfrac{3 + \sqrt{21}}{5}$

    2—T—Answer: 1.517

50. Use a calculator to approximate the number. Round to three decimal places.

    $(2.4)^{3/5}$

    2—T—Answer: 1.691

51. Use a calculator to approximate the number. Round to three decimal places.

    $(15.25)^{-1.2}$

    2—T—Answer: 0.380

52. Simplify: $\sqrt[3]{-625x^7y^5}$.

    (a) $5xy\sqrt[3]{-5x^4y^2}$ (b) $-5xy\sqrt[3]{5x^4y^2}$ (c) $-125x^2y\sqrt[3]{5xy^2}$

    (d) $-5x^2y\sqrt[3]{5xy^2}$ (e) Does not simplify

    1—Answer: d

53. Simplify: $\sqrt{75x^2y^{-4}}$.

    (a) $\dfrac{5\sqrt{3}x}{y^2}$ (b) $\dfrac{3\sqrt{5}|x|}{y^2}$ (c) $5\sqrt{3}|x|y^2$

    (d) $\dfrac{5\sqrt{3}|x|}{y^2}$ (e) None of these

    1—Answer: d

54. Simplify: $\sqrt{x^2y^2}$.

    (a) $\pm|xy|$ (b) $|xy|$ (c) xy

    (d) $-xy$ (e) None of these

    1—Answer: b

55. Simplify $\sqrt{a^4b^2}$ if a and b are both negative.

    (a) a^2 (b) $\pm|a^2b|$ (c) $-a^2b$

    (d) $\pm a^2b$ (e) None of these

    2—Answer: c

16   Chapter P   Prerequisites

56. Simplify: $\sqrt[3]{24x^4y^5}$.

(a) $3x^2y^2\sqrt[3]{6x^2y^3}$ (b) $8xy\sqrt[3]{3xy}$ (c) $2xy\sqrt[3]{6xy^2}$

(d) $2xy\sqrt[3]{3xy^2}$ (e) None of these

1—Answer: d

57. Perform the operation and simplify: $2^{3/2} \cdot 2^{5/2}$.

2—Answer: 16

58. Perform the operation and simplify: $\dfrac{5^{5/3}}{5^{2/3}}$.

2—Answer: 5

59. Perform the operation and simplify: $\dfrac{x^{4/3}y^{1/3}}{(xy)^{2/3}}$.

2—Answer: $\dfrac{x^{2/3}}{y^{1/3}}$

60. Perform the operation and simplify: $\dfrac{x^{-1/2} \cdot x^{1/3}}{x^2 \cdot x^{-3}}$.

2—Answer: $x^{5/6}$

61. Rationalize the denominator: $\dfrac{3}{\sqrt{7} + 2}$.

(a) $\sqrt{7} - 2$ (b) $\dfrac{3\sqrt{7} - 6}{5}$ (c) $\dfrac{3\sqrt{7} - 2}{3}$

(d) $\dfrac{3\sqrt{7} - 2}{5}$ (e) None of these

1—Answer: a

62. Rationalize the denominator: $\dfrac{5}{7 - \sqrt{2}}$.

(a) $\dfrac{35 + 5\sqrt{2}}{47}$ (b) $\dfrac{35 + \sqrt{2}}{47}$ (c) $7 + \sqrt{2}$

(d) $\dfrac{35 + \sqrt{2}}{3}$ (e) None of these

1—Answer: a

63. Rationalize the denominator: $\dfrac{2}{\sqrt[3]{2x}}$.

(a) $\dfrac{\sqrt[3]{2x}}{x}$ (b) $\dfrac{\sqrt[3]{4x^2}}{x}$ (c) $\dfrac{2\sqrt[3]{2x}}{2x}$

(d) $\sqrt[3]{4x}$ (e) None of these

2—Answer: b

Section P.2 Exponents and Radicals 17

64. Simplify by rationalizing the denominator: $\dfrac{6x}{5-\sqrt{2}}$.

(a) $\dfrac{6x(5+\sqrt{2})}{23}$ 	(b) $2x(5+\sqrt{2})$ 	(c) $\dfrac{30x-\sqrt{2}}{23}$

(d) $\dfrac{6x(5-\sqrt{2})}{21}$ 	(e) None of these

1—Answer: a

65. Simplify by rationalizing the denominator: $\dfrac{4}{\sqrt[3]{x}}$.

(a) $\dfrac{4\sqrt[3]{x}}{x}$ 	(b) $\dfrac{4\sqrt[3]{x^2}}{x}$ 	(c) 4

(d) $4\sqrt[3]{x}$ 	(e) None of these

1—Answer: b

66. Simplify by rationalizing the numerator: $\dfrac{6-\sqrt{2}}{5}$.

(a) $\dfrac{34}{5(6+\sqrt{2})}$ 	(b) $\dfrac{4}{30+\sqrt{2}}$ 	(c) $\dfrac{34}{5(6-\sqrt{2})}$

(d) $\dfrac{32}{5(6-\sqrt{2})}$ 	(e) None of these

1—Answer: a

67. Simplify by rationalizing the numerator: $\dfrac{\sqrt[3]{3}}{12}$.

(a) $\dfrac{1}{\sqrt[3]{4}}$ 	(b) $\dfrac{3}{4\sqrt[3]{3}}$ 	(c) $\dfrac{1}{4\sqrt[3]{3}}$

(d) $\dfrac{1}{4\sqrt[3]{9}}$ 	(e) None of these

1—Answer: d

68. Simplify by rationalizing the numerator: $\dfrac{\sqrt{2}-\sqrt{5}}{12}$.

(a) $\dfrac{1}{4(\sqrt{2}-\sqrt{5})}$ 	(b) $\dfrac{-1}{4(\sqrt{2}+\sqrt{5})}$ 	(c) $\dfrac{-7}{4(\sqrt{2}+\sqrt{5})}$

(d) $\dfrac{1}{4\sqrt{2}+\sqrt{5}}$ 	(e) None of these

1—Answer: b

69. Simplify by rationalizing the numerator: $\dfrac{5 + 2\sqrt{5}}{15}$.

(a) $\dfrac{-4}{5(1 - \sqrt{5})}$ (b) $\dfrac{2}{3(5 + 2\sqrt{5})}$ (c) $\dfrac{1}{5 - 2\sqrt{5}}$

(d) $\dfrac{1}{3(5 - 2\sqrt{5})}$ (e) None of these

1—Answer: d

70. Simplify by rationalizing the numerator: $\dfrac{\sqrt[3]{16}}{7}$.

(a) $\dfrac{16}{7\sqrt[3]{16}}$ (b) $\dfrac{1}{7\sqrt[3]{256}}$ (c) $\dfrac{4}{7\sqrt[3]{4}}$

(d) $\dfrac{2}{7\sqrt[3]{16}}$ (e) None of these

1—Answer: c

71. Rationalize the numerator: $\dfrac{3 - \sqrt{2}}{5}$.

1—Answer: $\dfrac{7}{5(3 + \sqrt{2})}$

72. Write as a single radical: $\sqrt[3]{\sqrt{2x}}$.

(a) $\sqrt[6]{2x}$ (b) $\sqrt[5]{2x}$ (c) $\sqrt[3]{2x}$

(d) $\sqrt[3]{4x^2}$ (e) None of these

1—M—Answer: a

73. Simplify: $\sqrt[3]{\sqrt{3x + 1}}$.

1—Answer: $\sqrt[6]{3x + 1}$

74. Simplify: $3\sqrt{2}\sqrt[3]{4}$.

(a) $6\sqrt[6]{2}$ (b) $3\sqrt[6]{8}$ (c) $6\sqrt{2}$

(d) $3\sqrt[6]{32}$ (e) None of these

2—M—Answer: a

75. Simplify: $(\sqrt[3]{81x^4y^9})(\sqrt[3]{2xy^2})$.

2—Answer: $3xy^3\sqrt{6x^2y^2}$

76. Simplify: $\sqrt{3x^2}\sqrt[3]{3x^2}$.

2—Answer: $\sqrt[6]{(3x^2)^5}$

Section P.2 Exponents and Radicals 19

77. Simplify: $\sqrt[6]{8x^3y^3}$.
    (a) $\sqrt{8xy}$ (b) $\sqrt[3]{2xy}$ (c) $\sqrt{2xy}$
    (d) $\sqrt[3]{8xy}$ (e) None of these

    1—Answer: c

78. Simplify: $\sqrt[6]{9x^2y^2}$.
    (a) $\sqrt{3xy}$ (b) $\sqrt[3]{3xy}$ (c) $\sqrt{9xy}$
    (d) $\sqrt[3]{9xy}$ (e) None of these

    1—Answer: b

79. Simplify: $3\sqrt[3]{4x^5y^3} + 7x\sqrt[3]{32x^2y^6}$.

    2—Answer: $(3 + 14y)xy\sqrt[3]{4x^2}$

80. Simplify: $4\sqrt{9x} - 2\sqrt{4x} + 7$.
    (a) $8x + 7$ (b) $8\sqrt{x} + 7$ (c) $2\sqrt{5x} + 7$
    (d) Does not simplify (e) None of these

    1—Answer: b

81. Simplify: $7\sqrt{25xy^2} - 4\sqrt{75xy^2} + 2\sqrt{12xy^2}$.
    (a) $35|y|\sqrt{x} - 16|y|\sqrt{3x}$ (b) $19|y|\sqrt{2x}$ (c) $35|y|\sqrt{x} - 6|y|\sqrt{2x}$
    (d) $5|y|\sqrt{38x}$ (e) None of these

    1—Answer: a

82. Simplify: $2x^2y\sqrt[3]{2x} + 7x^2\sqrt[3]{2xy^3} - 4\sqrt[3]{16x^7y^3}$.
    (a) $x^6y^3\sqrt[3]{2x}$ (b) $x^2y\sqrt[3]{2x}$ (c) $9x^2y\sqrt[3]{2x} - 8y\sqrt[3]{2x^7y}$
    (d) $2x^3y$ (e) None of these

    2—Answer: b

83. Write $\sqrt[3]{2x^2}$ in exponential form.
    (a) $2x^{3/2}$ (b) $2x^{2/3}$ (c) $(2x)^{2/3}$
    (d) $(2x^2)^{1/3}$ (e) None of these

    1—Answer: d

84. The period T, in seconds, of a pendulum is $T = 2\pi\sqrt{L/32}$ where L is the length of the pendulum in feet. Find the period of a pendulum whose length is $\frac{1}{2}$ foot.
    (a) π (b) $\dfrac{\pi}{2}$ (c) $\dfrac{\pi}{4}$
    (d) $\dfrac{\pi}{8}$ (e) None of these

    2—Answer: c

85. The period T, in seconds, of a pendulum is $T = 2\pi\sqrt{L/32}$ where L is the length of the pendulum in feet. Find the period of a pendulum whose length is 8 feet.

   (a) π (b) $\dfrac{\pi}{2}$ (c) $\dfrac{\pi}{4}$

   (d) $\dfrac{\pi}{8}$ (e) None of these

   2—Answer: a

86. Rewrite in scientific notation: 0.000004792.

   (a) 0.4792×10^5 (b) 4.792×10^{-6} (c) 4.792×10^{-5}

   (d) 4.792×10^6 (e) None of these

   1—Answer: b

87. Rewrite in decimal form: 3.75×10^{-7}.

   1—Answer: 0.000000375

88. Multiply: $(0.00000526)(72,000,000,000)^2$.

   (a) 3.7872×10^4 (b) 2.726784×10^{-20} (c) 2.726784×10^{16}

   (d) 2.726784×10^{-14} (e) None of these

   2—Answer: c

89. Simplify and write in decimal form: $\dfrac{(5.1 \times 10^{-5})(3 \times 10^6)}{1.7 \times 10^{-2}}$.

   2—Answer: 9000

90. Simplify and write in scientific notation: $\dfrac{(32,700,000,000,000)(72,000,000,000)^2}{0.0000000041}$.

   2—Answer: 4.13×10^{43}

91. Represent the number 0.0021367 in scientific notation.

   (a) 0.21367×10^{-2} (b) 21367×10^{-7} (c) 2.1367×10^{-2}

   (d) 2.1367×10^{-3} (e) None of these

   1—Answer: d

92. The highest peak in the Western hemisphere, located in Argentina, has an elevation of 22,831 feet. Represent this in scientific notation.

   (a) 22.831×10^4 (b) 2.2831×10^4 (c) 0.22831×10^5

   (d) 2.2831×10^3 (e) None of these

   1—Answer: b

93. The total area of Chile is approximately 2.9×10^5 square miles. Write this in decimal form.

   (a) 2900 (b) 29,000 (c) 290,000

   (d) 2,900,000 (e) None of these

   1—Answer: c

P.3 Polynomials and Special Products

1. Find the degree of the polynomial: $5x^4 - 2x^3 - 7x + 1$.

   (a) 4   (b) 5   (c) 8

   (d) 12   (e) None of these

   1—Answer: a

2. Find the degree of the polynomial: $5x^4 - 2x^2 + x$.

   (a) 7   (b) 6   (c) 5

   (d) 4   (e) None of these

   1—Answer: d

3. Find the degree of the polynomial: $4x^3 - 2x + 1$.

   (a) 2   (b) 3   (c) 4

   (d) 5   (e) This is not a polynomial.

   1—Answer: b

4. Find the degree of the polynomial: $5x^2 - 3x + 1$.

   (a) 2   (b) 3   (c) 4

   (d) 5   (e) This is not a polynomial.

   1—Answer: a

5. Identify any polynomials.

   (a) $4x^3 - 7\sqrt{x} + 3$ (b) $\dfrac{x^2 + 2x + 1}{x - 3}$ (c) $x^{-3} + 2x^{-2} + x$

   (d) None of these are polynomials.   (e) All of these are polynomials.

   1—Answer: d

6. Identify any polynomials.

   (a) $3x^2 - 2x + 1$ (b) $\dfrac{x + 1}{x - 1}$ (c) $x + \dfrac{1}{x}$

   (d) None of these are polynomials.   (e) All of these are polynomials.

   1—Answer: a

7. Identify any polynomials.

   (a) $\dfrac{3x + 1}{2x - 2}$ (b) $x + 2x^{-1} + 1$ (c) $\dfrac{1}{4}x^3 - \dfrac{2}{3}x - 2$

   (d) None of these are polynomials.   (e) All of these are polynomials.

   1—Answer: c

8. Identify any polynomials.

(a) $\frac{1}{2}x^3 + x - \frac{1}{3}$ (b) $7 + 4x^3 - 6x^7$ (c) $x + 1$

(d) None of these are polynomials.   (e) All of these are polynomials.

1—Answer: e

9. Simplify: $(6x^3 + 2x - 7) + (4x^2 + x + 1) - (x^3 + 3x^2 - 2)$.

(a) $5x^3 + 7x^2 + 3x - 8$ (b) $5x^3 + x^2 + 3x - 4$ (c) $5x^6 - x^4 + x - 8$

(d) $7x^6 - 8$ (e) None of these

1—Answer: b

10. Simplify: $(7x^4 - 2x^3 + 5x^2) + (7x^3 - 2x^2 + 5) - (6x^3 + 2x^2 - 12x)$.

(a) $7x^4 - x^3 + x^2 + 12x + 5$ (b) $7x^4 - x^3 + 5x^2 - 12x + 5$ (c) $-x^{10} + 5$

(d) $-x^9 + x^6 + 7x^4 - 12x + 5$ (e) None of these

1—Answer: a

11. Simplify: $(-4x^2 + 2x) - (5x^3 + 2x^2 - 1) + (x^2 + 1)$.

(a) $-9x^3 + 5x^2$ (b) $5x^3 - x^2 + 2x + 1$ (c) $6x^6$

(d) $-5x^3 - 5x^2 + 2x + 2$ (e) None of these

1—Answer: d

12. Simplify: $(2x^2 + 3x - 1) + (x^3 + x^2 + 5) - (2x^2 - 5x + 7)$.

(a) $x^3 - 2x + x^2 + 4$ (b) $x^3 + x^2 + 8x - 3$ (c) $8x^5 + 4$

(d) $63x^3 + 5x^6$ (e) None of these

1—Answer: b

13. Simplify: $(3x^2 - 2x) + (7x^3 - 2x^2 + 1) - (16x^2 - 7)$.

(a) $7x^3 - 15x^2 - 2x + 8$ (b) $7x^3 + 15x^2 - 6$ (c) $7x^3 - 15x^2 - 2x - 6$

(d) $15x^5 + 7x^3 - 2x - 6$ (e) None of these

1—Answer: a

14. Simplify: $(-2x^2 + 3x - 9) - (4x^2 - x + 2) + (x^3 - 2x^2 + 1)$.

(a) $x^3 - 8x^2 + 3x - 6$ (b) $x^3 - 8x^2 + 4x - 10$ (c) $x^3 - 2x^2 + 2x - 6$

(d) $x^3 - 4x^2 + 2x - 6$ (e) None of these

1—Answer: b

15. Simplify: $3x(5x + 2) - 14(2x^2 - x + 1)$.

(a) $28x^2 - 3x - 14$ (b) $13x^2 - 8x + 14$ (c) $-13x^2 + 20x - 14$

(d) $13x^2 - 20x + 14$ (e) None of these

1—Answer: c

16. Simplify: $3x(7x - 6) - 4x(x - 2)$.
    - (a) $17x^2 - 10x$
    - (b) $-84x^4 + 240x^3 - 144x^2$
    - (c) $21x^2 - 14x$
    - (d) $72x^3 - 96x^2$
    - (e) None of these

    1—Answer: a

17. Simplify: $(3x^4 - 7x^2) + 2x(x^2 - 1)(3x)$.
    - (a) $9x^4 - 13x^2$
    - (b) $9x^4 - 19x^3 - 2x$
    - (c) $3x^4 + 2x^3 - 7x^2 - 3x$
    - (d) $15x^7 - 50x^5 + 35x^3$
    - (e) None of these

    1—Answer: a

18. Simplify: $(5 - 2x)(3) - (3x + 2)(-2)$.
    - (a) $-6x^2 + 11x + 10$
    - (b) $-9x + 1$
    - (c) $-10x + 20$
    - (d) 19
    - (e) None of these

    1—Answer: d

19. Write in standard form: $(3x^2 + 2x) + x(1 - 7x) + (2x + 5)$.

    1—Answer: $-4x^2 + 5x + 5$

20. Write in standard form: $3x^2 - 2x(1 + 3x - x^2)$.

    1—Answer: $2x^3 - 3x^2 - 2x$

21. Multiply: $(2x^2 - 5)^2$.
    - (a) $4x^2 - 25$
    - (b) $4x^4 - 25$
    - (c) $4x^4 - 20x^2 + 25$
    - (d) $4x^4 - 10x^2 - 25$
    - (e) None of these

    1—Answer: c

22. Multiply: $(x - 2)(x^2 + 2x + 4)$.
    - (a) $x^3 - 8$
    - (b) $x^3 - 6x^2 + 12x - 8$
    - (c) $x^3 + 4x^2 - 8$
    - (d) $x^3 + 4x^2 + 8x - 8$
    - (e) None of these

    1—Answer: a

23. Multiply: $(x + 4y)^2$.
    - (a) $x^2 + 4xy + y^2$
    - (b) $x^2 + 16y^2$
    - (c) $x^2 + 8xy + 16y^2$
    - (d) $x^2 + 4y^2$
    - (e) None of these

    1—Answer: c

24. Multiply: $(x - 2\sqrt{3})(x + 2\sqrt{3})$.
    - (a) $x^2 - 12$
    - (b) $x^2 - 36$
    - (c) $x^2 - 6$
    - (d) $x^2 - 4\sqrt{3}$
    - (e) None of these

    1—Answer: a

25. Multiply: $(3x - 7)(2x + 9)$.
  (a) $6x^2 + 13x - 63$
  (b) $6x^2 - 63$
  (c) $6x^2 - 13x - 63$
  (d) $6x^2 + 63$
  (e) None of these

  1—Answer: a

26. Expand: $(2x - 1)^3$.

  2—Answer: $8x^3 - 12x^2 + 6x - 1$

27. Expand: $[(x - 1) + y]^2$.

  2—Answer: $x^2 - 2x + 1 + 2xy - 2y + y^2$

28. Multiply: $(x - 2)(x + 2)(x^2 + 4)$.

  2—Answer: $x^4 - 16$

29. Multiply: $(2x - y)(x + y)$.
  (a) $2x^2 - y^2$
  (b) $2x^2 + xy - y^2$
  (c) $-2x^2y^2$
  (d) $x^2y^2 - xy$
  (e) None of these

  1—Answer: b

30. Multiply: $[(x + 1) - y][(x + 1) + y]$.
  (a) $x^2 + 2x + 1 - y^2$
  (b) $x^2 + 1 - y^2$
  (c) $x^2y^2 + y^2$
  (d) $-x^2y^2 + 2xy^2 + y^2$
  (e) None of these

  1—Answer: a

31. Expand: $(3 + 2y)^3$.
  (a) $9 + 6y^3$
  (b) $27 + 8y^3$
  (c) $27 + 54y + 36y^2 + 8y^3$
  (d) $27 + 18y + 12y^2 + 8y^3$
  (e) None of these

  2—Answer: d

32. Represent the area of the region as a polynomial in standard form.
  (a) $-x^2 - 8x + 96$
  (b) $-3x^2 + 72$
  (c) 40
  (d) $3x^2 - 24x + 96$
  (e) None of these

  2—Answer: d

33. Represent the area of the region as a polynomial in standard form.
  (a) $84x$
  (b) $-10x^2 + 18x + 90$
  (c) $-6x^2 + 90x$
  (d) $10x + 36$
  (e) None of these

  2—Answer: c

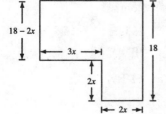

34. Represent the area of the region as a polynomial in standard form.

(a) $6x^2 - 72x$ (b) $48x - 18$

(c) $90x - 6x^2$ (d) $90x$

(e) None of these

2—Answer: a

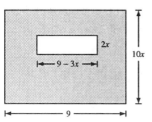

35. Represent the area of the region as a polynomial in standard form.

(a) $140x$ (b) $144x - 4x^2$

(c) $55x$ (d) $120x^2$

(e) None of these

2—Answer: b

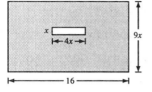

36. Represent the area of the region as a polynomial in standard form.

(a) $-2x + 18$ (b) $12x + 14$

(c) 14 (d) $11x + 14$

(e) None of these

2—Answer: d

37. The height, in feet, of a free-falling object after t seconds is given by the polynomial:

Height $= -16t^2 + 60t + 25$.

Determine the height of the object when $t = 2.5$ seconds.

(a) 215 feet (b) 165 feet (c) 39 feet

(d) 75 feet (e) None of these

1—Answer: d

38. The height, in feet, of a free-falling object after t seconds is given by the polynomial:

Height $= -16t^2 + 60t + 25$.

Determine the height of the object when $t = 3.5$ seconds.

(a) 61 feet (b) 79 feet (c) 39 feet

(d) 75 feet (e) None of these

1—Answer: c

39. The height, in feet, of a free-falling object after t seconds is given by the polynomial:

Height $= -16t^2 + 60t + 25$.

Determine the height of the object when $t = 1.5$ seconds.

(a) 61 feet (b) 79 feet (c) 39 feet

(d) 75 feet (e) None of these

1—Answer: b

40. The height, in feet, of a free-falling object after t seconds is given by the polynomial:

    Height $= -16t^2 + 60t + 25$.

    Determine the height of the object when $t = 3.0$ seconds.

    (a) 61 feet (b) 79 feet (c) 39 feet
    (d) 75 feet (e) None of these

    1—Answer: a

41. After t seconds, the height in feet of an object dropped from a hot air balloon is given by:

    Height $= 300 - 16t^2$.

    Find the height after 1.5 seconds.

    1—Answer: 264 feet

42. After t seconds, the height in feet of an object dropped from a hot air balloon is given by:

    Height $= 300 - 16t^2$.

    Find the height after 2.5 seconds.

    1—Answer: 200 feet

43. After t seconds, the height in feet of an object dropped from a hot air balloon is given by:

    Height $= 300 - 16t^2$.

    Find the height after 3.5 seconds.

    1—Answer: 104 feet

44. After t seconds, the height in feet of an object dropped from a hot air balloon is given by:

    Height $= 300 - 16t^2$.

    Find the height after 4.0 seconds.

    1—Answer: 44 feet

45. Find the area of the shaded region.

    2—Answer: $3x^2 + 4x$ square units

    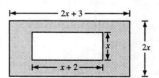

46. Find the area of the shaded region.

    2—Answer: $9x^2$ square units

    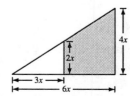

❏ P.4 Factoring

1. Factor: $3x^2 - 15x$.
   - (a) $45x^3$
   - (b) $3(x^2 - 5)$
   - (c) $3x(x - 5)$
   - (d) $3x(x - 5x)$
   - (e) None of these

   1—Answer: c

2. Factor: $(2x - 1)(x + 3) + (2x - 1)(2x + 1)$.

   2—Answer: $(2x - 1)(3x + 4)$

3. Factor: $(3x + 2)(x - 7) + (4x - 1)(x - 7)$.

   2—Answer: $(x - 7)(7x + 1)$

4. Factor: $8x^3 - 27$.
   - (a) $(2x - 3)^3$
   - (b) $(2x - 3)(4x^2 - 12x + 9)$
   - (c) $(2x - 3)(4x^2 + 6x + 9)$
   - (d) $(2x + 3)(4x^2 - 6x + 9)$
   - (e) None of these

   1—Answer: c

5. Factor: $x^3 + 216$.
   - (a) $(x + 6)^3$
   - (b) $(x + 6)(x^2 - 6x + 36)$
   - (c) $(x - 6)(x^2 + 6x - 36)$
   - (d) $(x + 6)(x^2 + 36)$
   - (e) None of these

   1—Answer: b

6. Factor: $y^3 - (x + 1)^3$.
   - (a) $(y - x - 1)(y^2 + xy + y + x^2 + 2x + 1)$
   - (b) $(y - x + 1)(y^2 + xy + y + x^2 + 1)$
   - (c) $(y + x - 1)(y^2 - xy - y + x^2 - 2x - 1)$
   - (d) $(y - x - 1)^3$
   - (e) None of these

   2—Answer: a

7. Factor: $(a - b)^2 - x^2$.
   - (a) $(a - b - x)^2$
   - (b) $(a - b - x)(a + b + x)$
   - (c) $(a - b - x)(a - b + x)$
   - (d) $a^2 - 2ab + b^2 - x^2$
   - (e) None of these

   1—Answer: c

8. Factor: $81 - 4x^2$.
   - (a) $(9 - 2x)^2$
   - (b) $(9 - 2x)(9 + 2x)$
   - (c) $(81 - 4x)^2$
   - (d) $(3 - 2x)^2(3 + 2x)^2$
   - (e) None of these

   1—Answer: b

9. Factor: $(x + 3)^2 - a^2$.

   2—Answer: $(x + 3 - a)(x + 3 + a)$

10. Factor: $9x^2 + 24xy + 16y^2$.

(a) $(9x + 4y)(x + 4y)$ (b) $(9x + 16y)(x + y)$ (c) $(3x - 4y)^2$

(d) $(3x + 4y)^2$ (e) None of these

1—Answer: d

11. Factor: $9x^2 - 24x + 16$.

1—Answer: $(3x - 4)^2$

12. Factor: $9x^2 - 42x + 49$.

(a) $(3x - 7)^2$ (b) $(9x - 7)^2$ (c) $(3x + 7)^2$

(d) $x(9x - 42 + 49)$ (e) None of these

1—Answer: a

13. Factor: $x^2 + 13x - 14$.

(a) $(x - 7)(x + 2)$ (b) $(x + 7)(x - 2)$ (c) $(x - 14)(x - 1)$

(d) $(x - 14)(x + 1)$ (e) None of these

1—Answer: e

14. Factor $x^2 + x - 12$.

(a) $(x - 4)(x - 3)$ (b) $(x + 4)(x - 3)$ (c) $(x + 6)(x - 2)$

(d) $(x - 12)(x - 1)$ (e) None of these

1—Answer: b

15. Factor: $x^2 - 2x - 15$.

(a) $(x - 3)(x + 5)$ (b) $(x - 15)(x + 1)$ (c) $(x + 3)(x - 5)$

(d) $(x + 3)(x + 5)$ (e) None of these

1—Answer: c

16. Factor: $3x^2 - 13x - 16$.

(a) $(3x - 16)(x - 1)$ (b) $(3x + 16)(x - 1)$ (c) $(x + 16)(3x - 1)$

(d) $(x - 16)(3x + 1)$ (e) None of these

1—Answer: e

17. Factor: $3x^2 - 19x - 14$.

(a) $(3x + 2)(x - 7)$ (b) $(3x - 7)(x + 2)$ (c) $(3x - 2)(x + 7)$

(d) $(3x + 7)(x - 2)$ (e) None of these

1—Answer: a

18. Factor: $14x^2 - 19x - 3$.

1—Answer: $(2x - 3)(7x + 1)$

19. Factor: $35x^2 + 9x - 2$.

1—Answer: $(5x + 2)(7x - 1)$

20. Factor: $2rs + 3rst - 8r - 12rt$.

(a) $r(2x + 3st - 8 - 12t)$ (b) $(rs - 4r)(2 + 3t)$ (c) $r(s - 4)(2 - 3t)$
(d) $r(s - 4)(2 + 3t)$ (e) None of these

2—Answer: d

21. Factor: $3rv - 2vt - 6rs + 4st$.

2—Answer: $(v - 2s)(3r - 2t)$

22. Factor: $3xz + 2yz - 6xw - 4yw$.

(a) $(3x + 2y)(z + 2w)$ (b) $(3x + 2y)(z - 2w)$ (c) $(3x - 2y)(z - 2w)$
(d) $(3xz + 2yz)(6xw + 4yw)$ (e) None of these

2—Answer: b

23. Factor: $4x^3 + 6x^2 - 10x$.

1—Answer: $2x(2x + 5)(x - 1)$

24. Factor into linear factors: $6x^3 + 33x^2y - 63xy^2$.

(a) $3x(2x - 3y)(x + 7y)$ (b) $(6x^2 - 9xy)(x + 7y)$ (c) $3x(2x + 3y)(x + 7y)$
(d) $(6x^2 + 9xy)(z + 7y)$ (e) None of these

1—Answer: a

25. Factor completely: $3x^4 - 48$.

(a) $3(x - 2)^2(x + 2)^2$ (b) $3(x - 2)^4$ (c) $3x^2(x - 4)^2$
(d) $3(x^2 + 4)(x + 2)(x - 2)$ (e) None of these

2—Answer: d

26. Factor: $3x - 24x^4$.

2—Answer: $3x(1 - 2x)(1 + 2x + 4x^2)$

27. Factor completely: $x^2(2x + 1)^3 - 4(2x + 1)^2$.

(a) $(2x + 1)^2(2x^3 - 3)$ (b) $(2x + 1)^2(2x^3 + x^2 - 4)$ (c) $(2x + 1)^3(x^2 - 4)$
(d) $(2x + 1)^3(x^2 - 8x - 4)$ (e) None of these

2—Answer: b

28. Factor: $4(x + 2)^2 - 3x(x + 2)^3$.

(a) $-(x + 2)^2(3x - 2)(x + 2)$ (b) $-(x + 2)^2(3x^2 + 6x - 4)$ (c) $(x + 2)^2(-3x^2 + 6)$
(d) $(x + 2)^2(-3x^2 + 6x + 4)$ (e) None of these

1—Answer: b

29. Factor: $9x(3x - 5)^2 + (3x - 5)^3$.

(a) $(3x - 5)^3(9x + 1)$ (b) $(3x - 5)^2(6x - 5)$ (c) $(3x - 5)^2(12x - 5)$
(d) $(3x - 5)(30x^2 - 70)$ (e) None of these

1—Answer: c

30. Factor: $6x(2x + 3)^{-4} - 24x^2(2x + 3)^{-5}$.

    (a) $-6x(2x - 3)(2x + 3)^{-5}$
    (b) $-6x(8x^2 + 12x - 1)(2x + 3)^{-4}$
    (c) $6x(2x + 3)^{-5}[(2x + 3)^{-1} + 4x]$
    (d) $6x(2x + 3)^{-4}(-8x^2 - 12x)$
    (e) None of these

    2—Answer: a

31. Factor: $(3x + 2)^{-3} - 9x(3x + 2)^{-4}$.

    (a) $(3x + 2)^{-4}[(3x + 2)^{-1} - 9x]$
    (b) $(-27x^2 - 18x)(3x + 2)^{-3}$
    (c) $-2(3x - 1)(3x + 2)^{-4}$
    (d) $-(27x^2 + 18x - 1)(3x + 2)^{-3}$
    (e) None of these

    2—Answer: c

32. Factor: $2x(4x - 1)^{-2} + (4x - 1)^{-1}$.

    (a) $(8x^2 - 1)(4x - 1)^{-1}$
    (b) $(4x - 1)^{-1}(8x^2 - 2x + 1)$
    (c) $(2x + 1)(4x - 1)^{-2}$
    (d) $(6x - 1)(4x - 1)^{-2}$
    (e) None of these

    1—Answer: d

33. The trinomial $x^2 - 4x + c$ will factor for which of the following values of c?

    (a) 3
    (b) -5
    (c) -12
    (d) All of these
    (e) None of these

    2—Answer: d

34. The trinomial $4x^2 + 3x + c$ will factor for which of the following values of c?

    (a) 6
    (b) 0
    (c) -1
    (d) Both a and b
    (e) Both b and c

    1—Answer: e

35. The total surface area of a right circular cylinder is found by using the formula $S = 2\pi r^2 + 2\pi rh$. Write the formula in factored form.

    1—Answer: $S = 2\pi r(r + h)$

36. The volume of the frustum of a cone can be found by using the formula $V = \frac{1}{3}(a^2\pi h + ab\pi h + b^2\pi h)$. Write the formula in factored form.

    1—Answer: $V = \frac{1}{3}\pi h(a^2 + ab + b^2)$

37. Find the perimeter, P, of the rectangle where the lengths of the sides can be obtained by factoring the expression for the area,

    Area $= x^2 - 64$.

    (a) $P = 2x$
    (b) $P = 4x$
    (c) $P = 4x + 32$
    (d) $P = 4x + 16$
    (e) None of these

    2—Answer: b

38. Find the perimeter, P, of the rectangle where the lengths of the sides can be obtained by factoring the expression for the area,

Area $= 2x^2 - 6x$.

(a) $P = 3x$ (b) $P = 3x - 3$ (c) $P = 4x - 3$

(d) $P = 6x - 6$ (e) None of these

2—Answer: d

39. The selling price, S, of a product is equal to the cost, C, plus the markup. The markup may be expressed as a percent, R, of the cost. Write an equation for the selling price in terms of the cost and markup, then factor that expression.

1—Answer: $S = C + RC \Rightarrow S = C(1 + R)$

40. The sale price, S, of a product is equal to the list price, L, minus the discount. The discount may be expressed as a percent, R, of the list price. Write an equation for the sale price in terms of the list price and discount, then factor that expression.

1—Answer: $S = L - RL \Rightarrow S = L(1 - R)$

41. Find the surface area of a compact disc with outside radius, R, and inside radius, r, then factor that expression.

1—Answer: $A = \pi R^2 - \pi r^2 \Rightarrow$

$A = \pi(R^2 - r^2)$

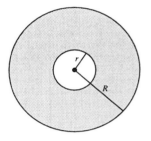

42. Write a polynomial and the correct factorization for the geometric factoring model.

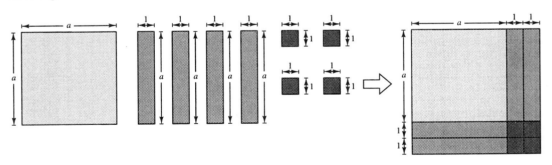

2—Answer: $a^2 + 4a + 4 = (a + 2)^2$

43. Write a polynomial and the correct factorization for the geometric factoring model.

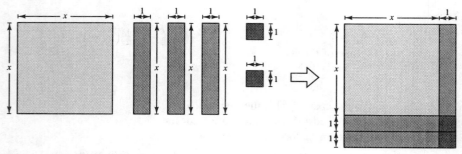

2—**Answer:** $x^2 + 3x + 2 = (x + 1)(x + 2)$

44. Write a polynomial and the correct factorization for the geometric factoring model.

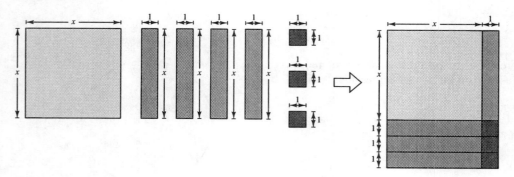

2—**Answer:** $x^2 + 4x + 3 = (x + 3)(x + 1)$

45. Write a polynomial and the correct factorization for the geometric factoring model.

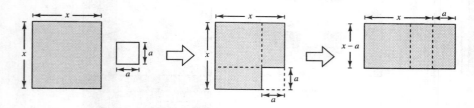

2—**Answer:** $x^2 - a^2 = (x + a)(x - a)$

❏ P.5 Fractional Expressions

1. Find the domain: $\frac{1}{2}x^2 + 2x + 1$.
   - (a) $(-\infty, \infty)$
   - (b) $[0, \infty)$
   - (c) $(-\infty, \frac{1}{2})(\frac{1}{2}, \infty)$
   - (d) $[\frac{1}{2}, \infty)$
   - (e) None of these

   1—Answer: a

2. Find the domain: $\dfrac{3x + 1}{x^2 - 2x}$.
   - (a) $(-\infty, 0)(0, \infty)$
   - (b) $(-\infty, -2)(0, \infty)$
   - (c) $(-\infty, 0)(0, 2)(2, \infty)$
   - (d) $\left(-\infty, \frac{1}{3}\right)\left(-\frac{1}{3}, 0\right)(0, 2)(2, \infty)$
   - (e) None of these

   1—Answer: c

3. Find the domain: $\dfrac{x + 1}{4 - x^2}$.
   - (a) $(-\infty, \infty)$
   - (b) $(-\infty, -1)(-1, 2)(2, \infty)$
   - (c) $(-\infty, -2)(2, \infty)$
   - (d) $(-\infty, -2)(-2, -1)(-1, 2)(2, \infty)$
   - (e) None of these

   1—Answer: e

4. Find the domain: $\sqrt{x + 2}$.
   - (a) $(-\infty, \infty)$
   - (b) $[0, \infty)$
   - (c) $[-2, \infty)$
   - (d) $(-\infty, -2)$
   - (e) None of these

   1—Answer: c

5. Find the domain: $\sqrt{3 - x}$.
   - (a) $(-\infty, 3]$
   - (b) $[3, \infty)$
   - (c) $(-\infty, 0]$
   - (d) $(-3, 3)$
   - (e) None of these

   1—Answer: a

6. Reduce: $\dfrac{x^2 - 8x + 12}{5x - 30}$.
   - (a) $\dfrac{x - 2}{5}$
   - (b) $\dfrac{2 + x}{-5}$
   - (c) $\dfrac{x + 2}{5}$
   - (d) $\dfrac{-x - 2}{5}$
   - (e) None of these

   1—Answer: a

7. Reduce to lowest terms: $\dfrac{2x^2 + 5x - 3}{6x - 3}$.

   1—Answer: $\dfrac{x + 3}{3}$

8. Reduce to lowest terms: $\dfrac{4x - 2x^2}{x^2 + x - 6}$.

   1—Answer: $\dfrac{-2x}{x + 3}$

9. Reduce: $\dfrac{x^2 - 7x + 10}{x^2 - 8x + 15}$.

   (a) $\dfrac{x + 2}{x + 3}$ (b) $\dfrac{x - 2}{x - 3}$ (c) $\dfrac{x + 3}{x + 2}$

   (d) $\dfrac{x - 2}{x + 3}$ (e) None of these

   1—Answer: b

10. Reduce: $\dfrac{x^2 + 3x - 10}{x^2 + 2x - 15}$.

    (a) $\dfrac{x - 2}{x - 3}$ (b) $\dfrac{x - 3}{x - 2}$ (c) $\dfrac{x + 2}{x - 3}$

    (d) $\dfrac{x + 2}{x + 3}$ (e) None of these

    1—Answer: a

11. Multiply: $\dfrac{1}{x + y}\left(\dfrac{x}{y} + \dfrac{y}{x}\right)$.

    (a) $\dfrac{1}{y} + \dfrac{1}{x}$ (b) 1   (c) $\dfrac{x + y}{xy}$

    (d) $\dfrac{x^2 + y^2}{xy(x + y)}$ (e) None of these

    1—Answer: d

12. Multiply, then simplify: $\dfrac{2 - x}{x^2 + 4} \cdot \dfrac{x + 2}{x^2 + 5x - 14}$.

    (a) $-\dfrac{x + 2}{(x^2 + 4)(x + 7)}$ (b) $\dfrac{1}{(x + 2)(x + 7)}$ (c) $\dfrac{x + 2}{(x^2 + 4)(x + 7)}$

    (d) $\dfrac{-1}{(x + 2)(x + 7)}$ (e) None of these

    1—Answer: a

13. Multiply, then simplify: $\dfrac{x^2 - 5x + 4}{x^2 + 4} \cdot \dfrac{x + 2}{x^2 + 3x - 4}$.

    1—Answer: $\dfrac{x^2 - 2x - 8}{(x^2 + 4)(x + 4)}$

14. Multiply, then simplify: $\dfrac{x^2 + 4x + 4}{x - 2} \cdot \dfrac{2 - x}{3x + 6}$.

   (a) $\dfrac{x - 2}{3}$ (b) $\dfrac{x + 2}{3}$ (c) $-\dfrac{x - 2}{3}$

   (d) $-\dfrac{x + 2}{3}$ (e) None of these

   1—Answer: d

15. Divide, then simplify: $\dfrac{x + 1}{x^2 - 1} \div \dfrac{x^2 + 1}{x - 1}$.

   1—Answer: $\dfrac{1}{x^2 + 1}$

16. Divide: $\dfrac{x + y}{x^3 - x^2} \div \dfrac{x^2 + y^2}{x^2 - x}$.

   (a) $\dfrac{1}{x(x + y)}$ (b) $\dfrac{x + y}{x(x^2 + y^2)}$ (c) $\dfrac{x(x^2 + y^2)}{x + y}$

   (d) $-x$ (e) None of these

   1—Answer: b

17. Divide: $\dfrac{4x - 16}{5x + 15} \div \dfrac{4 - x}{2x + 6}$.

   (a) 0 (b) $-\dfrac{4(x - 4)}{5(x + 3)^2}$ (c) $-\dfrac{8}{5}$

   (d) $-\dfrac{3}{10}$ (e) None of these

   1—Answer: c

18. Divide: $\dfrac{x^2 - 2x - 63}{x + 1} \div \dfrac{9 - x}{x^2 + x}$.

   (a) 585 (b) $-x(x + 7)$ (c) $x^2 + 7x$

   (d) $\dfrac{x^2 - 2x - 63}{9 - x}$ (e) None of these

   1—Answer: b

19. Subtract, then simplify: $\dfrac{1}{x} - \dfrac{x}{2y}$.

   (a) $\dfrac{2y - x}{2xy}$ (b) $\dfrac{1 - x}{2xy}$ (c) $\dfrac{1 - x}{x - 2y}$

   (d) $\dfrac{2y - x^2}{2xy}$ (e) None of these

   1—Answer: d

20. Subtract, then simplify: $\dfrac{2}{x-3} - \dfrac{1}{x+2}$.

(a) $\dfrac{1}{(x-3)(x+2)}$ (b) $\dfrac{x-1}{(x-3)(x+2)}$ (c) $\dfrac{x+7}{(x-3)(x+2)}$

(d) $\dfrac{x+1}{(x-3)(x+2)}$ (e) None of these

1—Answer: c

21. Subtract, then simplify: $\dfrac{3}{x} - \dfrac{9}{x+1}$.

1—Answer: $\dfrac{3(1-2x)}{x(x+1)}$

22. Add, then simplify: $\dfrac{4}{x+2} + \dfrac{7}{x-3}$.

(a) $\dfrac{11}{x-1}$ (b) $\dfrac{11}{2x-1}$ (c) $\dfrac{11x+2}{(x+2)(x-3)}$

(d) $\dfrac{11x+2}{x^2-6}$ (e) None of these

1—Answer: c

23. Add, then simplify: $\dfrac{2}{x^2-9} + \dfrac{5}{x^2-x-12}$.

(a) $\dfrac{7}{(x^2-9)(x^2-x-12)}$ (b) $\dfrac{7x^2-x-21}{(x^2-9)(x^2-x-12)}$ (c) $\dfrac{7x-7}{(x-3)(x-4)(x+3)}$

(d) $\dfrac{7x-23}{(x-3)(x+3)(x-4)}$ (e) None of these

2—Answer: d

24. Subtract, then simplify: $\dfrac{3}{x^2+2x+1} - \dfrac{1}{x+1}$.

(a) $\dfrac{4-x}{x^2+2x+1}$ (b) $\dfrac{-x^2+5x+2}{(x+1)(x^2+2x+1)}$ (c) $\dfrac{-x^2+x+2}{(x^2+2x+1)(x+1)}$

(d) $\dfrac{2-x}{x^2+2x+1}$ (e) None of these

2—Answer: d

25. Add, then simplify: $\dfrac{3}{x^2+x-2} + \dfrac{x}{x^2-x-6}$.

(a) $\dfrac{x^2+2x-9}{(x-3)(x+2)(x-1)}$ (b) $\dfrac{x^3+3x^2-5x-6}{(x^2+x-2)(x^2-x-6)}$ (c) $\dfrac{4x-10}{(x-3)(x+2)(x-1)}$

(d) $\dfrac{3+x}{2x^2-8}$ (e) None of these

2—Answer: a

26. Add, then simplify: $\dfrac{x+1}{x^2+x-2} + \dfrac{x+3}{x^2-4x+3}$.

   (a) $\dfrac{2x^2+3}{(x-1)(x-3)(x+2)}$
   (b) $\dfrac{2x^3+x^2-3}{(x^2+x-2)(x^2-4x+3)}$
   (c) $\dfrac{2x+4}{-3x+1}$

   (d) $\dfrac{2x^2+3x+3}{(x-1)(x-3)(x+2)}$
   (e) None of these

   2—Answer: d

27. Simplify: $\dfrac{\dfrac{2}{x+1} - \dfrac{x}{x+2}}{\dfrac{1}{x+1}}$.

   (a) $\dfrac{-x^2+3x+4}{x+2}$
   (b) $\dfrac{-x^2+x+4}{x+2}$
   (c) $\dfrac{2-x}{x+2}$

   (d) $\dfrac{-x^2+3x+3}{(x+1)(x+2)}$
   (e) None of these

   1—Answer: b

28. Simplify: $\dfrac{\dfrac{1}{x} - \dfrac{1}{x+1}}{\dfrac{1}{x^2+2x+1}}$.

   (a) $x+1$
   (b) $\dfrac{x}{x+1}$
   (c) $\dfrac{x+1}{x}$

   (d) $\dfrac{1}{x(x+1)(x^2+2x+1)}$
   (e) None of these

   2—Answer: c

29. Simplify: $\dfrac{\dfrac{1}{x} + \dfrac{7}{x+1}}{\dfrac{1}{x^2-1}}$.

   (a) $\dfrac{8x^2-7x-1}{x}$
   (b) $\dfrac{8x^2-1}{x}$
   (c) $\dfrac{7x^2-6x+1}{x}$

   (d) $\dfrac{8x+1}{x(x+1)(x^2-1)}$
   (e) None of these

   2—Answer: a

30. Simplify: $\dfrac{\dfrac{1}{x} - \dfrac{1}{y}}{xy}$.

   (a) $\dfrac{1}{x^2y^2}$
   (b) $\dfrac{y-x}{x^2y^2}$
   (c) $\dfrac{x-y}{x^2y^2}$

   (d) $y-x$
   (e) None of these

   1—Answer: b

38 Chapter P Prerequisites

31. Simplify: $\dfrac{(3x + 5)^{1/3} - \dfrac{x}{(3x + 5)^{2/3}}}{(3x + 5)^{2/3}}$.

(a) $\dfrac{x - 1}{(3x + 5)^{4/3}}$ (b) $\dfrac{9x^2 + 29x + 25}{(3x + 5)^{2/3}}$ (c) $\dfrac{2x + 5}{(3x + 5)^{2/3}}$

(d) $\dfrac{2x + 5}{(3x + 5)^{4/3}}$ (e) None of these

2—Answer: d

32. Simplify: $\dfrac{(7x + 2)^{1/3} - \dfrac{x}{(7x + 2)^{2/3}}}{(7x + 2)^{2/3}}$.

(a) $\dfrac{6x + 2}{(7x + 2)^{4/3}}$ (b) $\dfrac{6x + 2}{(7x + 2)^{2/3}}$ (c) $\dfrac{49x^2 + 13x + 4}{(7x + 2)^{4/3}}$

(d) $\dfrac{x + 4}{(7x + 2)^{4/3}}$ (e) None of these

2—Answer: a

33. Simplify: $\dfrac{(x + 2)^{1/2}}{(x + 2)^{1/2} - 4(x + 2)^{3/2}}$.

2—Answer: $-\dfrac{1}{4x + 7}$

34. Simplify: $\dfrac{\sqrt{x} + (6/\sqrt{x})}{\sqrt{x}}$.

(a) $\dfrac{6}{x}$ (b) $1 + 6\sqrt{x}$ (c) $\dfrac{x + 6\sqrt{x}}{x}$

(d) $\dfrac{x + 6}{x}$ (e) None of these

2—Answer: d

35. Simplify: $\dfrac{\sqrt{1 + x} - (x/\sqrt{1 + x})}{1 + x}$.

(a) $\dfrac{1 + 2\sqrt{1 + x}}{1 + x}$ (b) $\dfrac{-x + \sqrt{1 + x}}{(1 + x)\sqrt{1 + x}}$ (c) $\dfrac{1}{1 + x}$

(d) $\dfrac{\sqrt{1 + x}}{(1 + x)^2}$ (e) None of these

2—Answer: d

36. Simplify: $\dfrac{(3/\sqrt{x + 2}) - \sqrt{x + 2}}{5\sqrt{x + 2}}$.

2—Answer: $\dfrac{1 - x}{5(x + 2)}$

Section P.5 Fractional Expressions 39

37. Rationalize the denominator: $\dfrac{x}{3 - \sqrt{x+9}}$.

   2—Answer: $-(3 + \sqrt{x+9})$

38. The efficiency of a Carnot engine can be determined by using the formula $\text{Eff} = 1 - \dfrac{T_2}{T_1}$. Write this as a single fraction.

   (a) $\dfrac{T_1}{1 - T_2}$
   (b) $\dfrac{1 - T_2}{T_1}$
   (c) $\dfrac{T_1 - T_2}{T_1}$
   (d) $\dfrac{T_1 - T_2}{T_1 T_2}$
   (e) None of these

   1—Answer: c

39. After working for t hours on a common task, the fractional part of the job done by three workers is $t/2$, $t/3$, and $t/12$. What fractional part of the task has been completed?

   (a) $\dfrac{3t}{17}$
   (b) $\dfrac{11t}{12}$
   (c) $\dfrac{t^3}{72}$
   (d) $\dfrac{t}{4}$
   (e) None of these

   1—Answer: b

40. After working for t hours on a common task, the fractional part of the job done by three workers is $t/5$, $t/3$, and $2t/5$. What fractional part of the task has been completed?

   (a) $\dfrac{4t}{13}$
   (b) $\dfrac{4t}{15}$
   (c) $\dfrac{9t}{15}$
   (d) $\dfrac{14t}{15}$
   (e) None of these

   1—Answer: d

41. After working for t hours on a common task, the fractional part of the job done by three workers is $t/3$ and $t/2$. What fractional part of the task has *not* been completed?

   (a) $\dfrac{5t}{6}$
   (b) $\dfrac{3t}{5}$
   (c) $\dfrac{t}{6}$
   (d) $\dfrac{6 - t^2}{6}$
   (e) None of these

   1—Answer: c

42. After working for t hours on a common task, the fractional part of the job done by three workers is $t/5$ and $t/2$. What fractional part of the task has *not* been completed?

   (a) $\dfrac{7t}{10}$
   (b) $\dfrac{3t}{10}$
   (c) $\dfrac{4t}{5}$
   (d) $\dfrac{10 - t^2}{10}$
   (e) None of these

   1—Answer: b

43. Determine the average of the two real numbers $\dfrac{x}{10}$ and $\dfrac{x}{2}$.

   1—Answer: $\dfrac{3x}{10}$

44. Determine the three real numbers $\dfrac{x}{5}$, $\dfrac{x}{3}$, and $\dfrac{x}{6}$.

   1—Answer: $\dfrac{7x}{30}$

45. Find two real numbers that divide the real number line between $\dfrac{x}{9}$ and $\dfrac{x}{2}$ into three equal parts.

   2—Answer: $x_1 = \dfrac{13x}{54}, x_2 = \dfrac{10x}{27}$

46. Find three real numbers that divide the real number line between $\dfrac{x}{10}$ and $\dfrac{x}{5}$ into four equal parts.

   2—Answer: $x_1 = \dfrac{x}{8}, x_2 = \dfrac{3x}{20}$, and $x_3 = \dfrac{7x}{40}$

47. A marble is tossed in to a box whose base is shown. Find the probability that the marble will come to rest in the shaded portion of the box.

   2—Answer: $\dfrac{x}{4(x+2)}$

48. A marble is tossed in to a triangle whose base is shown. Find the probability that the marble will come to rest in the shaded portion of the triangle.

   2—Answer: $\dfrac{x+2}{4(x+3)}$

P.6 Errors and the Algebra of Calculus

1. Insert the required factor in the parentheses: $5x^3(7 - 2x^4)^5 = (\quad)(-40x^3)(7 - 2x^4)^5$.

   1—Answer: $-\frac{1}{8}$

2. Insert the required factor in the parentheses:
   $$\frac{3x^2 + 2x}{(2x^3 + 2x^2 - 1)^2} = (\quad)\frac{1}{(2x^3 + 2x^2 - 1)^2}(6x^2 + 4x)$$

   (a) $\frac{1}{2}$ (b) 2   (c) 4

   (d) $\frac{1}{4}$ (e) None of these

   1—Answer: a

3. Insert the required factor in the parentheses: $3x(2x + 1)^{1/2} + 5(2x + 1)^{3/2} = (2x + 1)^{1/2}(\quad)$.

   2—Answer: $13x + 5$

4. Insert the required factor in the parentheses: $5\sqrt{x + 3} - \frac{5x}{2\sqrt{x + 3}} = \frac{1}{2\sqrt{x + 3}}(\quad)$.

   (a) $2\sqrt{x + 3} - 5x$ (b) $5x + 30$ (c) 15

   (d) $5 - 5x$ (e) None of these

   1—Answer: b

5. Insert the required factor in the parentheses: $\frac{3}{4}(1 - x)^{2/3} + \frac{7}{8}(1 - x)^{5/3} = \frac{(1 - x)^{2/3}}{8}(\quad)$.

   2—Answer: $13 - 7x$

6. Insert the required factor in the parentheses: $-2x(3x^2 - 4)^{1/2} = (\quad)(3x^2 - 4)^{1/2}(6x)$.

   (a) $3x$ (b) $\left(-\frac{1}{3}\right)$ (c) $\left(\frac{1}{3}\right)$

   (d) 3   (e) None of these

   1—Answer: b

7. Insert the required factor in the parentheses: $\frac{-3}{16x^2 + 1} = (\quad)\frac{4}{16x^2 + 1}$.

   (a) $-\frac{3}{4}$ (b) $-\frac{4}{3}$ (c) $\frac{4}{3}$

   (d) $\frac{3}{4}$ (e) None of these

   1—Answer: a

8. Determine b^2: $\dfrac{4x^2}{7} + \dfrac{9y^2}{4} = \dfrac{x^2}{a^2} + \dfrac{y^2}{b^2}$.

   (a) $\dfrac{9}{4}$ (b) $\dfrac{4}{9}$ (c) 4

   (d) $\dfrac{1}{4}$ (e) None of these

   1—Answer: b

9. Determine a^2: $\dfrac{3x^2}{2} + \dfrac{7y^2}{9} = \dfrac{x^2}{a^2} + \dfrac{y^2}{b^2}$.

   (a) $\dfrac{7}{9}$ (b) $\dfrac{9}{7}$ (c) $\dfrac{3}{2}$

   (d) $\dfrac{2}{3}$ (e) None of these

   1–Answer: d

10. Determine a^2: $\dfrac{4x^2}{25} - \dfrac{y^2}{4} = \dfrac{x^2}{a^2} - \dfrac{y^2}{b^2}$.

    (a) $\dfrac{1}{4}$ (b) 4   (c) $\dfrac{25}{4}$

    (d) $\dfrac{4}{25}$ (e) None of these

    1—Answer: c

11. Rewrite $\dfrac{9x^2}{25} + 4y^2 = 1$ in the form $\dfrac{x^2}{a^2} + \dfrac{y^2}{b^2} = 1$.

    1—Answer: $\dfrac{x^2}{(5/3)^2} + \dfrac{y^2}{(1/2)^2} = 1$

12. Rewrite $12x^2 - 9y^2 = 16$ in the form $\dfrac{x^2}{a^2} - \dfrac{y^2}{b^2} = 1$.

    1—Answer: $\dfrac{x^2}{(2/\sqrt{3})^2} - \dfrac{y^2}{(4/3)^2} = 1$

13. Solve for a: $\dfrac{9x^2}{49} - \dfrac{4y^2}{25} = \dfrac{x^2}{a} - \dfrac{y^2}{b}$.

    (a) $\dfrac{9}{49}$ (b) $\dfrac{49}{9}$ (c) $\dfrac{3}{7}$

    (d) $\dfrac{7}{3}$ (e) None of these

    2—Answer: b

14. Rewrite the fraction as the sum of two terms: $\dfrac{4x^3 - 7x^2}{2x}$.

    1—Answer: $2x^2 - \dfrac{7}{2}x$

15. Rewrite the fraction as the sum of three terms: $\dfrac{3x^2 - 2x - 6}{3\sqrt{x}}$.

    1—Answer: $x^{3/2} - \dfrac{2}{3}x^{1/2} - 2x^{-1/2}$

16. Rewrite the fractions with no variables in the denominators: $\dfrac{1}{(2x)^2} - \dfrac{2}{\sqrt{x}} + \dfrac{5}{2x^3}$.

    1—Answer: $\dfrac{1}{4}x^{-2} - 2x^{-1/2} + \dfrac{5}{2}x^{-3}$

17. Write as a sum of terms: $\dfrac{x - 2x^2 + x^3}{\sqrt{x}}$.

    (a) $x - 2x^2 + x^3 - x^{1/2}$
    (b) $x^{-1/2} + 2x^{-2} - x^{-3} + x^{1/2}$
    (c) $x^{1/2} - 2x^{1/2} + x^{3/2}$
    (d) $x^{1/2} - 2x^{3/2} + x^{5/2}$
    (e) None of these

    1—Answer: d

18. Write as a sum of terms: $\dfrac{4x^3 - 3x^2 + 1}{x^{3/2}}$.

    (a) $4x^{9/2} - 3x^3 + x^{3/2}$
    (b) $4x^3 - 3x^2 + 1 - x^{3/2}$
    (c) $4x^{1/2} - 3x^{-1/2} + x^{-3/2}$
    (d) $4x^{-3/2} - 3x^{-1/2} + x^{-3/2}$
    (e) None of these

    1—Answer: e

19. Write as a sum of terms: $\dfrac{3x^4 - 2x^2 + 1}{\sqrt[3]{x}}$.

    (a) $3x^4 - 2x^2 + x^{-3}$
    (b) $3x^4 - 2x^2 + 1 - x^{1/3}$
    (c) $3x^{11/3} - 2x^{5/3} + x^{-1/3}$
    (d) $3x - 2^{-1} + x^{-1/3}$
    (e) None of these

    1—Answer: c

20. Simplify: $\dfrac{3}{4}x^2 - \dfrac{1}{2}x + \dfrac{5}{6}$.

    (a) $\dfrac{1}{6}(5x^2 - 3x + 5)$
    (b) $\dfrac{1}{12}(9x^2 - 6x + 10)$
    (c) $\dfrac{3}{4}\left(x^2 - \dfrac{1}{2}x + \dfrac{5}{6}\right)$
    (d) $12(9x - 5)(x - 2)$
    (e) None of these

    1—Answer: b

21. Simplify: $\dfrac{2}{3}x^3 + \dfrac{1}{7}x^2 - x$.

    (a) $14x^3 + 3x^2 - 21x$
    (b) $\dfrac{x(7x - 3)(2x + 7)}{21}$
    (c) $x\left(\dfrac{2}{3}x + 1\right)(x - 1)$
    (d) $\dfrac{x}{21}(14x^2 + 3x - 21)$
    (e) None of these

    1—Answer: d

22. Simplify: $\frac{5}{2}x^2 + \frac{1}{3}x$.

   (a) $x\left(\frac{5}{2}x - \frac{1}{3}\right)$ (b) $\frac{x}{6}(15x + 2)$ (c) $\frac{x}{6}(10x + 3)$

   (d) $\frac{1}{6}x(5x + 1)$ (e) None of these

   1—Answer: b

23. Simplify: $\frac{2}{3}x^3 + \frac{1}{8}x^2$.

   (a) $\frac{x}{24}(16x^2 + 3)$ (b) $\frac{x^2(16x + 3)}{24}$ (c) $\frac{x^2(2x + 1)}{24}$

   (d) $x^2\left(\frac{2}{3}x - \frac{1}{8}\right)$ (e) None of these

   1—Answer: b

24. Simplify: $\frac{8}{3}x^{5/3} + \frac{2}{3}x^{-1/3} - \frac{1}{3}x^{-4/3}$.

   (a) $\frac{8}{3}x^{5/3}\left(1 + \frac{1}{4}x^2 - \frac{1}{8}x^{1/3}\right)$ (b) $\frac{x^{5/3}}{3}(8 + 2x^{-2} - x^{-4})$ (c) $\frac{8x^3 + 2x - 1}{3x^{4/3}}$

   (d) $\frac{8x^{1/3} + 2x^{-5/3} - 1}{3x^{4/3}}$ (e) None of these

   2—Answer: c

25. Simplify: $\frac{1}{8}(3x + 1)^{3/2} + \frac{1}{4}(3x + 1)^{1/2}$.

   2—Answer: $\frac{3}{8}(3x + 1)^{1/2}(x + 1)$

26. Simplify: $\frac{(5x - 2)^{1/2}(2) - 2x\left(\frac{1}{2}\right)(5x - 2)^{-1/2}(5)}{(\sqrt{5x - 2})^2}$.

   (a) $\frac{2\sqrt{5x - 2} - 5x}{5x - 2}$ (b) $\frac{-4}{(5x - 2)}$ (c) $\frac{25x^2 + 2x + 8}{(5x - 2)^{3/2}}$

   (d) $\frac{5x - 4}{(5x - 2)^{3/2}}$ (e) None of these

   2—Answer: d

27. Simplify: $\frac{3x\left(\frac{5}{2}\right)(2x - 1)^{3/2} - (2x - 1)^{5/2}(3)}{(3x)^2}$.

   2—Answer: $-\frac{(x + 2)(2x - 1)^{3/2}}{6x^2}$

Section P.6 Errors and the Algebra of Calculus 45

28. Simplify: $\dfrac{(3x+1)^{3/2}(2) - 2x(\frac{3}{2})(3x+1)^{1/2}(3)}{[(3x+1)^{3/2}]^2}$.

   (a) $\dfrac{6x-7}{(3x+1)^{7/4}}$ (b) $\dfrac{6x-7}{(3x+1)^{5/2}}$ (c) $\dfrac{2-3x}{(3x+1)^{5/2}}$

   (d) $\dfrac{1-3x}{(3x+1)^2}$ (e) None of these

   2—Answer: c

29. Simplify: $7x(-5)(3-2x)^{-6}(-2) + (3-2x)^{-5}(7)$.

   (a) $\dfrac{70x+21}{(3-2x)^5}$ (b) $\dfrac{-140x^2+210x+7}{(3-2x)^6}$ (c) $\dfrac{68x+21}{(3-2x)^5}$

   (d) $\dfrac{56x+21}{(3-2x)^6}$ (e) None of these

   2—Answer: d

30. Simplify: $5x^2(-2)(3+2x)^{-2} + (3+2x)^{-1}(10x)$.

   (a) $\dfrac{10x(x+3)}{(3+2x)^2}$ (b) $\dfrac{2x(15-4x)}{(3+2x)^2}$ (c) $\dfrac{10x(1-x)}{(3+2x)}$

   (d) $\dfrac{-6}{3+2x}$ (e) None of these

   2—Answer: a

31. Simplify: $\dfrac{4}{3}x^3(7x+1)^{-2/3} + (7x+1)^{1/3}(12x^2)$.

   (a) $\dfrac{4x^3+12x^2(7x+1)^{1/3}}{3(7x+1)^{2/3}}$ (b) $\dfrac{x(91x+7)}{3(7x+1)^{2/3}}$ (c) $\dfrac{4x(64x+9)}{3(7x+1)^{2/3}}$

   (d) $\dfrac{4x^2(64x+9)}{3(7x+1)^{2/3}}$ (e) None of these

   2—Answer: c

32. Insert the required factor in the parentheses: $\dfrac{30x^2+18}{(5x^3+9x+2)^3} = (\quad)\dfrac{15x^2+9}{(5x^3+9x+2)^3}$.

   1—Answer: 2

33. Insert the required factor in the parentheses:

   $\frac{3}{2}x^2(3x+2)^{-3/4} + (3x+2)^{1/4}(6x^2) = \frac{9}{2}x^2(3x+2)^{-3/4}(\quad)$.

   (a) $4x+3$ (b) $4x+1$ (c) $12x+8$

   (d) $18x+12$ (e) None of these

   2—Answer: a

34. Insert the required factor: $\dfrac{\frac{3}{2}x^2 + \frac{1}{4}}{2} = (\quad)(6x^2 + 1)$.

    (a) $\dfrac{1}{6}$ (b) $\dfrac{1}{4}$ (c) $\dfrac{1}{8}$

    (d) $\dfrac{1}{2}$ (e) None of these

    1—Answer: c

35. Factor: $5x^2(x+3)^{1/2} + \frac{3}{2}(x+3)^{3/2}$.

    (a) $3(x+3)^{1/2}(5x^2 + 2x + 6)$ (b) $\frac{1}{3}(x+3)^{1/2}(15x^2 + 2x + 3)$

    (c) $3(x+3)^{1/2}(15x^2 + 2x + 6)$ (d) $\frac{1}{3}(x+3)^{1/2}(15x^2 + 2x + 6)$

    (e) None of these

    2—Answer: d

36. Factor: $\dfrac{2}{3}(x+5)^{4/3} - \dfrac{1}{5}(x+5)^{7/3}$.

    (a) $\dfrac{7}{15}(x+5)$ (b) $\dfrac{(x+5)^{4/3}}{15}(-5 - 3x)$ (c) $15(x+5)^{4/3}(5 - 3x)$

    (d) $15(x+5)^{4/3}(15 - 3x)$ (e) None of these

    2—Answer: b

❑ P.7 Graphical Representation of Data

1. The triangle shown in the figure has vertices at the points $(-1, -1)$, $(-1, 2)$, and $(1, 1)$. Shift the triangle 3 units to the right and 2 units down and find the vertices of the shifted triangle.

   (a) $(2, 1), (2, 4), (4, 3)$

   (b) $(-4, 1), (-4, 4), (-2, 3)$

   (c) $(-4, -3), (-4, 0), (-2, -1)$

   (d) $(2, -3), (2, 0), (4, -1)$

   (e) None of these

   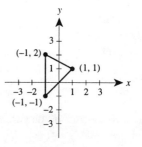

   1—Answer: d

2. The triangle shown in the figure has vertices at the points $(-1, 2)$, $(1, 2)$ and $(0, 0)$. Shift the triangle 2 units up and find the vertices of the shifted triangle.

   (a) $(1, 2), (3, 2), (2, 0)$

   (b) $(-1, 4), (1, 4), (0, 2)$

   (c) $(-1, 0), (1, 0), (0, -2)$

   (d) $(-3, 2), (-1, 2), (-2, 2)$

   (e) None of these

   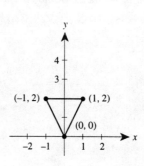

   1—Answer: b

3. The triangle shown in the figure has vertices at the points $(-1, 2)$, $(1, 2)$, and $(0, 0)$. Shift the triangle 3 units to the left and find the vertices of the shifted triangle.

   (a) $(-1, -1), (1, -1), (0, -3)$

   (b) $(-4, 2), (-2, 2), (-3, 0)$

   (c) $(-1, -1), (-1, 1), (-3, 0)$

   (d) $(2, 2), (-2, 2), (-3, 0)$

   (e) None of these

   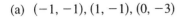

   1—Answer: b

4. Find the distance between the points $(3, 17)$ and $(-2, 5)$.

   (a) 13   (b) $\sqrt{145}$ (c) $\sqrt{485}$

   (d) $3\sqrt{51}$ (e) None of these

   1—Answer: a

5. Find the distance between the points $(-6, 10)$ and $(12, 2)$.

   (a) $2\sqrt{7}$ (b) $2\sqrt{97}$ (c) 10

   (d) $2\sqrt{65}$ (e) None of these

   1—Answer: b

6. Find the distance between the points $(3, -1)$ and $(7, 2)$.

   1—Answer: 5

7. Find the distance between the points $(3, 5)$ and $(-2, -1)$.

   1—Answer: $\sqrt{61}$

8. Find the midpoint of the line segment joining $(3, 7)$ and $(-6, 1)$.

   (a) $\left(-\frac{3}{2}, 4\right)$ (b) $\left(\frac{9}{2}, 3\right)$ (c) $(-3, 6)$

   (d) $(-3, 4)$ (e) None of these

   1—Answer: a

9. Find the midpoint of the line segment joining $(-3, 1)$ and $(5, -7)$.

   (a) $(-4, 4)$ (b) $(1, -3)$ (c) $(-4, -3)$

   (d) $(1, 4)$ (e) None of these

   1—Answer: b

10. Find the midpoint of the line segment joining $(-2, 1)$ and $(16, 3)$.

    (a) $(7, 2)$ (b) $(9, 1)$ (c) $(14, 4)$

    (d) $(-9, -1)$ (e) None of these

    1—Answer: a

11. Find the midpoint of the line segment joining $(6, 9)$ and $(-3, 1)$.

    1—Answer: $\left(\frac{3}{2}, 5\right)$

12. Find the midpoint of the line segment joining $(-6, -2)$ and $(5, -1)$.

    1—Answer: $\left(-\frac{1}{2}, -\frac{3}{2}\right)$

13. The point $(3, 2)$ is the midpoint of (x, y) and $(5, 1)$. Find the point (x, y).

    (a) $(3, 1)$ (b) $(1, 3)$ (c) $(10, 2)$

    (d) $\left(4, \frac{3}{2}\right)$ (e) None of these

    1—Answer: b

14. Find the distance between the origin and the midpoint of the two points $(3, 3)$ and $(3, 5)$.

    (a) $3\sqrt{2}$ (b) 7 (c) $\sqrt{34}$

    (d) 5 (e) None of these

    2—Answer: d

15. Find the distance between the origin and the midpoint of the two points $(2, 7)$ and $(6, 5)$.

    (a) 10 (b) $2\sqrt{13}$ (c) $4\sqrt{13}$

    (d) $2\sqrt{5}$ (e) None of these

    2—Answer: b

16. Find the distance between the origin and the midpoint of the two points $(5, 7)$ and $(-3, 1)$.

    (a) $\sqrt{17}$ (b) 5 (c) $\sqrt{10}$

    (d) 4 (e) None of these

    2—Answer: a

17. Identify the type of triangle that has $(-5, -1)$, $(2, 2)$, and $(0, -3)$ as vertices.

    (a) Scalene   (b) Right isosceles   (c) Equilateral

    (d) Isosceles   (e) None of these

    2—Answer: b

18. Identify the type of triangle that has $(1, 10)$, $(-3, -2)$, and $(3, 16)$ as vertices.

    (a) Isosceles   (b) Right   (c) Scalene

    (d) Equilateral   (e) These points do not form a triangle.

    2—Answer: e

19. Identify the type of triangle that has $(0, 0)$, $(4, 0)$ and, $(2, 2\sqrt{3})$ as vertices.

    (a) Scalene   (b) Right   (c) Isosceles

    (d) Equilateral   (e) These points do not form a triangle.

    2—Answer: d

Section P.7   Graphical Representation of Data   49

20. Identify the type of triangle that has (0, 0), (4, 0), and $(2, 4\sqrt{2})$ as vertices.

    (a) Scalene   (b) Right   (c) Isosceles

    (d) Equilateral   (e) These points do not form a triangle.

    2—Answer: c

21. Determine the quadrant in which the point (x, y) must be located if $x > 0$ and $y < 0$.

    (a) I   (b) II   (c) III

    (d) IV   (e) None of these

    1—Answer: d

22. Determine the quadrant in which the point (x, y) must be located if $x < 0$ and $y > 0$.

    (a) I   (b) II   (c) III

    (d) IV   (e) None of these

    1—Answer: b

23. Determine the quadrant(s) in which the point (x, y) must be located if $xy < 0$.

    (a) II   (b) II and III   (c) II and IV

    (d) I and III   (e) None of these

    1—Answer: c

24. Find the length of the hypotenuse of the right triangle determined by the points $(1, 1)$, $(-2, 1)$, and $(-2, 4)$.

    2—Answer: $3\sqrt{2}$

25. Find the length of the hypotenuse of the right triangle determined by the points $(-1, 1)$, $(3, 1)$, and $(3, -3)$.

    2—Answer: $4\sqrt{2}$

26. In a football game, the quarterback throws a pass from the 8-yard line, 15 yards from the sideline. The pass is caught on the 43-yard line, 3 yards from the same sideline. How long was the pass? (Assume the pass and the reception are on the same side of midfield.)

    (a) 40 yards   (b) 36.1 yards   (c) 39.4 yards

    (d) 37 yards   (e) None of these

    2—T—Answer: d

27. In a football game, the quarterback throws a pass from the 3-yard line, 10 yards from the sideline. The pass is caught on the 43-yard line, 40 yards from the same sideline. How long was the pass? (Assume the pass and the reception are on the same side of midfield.)

    (a) 60 yards   (b) 55 yards   (c) 50 yards

    (d) 45 yards   (e) None of these

    2—T—Answer: c

28. A homeowner needs to determine the distance y from the peak to the lower edge of the roof on the garage. He knows the distance from the ground to the peak is $14\frac{1}{2}$ feet and the distance from the lower edge of the roof to the ground is 11 feet. Find y if the garage is 20 feet wide.

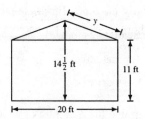

2—T—**Answer:** 10.6 feet

29. A homeowner needs to determine the distance y from the peak to the lower edge of the roof on his garage. He knows the distance from the ground to the peak is 13 feet and the distance from the lower edge of the roof to the ground is 10 feet. Find y if the garage is 18 feet wide. (Round to 1 decimal place.)

   (a) 10.1 feet   (b) 9.8 feet

   (c) 9.5 feet   (d) 9.3 feet

   (e) None of these

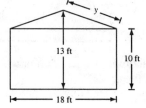

2—T—**Answer:** c

30. The accompanying figure gives the speed of a car (in mph) and the approximate stopping distance in feet.

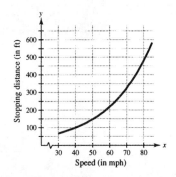

Complete the table by approximating the stopping distance.

Speed, x	30	40	50	60	70	80
Stopping distance, y						

2—**Answer:**

Speed, x	30	40	50	60	70	80
Stopping distance, y	65	95	140	205	300	450

31. The accompanying figure gives the normal Fahrenheit temperature, y, for Anchorage, Alaska, for each month, x, of the year where x = 1 represents January.

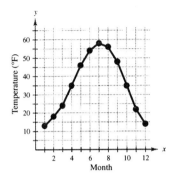

Complete the table by approximating the temperature.

Month, x	1	2	3	4	5	6	7	8	9	10	11	12
Temperature, y												

2—Answer:

Month, x	1	2	3	4	5	6	7	8	9	10	11	12
Temperature, y	13	18	24	35	46	54	58	56	48	35	22	14

CHAPTER ONE
Equations and Inequalities

❑ 1.1 Graphs and Graphing Utilities

1. Determine which of the following ordered pairs is not a solution of the equation $7x - 3y = 5$.

   (a) $(2, 3)$ (b) $\left(1, \frac{2}{3}\right)$ (c) $(4, 11)$

   (d) $\left(-\frac{1}{7}, -2\right)$ (e) All of these are solutions.

   1—Answer: c

2. Determine which of the following points does not lie on the graph of $y = \dfrac{1}{x^2 + 1}$.

   (a) $\left(-1, \dfrac{1}{2}\right)$ (b) $\left(-2, -\dfrac{1}{3}\right)$ (c) $\left(3, \dfrac{1}{10}\right)$

   (d) $\left(6, \dfrac{1}{37}\right)$ (e) All of these lie on the graph.

   1—Answer: b

3. Determine which of the following ordered pairs is a solution of the equation $y = x\sqrt{x + 1}$.

   (a) $(1, 2)$ (b) $(2, 6)$ (c) $(-1, 1)$

   (d) $(3, 6)$ (e) None of these are solutions.

   1—Answer: d

4. Complete the solutions table for $y = 4x^2 + 2$.

x	-2	0	2
y			

   1—Answer:

x	-2	0	2
y	18	2	18

5. Complete the following table for $y = 25x\left(\dfrac{4x}{x^2 + 5} + 2\right)$.

x	-2	0	2	4
y				

   1—Answer:

x	-2	0	2	4
y	$-\dfrac{500}{9}$	0	$\dfrac{1300}{9}$	$\dfrac{5800}{21}$

6. Use a graphing utility to graph $y = x^2 - 5x + 6$. Use a standard setting. Approximate any intercepts.

   2—T—Answer: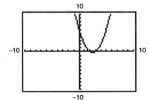

   x-intercepts: $(2, 0)$, $(3, 0)$; y-intercept: $(0, 6)$

7. Use a graphing utility to graph $y = x^2 - 2x - 3$. Use a standard setting. Approximate any intercepts.

   2—T—Answer: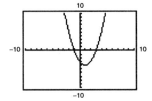

   x-intercepts: $(-1, 0)$, $(3, 0)$; y-intercept: $(0, -3)$

8. Use a graphing utility to graph $y = x^2 - 3x - 4$. Use a standard setting. Approximate any intercepts.

   2—T—Answer: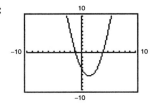

   x-intercepts: $(-1, 0)$, $(4, 0)$; y-intercept: $(0, -4)$

9. Use a graphing utility to graph $y = |x| - 2$. Use a standard setting. Approximate any intercepts.

   2—T—Answer: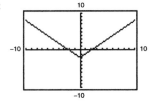

   x-intercepts: $(-2, 0)$, $(2, 0)$; y-intercept: $(0, -2)$

10. Use a graphing utility to graph $y = x(x + 6)$. Use a standard setting. Approximate any intercepts.

   2—T—Answer:
   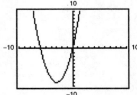

   x-intercepts: $(-6, 0), (0, 0)$; y-intercept: $(0, 0)$

11. Use a graphing utility to graph $y = \dfrac{2x}{x - 2}$. Use a standard setting. Approximate any intercepts.

   2—T—Answer:
   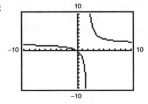

   x-intercept: $(0, 0)$; y-intercept: $(0, 0)$

12. Identify the type(s) of symmetry: $x^4 y^2 + 2x^2 y - 1 = 0$.
   (a) x-axis (b) y-axis (c) Origin
   (d) Both a and b (e) None of these
   1—Answer: b

13. Identify the type(s) of symmetry: $3x^4 + xy - 2 = 0$.
   (a) x-axis (b) y-axis (c) Origin
   (d) Both b and c (e) No symmetry
   1—Answer: c

14. Identify the type(s) of symmetry: $y = x^3 + 3x$.
   (a) x-axis (b) y-axis (c) Origin
   (d) Both a and b (e) None of these
   1—Answer: c

15. Identify the type(s) of symmetry: $y = |x| - 2$.
   1—Answer: Symmetric to the y-axis.

16. Identify the type(s) of symmetry: $x^2 + xy + y^2 = 0$.
   1—Answer: Symmetric to the origin.

17. Identify the graph of the equation: $y = |x + 7|$.

(a)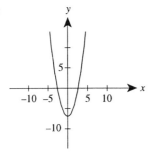

(b)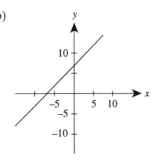

(c)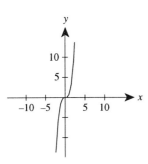

(d)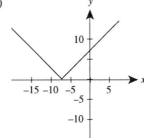

(e) None of these

1—Answer: d

18. Identify the graph of the equation: $y = \sqrt{2 - x}$.

(a)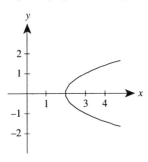

(b)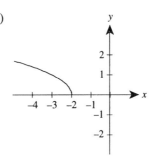

(c)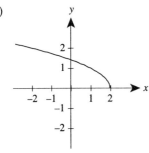

(d)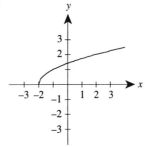

(e) None of these

1—Answer: c

19. Identify the graph of the equation: $y = \sqrt{x+3}$.

(a)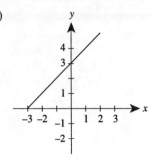

(b)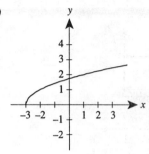

(c)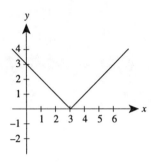

(d)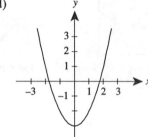

(e) None of these

1—Answer: b

20. Match the equation with the graph.

(a) $y = \sqrt{9 - x^2}$ (b) $y = |x^2 - 9|$

(c) $y = \sqrt{x^2 - 9}$ (d) $y = (9 - x)^2$

(e) None of these

1—Answer: a

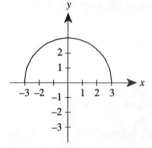

21. Match the equation with the graph.

(a) $y = \sqrt{x - 3}$ (b) $y = |x - 3|$

(c) $y = (x - 3)^2$ (d) $y = x - 3$

(e) None of these

1—Answer: b

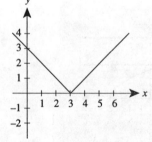

22. Describe the viewing rectangle of the graph of the equation: $y = 4x^2 - 1$.

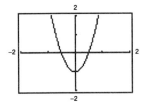

2—T—Answer:
Xmin = -2
Xmax = 2
Xscl = 1
Ymin = -2
Ymax = 2
Yscl = 1

23. Describe the viewing rectangle of the graph of the equation: $y = 2x^3 - x - 1$.

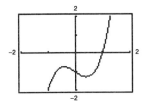

2—T—Answer:
Xmin = -2
Xmax = 2
Xscl = 1
Ymin = -2
Ymax = 2
Yscl = 1

24. Describe the viewing rectangle of the graph of the equation: $y = |x| + |x - 1|$.

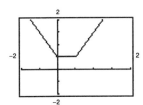

2—T—Answer:
Xmin = -2
Xmax = 4
Xscl = 1
Ymin = -2
Ymax = 4
Yscl = 1

25. Which of the following range settings was used to obtain the graph at the right?

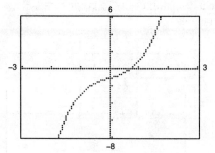

(a)
```
Xmin = -8
Xmax = 6
Xscl = 2
Ymin = -3
Ymax = 3
Yscl = 1
```

(b)
```
Xmin = -3
Xmax = 3
Xscl = 1
Ymin = -8
Ymax = 6
Yscl = 1
```

(c)
```
Xmin = -3
Xmax = 3
Xscl = 1
Ymin = -8
Ymax = 6
Yscl = 2
```

(d)
```
Xmin = -8
Xmax = 6
Xscl = 1
Ymin = -3
Ymax = 3
Yscl = 1
```

(e) None of these

1—T—Answer: c

26. Which of the following range settings was used to obtain the graph at the right?

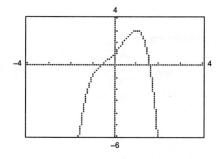

(a)
```
Xmin = -4
Xmax = 4
Xscl = 1
Ymin = -6
Ymax = 4
Yscl = 1
```

(b)
```
Xmin = -4
Xmax = 4
Xscl = 1
Ymin = -4
Ymax = 4
Yscl = 1
```

(c)
```
Xmin = -4
Xmax = 4
Xscl = 1
Ymin = -6
Ymax = 4
Yscl = 2
```

(d)
```
Xmin = -6
Xmax = 4
Xscl = 1
Ymin = -4
Ymax = 4
Yscl = 1
```

(e) None of these

1—T—Answer: a

27. Use a graphing utility to graph $y = 6 + 3x - 2x^2$. Use the standard viewing rectangle.

(a)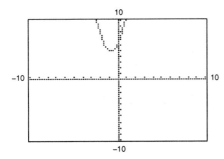

(b)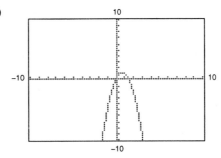

(c)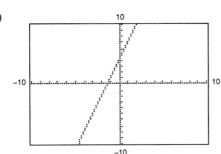

(d)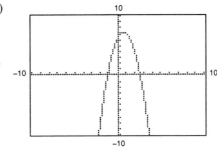

(e) None of these

1—T—Answer: d

28. Use a graphing utility to graph $y = x\sqrt{5 - x}$. Use the standard viewing rectangle.

(a)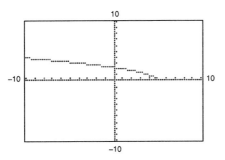

(b)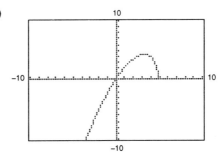

(c)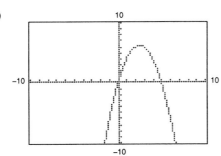

(d)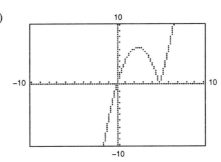

(e) None of these

1—T—Answer: b

29. Use a graphing utility to graph $y = x^3 + 4x$. Use the standard viewing rectangle.

(a)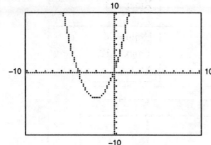

(b)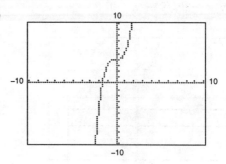

(c)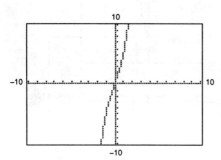

(d)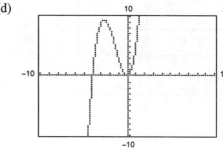

(e) None of these

1—T—Answer: c

30. Use a graphing utility and the specified viewing rectangle to graph $y = x^4 + 3x^3 - x + 4$.

```
Xmin = -5
Xmax = 3
Xscl = 1
Ymin = -4
Ymax = 12
Yscl = 2
```

(a)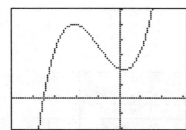

(b)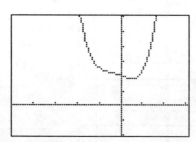

(c)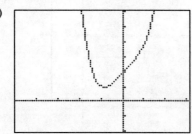

(d)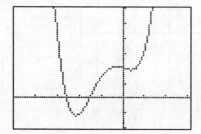

(e) None of these

1—T—Answer: d

31. Use a graphing utility and the specified viewing rectangle to graph $y = -30\sqrt{9-x}$.

```
Xmin = -5
Xmax = 3
Xscl = 1
Ymin = -4
Ymax = 12
Yscl = 2
```

(a)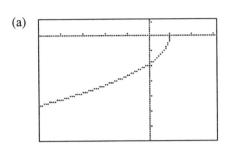

(b)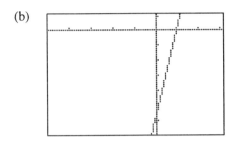

(c)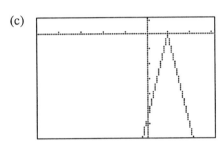

(d)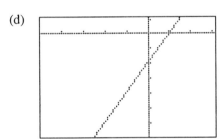

(e) None of these

1—T—Answer: a

32. Use a graphing utility and the specified viewing rectangle to graph $y = \frac{1}{4}(-x^2 - 5x + 3)$.

```
Xmin = -5
Xmax = 3
Xscl = 1
Ymin = -4
Ymax = 12
Yscl = 2
```

(a)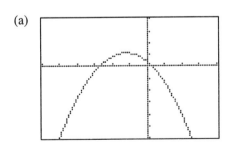

(b)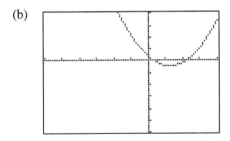

(c)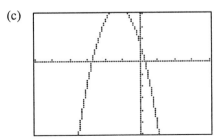

(d)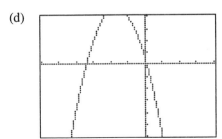

(e) None of these

1—T—Answer: a

33. Use a graphing utility and the specified viewing rectangle to graph $y = 7 - x + 2x^3$.

Xmin = -4
Xmax = 4
Xscl = 1
Ymin = -2
Ymax = 14
Yscl = 2

1—T—Answer:

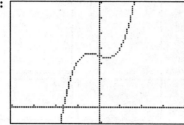

34. Use a graphing utility to graph $y = |2x + 1| - 3$.

(a)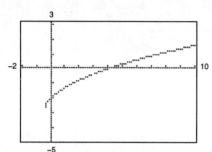

(b)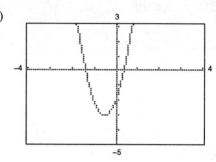

(c)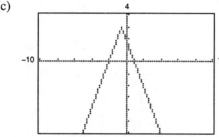

(d)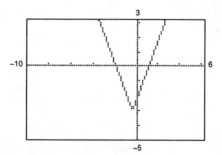

(e) None of these

1—T—Answer: d

35. Use a graphing utility to graph $y = 2x^3 + x^2 + x - 1$.

(a)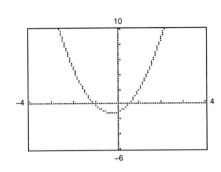

(b)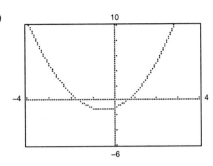

(c)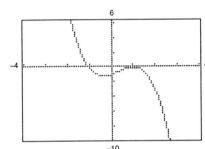

(d)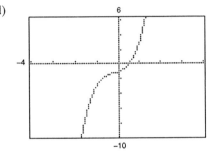

(e) None of these

1—T—Answer: d

36. Determine the standard equation of the circle with radius 3 and center $(3, -2)$.

   (a) $(x - 3)^2 + (y - 2)^2 = 3$
   (b) $(x - 3)(y - 2) = 9$
   (c) $(x - 3)^2 + (y + 2)^2 = \sqrt{3}$
   (d) $(x - 3)^2 + (y + 2)^2 = 9$
   (e) None of these

   1—Answer: d

37. Determine the standard equation of the circle with radius 5 and center $(-4, -3)$.

   (a) $(x - 4)^2 + (y - 3)^2 = 25$
   (b) $(x - 4)(x - 3) = 5$
   (c) $(x + 4)^2 + (y + 3)^2 = 5$
   (d) $(x + 4)^2 + (y - 3)^2 = 25$
   (e) None of these

   1—Answer: e

38. Determine the standard equation of the circle with radius 2 and center $(-1, 3)$.

   (a) $(x - 1)^2 + (y - 3)^2 = 2$
   (b) $(x - 1)^2 + (y - 3)^2 = 4$
   (c) $(x + 1)^2 + (y - 3)^2 = 2$
   (d) $(x + 1)^2 + (y - 3)^2 = 4$
   (e) None of these

   1—Answer: d

39. Determine the standard equation of the circle with radius 7 and center $(-2, -4)$.

   1—Answer: $(x + 2)^2 + (y + 4)^2 = 49$

40. Determine the standard equation of the circle with radius 16 and center $(0, \frac{1}{2})$.

    1—Answer: $x^2 + (y - \frac{1}{2})^2 = 256$

41. Find the center and radius of the circle: $(x - 2)^2 + (y + 4)^2 = 25$.
    - (a) Center: $(2, -4)$; radius: 5
    - (b) Center: $(-2, 4)$; radius: 5
    - (c) Center: $(2, -4)$; radius: 25
    - (d) Center: $(-2, 4)$; radius: 25
    - (e) None of these

    1—Answer: a

42. Find the center and radius of the circle: $(x + 4)^2 + (y - 1)^2 = 36$.
    - (a) Center: $(-4, 1)$; radius: 36
    - (b) Center: $(4, -1)$; radius: 36
    - (c) Center: $(-4, 1)$; radius: 6
    - (d) Center: $(4, -1)$; radius: 6
    - (e) None of these

    1—Answer: c

43. Find the center and radius of the circle: $(x + 1)^2 + (y + 2)^2 = 4$.
    - (a) Center: $(-1, -2)$; radius: 4
    - (b) Center: $(-1, -2)$; radius: 2
    - (c) Center: $(1, 2)$; radius: 4
    - (d) Center: $(1, 2)$; radius: 2
    - (e) None of these

    1—Answer: b

44. Use a graphing utility to graph $y_1 = \sqrt{64 - x^2}$ and $y_2 = -\sqrt{64 - x^2}$. Use a square setting. Identify the graph.

    2—T—Answer: a circle;

    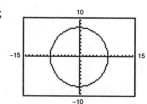

45. Use a graphing utility to graph $y_1 = \sqrt{16 - \frac{x^2}{9}}$ and $y_2 = -\sqrt{16 - \frac{x^2}{9}}$. Use a square setting. Identify the graph.

    2—T—Answer: an ellipse;

    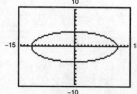

46. Use a graphing utility to graph $y_1 = 4 + \sqrt{25 - (x-1)^2}$ and $y_2 = 4 - \sqrt{25 - (x-1)^2}$. Use a square setting. Identify the graph.

2—T—Answer: a circle;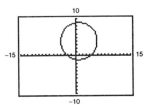

47. Find the center and radius of the circle: $\left(x - \dfrac{1}{2}\right)^2 + (y+3)^2 = \dfrac{4}{9}$.

1—T—Answer: Center: $\left(\dfrac{r}{2}, -3\right)$; radius: $\dfrac{2}{3}$

48. Find the center and radius of the circle: $(x-5)^2 + (y+4)^2 = 10$.

1—T—Answer: Center: $(5, -4)$; radius: $\sqrt{10}$

49. The earnings per share for a certain corporation from 1985 to 1990 can be approximated by the mathematical model $y = 1.23t + 0.25$ where y is the earnings and t represents the calendar year with $t = 0$ corresponding to the year 1985. Identify the graph of this equation.

(a)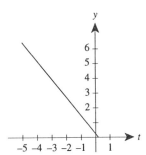

(b)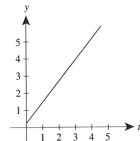

(c)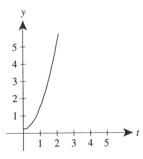

(d)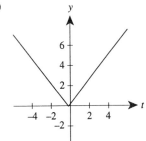

(e) None of these

1—Answer: b

50. The depreciated value y of a certain machine after t years is determined using the model $y = 36{,}000 - 4300t$, $0 \leq t \leq 5$. Sketch the graph of the equation over the given interval.

1—Answer:

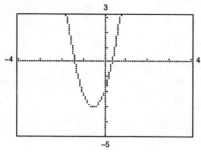

51. The depreciated value y of a certain machine after t years is determined using the model $y = 35{,}000 - 5000t$, $0 \leq t \leq 5$. Sketch the graph of the equation over the given interval.

(a)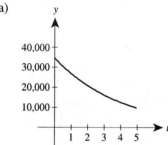

(b)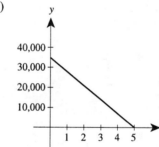

(c)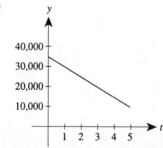

(d)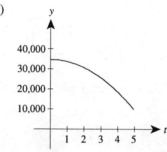

(e) None of these

1—Answer: c

52. The total cost C of a taxable item, in a state that has a 7% sales tax, is $C = 0.07p + p$ where p is the price of the item. Graph this equation: $0 \le p \le 100$.

(a)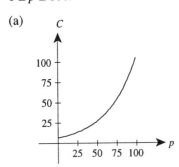

(b)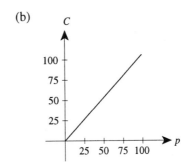

(c)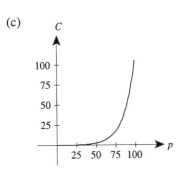

(d)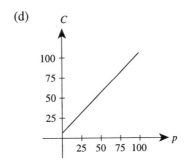

(e) None of these

1—Answer: b

53. The table gives the average attendance per professional basketball game in the United States.

Year	1983	1984	1985	1986	1987	1988	1989
Attendance per game	10,220	10,620	11,141	11,893	12,765	13,419	15,088

(Source: *National Basketball Association*)

A model for the average attendance per game between 1983 and 1989 is

$$y = 11x^4 - 252.6x^3 + 2143.1x^2 - 7279.5x + 18716.7,$$

where y is the average attendance per game and x is the year with $x = 3$ corresponding to 1983. Use a graphing utility to estimate the average attendance per game in 1986 according to the model.

(a) 14,092   (b) 12,091   (c) 11,885

(d) 11,463   (e) None of these

2—Answer: c

54. The total sales of food contractors in the United States between 1980 and 1990 can be approximated by the model

$$y = \tfrac{1}{2}(87x^2 + 604x + 13{,}634),\ 0 \le x \le 10$$

where y is the sales, in millions of dollars, and x is the year with $x = 0$ corresponding to 1980. Graph this equation. (Source: *National Restaurant Association*)

2—Answer:

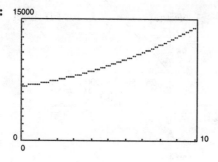

55. The formula to convert temperature in degrees Celsius to temperature in degrees Fahrenheit is $F = \tfrac{9}{5}C + 32$. Graph this equation for $0 \le C \le 100$.

1—Answer:

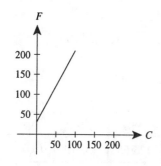

❑ 1.2 Linear Equations

1. Is $3x + 4(x - 2) = 10x$ a conditional equation or an identity?

(a) Conditional (b) Identity (c) Neither (d) Both

1—Answer: a

2. Is $3(x^2 + 2) = 5x - 9$ a conditional equation or an identity?

(a) Conditional (b) Identity (c) Neither (d) Both

1—Answer: a

3. Is $3x + 2 = 9x^2 - 1$ a conditional equation or an identity?

1—Answer: Conditional

4. Is $3(x^2 + 2x) = 7 + 3x^2$ a conditional equation or an identity?

(a) Conditional (b) Identity (c) Neither (d) Both

1—Answer: a

5. Is $2x + (3 - x)(3 + x) + 1 = 2 - (x - 4)(x + 2)$ a conditional equation or an identity?

   (a) Conditional   (b) Identity   (c) Neither   (d) Both

   1—Answer: b

6. Is $3 + \dfrac{3}{x - 1} = \dfrac{3x}{x - 1}$ a conditional equation or an identity?

   (a) Conditional   (b) Identity   (c) Neither   (d) Both

   1—Answer: b

7. Solve for x: $8x - 2 = 13 - 2x$.

   (a) $\frac{2}{3}$ (b) $\frac{3}{2}$ (c) $\frac{11}{6}$

   (d) $-\frac{2}{3}$ (e) None of these

   1—Answer: b

8. Solve the linear equation $7 - 3x + 3 = 4 + 2x - 6$.

   (a) $x = 2$ (b) $x = \frac{1}{2}$ (c) $x = \frac{12}{5}$

   (d) There is no solution.   (e) None of these

   1—Answer: c

9. Solve the linear equation $2 - 3x + 3 = 1 + 2x + 2$.

   (a) $x = -\frac{2}{5}$ (b) $x = \frac{5}{2}$ (c) $x = 0$

   (d) There is no solution.   (e) None of these

   1—Answer: e

10. Solve for x: $13x - 9 = 3x + 10$.

    (a) $\frac{19}{10}$ (b) $\frac{10}{19}$ (c) $\frac{1}{10}$

    (d) $\frac{19}{16}$ (e) None of these

    1—Answer: a

11. Solve for x: $7 - 3x + 2 = 4x - 1$.

    (a) 5   (b) 10   (c) $\frac{7}{10}$

    (d) $\frac{10}{7}$ (e) None of these

    1—Answer: d

12. Solve for x: $4 + 7x - 3x + 2 = 8x + 6$.

    (a) No solution   (b) 0   (c) 1

    (d) 2   (e) None of these

    1—Answer: b

13. Solve for x: $2 - 6 + 3x = 3x + 7$.

    (a) $x = -\frac{11}{6}$ (b) $x = \frac{11}{6}$ (c) $x = \frac{1}{6}$

    (d) There is no solution.   (e) None of these

    1—Answer: d

14. Solve for x: $0.3x - 4 = 2$.

(a) 2   (b) $\frac{9}{5}$ (c) 20

(d) $-\frac{20}{3}$ (e) None of these

1—Answer: c

15. Solve for x: $4x - 7(3x + 6) = 4x - 9$.

(a) $\frac{7}{11}$ (b) $-\frac{11}{7}$ (c) $-\frac{33}{13}$

(d) $-\frac{17}{7}$ (e) None of these

1—Answer: b

16. Solve for x: $5(3x - 2) + 5x - 7 = 16 + 2x$.

1—Answer: $\frac{11}{6}$

17. Solve for x: $2[x - (3x + 1)] = 4 - 2x$.

(a) -1 (b) -3 (c) 1

(d) $-\frac{1}{3}$ (e) None of these

2—Answer: b

18. Solve the equation $4(x + 3) + (6 - x) = 0$.

1—Answer: $x = -6$

19. Solve the equation $2(2 - x) = 3(x + 8)$.

1—Answer: $x = -4$

20. Solve the equation $3(x + 3) = 2 - (1 - 2x)$.

1—Answer: $x = -8$

21. Solve the equation $3(x - 6) = 2 + 2(x - 5)$.

1—Answer: $x = 10$

22. The solution to the linear equation $2 - 3[7 - 2(4 - x)] = 2x - 1$ is:

(a) $x = -\frac{3}{4}$ (b) $x = \frac{4}{3}$ (c) $x = \frac{3}{4}$

(d) $x = -\frac{4}{3}$ (e) None of these

1—Answer: c

23. The solution to the linear equation $3x - [x - 2(3 - 2x)] = -5$ is:

(a) $x = \frac{2}{11}$ (b) $x = \frac{11}{2}$ (c) $x = -\frac{2}{11}$

(d) $x = -\frac{11}{2}$ (e) None of these

1—Answer: b

24. The solution to the linear equation $3x - [2(3 - 2x) - x] = 4$ is:

(a) $x = \frac{5}{4}$ (b) $x = -\frac{4}{5}$ (c) $x = -\frac{5}{4}$

(d) $x = \frac{4}{5}$ (e) None of these

1—Answer: a

25. Solve the equation $3x - [5 - 2(1 - 2x)] = 7x - 5$.

1—Answer: $x = \frac{1}{4}$

26. Solve the equation $3x + [2(1 - 2x) + 5] = 5 - 7x$.

1—Answer: $x = -\frac{1}{3}$

27. Solve the equation $3x - [5 - 2(1 - 2x)] = 5 - 7x$.

1—Answer: $x = \frac{4}{3}$

28. Solve the equation $3x - [2(1 - 2x) - 5] = 5 - 7x$.

1—Answer: $x = \frac{1}{7}$

29. Determine which of the following is a solution of the equation $2(1 - x) - (4x + 3) = 11$.

(a) 2   (b) -2 (c) 1

(d) -3 (e) None of these

1—Answer: b

30. Determine which of the following is a solution of the equation $2(3x - 6) - 3(5 - x) = 9$.

(a) -2 (b) 5   (c) 2

(d) 4   (e) None of these

1—Answer: d

31. Solve for x: $0.15x + 0.10(30 - x) = 20$.

(a) -56 (b) 340   (c) 1695

(d) $\frac{850}{7}$ (e) None of these

2—Answer: b

32. Solve for r: $6390 = 6000(1 + r)$.

(a) 0.07   (b) 0.065   (c) 0.06

(d) 0.09   (e) None of these

1—Answer: b

33. Solve for x: $3[2x - (7x - 1)] = 5x + 13$.

(a) $\frac{4}{5}$ (b) -2 (c) $\frac{7}{13}$

(d) $-\frac{1}{2}$ (e) None of these

2—Answer: d

34. Solve for x: $\frac{3x}{5} + x = \frac{2}{3}$.

2—Answer: $\frac{5}{12}$

Chapter 1 Equations and Inequalities

35. Solve for x: $\dfrac{4x+1}{4} - \dfrac{2x+3}{3} = \dfrac{7}{12}$.

(a) 4   (b) $\dfrac{9}{2}$ (c) -2

(d) $\dfrac{3}{2}$ (e) None of these

2—Answer: a

36. Solve for x: $\dfrac{3x}{2} - \dfrac{x+1}{4} = 6$.

(a) 5   (b) $\dfrac{23}{5}$ (c) $\dfrac{35}{8}$

(d) $\dfrac{1}{2}$ (e) None of these

1—Answer: a

37. Solve for x: $\tfrac{3}{4}x - \tfrac{1}{2}(x+5) = 2$.

(a) -2 (b) 18   (c) 26

(d) 28   (e) None of these

1—Answer: b

38. Solve for x: $\dfrac{3x+2}{5} - \dfrac{6x+4}{3} = \dfrac{14}{3}$.

(a) $-\dfrac{44}{21}$ (b) 5   (c) $\dfrac{37}{12}$

(d) -4 (e) None of these

1—Answer: d

39. Solve for x: $\dfrac{2x-5}{x-3} = \dfrac{4x+1}{2x}$.

2—Answer: -3

40. Solve for x: $3 - \dfrac{4x+5}{x-2} = \dfrac{7x-9}{x-2}$.

(a) $-\dfrac{1}{4}$ (b) -1 (c) 1

(d) No solution   (e) None of these

2—Answer: a

41. Solve for x: $\dfrac{7x}{x-2} + \dfrac{2x}{x+2} = 9$.

(a) $-\dfrac{18}{5}$ (b) $\dfrac{2}{3}$ (c) $-\dfrac{2}{5}$

(d) $\dfrac{5}{18}$ (e) None of these

2—Answer: a

42. Solve for x: $\dfrac{4}{x} = \dfrac{7}{3}$.

(a) $\dfrac{7}{12}$ (b) $\dfrac{3}{28}$ (c) $\dfrac{28}{3}$

(d) $\dfrac{12}{7}$ (e) None of these

1—Answer: d

43. Solve for x: $8 = 3 + \dfrac{2}{x}$.

(a) $\dfrac{2}{5}$ (b) $\dfrac{5}{2}$ (c) $\dfrac{8}{5}$

(d) $\dfrac{4}{3}$ (e) None of these

1—Answer: a

44. Solve for x: $\dfrac{1}{x-2} + \dfrac{3}{x+3} = \dfrac{4}{x^2 + x - 6}$.

(a) $\dfrac{4}{7}$ (b) 3 (c) $\dfrac{7}{4}$

(d) 1 (e) None of these

1—Answer: c

45. Solve for x: $\dfrac{1}{x-3} - \dfrac{2}{x+3} = \dfrac{2x}{x^2 - 9}$.

(a) $-\dfrac{1}{2}$ (b) 3 (c) -3

(d) -3 and 3 (e) None of these

1—Answer: e

46. Solve $0.134x + 0.12(250 - x) = 50$ for x. Round your result to two decimal places.

(a) 1428.57 (b) 5624.30 (c) 3562.86

(d) −23.09 (e) None of these

1—T—Answer: a

47. Solve $1.93(x - 1) + 0.911x = 65.3$ for x. Round your results to two decimal places.

(a) 23.34   (b) 0.04   (c) 22.31

(d) 23.66   (e) None of these

1—T—Answer: d

48. Solve $2.25(x - 2) + 0.365x = 52.2$ for x. Round your results to two decimal places.

(a) 20.73   (b) 19.20   (c) 18.24

(d) 21.68   (e) None of these

2—T—Answer: d

49. Solve $0.55 + 13.9(2.1 - x) = 14$ for x. Round your results to two decimal places.

2—T—Answer: 1.13

50. Solve $0.68 + 12.5(3.2 - x) = 16$ for x. Round your results to two decimal places.

2—T—Answer: 1.97

51. Solve the equation and round your answer to two decimal places:

$$1.576x + 4 = 5.5.$$

1—T—Answer: $x = 0.95$

52. Solve the equation and round your answer to two decimal places:

$$315x - 267 = 25x + 765.$$

1—T—Answer: $x = 3.56$

53. Solve the equation and round your answer to two decimal places:

$$\frac{x}{5.25} = 3x + 10.25.$$

2—T—Answer: $x = -3.65$

54. Solve the equation and round your answer to two decimal places:

$$5x + \frac{5.2}{6} = \frac{3}{2}.$$

2—T—Answer: $x = 0.13$

55. Use a graphing utility to graph the equation and approximate any x-intercepts of $y = 3(x - 1) + 6$. Then set $y = 0$ and solve the resulting equation.

2—T—Answer: 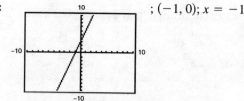 ; $(-1, 0)$; $x = -1$

56. Use a graphing utility to graph the equation and approximate any *x*-intercepts of $y = 4 - 4(x - 6)$. Then set $y = 0$ and solve the resulting equation.

2—T—Answer: 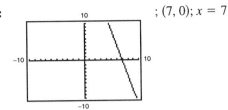 ; $(7, 0)$; $x = 7$

57. Use a graphing utility to graph the equation and approximate any *x*-intercepts of $y = 4[x - 2(x + 1)] + 4$. Then set $y = 0$ and solve the resulting equation.

2—T—Answer: 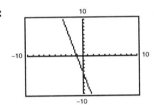 ; $(-1, 0)$; $x = -1$

58. Use a graphing utility to graph the equation and approximate any *x*-intercepts of $y = 5 + 3(2x - 1)$. Then set $y = 0$ and solve the resulting equation.

2—T—Answer: 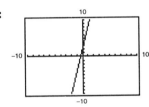 ; $\left(-\frac{1}{3}, 0\right)$; $x = -\frac{1}{3}$

59. Use a graphing utility to graph the equation and approximate any *x*-intercepts of $y = 20 + 3(2x + 1)$. Then set $y = 0$ and solve the resulting equation.

2—T—Answer: 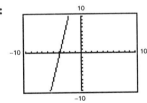 ; $(-3, 0)$; $x = -3$

60. Use a graphing utility to graph the equation and approximate any *x*-intercepts of $y = 2(-3x + 1) - 4$. Then set $y = 0$ and solve the resulting equation.

2—T—Answer: 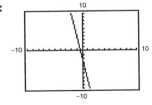 ; $(3, 0)$; $x = 3$

61. Use a graphing utility to graph the equation and approximate any x-intercepts of $y = 2 + 3(x - 1)$. Then set $y = 0$ and solve the resulting equation.

2—T—Answer: 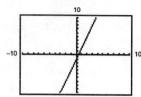 ; $(5.5, 0)$; $x = \frac{11}{2}$

62. The handicap, H, for a bowler with an average score, A, of less than 200 is determined using the formula $H = 0.8(200 - A)$. Find a bowler's average score if his handicap is 64.

    (a) 110   (b) 120   (c) 130
    (d) 140   (e) None of these

    1—Answer: b

63. The handicap, H, for a bowler with an average score, A, of less than 200 is determined using the formula $H = 0.8(200 - A)$. Find a bowler's average score if his handicap is 8.

    (a) 160   (b) 170   (c) 180
    (d) 190   (e) None of these

    1—Answer: d

64. The handicap, H, for a bowler with an average score, A, of less than 200 is determined using the formula $H = 0.8(200 - A)$. Find a bowler's average score if his handicap is 32.

    (a) 160   (b) 170   (c) 180
    (d) 190   (e) None of these

    1—Answer: a

65. The cost of x hundred pounds of fertilizer produced is given by $C = 0.55x + 210$. How many pounds of fertilizer cost $328.25?

    1—T—Answer: 215 pounds

66. The formula that converts Celsius temperature to Fahrenheit temperature is $F = \frac{9}{5}C + 32$. Find the Celsius temperature that corresponds to 98.6°F.

    2—T—Answer: 37°C

67. A new coffee shop has determined that its profit is growing according to the equation $P = 180t - 200$ where t is time in months with $t = 1$ corresponding to January. Determine the month its profit will reach $1780.

    2—Answer: $t = 11$; November

1.3 Modeling with Linear Equations

1. Let n represent a number. A variable expression in n, representing four added to the product of 6 and n, all divided by nine, is given by

   (a) $\dfrac{6n}{9} + 4$
   (b) $\dfrac{6n + 4}{9}$
   (c) $6n + \dfrac{4}{9}$
   (d) $\dfrac{6(n + 4)}{9}$
   (e) None of these

   1—Answer: b

2. Let n represent a number. A variable expression in n, representing the product of six and n divided by nine, all added to four, is given by

   (a) $\dfrac{6n}{9} + 4$
   (b) $\dfrac{6n + 4}{9}$
   (c) $6n + \dfrac{4}{9}$
   (d) $\dfrac{6(n + 4)}{9}$
   (e) None of these

   1—Answer: a

3. Let n represent a number. A variable expression in n, representing the product of six and the sum of n and four, all divided by nine, is given by

   (a) $\dfrac{6n}{9} + 4$
   (b) $\dfrac{6n + 4}{9}$
   (c) $6n + \dfrac{4}{9}$
   (d) $\dfrac{6(n + 4)}{9}$
   (e) None of these

   1—Answer: d

4. Let n represent a number. A variable expression in n, representing four-ninths added to the product of six and n is given by

   (a) $\dfrac{6n}{9} + 4$
   (b) $\dfrac{6n + 4}{9}$
   (c) $6n + \dfrac{4}{9}$
   (d) $\dfrac{6(n + 4)}{9}$
   (e) None of these

   1—Answer: c

5. The statement "the sum of ten and twice a number" translates into the algebraic expression:

   (a) $2(10) + n$
   (b) $2(n + 10)$
   (c) $2n + 10$
   (d) $10 + \dfrac{n}{2}$
   (e) None of these

   1—Answer: c

6. The statement "twice the sum of ten and a number" translates into the algebraic expression:

   (a) $2(10) + n$
   (b) $2(n + 10)$
   (c) $2n + 10$
   (d) $10 + \dfrac{n}{2}$
   (e) None of these

   1—Answer: b

7. The statement "twice ten added to a number" translates into the algebraic expression:

   (a) $2(10) + n$
   (b) $2(n + 10)$
   (c) $2n + 10$
   (d) $10 + \dfrac{n}{2}$
   (e) None of these

   1—Answer: a

8. Which of the following verbal expressions correctly represents the algebraic expression $\dfrac{5x + 4}{7}$?

   (a) Five-sevenths times the sum of x and four
   (b) Five times x, divided by seven, the result added to four
   (c) Five times the sum of one-seventh of x and four
   (d) The sum of four and the product of five and x, all divided by seven
   (e) Five times the sum of x and four-sevenths

   2—Answer: d

9. Which of the following verbal expressions correctly represents the algebraic expression $\dfrac{5(x + 4)}{7}$?

   (a) Five-sevenths times the sum of x and four
   (b) Five times x, divided by seven, the result added to four
   (c) Five times the sum of one-seventh of x and four
   (d) The sum of four and the product of five and x, all divided by seven
   (e) Five times the sum of x and four-sevenths

   2—Answer: a

10. Which of the following verbal expressions correctly represents the algebraic expression $\dfrac{5x}{7} + 4$?

    (a) Five-sevenths times the sum of x and four
    (b) Five times x, divided by seven, the result added to four
    (c) Five times the sum of one-seventh of x and four
    (d) The sum of four and the product of five and x, all divided by seven
    (e) Five times the sum of x and four-sevenths

    2—Answer: b

11. Which of the following verbal expressions correctly represents the algebraic expression $5\left(x + \dfrac{4}{7}\right)$?

    (a) Five-sevenths times the sum of x and four
    (b) Five times x, divided by seven, the result added to four
    (c) Five times the sum of one-seventh of x and four
    (d) The sum of four and the product of five and x, all divided by seven
    (e) Five times the sum of x and four-sevenths

    2—Answer: e

12. Without using a variable, write a verbal description of $3x - 7$.

    2—Answer: Seven less than the product of three and a number.
    NOTE: Verbal descriptions are not unique.

13. Without using a variable, write a verbal description of $3(x - 7)$.

    2—Answer: A number is diminished by seven, and the result tripled.
    NOTE: Verbal descriptions are not unique.

14. Without using a variable, write an verbal description of $\left(\dfrac{3}{x}\right) + 2$.

    2—Answer: Three is divided by a number and the result is added to 2.
    NOTE: Verbal descriptions are not unique.

15. Without using a variable, write a verbal description of $\dfrac{3}{x + 2}$.

    2—Answer: Three is divided by the sum of two and a number.
    NOTE: Verbal descriptions are not unique.

16. A telephone call costs $0.31 for the first minute plus $0.24 for each additional minute. Write an algebraic expression for the cost of a call lasting x minutes.

    (a) $0.31 + 0.24x$ (b) $0.24x$ (c) $0.31 + 0.24(x - 1)$

    (d) $0.31 + 0.24(x + 1)$ (e) None of these

    1—Answer: c

17. The cost of mailing a package is $0.29 for the first ounce plus $0.25 for each additional ounce. Write an algebraic expression for the cost of a package weighting x ounces.

    (a) $0.29 + 0.25x$ (b) $0.25x$ (c) $0.29 + 0.25(x + 1)$

    (d) $0.29 + 0.25(x - 1)$ (e) None of these

    1—Answer: d

18. The length of a rectangular room is 5 feet longer than the width. Write an algebraic expression for the perimeter of a room with width x feet.

    (a) $x(x + 5)$ (b) $2x + 2(x + 5)$ (c) $2x + 2(5x)$

    (d) $2x + 2x + 5$ (e) None of these

    1—Answer: b

19. The length of a rectangular room is 2 feet longer than the width. Write an algebraic expression for the perimeter of a room with width x feet.

    (a) $x(x + 2)$ (b) $4x + 2$ (c) $2x + 2(x + 2)$

    (d) $x + (x + 2)$ (e) None of these

    1—Answer: c

20. A jacket is discounted by 20%. Write an algebraic expression for the sale price of a jacket that originally sells for x dollars.

    (a) $x - 0.20x$ (b) $x - 0.20$ (c) $x + 0.20x$

    (d) $0.20x - x$ (e) None of these

    1—Answer: a

21. A stereo is discounted 40%. Write an algebraic expression for the sale price of a stereo that originally sells for x dollars.

    (a) $x - 0.40$ (b) $x - 0.40$ (c) $x + 0.40$

    (d) $0.40x$ (e) None of these

    1—Answer: b

22. Use an algebraic expression to represent the sum of the squares of two consecutive odd numbers.

    (a) $x^2 + (x + 1)^2$ (b) $x^2 + x^2 + 1$ (c) $x^2 + (x + 2)^2$

    (d) $(x^2 + 1) + (x^2 + 3)$ (e) None of these

    1—Answer: c

23. Use an algebraic expression to represent the sum of the squares of three consecutive numbers.

    1—Answer: $x^2 + (x + 1)^2 + (x + 2)^2$

24. A person rides a bicycle at a constant rate of 18 miles per hour for M miles. Write an expression for the time in hours elapsed.

    1—Answer: $\dfrac{M}{18}$

25. A person adds L liters of a fluid containing 65% antifreeze to a car radiator. Write an expression indicating the amount of antifreeze added.

    1—Answer: $0.65L$

26. A cash register contains x quarters and y dimes. Write an expression for the total amount of money in *cents*.

    1—Answer: $25x + 10y$

27. A cash register contains x quarters and y dimes. Write an expression for the total amount of money in *dollars*.

    1—Answer: $0.25x + 0.10y$

28. Write an algebraic expression for the distance traveled in t hours at an average speed of 55 miles per hour.

    1—Answer: $55t$

29. Write an algebraic expression for the distance traveled in 8 hours at an average speed of r miles per hour.

    1—Answer: $8r$

30. Write an algebraic expression for the average rate of speed when traveling 525 miles in *t* hours.

   1—Answer: $\dfrac{525}{t}$

31. Write an algebraic expression for the time to travel 525 miles at an average speed of *r* miles per hour.

   1—Answer: $\dfrac{525}{r}$

32. Write an expression for the area of the region.

   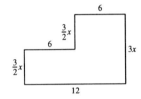

   2—Answer: $A = \left(\tfrac{3}{2}x\right)(6) + \left(\tfrac{3}{2}x\right)(12) = 27x$

33. Write an expression for the area of the region.

   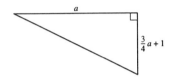

   2—Answer: $A = \tfrac{1}{2}a\left(\tfrac{3}{4}a + 1\right) = \tfrac{3}{8}a^2 + \tfrac{1}{2}a$

34. Find a number such that seven subtracted from twice this number is 75.

   (a) 41 (b) 69 (c) 90
   (d) 62 (e) None of these

   1—Answer: a

35. The sum of three consecutive odd integers is 75. Find the smallest of these integers.

   (a) 23 (b) 27 (c) 24
   (d) 25 (e) None of these

   1—Answer: a

36. The sum of two numbers is 27, and one number is twice the other. Find the numbers.

   1—Answer: 9, 18

37. Ten less than four times a number is 26. Find the number.

1—Answer: 9

38. The sum of three consecutive even integers is 24. Find the integers.

1—Answer: 6, 8, 10

39. The sum of two numbers is 26. The larger number is one less than twice the smaller number. Find the numbers.

1—Answer: 9, 17

40. David is three times as old as Maureen this year. In three years, David will be twice as old as Maureen. How old are David and Maureen now?

2—Answer: David is 9 years old, and Maureen is 3 years old.

41. A father is 42 years old and his son is 9. In how many years will the father be four times as old as the son?

2—Answer: Two years

42. Six years ago, Jay was four times as old as Hank. Now he is only two times as old as Hank. Find their present ages.

2—Answer: Hank is 9 years old, and Jay is 18 years old.

43. Paul is now twice as old as Richard. Five years ago, Paul was three times as old as Richard. Find their present ages.

2—Answer: Richard is 10 years old, and Paul is 20 years old.

44. 460 is what percent of 340?

   (a) 74%   (b) 0.74%   (c) 1.35%

   (d) 135.3%   (e) None of these

1—Answer: d

45. What is 0.17% of 432?

1—Answer: 0.7344

46. 29 is what percent of 37?

   (a) 7.8%   (b) 78.4%   (c) 12.76%

   (d) 127.6%   (e) None of these

1—Answer: b

47. Eleanor invests $18,000 in two funds paying $9\frac{1}{4}\%$ and $10\frac{1}{2}\%$ simple interest. How much is invested at $9\frac{1}{4}\%$ if the total yearly interest is $1827.50?

   (a) $1300.00   (b) $5000.00   (c) $8000.00

   (d) $10,000.00   (e) None of these

2—Answer: b

48. Maria inherited $15,000. She decided to invest it in two funds, one paying $9\frac{1}{4}\%$ simple interest, the other paying $11\frac{1}{2}\%$ simple interest. Her annual income from these investments will total $1623.75. How much did she invest in the fund that pays $9\frac{1}{4}\%$ simple interest?

(a) $4500 (b) $10,500 (c) $488

(d) $14,512 (e) None of these

2—Answer: a

49. Ann invested $8000 in a fund that pays $2\frac{1}{2}\%$ more simple interest per year than a similar fund in which her husband had invested $10,000. At the end of a year their interest totaled $1690.00. What rate of interest did Ann receive?

2—Answer: $10\frac{7}{9}\%$

50. Two trains traveling the same speed leave the city. The southbound train reaches its destination in 45 minutes. The eastbound train reaches its destination in 1 hour. How fast were the trains traveling if their destinations are 88 miles apart?

(a) 70.4 mph (b) 1.2 mph (c) 2 mph

(d) 49.6 mph (e) None of these

2—Answer: a

51. Two brothers, Bob and Bill, live 450 miles apart. Starting at the same time they plan to drive until they meet. Bill averages 10 miles per hour faster than Bob who averages 50 mph. How long will it take them to meet?

(a) $3\frac{2}{15}$ hours (b) $3\frac{9}{11}$ hours (c) $4\frac{1}{11}$ hours

(d) $4\frac{1}{2}$ hours (e) None of these

2—Answer: c

52. Two friends living 216 miles apart in bordering states are planning to meet at the state line. The speed limit in one state is 55 mph and 65 mph in the other. Assuming each will drive the speed limit and each will travel the same length of time, determine how far from the state line the person who is traveling at 65 mph lives.

(a) 99 miles (b) 117 miles (c) 108 miles

(d) 180 miles (e) None of these

2—Answer: b

53. A truck driver averaged 60 mph on a 600 mile trip and averaged 40 mph on the return trip. What was the average speed for the round trip?

(a) 45 mph (b) 48 mph (c) 50 mph

(d) 55 mph (e) None of these

2—Answer: b

54. A truck driver averaged 60 mph on a 600 mile trip and averaged 40 mph on the return trip. What was the average speed for the round trip?

(a) 48 mph (b) 50 mph (c) 55 mph

(d) 45 mph (e) None of these

2—Answer: a

84   Chapter 1   Equations and Inequalities

55. A truck driver averaged 60 mph on a 600 mile trip and averaged 40 mph on the return trip. What was the average speed for the round trip?

   (a) 50 mph   (b) 55 mph   (c) 45 mph

   (d) 48 mph   (e) None of these

   2—Answer: d

56. A truck driver averaged 60 mph on a 600 mile trip and averaged 40 mph on the return trip. What was the average speed for the round trip?

   (a) 55 mph   (b) 45 mph   (c) 48 mph

   (d) 50 mph   (e) None of these

   2—Answer: c

57. Two cars, starting together, travel in opposite directions on a highway, one at 55 mph and the other at 45 mph. How far apart are they after $2\frac{1}{2}$ hours?

   1—Answer: 250 miles

58. Two cars, starting together, travel in opposite directions on a highway, one at 55 mph and the other at 45 mph. How far apart are they after one hour and 45 minutes?

   1—Answer: 175 miles

59. Two cars, starting together, travel in opposite directions on a highway, one at 55 mph and the other at 45 mph. How far apart are they after three hours and 12 minutes?

   1—Answer: 320 miles

60. Find the original price of a television set that was reduced 40% and is now priced at $285.50.

   1—Answer: $475.83

61. Your weekly gross income after a 2.3% cost of living raise is $593.34. What was your income before the raise?

   (a) $575   (b) $590   (c) $580

   (d) $585   (e) None of these

   1—Answer: c

62. The price of a new car including a 6% sales tax is $17,362.80. What was the price of the car before the sales tax was added?

   (a) $16,380   (b) $17,040   (c) $16,321

   (d) $16,870   (e) None of these

   1—Answer: a

63. The list price of your purchase is $572.75, but you receive a 20% discount. The sales tax is 6%. How much do you owe?

   2—Answer: $485.69

64. The list price of your purchase is $744.12, but you receive a 16% discount. The sales tax is 6.5%. How much do you owe?

   2—Answer: $665.69

Section 1.3   Modeling with Linear Equations   85

65. The list price of your purchase is $492.95, but you receive an 18% discount. The sales tax is 5.5%. How much do you owe?

    2—Answer: $426.45

66. The annual snowfall decreased by 32% from last year's 85.2 inches. This year's snowfall to the nearest one-tenth of an inch was:

    (a) 57.9   (b) 27.2   (c) 64.5

    (d) 53.2   (e) None of these

    1—Answer: a

67. The annual snowfall decreased by 12% from last year's 68.7 inches. This year's snowfall to the nearest one-tenth of an inch was:

    (a) 27.2   (b) 64.5   (c) 53.2

    (d) 60.5   (e) None of these

    1—Answer: d

68. A 15-quart radiator contains a 40% concentration of antifreeze. How much of the solution must be drained and replaced by 100% antifreeze to bring the solution up to 70%?

    (a) 4.5 quarts   (b) 5 quarts   (c) 10.5 quarts

    (d) 7.5 quarts   (e) None of these

    2—Answer: d

69. Determine the number of milliliters of a 70% sulfuric acid solution that must be added to 160 milliliters of a 25% solution to make a 30% solution.

    (a) 40   (b) 35   (c) 25

    (d) 20   (e) None of these

    2—Answer: d

70. How many ounces of pure antifreeze must be added to 100 ounces of 40% antifreeze solution to obtain a 50% solution?

    1—Answer: 20 ounces

71. How many ounces of pure antifreeze must be added to 100 ounces of 40% antifreeze solution to obtain a 60% solution?

    1—Answer: 50 ounces

72. How many ounces of water must be added to 100 ounces of 40% antifreeze solution to obtain a 25% solution?

    1—Answer: 60 ounces

73. How many ounces of water must be added to 100 ounces of 40% antifreeze solution to obtain a 16% solution?

    1—Answer: 150 ounces

74. A grocer mixes four pounds of nuts, costing $9 per pound, with sixteen pounds of nuts, costing $4 per pound. What price per pound is the mixture worth?

   (a) $13.00 (b) $5.00 (c) $6.50
   (d) 20 pounds (e) None of these

   1—Answer: b

75. A grocer mixes four pounds of nuts, costing $9 per pound, with sixteen pounds of nuts, costing $4 per pound. What price per pound is the mixture worth?

   (a) 20 pounds (b) $13.00 (c) $5.00
   (d) $6.50 (e) None of these

   1—Answer: c

76. A grocer mixes four pounds of nuts, costing $9 per pound, with sixteen pounds of nuts, costing $4 per pound. What price per pound is the mixture worth?

   (a) $6.50 (b) 20 pounds (c) $13.00
   (d) $5.00 (e) None of these

   1—Answer: d

77. A grocer wants to mix cashew nuts worth $8 per pound with peanuts worth $3 per pound. She wants to obtain a mixture to sell for $4 per pound. If ten pounds of peanuts are used, what is the total weight of the mixture?

   2—Answer: 12.5 pounds

78. A grocer wants to mix cashew nuts worth $8 per pound with peanuts worth $3 per pound. She wants to obtain a mixture to sell for $4 per pound. If ten pounds of peanuts are used, what is the total value in dollars of the mixture?

   2—Answer: $50

79. You want to measure the height of a steeple. To do this, you measure the steeple's shadow and find it is $8\frac{3}{4}$ feet long. You also measure the shadow of a 4 foot steak and find that its shadow is $2\frac{1}{3}$ long. Determine the height of the steeple.

   (a) 12 ft (b) 15 ft (c) 19 ft
   (d) 22 ft (e) None of these

   1—Answer: b

80. You want to make a scale model of a rectangular room that is 10 feet wide and 15 feet long. If the width of the model is 9 inches, what will be the length?

   (a) $10\frac{1}{4}$ in. (b) 12 in. (c) 13 in.
   (d) $13\frac{1}{2}$ in. (e) None of these

   2—Answer: d

81. You want to measure the height of a tree. To do this, you measure the tree's shadow and find that it is 52.5 feet long. You also measure the shadow of a 4 foot stake and find that its shadow is 6 feet long. How tall is the tree?

(a) 35 ft   (b) $78\frac{3}{4}$ ft   (c) 42 ft

(d) $60\frac{1}{3}$ ft   (e) None of these

1—Answer: a

82. A homeowner wants to fence in a section of his backyard. The fenced-in area will be rectangular with the length $1\frac{1}{3}$ times the width. How many feet of fencing will he need to buy if the width is 27 feet long?

(a) 36 ft   (b) 63 ft   (c) 126 ft

(d) 972 ft   (e) None of these

2—Answer: c

83. A company has fixed costs of $10,000 per month and variable costs of $9.75 per unit manufactured. The company has $96,190 available to cover the monthly costs. How many units can the company manufacture?

(a) 9866   (b) 8840   (c) 10,891

(d) 9463   (e) None of these

2—Answer: b

84. A company has fixed cost of $8000 per month and variable costs of $8.65 per unit manufactured. The company has $64,000 available to cover the monthly costs. How many units can the company manufacture?

(a) 6474   (b) 7399   (c) 8324

(d) 6920   (e) None of these

2—Answer: a

85. Solve for P: $A = P + PRT$.

(a) $A - PRT$ (b) $A - RT$ (c) $\dfrac{A}{RT}$

(d) $\dfrac{A}{1 + RT}$ (e) None of these

1—Answer: d

86. Solve for h: $V = \dfrac{\pi}{3}r^2h$.

(a) $\dfrac{V}{3\pi r^2}$ (b) $\dfrac{\pi r^2 V}{3}$ (c) $\dfrac{\pi r^2}{3V}$

(d) $\dfrac{3V}{\pi r^2}$ (e) None of these

1—Answer: d

87. Solve for h: $A = \frac{1}{2}(a + b)h$.

    (a) $\dfrac{A}{2(a + b)}$
    (b) $\dfrac{2A}{a + b}$
    (c) $2A(a + b)$
    (d) $A - \dfrac{1}{2}(a + b)$
    (e) None of these

    1—Answer: b

88. Solve for p: $g = \dfrac{4\pi^2 p}{r^2}$.

    1—Answer: $\dfrac{gr^2}{4\pi^2}$

89. Solve for b: $A = \dfrac{1}{2}bh$.

    1—Answer: $\dfrac{2A}{h}$

❏ 1.4 Quadratic Equations and Applications

1. Write the quadratic equation in standard form: $x^2 - 4x = x + 2$.
   (a) $x^2 - 3x = 2$
   (b) $x^2 - 5x = 2$
   (c) $x^2 - 5x - 2 = 0$
   (d) $x^2 = 3x - 2$
   (e) None of these

   1—Answer: c

2. Write the quadratic equation in standard form: $\dfrac{3}{x + 7} - \dfrac{4}{x + 2} = 6$.
   (a) $6x^2 + 55x + 106 = 0$
   (b) $6x^2 + 55x + 72 = 0$
   (c) $6x^2 - x + 106 = 0$
   (d) $6x^2 - x = 72$
   (e) None of these

   2—Answer: a

3. Write the quadratic equation in standard form: $x^2 = 64x$.
   (a) $x^2 = 64x$
   (b) $x^2 - 64x = 0$
   (c) $x - 64 = 0$
   (d) $64 - x^2 = 0$
   (e) None of these

   1—Answer: b

4. Write the quadratic equation in standard form: $\dfrac{1}{x + 1} - \dfrac{3}{x - 2} = 5$.
   (a) $x = -5$
   (b) $2x - 6 = 0$
   (c) $5x^2 + 2x - 5 = 0$
   (d) $5x^2 - 3x - 5 = 0$
   (e) None of these

   1—Answer: d

5. Write the quadratic equation in standard form: $x^2 + 1 = \dfrac{3x - 2}{5}$.

   (a) $x^2 - \dfrac{3}{5}x + 3 = 0$ (b) $5x^2 - 3x + 15 = 0$ (c) $5x^2 - 3x + 7 = 0$

   (d) $5x^2 - 3x + 3 = 0$ (e) None of these

   1—Answer: c

6. Solve for x: $5x^2 - 2 = 3x$.

   (a) $\frac{2}{5}, -1$ (b) $-\frac{1}{5}, 2$ (c) $-\frac{2}{5}, 1$

   (d) $\frac{1}{5}, -2$ (e) None of these

   2—Answer: c

7. Solve for x: $2x^2 + 4x = 9x + 18$.

   (a) $-2, \frac{9}{2}$ (b) $2, -\frac{9}{2}$ (c) $\frac{9}{2}$

   (d) $-\frac{9}{2}$ (e) None of these

   1—Answer: a

8. Solve for x: $4x^2 + 12x + 9 = 0$.

   1—Answer: $-\frac{3}{2}$

9. Solve for x: $3x^2 + 19x - 14 = 0$.

   1—Answer: $-7, \frac{2}{3}$

10. Solve for x: $2x^2 + x = 3$.

    1—Answer: $-\frac{3}{2}, 1$

11. Solve for x: $2x^2 = 162$.

    (a) 9 (b) 9 (c) $-9, 9$

    (d) 81 (e) None of these

    1—Answer: c

12. Solve for x: $7(x + 2)^2 = 12$.

    1—Answer: $\frac{1}{7}\left(-14 \pm 2\sqrt{21}\right)$

13. Solve for x: $(x + 7)^2 = 5$.

    1—Answer: $-7 \pm \sqrt{5}$

14. Solve for x: $3x^2 = 192$.

    (a) ± 8 (b) 8 (c) -1

    (d) 64 (e) None of these

    1—Answer: a

Chapter 1 Equations and Inequalities

15. Solve for x: $(2x + 3)^2 = 4$.

(a) $\frac{1}{2}, \frac{5}{2}$ (b) $-\frac{1}{2}$ (c) $-\frac{5}{2}, -\frac{1}{2}$

(d) ± 2 (e) None of these

1—Answer: c

16. Solve by completing the square: $x^2 - 8x + 2 = 0$.

1—Answer: $4 \pm \sqrt{14}$

17. Solve by completing the square: $x^2 + 4x - 2 = 0$.

(a) $2 \pm \sqrt{6}$ (b) $2 \pm \sqrt{2}$ (c) $-2 \pm \sqrt{2}$

(d) $-2 \pm \sqrt{6}$ (e) None of these

1—Answer: d

18. Solve by completing the square: $x^2 - 6x + 1 = 0$.

(a) $3 \pm \sqrt{26}$ (b) $3 \pm \sqrt{10}$ (c) $3 \pm \sqrt{17}$

(d) $3 \pm 2\sqrt{2}$ (e) None of these

1—Answer: d

19. Solve by completing the square: $1 - x = x(x + 3)$.

(a) $x = -\frac{3}{2} \pm \frac{\sqrt{13}}{2}$ (b) $-2 \pm \sqrt{5}$ (c) $-1 \pm \sqrt{2}$

(d) $\frac{3}{2} \pm \frac{\sqrt{11}}{2}$ (e) None of these

1—Answer: b

20. Solve by completing the square: $6x - x^2 = 3$.

(a) $-3 \pm 2\sqrt{3}$ (b) $3 \pm \sqrt{6}$ (c) $3 \pm 2\sqrt{3}$

(d) $3, 9$ (e) None of these

1—Answer: b

21. Solve by completing the square: $2x^2 - 8x + 5 = 0$.

(a) $2 \pm \frac{\sqrt{6}}{2}$ (b) $-2 \pm \frac{\sqrt{3}}{2}$ (c) $4 \pm \frac{\sqrt{19}}{2}$

(d) $8 \pm \frac{\sqrt{5}}{2}$ (e) None of these

2—Answer: a

22. Solve by completing the square: $3x^2 + 18x - 22 = 0$

(a) $-18 \pm \frac{\sqrt{22}}{3}$ (b) $-3 \pm \frac{\sqrt{15}}{3}$ (c) $-3 \pm \frac{4\sqrt{2}}{3}$

(d) $6 \pm \frac{3\sqrt{3}}{5}$ (e) None of these

2—Answer: e

23. Solve by completing the square: $7x^2 - 14x + 6 = 0$.

(a) $1 \pm \dfrac{\sqrt{-5}}{7}$ (b) $14 \pm \dfrac{\sqrt{6}}{7}$ (c) $1 \pm \dfrac{5\sqrt{2}}{7}$

(d) $1 \pm \dfrac{\sqrt{7}}{7}$ (e) None of these

2—Answer: d

24. Solve by completing the square: $4x^2 - 16x - 13 = 0$.

(a) $2 \pm \dfrac{\sqrt{17}}{2}$ (b) $4 \pm \dfrac{\sqrt{17}}{4}$ (c) $2 \pm \dfrac{\sqrt{3}}{2}$

(d) $-2 \pm \dfrac{\sqrt{-9}}{2}$ (e) None of these

2—Answer: e

25. Solve by completing the square: $2x^2 + 12x + 13 = 0$.

(a) $-6 \pm \dfrac{\sqrt{13}}{2}$ (b) $-3 \pm \dfrac{\sqrt{10}}{2}$ (c) $\dfrac{3 \pm \sqrt{5}}{2}$

(d) $\pm \dfrac{\sqrt{23}}{2}$ (e) None of these

2—Answer: b

26. Solve by completing the square: $2x^2 - 7x - 12 = 0$.

2—Answer: $\frac{1}{4}\left(7 \pm \sqrt{145}\right)$

27. Solve for x: $\dfrac{3x + 25}{x + 7} - 5 = \dfrac{3}{x}$.

(a) $\dfrac{3}{2}, 7$ (b) $\dfrac{7}{2}, 3$ (c) $-\dfrac{3}{2}, -7$

(d) $-\dfrac{7}{2}, -3$ (e) None of these

1—Answer: d

28. Complete the square: $2x^2 + 9x - 4$.

(a) $2\left(x + \dfrac{9}{4}\right)^2 - \dfrac{113}{8}$ (b) $2\left(x + \dfrac{9}{4}\right)^2 - \dfrac{145}{16}$ (c) $2\left(x + \dfrac{9}{4}\right)^2 + \dfrac{49}{8}$

(d) $2\left(x + \dfrac{9}{4}\right)^2 + \dfrac{17}{16}$ (e) None of these

2—Answer: a

29. Complete the square: $3x^2 - 2x + 1$.

(a) $(3x - 1)^2$ (b) $3(x - 1)^2 + 1$ (c) $3\left(x - \dfrac{1}{3}\right)^2 + \dfrac{1}{9}$

(d) $3\left(x - \dfrac{1}{3}\right)^2 + \dfrac{2}{3}$ (e) None of these

2—Answer: d

30. Complete the square: $2x^2 - 6x + 9$.

2—Answer: $2\left(x - \frac{3}{2}\right)^2 + \frac{9}{2}$

31. Complete the square in the denominator: $\dfrac{3}{4x^2 + 10x - 7}$.

(a) $\dfrac{3}{4\left(x + \frac{5}{4}\right)^2 - \frac{53}{4}}$
(b) $\dfrac{3}{4\left(x + \frac{5}{2}\right)^2 - 32}$
(c) $\dfrac{3}{4\left(x + \frac{5}{4}\right)^2 - \frac{3}{4}}$

(d) $\dfrac{3}{4\left(x + \frac{5}{2}\right)^2 + 18}$
(e) None of these

2—Answer: a

32. Complete the square in the denominator: $\dfrac{5}{2x^2 - 3x + 1}$.

(a) $\dfrac{5}{2\left(x - \frac{3}{2}\right)^2 + \frac{13}{4}}$
(b) $\dfrac{5}{2\left(x - \frac{3}{2}\right)^2 + 1}$
(c) $\dfrac{5}{2\left(x - \frac{3}{4}\right)^2 + \frac{11}{2}}$

(d) $\dfrac{5}{2\left(x - \frac{3}{2}\right) - \frac{5}{4}}$
(e) None of these

2—Answer: e

33. Identify the Quadratic Formula.

(a) $x = -b \pm \dfrac{\sqrt{b^2 - 4ac}}{2a}$
(b) $x = \dfrac{-b \pm \sqrt{b^2 - 4a}}{2c}$
(c) $x = \dfrac{-b \pm \sqrt{b^2 - 4ac}}{2a}$

(d) $x = \dfrac{-b \pm \sqrt{b^2 - 4ac}}{2}$
(e) None of these

1—Answer: c

34. Use the discriminant to determine the number of real solutions: $4x^2 - 2x - 7 = 0$.

(a) 0 (b) 1 (c) 2

(d) 3 (e) None of these

1—Answer: c

35. Use the discriminant to determine the number of real solutions: $7x^2 - 3x + 15 = 0$.

(a) 0 (b) 1 (c) 2

(d) 3 (e) None of these

1—Answer: a

36. Use the discriminant to determine the number of real solutions: $\frac{1}{3}x^2 - 2x + 3 = 0$.

(a) 0 (b) 1 (c) 2

(d) 3 (e) None of these

1—Answer: b

37. Use the discriminant to determine the number of real solutions: $10x = x^2 - 14x + 50$.

  1—Answer: 2

38. Use the discriminant to determine the number of real solutions: $5x^2 - 7x + 16 = 0$.

  1—Answer: 0

39. Use the discriminant to determine which quadratic equation has one (repeated) real number solution.

  (a) $3x^2 + 4x + 2 = 0$ (b) $x^2 - 5x - 4 = 0$ (c) $9x^2 - 6x + 1 = 0$
  (d) $7x^2 + 2x - 1 = 0$ (e) None of these

  1—Answer: c

40. Use the discriminant to determine which quadratic equation has no real solutions.

  (a) $x^2 + 4x + 2 = 0$ (b) $9x^2 - 6x + 1 = 0$ (c) $x^2 - 5x + 4 = 0$
  (d) $7x^2 + 2x - 1 = 0$ (e) None of these

  1—Answer: e

41. Solve for x: $x^2 - 3x + \dfrac{3}{2} = 0$.

  1—Answer: $\dfrac{3 \pm \sqrt{3}}{2}$

42. Solve for x: $-3x^2 + 4x + 6 = 0$.

  1—Answer: $\dfrac{2 \pm \sqrt{22}}{3}$

43. Solve for x: $(x + 2)^2 = -16x$.

  (a) $-8 \pm 2\sqrt{15}$ (b) $-10 \pm 4\sqrt{6}$ (c) $-10 \pm 2\sqrt{26}$
  (d) $-8 \pm 4\sqrt{15}$ (e) None of these

  1—Answer: b

44. Solve for x: $\dfrac{1}{x - 1} + \dfrac{x}{x + 2} = 2$.

  1—Answer: $-1 \pm \sqrt{7}$

45. Solve: $3x^2 - 6x + 2 = 0$.

  (a) $\dfrac{3 \pm \sqrt{3}}{3}$ (b) $1 \pm \sqrt{3}$ (c) $\dfrac{3 \pm \sqrt{15}}{3}$
  (d) $\dfrac{1}{3}, 2$ (e) None of these

  1—Answer: a

46. Solve: $(x - 1)^2 = 3x + 5$.

(a) 1, 4
(b) $\dfrac{5 \pm \sqrt{39}}{2}$
(c) $\dfrac{5 \pm \sqrt{41}}{2}$
(d) $-1, 6$
(e) None of these

2—Answer: c

47. Solve: $4x^2 + 12x = 135$.

(a) $-\dfrac{9}{2}, \dfrac{15}{2}$
(b) $-\dfrac{5}{2}, \dfrac{3}{2}$
(c) $-\dfrac{15}{2}, \dfrac{9}{2}$
(d) $\dfrac{-3 \pm \sqrt{6}}{2}$
(e) None of these

2—Answer: c

48. Use your calculator to solve: $2.5x^2 + 3.267x - 8.97 = 0$. Round your answers to three decimal places.

(a) $-6.643, 3.376$
(b) $-5, 271, -1.263$
(c) $-8.276, 1.742$
(d) $-2.657, 1.350$
(e) None of these

2—T—Answer: d

49. Use your calculator to solve: $1.37x^2 - 2.4x - 5.41 = 0$. Round your answers to three decimal places.

(a) $0.228, 4.572$
(b) $-1.296, 3.048$
(c) $-1.775, 4.175$
(d) $-5.720, 2.432$
(e) None of these

2—T—Answer: b

50. Use a calculator to solve: $3x^2 - 0.24x - 0.57 = 0$.

(a) $0.478, -0.398$
(b) $0.474, -3.94$
(c) $1.434, -1.194$
(d) $1.422, -1.182$
(e) None of these

1—T—Answer: a

51. Use a calculator to solve: $27x^2 - 3.2x - 71 = 0$.

(a) $1.680, -1.561$
(b) $1.682, -1.563$
(c) $-1.678, 1.623$
(d) $-1.680, 1.569$
(e) None of these

2—T—Answer: b

52. Use a calculator to solve: $3x^2 - 0.482x - 1.2 = 0$.

(a) $-0.557, 0.718$
(b) $-0.718, 0.557$
(c) $-1.671, 2.154$
(d) $-0.547, 0.708$
(e) None of these

2—T—Answer: a

53. Use a calculator to solve: $62x^2 - 78.2x + 5.1 = 0$.

(a) $-0.062, 1.323$
(b) $-3.244, 1.983$
(c) $0.069, 1.192$
(d) $4.277, 73.903$
(e) None of these

2—T—Answer: c

Section 1.4 Quadratic Equations and Applications 95

54. Use a calculator to solve: $3.2x^2 + 0.61x - 7.4 = 0$.

   (a) $-1.619, 1.428$ (b) $-0.842, 2.748$ (c) $-5.181, 4.570$

   (d) $-0.617, 1.529$ (e) None of these

   2—T—Answer: a

55. Solve for x: $2x^2 - 5(x - 1) = 3(x + 5)$.

   (a) $\dfrac{1 \pm \sqrt{39}}{2}$ (b) $\pm \dfrac{\sqrt{15}}{3}$ (c) $0, 1, -5$

   (d) $-1, 5$ (e) None of these

   2—Answer: d

56. Solve for x: $(2x + 5)^2 = 9$.

   (a) $-4, -1$ (b) -1 (c) 38

   (d) $14, 8$ (e) None of these

   1—Answer: a

57. Solve for x: $\dfrac{2}{5}x^2 - \dfrac{1}{2} = 0$.

   (a) $\pm \dfrac{\sqrt{2}}{5}$ (b) $\pm \dfrac{\sqrt{5}}{5}$ (c) $\pm \dfrac{5\sqrt{2}}{4}$

   (d) $\pm \dfrac{1}{2}\sqrt{5}$ (e) None of these

   1—Answer: d

58. Solve for x: $4x + 3(x^2 - 1) = 2 + 3x^2$.

   (a) $\pm \dfrac{\sqrt{5}}{2}$ (b) $\dfrac{5}{4}$ (c) $\dfrac{2 \pm \sqrt{34}}{12}$

   (d) $\dfrac{-2 \pm \sqrt{10}}{6}$ (e) None of these

   1—Answer: b

59. Solve for x: $(x - 2)(x + 1) = (x - 3)^2$.

   (a) $\dfrac{11}{5}$ (b) $2, 9$ (c) 11

   (d) $7, -2$ (e) None of these

   1—Answer: a

60. Solve for x: $2x^2 - 5x = x^2 + 1$.

   (a) $0, 5$ (b) $\dfrac{4}{11}$ (c) $\dfrac{5 \pm \sqrt{29}}{2}$

   (d) $\dfrac{-5 \pm \sqrt{21}}{2}$ (e) None of these

   1—Answer: c

61. Solve for x: $(x + 1)^2 + x^2 = 9$.

(a) 8   (b) ± 2 (c) $\dfrac{1 \pm \sqrt{15}}{2}$

(d) $\dfrac{-1 \pm \sqrt{17}}{2}$ (e) None of these

1—Answer: d

62. Solve for x: $(x - 1)^2 + (x + 1)^2 = 100$.

(a) $\pm 5\sqrt{2}$ (b) $\pm 2\sqrt{5}$ (c) ± 7

(d) $-4, 6$ (e) None of these

1—Answer: c

63. Two airplanes leave simultaneously from the same airport, one flying due east, and the other flying due north. The eastbound plane is flying 50 miles per hour slower than the northbound one. If after 4 hours they are 1000 miles apart, how fast is the northbound plane traveling?

(a) 150 mph   (b) 200 mph   (c) 100 mph

(d) 300 mph   (e) None of these

2—Answer: b

64. An open box is to be constructed from a square piece of material by cutting a 3-inch square from each corner. Find the dimensions of the square piece of material if the box is to have a volume of 363 cubic inches.

(a) 14" by 14"   (b) 17" by 17"   (c) 20" by 20"

(d) 23" by 23"   (e) None of these

2—Answer: b

65. Find two consecutive positive integers m and n such that $n^2 - m^2 = 27$.

2—Answer: 13, 14

66. The Curriers have decided to fence in part of their back yard to form a rectangular region with an area of 1248 square feet. The fence will extend 2 feet on each side of their 48-foot wide house. How many feet of fencing will they need to enclose the play area? (There is no fence along the house wall.)

2—Answer: 104 feet

67. Use the cost equation, $C = 0.5x^2 + 15x + 6000$, to find the number of units, x, that manufacturer can produce with a total cost, C, of $13,700.

(a) 140   (b) 110   (c) 94,056,500

(d) 47,134,000   (e) None of these

2—T—Answer: b

68. The daily cost in dollars, C, of producing x chairs is given by the quadratic equation $C = x^2 - 120x + 4200$. How many chairs are produced each day if the daily cost is $600?

(a) 60 (b) 600 (c) 90

(d) 40 (e) None of these

2—T—Answer: a

69. You plan to stabilize a T.V. antenna with two guy wires. The guy wires are attached to the antenna 30 feet from the base. How much wire will you need if each of the wires is secured 20 feet from the base of the antenna?

(a) $10\sqrt{13}$ ft (b) $20\sqrt{5}$ ft (c) $20\sqrt{13}$ ft

(d) $10\sqrt{5}$ ft (e) None of these

2—Answer: c

70. Find the smaller of two consecutive positive integers such that the one number times twice the other equals 612.

(a) -18 (b) 12 (c) 18

(d) 17 (e) None of these

2—Answer: d

71. Find the two consecutive *odd* numbers whose product is 483.

1—Answer: 21 and 33

72. Find two consecutive *even* numbers whose product is 288.

1—Answer: 16 and 18

73. Find two consecutive *odd* numbers, the sum of whose squares is 394.

1—Answer: 13 and 15

74. Find two consecutive *even* numbers, the sum of whose squares is 724.

1—Answer: 18 and 20

75. The area of a rectangle is 56 square yards. Its length is 2 yards more than three times its width. Find its dimensions.

1—Answer: 4 yards by 14 yards

76. The area of a rectangle is 221 square feet. Its perimeter is 60 feet. Find its dimensions.

1—Answer: 17 feet by 13 feet

77. The area of a rectangle is 270 square feet. Its perimeter is 66 feet. Find its dimensions.

1—Answer: 18 feet by 15 feet

78. The area of a rectangle is 434 square yards. Its length is 3 yards more than twice its width. Find its dimensions.

1—Answer: 14 yards by 31 yards

79. The area of a triangle is given by $A = \frac{1}{2}bh$ where A is in square inches when b and h are in inches. The height of a triangle is 4 inches longer than the base and its area is 336 square inches. Find the base and height.

1—Answer: $b = 24$ inches and $h = 28$ inches

80. The area of a triangle is given by $A = \frac{1}{2}bh$ where A is in square inches when b and h are in inches. The height of a triangle is 4 inches shorter than the base and its area is 198 square inches. Find the base and height.

1—Answer: $b = 22$ inches and $h = 18$ inches

81. The area of a triangle is given by $A = \frac{1}{2}bh$ where A is in square inches when b and h are in inches. The height of a triangle is 2 inches longer than the base and its area is 18 square inches. Find the base (two decimal places).

1—T—Answer: $b \approx 5.08$ inches

82. The height of an object dropped from an initial height of 350 feet is given by $h = 350 - 16t^2$, where t is in seconds and h is in feet. How many seconds (to two decimal places) has the object been falling when it strikes the ground?

1—T—Answer: 4.68 seconds

83. The height of an object dropped from an initial height of 450 feet is given by $h = 450 - 16t^2$, where t is in seconds and h is in feet. How many seconds (to two decimal places) has the object been falling when it strikes the ground?

1—Answer: 5.30 seconds

84. A train makes a round trip between cities 300 miles apart. On the return half, the average speed is 25 mph faster than the average speed on the trip out and takes 1 hour less time. Find the time required on the trip out.

1—Answer: 4 hours

85. A train makes a round trip between cities 300 miles apart. On the return half, the average speed is 25 mph faster than the average speed on the trip out and takes 1 hour less time. Find the time required on the return trip.

1—Answer: 3 hours

86. A group of people could rent a social hall for $600. When 5 more people join the venture the cost per person is decreased by $10. How many people are in the larger group?

1—Answer: 20

87. A group of people could rent a social hall for $500. When 15 more people join the venture the cost per person is decreased by $7.50. How many people are in the larger group?

1—Answer: 40

1.5 Complex Numbers

1. Find b so that the equation is true: $(a + 3) + (2b - 1)i = 5 + 9i$.
   - (a) 2
   - (b) 4
   - (c) 5
   - (d) 8
   - (e) None of these

   1—Answer: c

2. Find a so that the equation is true: $(a + 6) + (3b + 1)i = 4 + 3i$.
   - (a) -2
   - (b) $\frac{2}{3}$
   - (c) 10
   - (d) $\frac{3}{2}$
   - (e) None of these

   1—Answer: a

3. Find b so that the equation is true: $(2a + 1) + (b - 2)i = 3 + 7i$.
   - (a) -1
   - (b) 1
   - (c) 5
   - (d) 9
   - (e) None of these

   1—Answer: d

4. Find b so that the equation is true: $(3a + 1) + (b - 6)i = 4 - 5i$.
   - (a) -1
   - (b) 1
   - (c) 7
   - (d) 5
   - (e) None of these

   1—Answer: b

5. Find b so that the equation is true: $(2a + 1) + (2b + 3)i = 4 - 7i$.
   - (a) $\frac{2}{3}$
   - (b) 5
   - (c) -5
   - (d) -2
   - (e) None of these

   1—Answer: c

6. Write in standard form: $3 + \sqrt{-9} - 16 + 2i^2$.
   - (a) $-15 - 3i$
   - (b) $-15 + 3i$
   - (c) -18
   - (d) $-17 + 3i$
   - (e) None of these

   1—Answer: b

7. Write in standard form: $2i^3 - 3\sqrt{-16} + 2$.
   - (a) $2 - 14i$
   - (b) $14 - 2i$
   - (c) $2 + 10i$
   - (d) $2 - 4i$
   - (e) None of these

   1—Answer: a

8. Write in standard form: $4 - \sqrt{-8}$.
   - (a) $4 - 2i$
   - (b) $4 + 2\sqrt{2}i$
   - (c) $4 - 2\sqrt{2}i$
   - (d) 8
   - (e) None of these

   1—Answer: c

9. Write in standard form: $2i^4 + 7i^3$.

   (a) $2 - 7i$
   (b) $-2 + 7i$
   (c) $7 - 2i$
   (d) $-7 + 2i$
   (e) None of these

   1—Answer: a

10. Write in standard form: $(16 + 2i) - (3 + 4i^2)$.

    (a) $9 + 2i$
    (b) $13 - 2i$
    (c) $15 + 2i$
    (d) $17 + 2i$
    (e) None of these

    1—Answer: d

11. Simplify, then write your result in standard form: $(3 + 6i) - 2(i + 7) - \sqrt{-4}$.

    (a) $1 + 4i$
    (b) $-11 + 6i$
    (c) $-11 + 2i$
    (d) $3 + 4i$
    (e) None of these

    1—Answer: c

12. Simplify, then write your result in standard form: $(6 + \sqrt{-9}) - 2i + 10 - \sqrt{16}$.

    (a) $16 - 3i$
    (b) $13 - 6i$
    (c) $9 - 2i$
    (d) $12 + i$
    (e) None of these

    1—Answer: d

13. Simplify, then write your result in standard form: $(4 - \sqrt{-1}) - 2(3 + 2i)$.

    (a) $-2 - 3i$
    (b) $-2 + 5i$
    (c) $-2 - 5i$
    (d) $-2 + i$
    (e) None of these

    1—Answer: c

14. Simplify, then write your result in standard form: $(4 - \sqrt{25}) + 2\sqrt{-9} - 4i + 7$.

    (a) $11 - 3i$
    (b) $13 - 12i$
    (c) $15 - 4i$
    (d) $13 - 2i$
    (e) None of these

    1—Answer: a

15. Simplify, then write your result in standard form: $3(2 - \sqrt{-9}) + 2i(4i - 7)$.

    1—Answer: $-2 - 23i$

16. Write the conjugate: $4 - \sqrt{-3}$.

    (a) $-4 + \sqrt{3}i$
    (b) $16 - 3i$
    (c) $4 + \sqrt{3}i$
    (d) $4 - \sqrt{3}i$
    (e) None of these

    1—Answer: c

17. Write the conjugate: $6 + \sqrt{-16}$.

    (a) $6 - 4i$
    (b) $6 + 4i$
    (c) $-6 - 4i$
    (d) $-6 + 4i$
    (e) None of these

    1—Answer: a

18. Write the conjugate: $3 - \sqrt{-1}$.

(a) $-3 - i$ (b) $-3 + i$ (c) 4

(d) $3 + i$ (e) None of these

1—Answer: d

19. Write the conjugate: $\frac{1}{2} + 4i$.

(a) $\frac{1}{2} - 4i$ (b) $2 - \frac{1}{4}i$ (c) $-\frac{1}{2} + 4i$

(d) $-2 - 4i$ (e) None of these

1—Answer: a

20. Write the conjugate: $\frac{3 + 4i}{16}$.

(a) $\frac{3}{16} + \frac{1}{4}i$ (b) $\frac{16}{3 + 4i}$ (c) $\frac{3}{16} - \frac{1}{4}i$

(d) $-\frac{3 + 4i}{16}$ (e) None of these

1—Answer: c

21. Multiply: $(3 + 7i)(6 - 2i)$.

(a) $18 + 22i$ (b) $4 + 48i$ (c) $4 + 36i$

(d) $32 + 36i$ (e) None of these

1—Answer: d

22. Multiply: $(3 - \sqrt{-4})(7 + \sqrt{-9})$.

(a) $15 + 23i$ (b) $27 - 5i$ (c) $27 + 5i$

(d) $15 + 5i$ (e) None of these

1—Answer: b

23. Multiply: $(4 - \sqrt{-9})^2$.

(a) 7 (b) $7 - 24i$ (c) $25 - 24i$

(d) $7 - 12i$ (e) None of these

1—Answer: b

24. Divide then write your answer in standard form: $\frac{6 + 10i}{2i}$.

(a) $\frac{3}{i} + 5$ (b) $5 - 3i$ (c) $5 + 3i$

(d) $3 + 5i$ (e) None of these

1—Answer: b

25. Divide then write the result in standard form: $\dfrac{-4+i}{1+4i}$.

(a) $-\dfrac{8}{17}+i$ (b) $-i$ (c) i

(d) $\dfrac{8}{17}-i$ (e) None of these

2—Answer: c

26. $\dfrac{3-2i}{2+4i}$ written in standard form, is:

(a) $\dfrac{1}{10}-\dfrac{4}{5}i$ (b) $-\dfrac{1}{10}+\dfrac{4}{5}i$ (c) $-\dfrac{1}{10}-\dfrac{4}{5}i$

(d) $\dfrac{1}{10}+\dfrac{4}{5}i$ (e) None of these

2—Answer: c

27. $\dfrac{2+4i}{3-2i}$ written in standard form, is:

(a) $-\dfrac{2}{13}-\dfrac{16}{13}i$ (b) $\dfrac{2}{13}-\dfrac{16}{13}i$ (c) $\dfrac{2}{13}+\dfrac{16}{13}i$

(d) $-\dfrac{2}{13}+\dfrac{16}{13}i$ (e) None of these

2—Answer: d

28. Divide, then write the result in standard form: $\dfrac{3+7i}{3-7i}$.

2—Answer: $-\dfrac{20}{29}+\dfrac{21}{29}i$

29. Divide, then write the result in standard form: $\dfrac{3-7i}{3+7i}$.

2—Answer: $-\dfrac{20}{29}-\dfrac{21}{29}i$

30. Divide, then write the result in standard form: $\dfrac{1+4i}{4-i}$.

2—Answer: i

31. Divide, then write the result in standard form: $\dfrac{4-i}{1+4i}$.

2—Answer: $-i$

32. Divide, then write the result in standard form: $\dfrac{2+3i}{1-i}$.

2—Answer: $\dfrac{-1}{2}+\dfrac{5}{2}i$

33. Solve for x: $3x^2 = 4x - 2$.

    (a) $\dfrac{2 \pm 2\sqrt{2}i}{3}$ (b) $\dfrac{2 \pm 2\sqrt{10}}{3}$ (c) $\dfrac{2 \pm \sqrt{2}i}{3}$

    (d) $\dfrac{2 \pm \sqrt{10}}{3}$ (e) None of these

    1—Answer: c

34. Solve for x: $3x^2 = x + 14$.

    (a) $-7, \dfrac{2}{3}$ (b) $\dfrac{7}{3}, -2$ (c) $\dfrac{x+14}{3x}$

    (d) $-\dfrac{1}{6} \pm \dfrac{\sqrt{167}}{6} i$ (e) None of these

    1—Answer: b

35. Solve for x: $8x^2 = 2x - 3$.

    (a) $\dfrac{2x-3}{8}$ (b) $\dfrac{1}{16} + \dfrac{\sqrt{23}}{16} i$ (c) $2 \pm \dfrac{\sqrt{23}}{16} i$

    (d) $\dfrac{1 \pm \sqrt{23}i}{8}$ (e) None of these

    1—Answer: d

36. Solve for x: $x^2 - 2x + 10 = 0$.

    (a) $7, -1$ (b) $1 + 3i, -1 + 3i$ (c) $1 + 3i, 1 - 3i$

    (d) $4, -2$ (e) None of these

    1—Answer: c

37. Use the Quadratic Formula to solve for x: $5x^2 - 2x + 6 = 0$.

    1—Answer: $\tfrac{1}{5}(1 \pm \sqrt{29}i)$

38. Use the Quadratic Formula to solve for x: $2x^2 - 4x + 3 = 0$.

    1—Answer: $\tfrac{1}{2}(2 \pm \sqrt{2}i)$

39. Use the Quadratic Formula to solve for x: $3x^2 - 2x + 1 = 0$.

    1—Answer: $\dfrac{1}{3} \pm \dfrac{\sqrt{2}}{3} i$

40. Use the Quadratic Formula to solve for x: $4x^2 - 4x + 3 = 0$.

    1—Answer: $\dfrac{1}{2} \pm \dfrac{\sqrt{2}}{2} i$

41. Use the Quadratic Formula to solve for x: $2x^2 - 5x + 6 = 0$.

    1—Answer: $\dfrac{5}{4} \pm \dfrac{\sqrt{23}}{4} i$

Chapter 1 Equations and Inequalities

42. $(i)^{26} =$
   (a) 1   (b) i (c) -1
   (d) $-i$ (e) None of these
   1—Answer: c

43. $(i)^{31} =$
   (a) 1   (b) i (c) -1
   (d) $-i$ (e) None of these
   1—Answer: d

44. $(i)^{44} =$
   (a) 1   (b) i (c) -1
   (d) $-i$ (e) None of these
   1—Answer: a

45. $(i)^{33} =$
   (a) 1   (b) i (c) -1
   (d) $-i$ (e) None of these
   1—Answer: b

46. Simplify: $(\sqrt{-25})^3$.
   (a) -125 (b) $125i$ (c) $-125i$
   (d) 125 (e) None of these
   1—Answer: c

47. Simplify: $4i^{16} + 3i^{12}$.
   (a) 7 (b) $4 - 3i$ (c) $7i$
   (d) $-3 + 4i$ (e) None of these
   1—Answer: a

48. Simplify: $6i^{17} + 4i^{20}$.
   (a) $6 - 4i$ (b) $4 - 6i$ (c) $-6 + 4i$
   (d) $4 + 6i$ (e) None of these
   1—Answer: d

49. Use a graphing utility to graph $y = x^2 + x + 1$. Use the graph to approximate the x-intercepts. Then set $y = 0$ and solve the equation. Compare the result with the x-intercepts of the graph.

2—T—Answer: 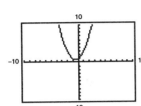 ; $x = -\dfrac{1}{2} \pm \dfrac{\sqrt{3}}{2} i$; no x-intercepts

50. Use a graphing utility to graph $y = x^2 - 2x + 5$. Use the graph to approximate the x-intercepts. Then set $y = 0$ and solve the equation. Compare the result with the x-intercepts of the graph.

2—T—Answer: 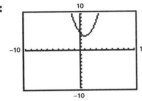 ; $x = 1 \pm 2i$; no x-intercepts

51. Use a graphing utility to graph $y = x^2 - 3x + 6$. Use the graph to approximate the x-intercepts. Then set $y = 0$ and solve the equation. Compare the result with the x-intercepts of the graph.

2—T—Answer: 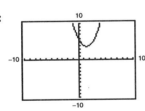 ; $x = \dfrac{3}{2} \pm \dfrac{\sqrt{15}}{2} i$; no x-intercepts

1.6 Other Types of Equations

1. Solve for x: $3x^3 = 27x$.
   - (a) 3
   - (b) $-3, 3$
   - (c) $-3, 0, 3$
   - (d) $0, 3i, -3i$
   - (e) None of these

   1—Answer: c

2. Solve for x: $3x^3 - 24x^2 + 21x = 0$.
   - (a) 7, 1
   - (b) $-7, -1$
   - (c) 0, 1, 7
   - (d) $0, -1, -7$
   - (e) None of these

   1—Answer: c

3. Solve for x: $20x^3 - 500x = 0$.

   1—Answer: $0, \pm 5$

4. Solve for x: $7x^3 = 252x$.

   1—Answer: $\pm 6, 0$

5. Solve: $x^3 + x^2 - 2x = 2$.
   - (a) $-1, \pm\sqrt{2}$
   - (b) $1 \pm \sqrt{3}$
   - (c) $0, 1 \pm \sqrt{3}$
   - (d) $-2, 1$
   - (e) None of these

   2—Answer: a

6. Solve: $x^3 - 5x - 2x^2 + 10 = 0$.
   - (a) $-2, \pm\sqrt{5}$
   - (b) $\pm\sqrt{5}$
   - (c) $2, \sqrt{5}$
   - (d) $2, \pm\sqrt{5}$
   - (e) None of these

   2—Answer: d

7. Solve for x: $x^4 + 5x^2 - 36 = 0$.
   - (a) $\pm 3, \pm 2$
   - (b) 9, 4
   - (c) $\pm 2, \pm 3i$
   - (d) $-9, 4$
   - (e) None of these

   1—Answer: c

8. Solve for x: $2x^4 - 7x^2 + 5 = 0$.

   1—Answer: $\pm 1, \pm \dfrac{\sqrt{10}}{2}$

9. Solve: $9x^4 - 24x^2 + 16 = 0$.
   - (a) $\pm \dfrac{2}{\sqrt{3}}$
   - (b) $\dfrac{2}{\sqrt{3}}$
   - (c) $0, \pm \dfrac{2}{\sqrt{3}}$
   - (d) $\dfrac{4}{3}$
   - (e) None of these

   1—Answer: a

10. Solve for x: $3x - 2\sqrt{x} - 5 = 0$.

  (a) $\frac{5}{3}$ (b) $-1, \frac{5}{3}$ (c) $1, \frac{25}{9}$

  (d) $\frac{25}{9}$ (e) None of these

  1—Answer: d

11. Solve for x: $x^{2/3} - 6x^{1/3} = 7$.

  2—Answer: $343, -1$

12. Solve for x: $\sqrt{2 - 5x} = 5x$.

  (a) $\frac{1}{5}$ (b) $-\frac{2}{5}$ (c) $\frac{1}{5}, -\frac{2}{5}$

  (d) $\frac{1}{10}$ (e) None of these

  1—Answer: a

13. Solve for x: $3x + 5 = \sqrt{2 - 2x}$.

  1—Answer: -1

14. Solve for x: $\sqrt[3]{4x - 1} = 3$.

  1—Answer: 7

15. Solve for x: $\sqrt{15x + 4} = 4 - \sqrt{2x + 3}$.

  (a) 3 (b) $\frac{11}{169}$ (c) $3, \frac{11}{169}$

  (d) $-3, -\frac{11}{169}$ (e) None of these

  2—Answer: b

16. Solve for x: $\sqrt{x + 16} = 3 + \sqrt{x - 2}$.

  (a) 3 (b) $\frac{17}{4}$ (c) $\frac{1}{2}$

  (d) No solution (e) None of these

  2—Answer: b

17. Solve for x: $\sqrt{x + 1} = 9 - \sqrt{x}$.

  2—Answer: $\frac{1600}{81}$

18. Solve for x: $(x^2 + 4)^{2/3} = 25$.

  (a) $-5.8, 5.8$ (b) $-4.6, 4.6$ (c) 21

  (d) $-11, 11$ (e) None of these

  1—Answer: d

19. Solve for x: $(x^2 - 9x + 2)^{3/2} = 216$.

  (a) $9, -\frac{1}{2}$ (b) $\frac{9}{2} \pm 3i$ (c) $\frac{9}{2} \pm \frac{\sqrt{217}}{2}$

  (d) $9 \pm \frac{\sqrt{217}}{2}$ (e) None of these

  2—Answer: c

108 Chapter 1 Equations and Inequalities

20. Solve: $(x^2 - 2x + 5)^{2/3} = 4$.

(a) $-3, 1$ (b) $-1 \pm \sqrt{13}$ (c) $-1, 3$

(d) 1 (e) None of these

2—Answer: c

21. Solve: $\dfrac{1}{x} - \dfrac{1}{x+1} = 1$.

(a) $-1, 0$ (b) $\dfrac{-1-\sqrt{5}}{2}, \dfrac{-1+\sqrt{5}}{2}$ (c) $-1 - \dfrac{\sqrt{5}}{2}, -1 + \dfrac{\sqrt{5}}{2}$

(d) $-1, 1$ (e) None of these

2—Answer: b

22. Solve: $\dfrac{1}{x-1} + \dfrac{x}{x+2} = 2$.

(a) $-2, 1$ (b) $-2, 0, 1$ (c) $-1 \pm \sqrt{7}$

(d) $-1 \pm \sqrt{3}$ (e) None of these

2—Answer: c

23. Solve for x: $\dfrac{2}{x^2 - 1} + \dfrac{1}{x+1} = 5$.

(a) $\dfrac{6}{5}$ (b) $-1, \dfrac{6}{5}$ (c) $\dfrac{1 \pm \sqrt{41}}{10}$

(d) $\pm \dfrac{\sqrt{2}}{2}$ (e) None of these

1—Answer: a

24. Solve for x: $\dfrac{4}{x} - \dfrac{3}{x+1} = 7$.

(a) $-1, 5$ (b) 3 (c) $\dfrac{-3 \pm \sqrt{37}}{7}$

(d) $\dfrac{-7 \pm \sqrt{77}}{14}$ (e) None of these

1—Answer: c

25. Solve for x: $\dfrac{x}{x^2 - 9} + \dfrac{2}{x+3} = 3$.

(a) $\dfrac{3}{7}, -\dfrac{1}{5}$ (b) $\dfrac{-1 \pm \sqrt{17}}{2}$ (c) $-3, 7$

(d) $\dfrac{1 \pm \sqrt{5}}{2}$ (e) None of these

1—Answer: e

26. Solve for x: $|3x + 10| = 13$.

(a) 1   (b) 1, −1   (c) 1, $-\frac{23}{3}$

(d) 1, $\frac{23}{3}$ (e) None of these

1—Answer: c

27. Solve for x: $|2 - 4x| = 12$.

(a) $-\frac{5}{2}, \frac{7}{2}$ (b) $-\frac{5}{2}, -\frac{7}{2}$ (c) $\frac{5}{2}, -\frac{5}{2}$

(d) $-\frac{5}{2}$ (e) None of these

1—Answer: a

28. Solve: $|x^2 - 2x| = 2x - 3$.

(a) $1, 3, \pm\sqrt{3}$ (b) $\sqrt{3}, 3$ (c) 1, 3

(d) 3   (e) None of these

2—Answer: b

29. Solve: $|x^2 - 2x| = x$.

(a) 0   (b) 0, ±1   (c) 0, ±3

(d) 0, 1, 3   (e) None of these

1—Answer: d

30. Solve: $|x^2 - 2x| = 3x - 6$.

(a) 2   (b) 2, ±3   (c) 2, 3

(d) ±3   (e) None of these

2—Answer: c

31. Use a graphing utility to approximate the solutions of $x^3 + 3x^2 = 5x$.

(a) −4.19, 1.19   (b) 0, ±4.19   (c) 0, ±1.19

(d) 0, −4.19, 1.19   (e) None of these

1—T—Answer: d

32. Use a graphing utility to approximate the solutions of $9x^4 - 24x^2 + 16 = 0$.

(a) ±1.15   (b) 1.15   (c) 0, ±1.15

(d) 1.33   (e) None of these

1—T—Answer: a

33. Use a graphing utility to approximate the solutions of $\sqrt{x + 16} = 3 + \sqrt{x - 2}$.

(a) 3   (b) 4.25   (c) 0.5

(d) No solution   (e) None of these

2—T—Answer: b

110 Chapter 1 Equations and Inequalities

34. Use a graphing utility to approximate the solution of $(x^2 - 9x + 2)^{3/2} = 216$.
    (a) $-0.5, 9$
    (b) $\pm 3, 4.5$
    (c) $-2.87, 11.87$
    (d) $1.63, 16.37$
    (e) None of these

    2—T—Answer: c

35. Use a graphing utility to approximate the solution of $\dfrac{1}{x} - \dfrac{1}{x+1} = 1$.
    (a) $-1, 0$
    (b) $-1.62, 0.62$
    (c) $-2.12, 0.12$
    (d) $-1, 1$
    (e) None of these

    2—T—Answer: b

36. Use a graphing utility to approximate the solutions of $\dfrac{x}{x^2 - 9} + \dfrac{2}{x+3} = 3$.
    (a) $-0.20, 0.43$
    (b) $-2.56, 1.56$
    (c) $-3, 7$
    (d) $-0.62, 1.62$
    (e) None of these

    2—T—Answer: e

37. Use a graphing utility to approximate the solutions of $|3x + 10| = 13$.
    (a) 1
    (b) $-1, 1$
    (c) $-7.67, 1$
    (d) $-1, 7.67$
    (e) None of these

    2—T—Answer: c

38. Use a graphing utility to graph $y = x^3 - 7x^2 + 12x$. Approximate any x-intercepts. Set $y = 0$ and solve the equation.

    2—T—Answer: 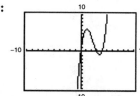 ; $(0, 0), (3, 0), (4, 0)$; $x = 0, 3, 4$

39. Use a graphing utility to graph $y = \sqrt{2x - 1} - 2$. Approximate any x-intercepts. Set $y = 0$ and solve the equation.

    2—T—Answer: 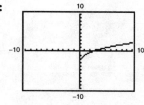 ; $\left(\tfrac{1}{2}, 0\right)$; $x = \tfrac{1}{2}$

40. Use a graphing utility to graph $y = |x^2 - 1| - 5$. Approximate any x-intercepts. Set $y = 0$ and solve the equation.

2—T—Answer: 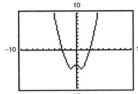 ; $(2.45, 0), (-2.45, 0)$; $x = \pm\sqrt{6}$

41. Use a graphing utility to graph $y = \dfrac{1}{x} + \dfrac{1}{x+1} - 2$. Approximate any x-intercepts. Set $y = 0$ and solve the equation.

2—T—Answer: 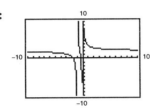 ; $(-0.7, 0), (0.7, 0)$; $x = \pm\dfrac{\sqrt{2}}{2}$

42. A church youth group decides to go bowling. They can rent three lanes for two hours for a total of $60. The cost per person will drop by $0.60 if they can get 5 visitors to attend also. How many people are in the youth group?

(a) 18 (b) 20 (c) 22

(b) 24 (e) None of these

2—Answer: b

43. During a one week leave, three military personnel decide to rent a car and share equally in the cost. By adding a fourth person to the group, each person could save $12.25. How much is the weekly rental for the car?

(a) $49 (b) $163 (c) $189

(d) $147 (e) None of these

2—Answer: d

44. The demand equation for a certain product is $p = 25(20 - \sqrt{x})$ where x is the number of units demanded per day and p is the price per unit. Find the demand if the price is set at $250.

(a) 100 (b) 10 (c) 19

(d) 14 (e) None of these

2—Answer: a

112 Chapter 1 Equations and Inequalities

45. The demand equation for a certain product is $p = 40 - \sqrt{0.0001x + 1}$ where x is the number of units demanded per day and p is the price per unit. Find the demand if the price is set at $10.50.

   (a) 92,316,410 (b) 4,765,180 (c) 8,692,500
   (d) 576,910 (e) None of these

   2—Answer: c

46. Find the height of the rectangular solid if the volume is $80x$ cubic units.

   (a) 5 (b) 10
   (c) 3 (d) 6
   (e) None of these

   1—Answer: c

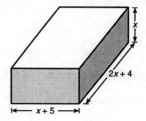

❑ 1.7 Linear Inequalities

1. Write an inequality to represent the interval.

   (a) $-3 \leq x < 5$ (b) $-3 \geq x < 5$ (c) $-3 < x \leq 5$
   (d) $5 > x \leq -3$ (e) None of these

   1—Answer: c

2. Write an inequality to represent the interval.

   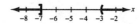

   (a) $x < -7$ or $x \geq -3$ (b) $-7 \leq x > -3$ (c) $-7 < x < -3$
   (d) $x \leq -7$ or $x > -3$ (e) None of these

   1—Answer: d

3. Graph the solution: $3 - 2x < 15$.

   (a) ![number line with open circle at -6, shaded right]
   (b) ![number line with open circle at -6, shaded left]
   (c) ![number line with open circle at 6, shaded right]
   (d) ![number line with open circle at 6, shaded left]
   (e) None of these

   1—Answer: a

4. Solve the inequality: $14 - 2x \leq 5$.
   - (a) $\left(-\infty, \frac{9}{2}\right)$
   - (b) $\left(-\infty, \frac{9}{2}\right]$
   - (c) $\left(\frac{9}{2}, \infty\right)$
   - (d) $\left[\frac{9}{2}, \infty\right)$
   - (e) None of these

   1—Answer: d

5. Solve: $3 - 2x \leq 9$.
   - (a) $(-\infty, -3]$
   - (b) $(-\infty, 3]$
   - (c) $[-3, \infty)$
   - (d) $[3, \infty)$
   - (e) None of these

   1—Answer: c

6. Graph the solution: $2x - 4 < 8$.
   - (a)
   - (b)
   - (c)
   - (d)
   - (e) None of these

   1—Answer: d

7. Graph the solution: $-3 \leq 2x + 1 \leq 5$.
   - (a)
   - (b)
   - (c)
   - (d)
   - (e) None of these

   1—Answer: b

8. Graph the solution: $-6 < 7x + 2 \leq 5$.
   - (a)
   - (b)
   - (c)
   - (d)
   - (e) None of these

   1—Answer: b

9. Graph the solution: $\frac{1}{2} < 3 - x < 5$.

   1—Answer:

10. Graph the solution: $-16 \leq 7 - 2x < 5$.

    1—Answer:

11. Solve: $5x + 6 > 7x + 9$.

(a) $\left(-\frac{3}{2}, \infty\right)$ (b) $\left(\frac{6}{5}, \frac{9}{7}\right)$ (c) $\left(-\infty, -\frac{3}{2}\right)$
(d) $\left(\frac{3}{2}, \infty\right)$ (e) None of these

1—Answer: c

12. Solve the inequality: $4 - 3x \geq 5x + 12$.

(a) $(-\infty, -1]$ (b) $(-\infty, 8]$ (c) $[-1, \infty)$
(d) $(-\infty, -2)$ (e) None of these

1—Answer: a

13. Solve the inequality: $6 - 5x \leq x - 6$.

(a) $[0, \infty)$ (b) $[2, \infty)$ (c) $(-\infty, 2]$
(d) $(-\infty, 0)$ (e) None of these

1—Answer: b

14. Solve: $-2 < 3x + 1 < 7$.

(a) $-1 \leq x \leq 2$ (b) $-\frac{1}{3} < x < \frac{8}{3}$ (c) $x > 2$
(d) $-1 < x < 2$ (e) None of these

1—Answer: d

15. Solve: $-6 < 7x + 2 \leq 5$.

(a) $\left(-\infty, -\frac{8}{7}\right)$ or $\left(\frac{3}{7}, \infty\right)$ (b) $\left(-\frac{8}{7}, \frac{3}{7}\right]$ (c) $\left(-\frac{8}{7}, \infty\right)$
(d) $\left(-\infty, \frac{3}{7}\right)$ (e) None of these

1—Answer: b

16. Solve the inequality algebraically: $-16 \leq 7 - 2x \leq 5$.

(a) $x \leq 1$ or $x \geq \frac{23}{2}$ (b) $-1 \leq x \leq \frac{23}{3}$ (c) $1 \leq x \leq \frac{23}{2}$
(d) $-\frac{23}{2} \leq x \leq 1$ (e) None of these

1—Answer: c

17. Solve the inequality algebraically: $-2 < 3x + 1 < 7$.

(a) $-1 \leq x \leq 2$ (b) $-\frac{1}{3} < x < \frac{8}{3}$ (c) $x > 2$
(d) $-1 < x < 2$ (e) None of these

1—Answer: d

18. Solve the inequality algebraically: $14 - 2x \leq 5$.

(a) $\left(-\infty, \frac{9}{2}\right)$ (b) $\left(-\infty, \frac{9}{2}\right]$ (c) $\left(\frac{9}{2}, \infty\right)$
(d) $\left[\frac{9}{2}, \infty\right)$ (e) None of these

1—Answer: d

19. Solve the inequality algebraically: $3 - 2x \leq 9$.
    (a) $(-\infty, -3]$ (b) $(-\infty, 3]$ (c) $[-3, \infty)$
    (d) $[3, \infty)$ (e) None of these
    1—Answer: c

20. Solve the inequality algebraically: $5x + 6 > 7x + 9$.
    (a) $\left(-\frac{3}{2}, \infty\right)$ (b) $\left(\frac{6}{5}, \frac{9}{7}\right)$ (c) $\left(-\infty, -\frac{3}{2}\right)$
    (d) $\left(\frac{3}{2}, \infty\right)$ (e) None of these
    1—Answer: c

21. Solve the inequality algebraically: $4 - 3x \geq 5x + 12$.
    (a) $(-\infty, -1]$ (b) $(-\infty, 8]$ (c) $[-1, \infty)$
    (d) $(-\infty, -2)$ (e) None of these
    1—Answer: a

22. Solve the inequality algebraically: $-16 \leq 7 - 2x \leq 5$.
    (a) $x \leq 1$ or $x \geq \frac{23}{2}$ (b) $-1 \leq x \leq \frac{23}{2}$ (c) $1 \leq x \leq \frac{23}{2}$
    (d) $-\frac{23}{2} \leq x \leq 1$ (e) None of these
    1—Answer: c

23. Solve the inequality algebraically: $-2 < 3x + 1 < 7$.
    (a) $-1 \leq x \leq 2$ (b) $-\frac{1}{3} < x < \frac{8}{3}$ (c) $x > 2$
    (d) $-1 < x < 2$ (e) None of these
    1—Answer: d

24. Use a graphing utility to solve the inequality: $-6 < 7x + 2 \leq 5$.
    (a) $x < -\frac{8}{7}$ or $x > \frac{3}{7}$ (b) $-\frac{8}{7} < x \leq \frac{3}{7}$ (c) $x > -\frac{8}{7}$
    (d) $x < \frac{3}{7}$ (e) None of these
    1—T—Answer: b

25. Use a graphing utility to solve the inequality: $6 - 5x \leq x - 6$.
    (a) $[0, \infty)$ (b) $[2, \infty)$ (c) $(-\infty, 2]$
    (d) $(-\infty, 0)$ (e) None of these
    1—T—Answer: b

26. Use a graphing utility to solve the inequality: $5 + 2x > 4x + 7$.
    (a) $x < 1$ (b) $x > 1$ (c) $x > -1$
    (d) $x < -1$ (e) None of these
    1—T—Answer: d

27. Find the interval for which the radicand is nonnegative: $\sqrt{5 - 4x}$.

(a) $\left(-\infty, -\frac{4}{5}\right]$ (b) $\left[\frac{5}{4}, \infty\right)$ (c) $\left(-\infty, \frac{4}{5}\right]$

(d) $\left(-\infty, \frac{5}{4}\right]$ (e) None of these

1—Answer: d

28. Find the interval for which the radicand is nonnegative: $\sqrt{2 - 3x}$.

(a) $\left[-\infty, -\frac{3}{2}\right]$ (b) $\left(-\infty, \frac{2}{3}\right]$ (c) $\left[\frac{2}{3}, \infty\right)$

(d) $\left[\frac{3}{2}, \infty\right)$ (e) None of these

1—Answer: b

29. Find the intervals for which the radicand is nonnegative: $\sqrt{7 + 5x}$.

(a) $\left(-\infty, \frac{5}{7}\right]$ (b) $\left(-\infty, -\frac{7}{5}\right]$ (c) $\left[-\frac{5}{7}, \infty\right)$

(d) $\left[\frac{7}{5}, \infty\right)$ (e) None of these

1—Answer: e

30. Find the interval for which the radicand is positive: $\dfrac{1}{\sqrt{2 + 5x}}$.

(a) $\left(-\infty, -\frac{2}{5}\right)$ (b) $\left(-\infty, -\frac{5}{2}\right)$ (c) $\left(-\frac{5}{2}, \infty\right)$

(d) $\left(-\frac{2}{5}, \infty\right)$ (e) None of these

1—Answer: d

31. Find the interval for which the radicand is positive: $\dfrac{1}{\sqrt[4]{3 - 4x}}$.

(a) $\left(\frac{4}{3}, \infty\right)$ (b) $\left(-\frac{3}{4}, \infty\right)$ (c) $\left(-\infty, \frac{3}{4}\right)$

(d) $\left(-\infty, -\frac{4}{3}\right)$ (e) None of these

1—Answer: c

32. Match the inequality with the graph.

```
 (——+——+——+——+——+——+——+——)→ x
 -7  -6  -5  -4  -3  -2  -1   0   1
```

(a) $|x - 7| < 4$ (b) $|x - 1| < 3$ (c) $|x + 3| < 7$

(d) $|x + 3| < 4$ (e) None of these

1—Answer: d

33. Match the inequality with the graph.

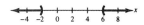

(a) $|x + 4| < -2$ (b) $|x - 2| > 4$ (c) $|x - 2| < 4$
(d) $|x - 6| > 2$ (e) None of these
1—Answer: b

34. Match the inequality with the graph.

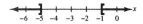

(a) $|x + 2| > 5$ (b) $|x + 3| \geq 2$ (c) $|x - 1| \leq 5$
(d) $|x - 5| \geq 2$ (e) None of these
1—Answer: b

35. Graph the solution: $|3x - 1| \geq 5$.

(a) [graph from -1 to 2]
(b) [graph from -2 to -1, and from 2 to right]
(c) [graph from -2 to 1]
(d) [graph from -3 to -1, and from 1 to right]
(e) None of these
1—Answer: b

36. Graph the solution: $|x + 2| < 9$.

(a) [graph from -15 to 5, open]
(b) [graph to -5, and from 5 to right]
(c) [graph from -10 to 10, open]
(d) [graph to 0, and from 10 to right]
(e) None of these
1—Answer: a

37. Graph the solution: $|3x - 1| > 9$.

1—Answer: [graph with open endpoints at $-\frac{8}{3}$ and $\frac{10}{3}$]

38. Solve the inequality algebraically: $|3x - 1| > 2$.

(a) $\left(-\frac{1}{3}, 1\right)$ (b) $\left[-\frac{1}{3}, 1\right]$ (c) $\left(-\infty, -\frac{1}{3}\right), (1, \infty)$
(d) $\left(-\infty, -\frac{1}{3}\right], [1, \infty)$ (e) None of these
1—Answer: c

39. Solve the inequality algebraically: $|x + 5| \leq 2$.

1—Answer: $-7 \leq x \leq -3$

40. Use a graphing utility to solve the inequality: $|2x + 5| > 3$.

(a) $x < -4$ or $x > -1$ (b) $-4 < x < -1$ (c) $x < 1$ or $x > 4$

(d) $x < -1$ or $x > 4$ (e) None of these

1—T—Answer: a

41. Use the absolute value notation to define all real numbers on the real number line within 6 units of 10.

(a) $|x - 6| \geq 10$ (b) $|x - 10| \geq 6$ (c) $|x - 6| \geq 10$

(d) $|x - 10| \leq 6$ (e) None of these

1—Answer: d

42. Use absolute value notation to denote all real numbers x that are at least 3 units from the number 5.

(a) $|x - 3| < 5$ (b) $|x - 3| \geq 5$ (c) $|x - 5| < 3$

(d) $|x - 5| \geq 3$ (e) None of these

1—M—Answer: d

43. Use absolute value to define the interval: $-7 \leq x \leq 7$.

1—Answer: $|x| \leq 7$

44. Use absolute value to define the interval: $x < -3$ or $x > 3$.

1—Answer: $|x| > 3$

45. Use absolute value to define the interval: $-7 < x < 3$.

2—Answer: $|x + 2| < 5$

46. Use absolute value notation to define the interval on the real number line.

1—Answer: $|x - 3| < 2$

47. Use absolute value notation to define the interval on the real number line.

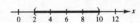

1—Answer: $|x - 6| < 4$

48. Use absolute value notation to define the interval on the real number line.

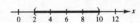

1—Answer: $|x + 2| \leq 4$

Section 1.7 Linear Inequalities 119

49. Use absolute value notation to define the interval on the real number line.

1—Answer: $\left|x - \frac{3}{2}\right| < \frac{15}{2}$

50. The revenue for selling x units of a product is $R = 35.95x$. The cost of producing x units is $14.75x + 848$. In order to obtain a profit, the revenue must be greater than the cost. For what values of x will this product return a profit?

   (a) $x > 123$ (b) $x > 117$ (c) $x > 52$
   (d) $x > 40$ (e) None of these

   2—Answer: d

51. The revenue for selling x units of a product is $R = 257x$. The cost of producing x units is $193x + 5248$. In order to obtain a profit, the revenue must be greater than the cost. For what values of x will this product return a profit?

   (a) $x > 103$ (b) $x > 82$ (c) $x > 77$
   (d) $x > 12$ (e) None of these

   2—Answer: b

52. The revenue for selling x units of a product is $R = 4.50x$ The cost of producing x units is $3x + 3717$. In order to obtain a profit, the revenue must be greater than the cost. For what values of x will this product return a profit?

   (a) $x > 1893$ (b) $x > 1239$ (c) $x > 2478$
   (d) $x > 496$ (e) None of these

   2—Answer: c

53. You buy a bag of candy that costs $2.90 per pound. The weight that is listed on the bag is 1.10 pounds. If the scale that weighed the candy is only accurate to within 0.125 of a pound, how much money might you have been overcharged or undercharged?

   (a) $22\frac{1}{8}$¢   (b) $17\frac{3}{4}$¢   (c) $36\frac{1}{4}$¢
   (d) $31\frac{1}{2}$¢   (e) None of these

   2—Answer: c

54. You buy a bag of candy that costs $3.15 per pound. The weight that is listed on the bag is 0.90 pounds. If the scale that weighed the candy is only accurate to within 0.125 of a pound, how much money might you have been overcharged or undercharged?

   (a) $39\frac{3}{8}$¢   (b) $29\frac{1}{4}$¢   (c) $16\frac{2}{3}$¢
   (d) $51\frac{1}{2}$¢   (e) None of these

   2—Answer: a

1.8 Other Types of Inequalities

1. Solve the inequality: $(x-2)^2 \leq 9$.

   1—Answer: $[-1, 5]$

2. Solve the inequality: $(x+3)^2 \geq 4$.

   (a) $[1, 5]$ (b) $(-\infty, 5]$ (c) $(-\infty, -5] \cup [-1, \infty)$

   (d) $[-5, -1]$ (e) None of these

   1—Answer: c

3. Solve the inequality: $(x-1)^2 \leq 25$.

   (a) $[-4, 6]$ (b) $(-\infty, -4] \cup [6, \infty)$ (c) $(-\infty, -6] \cup [4, \infty)$

   (d) $[-6, 4]$ (e) None of these

   1—Answer: a

4. Use a graphing utility to solve $(x+1)^2 \geq 9$.

   (a) $-4, 2$ (b) $(-\infty, -4] \cup [2, \infty)$ (c) $[-2, 4]$

   (d) $(-\infty, -2) \cup (4, \infty)$ (e) None of these

   1—T—Answer: b

5. Solve the inequality: $x^2 - x > 6$.

   1—Answer: $(-\infty, -2) \cup (3, \infty)$

6. Solve the inequality: $3x^3 - 6x^2 > 0$.

   (a) $(-\infty, 0) \cup (2, \infty)$ (b) $(0, 2)$ (c) $(-\infty, 0)$

   (d) $(2, \infty)$ (e) None of these

   1—Answer: d

7. Use a graphing utility to solve $2x^2 + 3x \geq 5$.

   (a) $[-2.5, 1]$ (b) $[-2.5, \infty)$ (c) $(-\infty, -2.5] \cup [1, \infty)$

   (d) $(-\infty, -1.5) \cup [5, \infty)$ (e) None of these

   1—T—Answer: c

8. Solve: $2x^2 + 3x < 9$.

   (a) $\left(-3, \frac{3}{2}\right)$ (b) $(-\infty, -3) \cup \left(\frac{3}{2}, \infty\right)$ (c) $\left[-3, \frac{3}{2}\right]$

   (d) $(-\infty, 3) \cup (9, \infty)$ (e) None of these

   1—Answer: a

9. Use a graphing utility to solve $2x^3 \leq 4x^4$.

   (a) $(-\infty, \infty)$ (b) $\left[0, \frac{1}{2}\right]$ (c) $(-\infty, 0] \cup \left[\frac{1}{2}, \infty\right)$

   (d) $\left[\frac{1}{2}, \infty\right)$ (e) None of these

   1—T—Answer: c

10. Solve: $x^2 + 1 \geq 0$.
    - (a) $(-\infty, \infty)$
    - (b) $[-1, 1]$
    - (c) $(-\infty, -1] \cup [1, \infty)$
    - (d) $(-1, 1)$
    - (e) None of these

    2—Answer: a

11. Use a graphing utility to solve: $x^2 + 4x + 2 \leq 0$.
    - (a) $[-3.41, \infty)$
    - (b) $[-3.41, -0.59]$
    - (c) $(-\infty, -3.41] \cup [-0.59, \infty)$
    - (d) $[-2, 2]$
    - (e) None of these

    2—T—Answer: b

12. Solve: $2x^2 - 5x > 3$.
    - (a) $\left(-\frac{1}{2}, 3\right)$
    - (b) $\left(-\infty, -\frac{1}{2}\right) \cup (3, \infty)$
    - (c) $(-\infty, -3) \cup \left(\frac{1}{2}, \infty\right)$
    - (d) $\left(-\frac{1}{2}, \infty\right)$
    - (e) None of these

    1—Answer: b

13. Solve: $x^2 + 6x + 1 \geq 0$.
    - (a) $(-\infty, -3 - 2\sqrt{2}]$ or $[-3 + 2\sqrt{2}, \infty)$
    - (b) $[-3 - 2\sqrt{2}, -3 + 2\sqrt{2}]$
    - (c) $(-\infty, -3 - \sqrt{10}]$ or $[-3 + \sqrt{10}, \infty)$
    - (d) $[-3 - \sqrt{10}, -3\sqrt{10}]$
    - (e) None of these

    2—Answer: a

14. Solve: $x^2 + x - 3 < 0$.
    - (a) $(-\infty, -3)$ or $(1, \infty)$
    - (b) $\left(-\infty, \frac{-1 - \sqrt{13}}{2}\right)$ or $\left(\frac{-1 + \sqrt{13}}{2}, \infty\right)$
    - (c) $\left(\frac{-1 - \sqrt{13}}{2}, \frac{-1 + \sqrt{13}}{2}\right)$
    - (d) $(-3, 1)$
    - (e) None of these

    2—Answer: c

15. Solve: $x^2 + 6x + 1 \geq 0$.
    - (a) $\left(-\infty, -3 - 2\sqrt{2}\right] \cup \left[-3 + 2\sqrt{2}, \infty\right)$
    - (b) $\left[-3 - 2\sqrt{2}, -3 + 2\sqrt{2}\right]$
    - (c) $\left(-\infty, -3 - \sqrt{10}\right] \cup \left[-3 + \sqrt{10}, \infty\right)$
    - (d) $\left[-3 - \sqrt{10}, -3 + \sqrt{10}\right]$
    - (e) None of these

    2—Answer: a

16. Use a graphing utility to solve $x^2 + x - 3 < 0$.
    - (a) $(-\infty, -3) \cup (1, \infty)$
    - (b) $(-\infty, -2.30) \cup (1.30, \infty)$
    - (c) $(-2.30, 1.30)$
    - (d) $(-3, 1)$
    - (e) None of these

    2—T—Answer: c

17. Use a graphing utility to solve $x^2 - 4x + 2 < 0$.

    (a) $(-3.41, -0.59)$ (b) $(0.59, 3.41)$ (c) $(-\infty, -3.41) \cup (-0.59, \infty)$

    (d) $(-\infty, 0.59) \cup (3.41, \infty)$ (e) None of these

    2—T—Answer: b

18. Solve: $x^2 + 3x + 4 > 0$.

    (a) $(-\infty, \infty)$ (b) $(-\infty, 0) \cup (0, \infty)$ (c) Empty set

    (d) $\left(\dfrac{-3 - \sqrt{7}}{2}, \dfrac{-3 + \sqrt{7}}{2}\right)$ (e) None of these

    2—Answer: a

19. Solve: $x^2 - 3x + 4 < 0$.

    (a) $(-\infty, \infty)$ (b) $(-\infty, 0) \cup (0, \infty)$ (c) Empty set

    (d) $\left(\dfrac{3 - \sqrt{7}}{2}, \dfrac{3 + \sqrt{7}}{2}\right)$ (e) None of these

    2—Answer: c

20. Solve: $x^2 + 6x + 9 \leq 0$.

    (a) $(-\infty, -3]$ or $[-3, \infty)$ (b) -3 (c) $(-\infty, \infty)$

    (d) Empty set   (e) None of these

    2—Answer: b

21. Use a graphing utility to solve $x^2 + 3x + 9 \geq 0$.

    (a) $(-\infty, -3] \cup [3, \infty)$ (b) $[-3, 3]$ (c) $(-\infty, \infty)$

    (d) Empty set   (e) None of these

    2—T—Answer: c

22. Solve: $x^2 + 4 < 2x$.

    (a) $(-\infty, \infty)$ (b) $(-\infty, -2) \cup (2, \infty)$ (c) $(-2, 2)$

    (d) Empty set   (e) None of these

    2—Answer: d

23. Solve the inequality: $\dfrac{x + 16}{3x + 2} \leq 5$.

    (a) $\left(-\infty, -\dfrac{2}{3}\right] \cup \left[\dfrac{3}{7}, \infty\right)$ (b) $\left[-\dfrac{2}{3}, \dfrac{3}{7}\right]$ (c) $\left(-\infty, -\dfrac{2}{3}\right) \cup \left[\dfrac{3}{7}, \infty\right)$

    (d) $\left(-\dfrac{2}{3}, \dfrac{3}{7}\right]$ (e) None of these

    2—Answer: c

24. Use a graphing utility to solve $\dfrac{x+7}{3x-1} < 1$.

   (a) $(0.33, 4)$ (b) $[0.33, 4]$ (c) $(-\infty, 0.33) \cup (4, \infty)$

   (d) $(-\infty, 0.33] \cup [4, \infty)$ (e) None of these

   2—T—Answer: c

25. Solve the inequality: $\dfrac{3x-7}{x+2} < 1$.

   2—Answer: $\left(-2, \dfrac{9}{2}\right)$

26. Solve the inequality: $\dfrac{2}{x-1} < 5$.

   (a) $(-\infty, 1)$ (b) $\left(\dfrac{7}{5}, \infty\right)$ (c) $\left(-\infty, -\dfrac{3}{5}\right) \cup \left(\dfrac{7}{5}, \infty\right)$

   (d) $(-\infty, 1) \cup \left(\dfrac{7}{5}, \infty\right)$ (e) None of these

   2—Answer: d

27. Use a graphing utility to solve $\dfrac{2}{x+1} \geq 5$.

   (a) $(-\infty, -1) \cup \left(-\dfrac{3}{5}, \infty\right)$ (b) $\left(-\infty, -\dfrac{3}{5}\right]$ (c) $\left(-1, -\dfrac{3}{5}\right]$

   (d) $\left(-\infty, \dfrac{1}{5}\right]$ (e) None of these

   2—T—Answer: c

28. Solve the inequality: $\dfrac{4}{x+1} \leq \dfrac{3}{x+2}$.

   (a) $(-\infty, -5] \cup (-2, -1)$ (b) $(-5, -2) \cup [-1, \infty)$ (c) $(-\infty, -5) \cup (-2, -1)$

   (d) $(-5, -2) \cup [-1, \infty)$ (e) None of these

   2—Answer: a

29. Solve the inequality: $\dfrac{3}{x-2} \leq \dfrac{5}{x+2}$.

   (a) $(-\infty, -2) \cup (2, 6)$ (b) $(-2, 2) \cup [6, \infty)$ (c) $[8, \infty)$

   (d) $(-2, 2) \cup [8, \infty)$ (e) None of these

   2—Answer: d

30. Solve the inequality: $\dfrac{2}{x-1} \leq \dfrac{3}{x+1}$.

   (a) $(-1, 1) \cup [5, \infty)$ (b) $(-\infty, -1) \cup (1, 5]$ (c) $[5, \infty)$

   (d) Empty set   (e) None of these

   2—Answer: a

Chapter 1 Equations and Inequalities

31. Use a graphing utility to solve $\dfrac{3}{x-1} \leq \dfrac{2}{x+1}$.

(a) $(-\infty, -5]$ (b) $(-\infty, -5] \cup (-1, 1)$ (c) $[-5, -1) \cup (1, \infty)$

(d) $(-\infty, -2] \cup (1, \infty)$ (e) None of these

2—T—Answer: b

32. Solve the inequality; $\dfrac{2}{x+2} \geq \dfrac{3}{x-1}$.

(a) $[-8, \infty)$ (b) $[-8, -2) \cup (1, \infty)$ (c) $(-\infty, -8] \cup (-2, 1)$

(d) $(-\infty, -8]$ (e) None of these

2—Answer: c

33. Find the domain of $\sqrt{x^2 - 7x - 8}$.

(a) $(-\infty, -1] \cup [8, \infty)$ (b) $(-\infty, -1) \cup (8, \infty)$ (c) $[-1, 8]$

(d) $(-1, 8)$ (e) None of these

1—Answer: a

34. Find the domain of $\sqrt{169 - 9x^2}$.

(a) $\left(-\dfrac{13}{3}, \dfrac{13}{3}\right)$ (b) $\left[-\dfrac{13}{3}, \dfrac{13}{3}\right]$ (c) $\left(-\infty, -\dfrac{13}{3}\right] \cup \left(\dfrac{13}{3}, \infty\right)$

(d) $\left(-\infty, -\dfrac{13}{3}\right) \cup \left(\dfrac{13}{3}, \infty\right)$ (e) None of these

1—Answer: b

35. Find the domain of $\sqrt{36 - x^2}$.

1—Answer: $[-6, 6]$

36. Find the domain of $\sqrt{16 - 4x^2}$.

1—Answer: $[-2, 2]$

37. Find the domain of $\dfrac{1}{\sqrt{x^2 - 7x - 8}}$.

(a) $[-1, 8]$ (b) $(-\infty, -1] \cup [8, \infty)$ (c) $(-\infty, -1) \cup (8, \infty)$

(d) $(-1, 8)$ (e) None of these

1—Answer: c

38. P dollars, invested at interest rate, r, compounded annually, increases to an amount $A = P(1 + r)^2$ in two years. If an investment of \$750 is to increase to an amount greater than \$883 in two years, then the interest rate must be greater than what percentage?

(a) 8.86%   (b) 8.5%   (c) 17.7%

(d) 5.6%   (e) None of these

2—T—Answer: b

39. P dollars, invested at interest rate, r, compounded annually, increases to an amount $A = P(1 + r)^2$ in two years. If an investment of $970 is to increase to an amount greater than $1100 in two years, then the interest rate must be greater than what percentage?

(a) 5.5%   (b) 6.0%   (c) 6.5%

(d) 7.0%   (e) None of these

2—T—Answer: c

40. Solve the inequality $0.2x^2 + 3.6 < 10.6$. Round each number in your solution to two decimal places.

2—T—Answer: $(-5.92, 5.92)$

41. Solve the inequality $-0.5x^2 + 4.26 \geq 2.56$. Round each number in your solution to two decimal places.

2—T—Answer: $[-1.84, 1.84]$

42. Solve the inequality $-6.26x^2 + 7.10 \leq 2.4x$. Round each number in your solution to two decimal places.

2—T—Answer: $(-\infty, -1.27] \cup [0.89, \infty)$

43. Solve the inequality $1.4x^2 + 5.6x - 10.75 \geq 0$. Round each number in your solution to two decimal places.

2—T—Answer: $(-\infty, -5.42] \cup [1.42, \infty)$

44. Solve the inequality $\dfrac{1}{3.2x - 6} > 2.4$. Round each number in your solution to two decimal places.

2—T—Answer: $(1.88, 2.00)$

45. A projectile is fired straight upward from ground level with an initial velocity of 64 feet per second. When will the height be less than 48 feet?

2—T—Answer: $(0, 1) \cup (3, 4)$

46. A projectile is fired straight upward from ground level with an initial velocity of 64 feet per second. When will the height be more than 48 feet?

2—T—Answer: $(1, 3)$

47. A projectile is fired straight upward from ground level with an initial velocity of 96 feet per second. When will the height be less than 84 feet?

2—T—Answer: $(0, 1.06) \cup (4.94, 6)$

48. A projectile is fired straight upward from ground level with an initial velocity of 96 feet per second. When will the height be more than 84 feet?

2—T—Answer: $(1.06, 4.94)$

CHAPTER TWO
Functions and Their Graphs

❑ 2.1 Lines in the Plane and Slope

1. Determine the slope of the line.

    (a) 2 \qquad (b) $\frac{1}{2}$

    (c) -2 \qquad (d) $-\frac{1}{2}$

    (e) None of these

    1—Answer: a

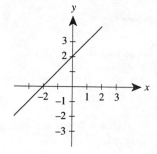

2. Determine the slope of the line.

    (a) $\frac{5}{2}$ \qquad (b) $-\frac{5}{2}$

    (c) $\frac{2}{5}$ \qquad (d) $-\frac{2}{5}$

    (e) None of these

    1—Answer: b

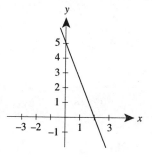

3. Determine the slope of the line shown at the right.

    (a) $\frac{2}{3}$ \qquad (b) $-\frac{2}{3}$

    (c) $\frac{3}{2}$ \qquad (d) $-\frac{3}{2}$

    (e) None of these

    1—Answer: c

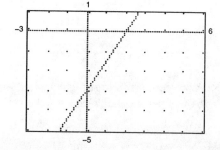

4. Determine the slope of the line shown at the right.

    (a) $-\frac{1}{2}$ \qquad (b) $\frac{1}{2}$

    (c) 2 \qquad (d) -2

    (e) None of these

    1—Answer: a

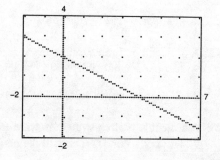

5. Determine the slope of the line.

    (a) $\frac{1}{5}$
    (b) $-\frac{1}{5}$
    (c) 5
    (d) -5
    (e) None of these

    1—Answer: d

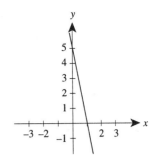

6. Find the slope of the line passing through $(6, 10)$ and $(-1, 4)$.

    (a) $\frac{7}{6}$
    (b) $-\frac{7}{6}$
    (c) $\frac{6}{7}$
    (d) $-\frac{6}{7}$
    (e) None of these

    1—Answer: c

7. Find the slope of the line passing through $(-1, 16)$ and $(4, 2)$.

    (a) $-\frac{5}{14}$
    (b) $-\frac{14}{5}$
    (c) $\frac{5}{14}$
    (d) $\frac{14}{5}$
    (e) None of these

    1—Answer: b

8. Find the slope of the line passing through $(3, -2)$ and $(5, 7)$.

    (a) $-\frac{9}{7}$
    (b) $\frac{9}{2}$
    (c) $\frac{5}{2}$
    (d) $\frac{2}{9}$
    (e) None of these

    1—Answer: b

9. Find the slope of the line passing through $(5, 9)$ and $(-1, -3)$.

    1—Answer: 2

10. Find the slope of the line passing through $(3, 7)$ and $(-1, -2)$.

    1—Answer: $\frac{9}{4}$

11. Which of the following points does **not** lie on the line that contains the point $(7, 7)$ and has slope $\frac{2}{7}$.

    (a) $\left(4, \frac{43}{7}\right)$
    (b) $(0, 5)$
    (c) $(-7, 2)$
    (d) $(-14, 1)$
    (e) All of these lie on the line.

    2—Answer: c

12. Determine which points lie on the line that contains the point $(5, 7)$ with slope 0.

    (a) $(5, 0)$
    (b) $(0, 7)$
    (c) $(7, 5)$
    (d) All of these lie on the line.
    (e) None of these lie on the line.

    1—Answer: b

13. Determine which points lie on the vertical line that contains the point (5, 1).
    (a) (5, 0)   (b) (0, 1)   (c) (1, 5)
    (d) All of these points lie on the line.   (e) None of these points lie on the line.

    1—Answer: a

14. Determine which points lie on the line that contains the point (2, −3) and has a slope of $-\frac{7}{4}$.
    (a) $\left(4, -\frac{13}{2}\right)$ (b) (−2, 4)   (c) $\left(0, \frac{1}{2}\right)$
    (d) All of these points lie on the line.   (e) None of these points lie on the line.

    2—Answer: d

15. Use the slope to describe the behavior of the line that passes through (3, 0) and (9, −2).
    (a) Rises from left to right   (b) Falls from left to right   (c) Horizontal
    (d) Vertical   (e) None of these

    2—Answer: b

16. Use the slope to describe the behavior of the line that passes through (5, 6) and (−1, 6).
    (a) Rises from left to right   (b) Falls from left to right   (c) Horizontal
    (d) Vertical   (e) None of these

    2—Answer: c

17. What is the slope of a line that is perpendicular to the line given by $2x + 3y + 9 = 0$?
    (a) $\frac{2}{3}$ (b) $-\frac{2}{3}$ (c) $\frac{3}{2}$
    (d) $-\frac{3}{2}$ (e) None of these

    1—Answer: c

18. What is the slope of the line parallel to the line $4x - 2y = 9$.
    (a) $\frac{9}{2}$ (b) $\frac{9}{4}$ (c) $-\frac{2}{4}$
    (d) 2   (e) None of these

    1—Answer: d

19. What is the slope of the line perpendicular to the line $y = 7$?
    (a) 0   (b) Undefined   (c) $\frac{1}{7}$
    (d) $-\frac{1}{7}$ (e) None of these

    1—Answer: b

20. Find the slope of the line perpendicular to the line $3x - 4y = 12$.
    (a) Undefined   (b) 0   (c) $\frac{4}{3}$
    (d) $-\frac{3}{4}$ (e) None of these

    1—Answer: e

21. Find the slope of the line $7x - 2y = 12$.

(a) $\frac{7}{2}$ (b) $-\frac{2}{7}$ (c) $\frac{12}{7}$

(d) -6 (e) None of these

1—Answer: a

22. Find the slope of the line $y = \frac{4x - 13}{7}$.

(a) 4 (b) $\frac{4}{7}$ (c) $-\frac{13}{7}$

(d) $\frac{1}{7}$ (e) None of these

1—Answer: b

23. Find the slope of the line $y = 3$.

(a) 3 (b) 0 (c) $\frac{1}{3}$

(d) Undefined   (e) None of these

1—Answer: b

24. Find the slope of the line $x = \frac{1}{2}$.

(a) $\frac{1}{2}$ (b) 0 (c) -2

(d) Undefined   (e) None of these

1—Answer: d

25. Find the equation of the line that has a slope of $-\frac{3}{4}$ and passes through $(1, 2)$.

(a) $3x - 4y - 7 = 0$ (b) $3x - 4y - 11 = 0$ (c) $3x + 4y - 11 = 0$

(d) $3x + 4y + 11 = 0$ (e) None of these

1—Answer: c

26. Find the equation of the vertical line that passes through $(2, 5)$.

(a) $y = 2$ (b) $y = 5$ (c) $x = 2$

(d) $x = 5$ (e) None of these

1—Answer: c

27. Find the equation of the line that passes through $(0, 0)$ and has a slope that is undefined.

(a) $y = 0$ (b) $x = 0$ (c) $x + y = 0$

(d) $x = y$ (e) None of these

1—Answer: b

28. Find the equation of the line that passes through $(-1, 5)$ and has a slope of 2.

1—Answer: $2x - y + 7 = 0$

29. Find the equation of the line that is perpendicular to $2x + 3y = 12$ but has the same y-intercept.

    (a) $2x + 3y = 8$ (b) $2x - 3y = 12$ (c) $2x + 3y = -12$

    (d) $3x - 2y = -8$ (e) None of these

    2—Answer: d

30. Find an equation of the line shown at the right.

    (a) $y = \frac{2}{3}x - \frac{11}{3}$ (b) $y = \frac{4}{5}x + 4$

    (c) $y = -\frac{2}{3}x - 4$ (d) $y = -\frac{4}{5}x - \frac{7}{2}$

    (e) None of these

    2—Answer: a

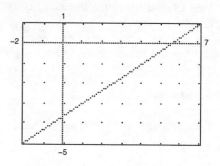

31. Find the equation of the line shown at the right.

    (a) $y = \frac{3}{4}x + 2$ (b) $y = -\frac{1}{3}x + \frac{5}{4}$

    (c) $y = -\frac{1}{2}x + \frac{2}{3}$ (d) $y = -\frac{3}{5}x + \frac{7}{5}$

    (e) None of these

    2—Answer: d

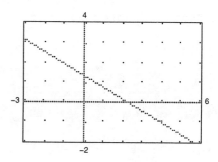

32. Rewrite the equation of the line $2x - 5y = 20$ in slope-intercept form.

    (a) $y = -\frac{5}{2}x + 4$ (b) $y = \frac{5}{2}x - 4$ (c) $y = \frac{2}{5}x - 4$

    (d) $y = -\frac{2}{5}x + 4$ (e) None of these

    1—Answer: c

33. Rewrite the equation of the line $x + 7y = 35$ in slope-intercept form.

    (a) $y = \frac{1}{7}x - 5$ (b) $y = -\frac{1}{7}x + 5$ (c) $y = 7x + 5$

    (d) $y = -7x + 5$ (e) None of these

    1—Answer: b

34. Describe the graph of $6x - y = 12$.

    (a) Rises from left to right   (b) Falls from left to right   (c) Horizontal

    (d) Vertical   (e) None of these

    2—Answer: a

35. Find an equation of the line that passes through $(3, 10)$ and is parallel to the line $x - 3y = 1$.

    (a) $y = \frac{1}{3}x + 9$ (b) $y = 3x + 1$ (c) $y = -3x + 19$

    (d) $y = -\frac{1}{3}x + 11$ (e) None of these

    2—Answer: a

36. Find an equation of the line that passes through $(-1, -3)$ and is parallel to the line $2x + y = 19$.

    (a) $y = -2x - 3$
    (b) $y = -2x - 5$
    (c) $y = 2x - 1$
    (d) $y = -\frac{1}{2}x - \frac{7}{2}$
    (e) None of these

    2—Answer: b

37. Find an equation of the line that passes through $(6, 2)$ and is perpendicular to the line $3x + 2y = 2$.

    (a) $y = -\frac{3}{2}x + 11$
    (b) $y = -\frac{2}{3}x + 6$
    (c) $y = \frac{3}{2}x - 7$
    (d) $y = \frac{2}{3}x - 2$
    (e) None of these

    2—Answer: d

38. Find an equation of the line that passes through $(8, 17)$ and is perpendicular to the line $x + 2y = 2$.

    2—Answer: $y = 2x + 1$

39. Find an equation of the line that passes through $(3, 5)$ and is perpendicular to the line $x + 3y = 6$.

    2—Answer: $y = 3x - 4$

40. Find the equation of the line that passes through $(1, 3)$ and is perpendicular to the line $2x + 3y + 5 = 0$.

    (a) $3x - 2y + 3 = 0$
    (b) $2x + 3y - 11 = 0$
    (c) $2x + 3y - 9 = 0$
    (d) $3x - 2y - 7 = 0$
    (e) None of these

    2—Answer: a

41. Find the equation of the line that passes through $(2, -1)$ and is parallel to the line $2x + 7y = 5$.

    (a) $2x - 7y - 11 = 0$
    (b) $2x + 7y + 3 = 0$
    (c) $2x + 7y - 12 = 0$
    (d) $7x - 2y - 16 = 0$
    (e) None of these

    2—Answer: b

42. Find the equation of the line that passes through $(-3, -2)$ and is parallel to the line $3x + 2y - 5 = 0$.

    2—Answer: $3x + 2y + 13 = 0$

43. Write the equation of the line that passes through the points $(3, -1), (-4, -1)$.

    (a) $x = -1$
    (b) $y = -1$
    (c) $4x - 3y = 1$
    (d) $7x + 2y = 19$
    (e) None of these

    1—Answer: b

44. Write the equation of the line that passes through the points $(2, -1), (2, -6)$.

    (a) $7x + 4y = 10$
    (b) $x + y = -7$
    (c) $x = 2$
    (d) $y = 2$
    (e) None of these

    1—Answer: c

Chapter 2 Functions and Their Graphs

45. Which viewing rectangle should be used to graph $y = \frac{1}{3}x + 2$ and $y = -3x - 1$ so that the lines appear perpendicular?

(a)
```
Xmin = -7
Xmax = 2
Xscl = 1
Ymin = -2
Ymax = 4
Yscl = 1
```

(b)
```
Xmin = -9
Xmax = 3
Xscl = 1
Ymin = -3
Ymax = 5
Yscl = 1
```

(c)
```
Xmin = -9
Xmax = 6
Xscl = 1
Ymin = -5
Ymax = 5
Yscl = 1
```

(d) All of these (e) None of these

2—T—Answer: d

46. Which viewing rectangle should be used to graph $3x - 4y = -4$ and $4x + 3y = -3$ so that the lines appear perpendicular?

(a)
```
Xmin = -4
Xmax = 4
Xscl = 1
Ymin = -3
Ymax = 3
Yscl = 1
```

(b)
```
Xmin = -4
Xmax = 4
Xscl = 1
Ymin = -4
Ymax = 4
Yscl = 1
```

(c)
```
Xmin = -5
Xmax = 4
Xscl = 1
Ymin = -3
Ymax = 3
Yscl = 1
```

(d) All of these (e) None of these

2—T—Answer: c

47. Which viewing rectangle should be used to graph $4x - y - 1 = 0$ and $x + 4y - 4 = 0$ so that the lines appear perpendicular?

(a)
```
Xmin = -4
Xmax = 4
Xscl = 1
Ymin = -4
Ymax = 4
Yscl = 1
```

(b)
```
Xmin = -5
Xmax = 5
Xscl = 1
Ymin = -3
Ymax = 3
Yscl = 1
```

(c)
```
Xmin = -10
Xmax = 10
Xscl = 1
Ymin = -3
Ymax = 3
Yscl = 1
```

(d) All of these (e) None of these

1—T—Answer: b

48. An employee is paid $200 plus 8% of her net sales. Write a linear equation of her wages, *W*, in terms of her net sales, *x*.

1—Answer: $W = 0.08x + 200$

49. A new vehicle worth $15,000 depreciates $2500 every year after it is purchased. Write a linear equation of its value, *V*, in terms of the number of years, *t*, since it was purchased.

1—Answer: $V = -2500t + 15{,}000$

50. A lawyer charges an initial fee of $500 plus $50 per hour. Write a linear equation of the fee, *F*, in terms of the number of hours of service, *t*.

1—Answer: $F = 50t + 500$

51. Morgan Sporting Goods had net sales of $150,000 in January of this past year. In March, their net sales were $300,000. Assuming that their sales are increasing linearly, write an equation of net sales, S, in terms of the month using $t = 1$ for January.

   2—Answer: $S = 75,000t + 75,000$

52. Curtis Area Schools had an enrollment of 2800 students in 1980 and 12,600 in 1988. Assuming the growth is linear, write an equation of the enrollment, E, in terms of the year using $t = 0$ for 1980.

   2—Answer: $E = 1225t + 2800$

53. A new Audi purchased in 1987 was valued at $28,000. In 1993, it was worth $16,000. Assuming the depreciation is linear, write an equation of its value, V, in terms of the number of years since its purchase using $t = 0$ for 1987.

   2—Answer: $V = -2000t + 28,000$

54. The cost of parts of your automobile repair bill was $152. The cost for labor was $30 per hour. Write a linear equation giving the total cost C, in terms of t, the number of hours.

   1—Answer: $C = 152 + 30t$

55. An employee is paid $15 per hour plus $2 for each unit produced per hour. Match this with the appropriate graph.

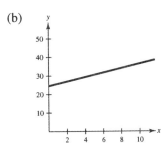

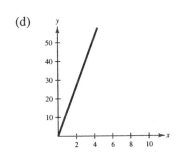

   (e) None of these

   1—Answer: a

56. A sales representative receives $25 per day for food and $0.30 for each mile traveled. Match this with the appropriate graph.

(a)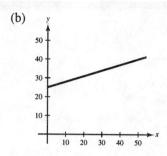

(b)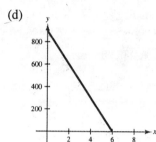

(c)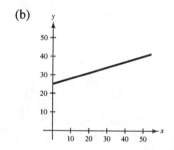

(d)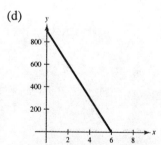

(e) None of these

1—**Answer:** b

57. A word processor that was purchased for $900 depreciates $150 per year. Match this with the appropriate graph.

(a)

(b)

(c)

(d)

(e) None of these

1—**Answer:** d

58. The radio advertising expenditures in the United States were about $3700 million in 1980 and $7800 million in 1988. Predict the expenditures in 1996, assuming the expenditures follow a linear growth pattern.
    (Source: *McCann-Erickson, Inc.*)
    (a) $15,600 million   (b) $11,900 million   (c) $11,500 million
    (d) $10,400 million   (e) None of these

    2—Answer: b

59. Your salary was $18,000 in 1983 and $27,900 in 1992. If your salary follows a linear growth pattern, what will it be in 1995?
    (a) $29,300   (b) $29,900   (c) $31,200
    (d) $37,200   (e) None of these

    2—Answer: c

60. The population of Morgan Falls in 1990 was 12,500 and in 1995 it was 18,750. If the population follows a linear growth pattern, what will it be in 1998?
    (a) 20,000   (b) 25,000   (c) 27,500
    (d) 22,500   (e) None of these

    1–Answer: d

61. A school's growth in enrollment is approximately linear. In 1970 it had 2000 students and in 1990 there were 3000. Estimate the approximate number of students in 1982.
    (a) 2500   (b) 2600   (c) 2650
    (d) 2550   (e) None of these

    1—Answer: b

62. A school's growth in enrollment is approximately linear. In 1970 it had 2000 students and in 1990 there were 3000. Estimate the approximate number of students in 1978.
    (a) 2400   (b) 2350   (c) 2500
    (d) 2450   (e) None of these

    1—Answer: a

63. A school's growth in enrollment is approximately linear. In 1970 it had 2000 students and in 1990 there were 3000. Estimate the approximate number of students in 1983.
    (a) 2700   (b) 2550   (c) 2600
    (d) 2650   (e) None of these

    1—Answer: d

64. The growth of the number of employees hired by a company has been linear for the past ten years, starting at 323 ten years ago and currently standing at 393. Estimate the number of employees that will be employed three years from now (by linear extrapolation).

    1—Answer: 414

65. The growth of the number of employees hired by a company has been linear for the past ten years, starting at 287 ten years ago and currently standing at 347. Estimate the number of employees that will be employed three years from now (by linear extrapolation).

    1—Answer: 365

66. The growth of the number of employees hired by a company has been linear for the past ten years, starting at 415 ten years ago and currently standing at 495. Estimate the number of employees that will be employed two years from now (by linear extrapolation).

    1—Answer: 511

67. The growth of the number of employees hired by a company has been linear for the past ten years, starting at 242 ten years ago and currently standing at 312. Estimate the number of employees that will be employed three years from now (by linear extrapolation).

    1—Answer: 333

❑ 2.2 Functions

1. Given $A = \{-2, -1, 0, 1\}$ and $B = \{1, 2, 3\}$, determine which of the sets of ordered pairs represents a function from A to B.

    (a) $\{(-2, 1), (-1, 1), (0, 1), (1, 1)\}$ (b) $\{(-2, 1), (-2, 2), (-2, 3)\}$

    (c) $\{(-2, 1), (-1, 2), (-1, 3)\}$ (d) All of these

    (e) None of these

    1—Answer: a

2. Given $A = \{-1, 0, 1, 2\}$ and $B = \{1, 2, 3\}$, determine which of the sets of ordered pairs represents a function from A to B.

    (a) $\{(-1, 1), (0, 2), (1, 2), (2, 3)\}$ (b) $\{(-1, 2), (0, 1), (2, 3)\}$

    (c) $\{(-1, 2), (0, 2), (1, 3), (2, 3)\}$ (d) All of these

    (e) None of these

    1—Answer: d

3. Given $A = \{1, 2, 3\}$ and $B = \{-2, -1, 0, 1\}$, determine which of the sets of ordered pairs represents a function from A to B.

    (a) $\{(1, -2), (2, -2), (3, -1), (2, 0), (2, 1)\}$ (b) $\{(1, -2), (2, -1), (2, 0), (3, 1)\}$

    (c) $\{(1, -2), (2, -1), (3, 0), (1, 1)\}$ (d) All of these

    (e) None of these

    1—Answer: e

4. In which of the following equations is y a function of x?

    (a) $3y + 2x - 9 = 17$ (b) $2x^2 + x = 4y$

    (c) Both a and b   (d) Neither a nor b

    1—Answer: c

5. In which of the following equations is y a function of x?

   (a) $2x + 3y - 1 = 0$ (b) $x^2 + 3y^2 = 7$ (c) $2x^2y = 7$

   (d) Both a and b   (e) Both a and c

   1—Answer: e

6. In which of the following equations is y a function of x?

   (a) $3y + 2x - 7 = 0$ (b) $5x^2y = 9 - 2x$ (c) $3x^2 - 4y^2 = 9$

   (d) $x = 3y^2 - 1$ (e) Both a and b

   1—Answer: e

7. Let $A = \{0, 1, 2, 3\}$ and $B = \{-4, -2, 0, 2, 4\}$. Which set of ordered pairs represents a function from A to B?

   (a) $(1, -4), (1, -2), (2, 0), (2, 4)$ (b) $(0, -3), (1, -1), (2, 1), (3, 3)$

   (c) $(0, 4), (1, 0), (2, -2), (3, -4)$ (d) $(2, -2), (2, 0), (2, 2), (2, 4)$

   (e) None of these

   1—Answer: c

8. Let $A = \{0, 1, 2, 3\}$ and $B = \{-4, -2, 0, 2, 4\}$. Which set of ordered pairs represents a function from A to B?

   (a) $(1, -4), (1, -2), (2, 0), (2, 4)$ (b) $(0, 4), (1, 2), (2, 0), (3, 2)$

   (c) $(0, 3), (1, 1), (2, -1), (3, -3)$ (d) $(1, 4), (1, 2), (2, -2), (2, -4)$

   (e) None of these

   1—Answer: b

9. Let $A = \{0, 1, 2, 3\}$ and $B = \{-4, -2, 0, 2, 4\}$. Which set of ordered pairs represents a function from A to B?

   (a) $(0, -2), (0, 0), (0, 2), (0, 4)$ (b) $(1, -2), (1, 0), (1, 2), (1, 4)$

   (c) $(2, -2), (2, 0), (2, 2), (2, 4)$ (d) $(3, -2), (3, 0), (3, 2), (3, 4)$

   (e) None of these

   1—Answer: e

10. Let $A = \{0, 1, 2, 3\}$ and $B = \{-4, -2, 0, 2, 4\}$. Which set of ordered pairs represents a function from A to B?

    (a) $(1, 4), (2, 2), (3, 4), (3, 0)$ (b) $(0, -4), (1, -2), (1, 0), (2, 2)$

    (c) $(0, 0), (0, 2), (1, -2), (1, -4)$ (d) $(0, 4), (1, -4), (2, 2), (3, -2)$

    (e) None of these

    1—Answer: d

11. Let $A = \{0, 2, 4\}$ and $B = \{1, 3, 5\}$. Fill in the missing number so that the set of ordered pairs represents a function from A to B.

    $(0, 3), (\ \ , 5), (2, 1)$

    1—Answer: 4

12. Let $A = \{0, 2, 4\}$ and $B = \{1, 3, 5\}$. Fill in the missing number so that the set of ordered pairs represents a function from A to B.

   $(0, 5), (4, 5), (, 3)$

   1—Answer: 2

13. In which of the following equations is y a function of x?

   (a) $y = 3x^2 - 9$ (b) $3x + 2y = 5$ (c) $|x| = y$

   (d) All of these   (e) None of these

   1—Answer: d

14. Determine which equation represents y as a function of x.

   (a) $|x| + |y| = 1$ (b) $x^2 + y^2 = 1$ (c) $y^2 = 7 - x$

   (d) $2x - 3y = 7$ (e) None of these

   1—Answer: d

15. Determine which equation represents y as a function of x.

   (a) $x + y^2 = 4$ (b) $|y| = x + 1$ (c) $y = 3x^2 - 2x$

   (d) $x^2 + y^2 = 4$ (e) None of these

   1—Answer: c

16. Determine which equation represents y as a function of x.

   (a) $x^2 + 2x - y + 3 = 0$ (b) $y^2 + 2x + 4 = 0$ (c) $|y| - x = 1$

   (d) $x^2 + y^2 = 25$ (e) None of these

   1—Answer: a

17. Determine which equation represents y as a function of x.

   (a) $x + 4y^2 - 7 = 0$ (b) $x + 4y - 7 = 0$ (c) $x + 4|y| - 7 = 0$

   (d) $|x| + 4|y| - 7 = 0$ (e) None of these

   1—Answer: b

18. Select which of the two equations represents y as a function of x and specify why the other does not.

   (a) $x^2 + 5y - x = 7$ (b) $y^2 - 5x = 7$

   E—Answer: (a); in (b), some values of x determine more than one value of y.

19. Select which of the two equations represents y as a function of x and specify why the other does not.

   (a) $|x| + y = 4$ (b) $x + |y| = 4$

   1—Answer: (a); in (b), some values of x determine more than one value of y.

20. Select which of the two equations represents y as a function of x and specify why the other does not.

   (a) $x^2 + y^2 = 4$ (b) $x^2 + y = 4$

   1—Answer: (b); in (a), some values of x determine more than one value of y.

21. Select which of the two equations represents y as a function of x and specify why the other does not.

    (a) $y^2 - 4y = x$ (b) $x^2 - 4x = y$

    1—**Answer:** (b); in (a), some values of x determine more than one value of y.

22. Given $f(x) = 4x^2 - 2$, find $f(-2)$.

    (a) 62 (b) -66 (c) 14

    (d) -18 (e) None of these

    1—**Answer:** c

23. Given $g(x) = 3 - 4x^2$, find $g(-2)$.

    (a) -13 (b) 19 (c) -61

    (d) -3 (e) None of these

    1—**Answer:** a

24. Given $h(x) = 10x + 2x^2$, find $h(-2)$.

    (a) 6 (b) -22 (c) -28

    (d) -12 (e) None of these

    1—**Answer:** d

25. Given $f(x) = \begin{cases} 7x - 10, & x \leq 2 \\ x^2 + 6, & x > 2 \end{cases}$, find $f(0)$.

    (a) -10 (b) 0 (c) -4

    (d) 6 (e) None of these

    1—**Answer:** a

26. Given $f(x) = \begin{cases} 3x + 4, & x \leq 2 \\ x^2 + 1, & x > 2 \end{cases}$, find $f(3)$.

    (a) 13 (b) 10 (c) 5

    (d) 3 (e) None of these

    1—**Answer:** b

27. Given $f(x) = \begin{cases} 2x - 1, & x \leq -2 \\ x + 6, & x > -2 \end{cases}$, find $f(-6)$.

    (a) -11 (b) -13 (c) 0

    (d) 11 (e) None of these

    1—**Answer:** b

28. Given $f(x) = x^2 - 3x + 4$, find $f(x + 2) - f(2)$.

    (a) $x^2 - 3x + 4$ (b) $x^2 + x$ (c) $x^2 + x - 8$

    (d) $x^2 - 3x - 4$ (e) None of these

    2—**Answer:** b

29. Given $f(x) = |x - 3| - 5$, find $f(1) - f(5)$.

   (a) 0
   (b) -4
   (c) 14
   (d) -14
   (e) None of these

   1—Answer: a

30. Given $f(x) = |3x + 1| - 5$, find $f(x + 1) - f(x)$.

   (a) 3
   (b) -5
   (c) $|3x + 4| - |3x + 1| - 10$
   (d) $|3x + 4| - |3x + 1|$
   (e) None of these

   2—Answer: d

31. Given $f(x) = 3x - 7$, find $f(x + 1) + f(2)$.

   2—Answer: $3x - 5$

32. Given $g(x) = \dfrac{x}{2x + 1}$, find $g(k - 2)$.

   2—Answer: $\dfrac{k - 2}{2k - 3}$

33. Given $h(x) = \dfrac{1}{2}x^3$, find $h(t + 1) - h(1)$.

   2—Answer: $\dfrac{1}{2}(t^3 + 3t^2 + 3t)$

34. $F(x) = \begin{cases} 3 - x^2, & \text{if } x \geq 0 \\ 3 + 2x, & \text{if } x < 0 \end{cases}$ $F(2) - F(-2) =$

   (a) 0
   (b) 1
   (c) -2
   (d) $F(4)$
   (e) None of these

   1—Answer: a

35. $F(x) = \begin{cases} 3 - x^2, & \text{if } x \geq 0 \\ 3 + 2x, & \text{if } x < 0 \end{cases}$ $F(1) - F(-1) =$

   (a) 0
   (b) 1
   (c) 3
   (d) $F(2)$
   (e) None of these

   1—Answer: b

36. $F(x) = \begin{cases} 3 - x^2, & \text{if } x \geq 0 \\ 3 + 2x, & \text{if } x < 0 \end{cases}$ $F(2) - F(-1) =$

   (a) -6
   (b) 0
   (c) -2
   (d) $F(3)$
   (e) None of these

   1—Answer: c

37. $F(x) = \begin{cases} 3 - x^2, & \text{if } x \geq 0 \\ 3 + 2x, & \text{if } x < 0 \end{cases}$ $F(3) - F(-3) =$

   (a) 0   (b) 3   (c) $F(6)$

   (d) -3 (e) None of these

   1—Answer: d

38. Given $f(x) = \begin{cases} x^2 + 1, & x < 4 \\ 6x - 7, & x \geq 4 \end{cases}$, find $f(-2)$.

   (a) -19 (b) 5   (c) 4

   (d) -5 (e) None of these

   1—Answer: b

39. Given $f(x) = \begin{cases} \sqrt{-x}, & x \leq 0 \\ 6x, & x > 0 \end{cases}$, find $f(4)$.

   (a) 2   (b) -2 (c) 10

   (d) 24   (e) None of these

   1—Answer: d

40. Given $f(x) = \begin{cases} x - 19, & x < -5 \\ |x + 3|, & x \geq -5 \end{cases}$, find $f(-4)$.

   (a) -15 (b) 1   (c) -1

   (d) -23 (e) None of these

   1—Answer: b

41. Given $f(x) = \begin{cases} |x - 2|, & x \leq 1 \\ |2x - 5|, & x > 1 \end{cases}$, find $f(1)$.

   (a) 1   (b) -1 (c) 3

   (d) -3 (e) None of these

   1—Answer: a

42. $F(x) = -1 + 2x - x^2$. Find $F(k + 1)$ and simplify.

   2—Answer: $-k^2$

43. $F(x) = 5 + 2x - x^2$. Find $F(k + 1) - F(k)$ and simplify.

   2—Answer: $1 - 2k$

44. $F(x) = 5 + 2x - x^2$. Find $F(k - 1) - F(k)$ and simplify.

   2—Answer: $2k - 3$

45. $F(x) = 5 + 2x - x^2$. Find $F(k + 1) - F(k - 1)$ and simplify.

   2—Answer: $4 - 4k$

46. Given $f(x) = 2 - x^2$, find $\dfrac{f(x + \Delta x) - f(x)}{\Delta x}$.

   (a) $\dfrac{x^2 - \Delta x - (\Delta x)^2}{\Delta x}$ (b) $\dfrac{-2x^2 - \Delta x^2}{\Delta x}$ (c) $-2x - \Delta x$

   (d) $\dfrac{1}{2}$ (e) None of these

   2—Answer: c

47. Given $f(x) = 2 - 3x^2$, find $\dfrac{f(x + \Delta x) - f(x)}{\Delta x}$.

   (a) 2   (b) $-6x - 3\Delta x$ (c) $2x + \Delta x$

   (d) $\dfrac{6x^2 + 6x\Delta x + 9(\Delta x)^2}{\Delta x}$ (e) None of these

   2—Answer: b

48. Given $f(x) = 1 - 2x^2$, find $\dfrac{f(x + \Delta x) - f(x)}{\Delta x}$.

   (a) $-4x - 2\Delta x$ (b) $2x + \Delta x$ (c) $\dfrac{4x^2 + 2x\Delta x + (\Delta x)^2}{\Delta x}$

   (d) -9 (e) None of these

   2—Answer: a

49. Given $f(x) = 2 - 4x^2$, find $\dfrac{f(x + \Delta x) - f(x)}{\Delta x}$.

   2—Answer: $-8x - 4\Delta x$

50. Given $f(x) = 3 - 2x^2$, find $\dfrac{f(x + \Delta x) - f(x)}{\Delta x}$.

   2—Answer: $-4x - 2\Delta x$

51. Find all real values of x such that $f(x) = 0$: $f(x) = \dfrac{x - 3}{x + 4}$.

   (a) $\dfrac{3}{4}$ (b) 3   (c) -4

   (d) $3, -4$ (e) None of these

   1—Answer: b

52. Find all real values of x such that $f(x) = 0$: $f(x) = \dfrac{x - 6}{x + 5}$.

   (a) 6   (b) -5 (c) $-\dfrac{6}{5}$

   (d) $6, -5$ (e) None of these

   1—Answer: a

53. Find all real values of x such that $f(x) = 0$; $f(x) = |x + 3|$.

(a) ± 3 (b) -3 (c) 3

(d) $-\frac{1}{3}$ (e) None of these

1—Answer: b

54. Find all real values of x such that $f(x) = 0$: $f(x) = x\sqrt{x^2 - 9}$.

(a) ± 3 (b) 3 (c) $0, \pm 3$

(d) $0, 3$ (e) None of these

1—Answer: c

55. Find all real values of x such that $f(x) = 0$: $f(x) = x^3 - 2x^2 - 5x$.

2—Answer: $x = 0, 1 \pm \sqrt{6}$

56. Find all real values of x such that $f(x) = 0$: $f(x) = x^3 - 2x^2 - 4x$.

2—Answer: $x = 0, 1 \pm \sqrt{5}$

57. Which of the functions fits the data?

x	-4	-2	0	2	4	6
y	8	4	0	4	8	12

(a) $f(x) = 2|x|$ (b) $f(x) = 2x^2$ (c) $f(x) = 2x$

(d) $f(x) = 2\sqrt{x}$ (e) None of these

1—Answer: a

58. Which of the functions fits the data?

x	-2	0	1	3	5	10
y	-6	0	3	9	15	30

(a) $f(x) = x^3$ (b) $f(x) = \sqrt[3]{x}$ (c) $f(x) = |x|^3$

(d) $f(x) = 3x$ (e) None of these

1—Answer: d

59. Which of the functions fits the data?

x	-4	-2	0	2	8
y	16	4	0	4	64

(a) $f(x) = 4x$ (b) $f(x) = x^2$ (c) $f(x) = \frac{1}{2}x^2$

(d) $f(x) = 2|x|$ (e) None of these

1—Answer: b

60. For what values of x does $f(x) = g(x)$? $f(x) = 3x + 1$ $g(x) = x^2 - 3$

(a) 0 (b) $4, 1$ (c) $-4, -1$

(d) $4, -1$ (e) None of these

1—Answer: d

61. Find all real values of x for which $f(x) = g(x)$: $f(x) = x^4 + 3x^2$ $g(x) = 7x^2$

   (a) $-4, 0, 4$ (b) $0, 2$ (c) 2

   (d) $0, 2, -2$ (e) None of these

   1—Answer: d

62. Find all real values of x for which $f(x) = g(x)$: $f(x) = x - 2$ $g(x) = \sqrt{x}$

   (a) $\dfrac{1 \pm \sqrt{17}}{2}$ (b) $\dfrac{-1 \pm \sqrt{15}}{2}$ (c) $1, 4$

   (d) $-1, 3$ (e) None of these

   2—Answer: c

63. Find all real values of x for which $f(x) = g(x)$: $f(x) = x^2 - 2$ $g(x) = 2x + 1$

   (a) $1, -3$ (b) $1, 3$ (c) No real values of x

   (d) $-1, 3$ (e) None of these

   1—Answer: d

64. Find all real values of x for which $f(x) = g(x)$: $f(x) = x^2 - 5$ $g(x) = x + 1$

   1—Answer: $-2, 3$

65. Find all real values of x for which $f(x) = g(x)$: $f(x) = x^2$ $g(x) = x + 9$

   2—Answer: $\dfrac{1 \pm \sqrt{37}}{2}$

66. Find the domain of the function $\{(-2, 1), (-1, 0), (0, -3), (1, -8)\}$.

   (a) $[-2, 1]$ (b) $(-\infty, \infty)$ (c) $\{-2, -1, 0, 1\}$

   (d) $\{1, 0, -3, -8\}$ (e) $\{-8, -3, -2, -1, 0, 1\}$

   1—Answer: c

67. Find the domain of the function $\{(1, 1), (2, 4), (3, 9)\}$.

   (a) $[1, 9]$ (b) $\{1, 2, 3, 4, 9\}$ (c) $\{1, 4, 9\}$

   (d) $\{1, 2, 3\}$ (e) None of these

   1—Answer: d

68. Find the domain of the function: $f(x) = \dfrac{9}{x}$.

   (a) $(9, \infty)$ (b) $(-\infty, \infty)$ (c) All real numbers $x \neq 0$

   (d) All real numbers $x \neq 9$ (e) None of these

   1—Answer: c

69. Find the domain of the function: $g(x) = \dfrac{x}{x^2 + 1}$.

   (a) All real $x \neq 1$ (b) All real $x \neq -1$ (c) All real $x \neq 0$

   (d) All real x (e) None of these

   1—Answer: d

70. Find the domain of the function $h(x) = \dfrac{x + 4}{x(x - 5)}$.

1—Answer: All real $x \neq 0$, $x \neq 5$

71. Find the domain of the function $g(x) = \dfrac{5x}{x^2 - 7x + 12}$.

1—Answer: All real $x \neq 3$, $x \neq 4$

72. Find the domain of the function $f(x) = \dfrac{2x - 1}{2x + 1}$.

1—Answer: All real $x \neq -\frac{1}{2}$

73. Find the domain of the function: $y = \dfrac{1}{x}$.

(a) $(-\infty, \infty)$ (b) $(-\infty, 0)(0, \infty)$ (c) $(-\infty, 0)$

(d) $(0, \infty)$ (e) None of these

1—Answer: b

74. Find the domain of the function: $f(x) = \sqrt{2x + 3}$.

(a) $[0, \infty)$ (b) $(0, \infty)$ (c) $\left[-\frac{3}{2}, \infty\right)$

(d) $\left(-\frac{3}{2}, \infty\right)$ (e) None of these

1—Answer: c

75. Find the domain of the function: $f(x) = \dfrac{1}{x + 2}$.

1—Answer: Domain: $(-\infty, -2), (-2, \infty)$

76. Find the domain of the function: $f(x) = \dfrac{1}{\sqrt{x^2 + 1}}$.

(a) $(-\infty, -1), (-1, 1), (1, \infty)$ (b) $(-\infty, 0), (0, \infty)$ (c) $(-\infty, \infty)$

(d) $(-\infty, -1), (-1, \infty)$ (e) None of these

1—Answer: c

77. Find the domain of the function $f(x) = \dfrac{1}{x^2 - 3x + 2}$.

(a) $(-\infty, -2), (-2, 1), (1, \infty)$ (b) $(-\infty, 1), (1, 2), (2, \infty)$ (c) $(-\infty, \infty)$

(d) $\left(-\infty, \frac{1}{2}\right), \left(\frac{1}{2}, \infty\right)$ (e) None of these

2—Answer: b

78. The domain of the function $f(x) = \dfrac{1}{\sqrt{x - 4}}$ is:

(a) All real $x \geq 4$ (b) All real x (c) All real $x > 4$

(d) All positive real numbers   (e) None of these

1—Answer: c

79. The domain of the function $f(x) = \dfrac{3-x}{4+x}$ is:

    (a) All real $-x \neq 4$ (b) All real x (c) $-4 < x \leq 3$

    (d) All real $x \geq 3$ and $x < -4$ (e) None of these

    1—Answer: a

80. Find the domain of the function $f(x) = \dfrac{x-2}{(x-5)(x+3)}$.

    (a) All real $x \neq 5$ (b) All real $x \neq -3$ (c) All real $x \neq 5, x \neq -3$

    (d) All real $x \neq 5, x \neq -3$ and $x \neq 2$ (e) None of these

    1—Answer: c

81. The length of a rectangle is four times the width. Express the perimeter, P, as a function of the width, w.

    (a) $P(w) = 6w$ (b) $P(w) = 10w$ (c) $P(w) = 5w$

    (d) $P(w) = 8w$ (e) None of these

    1—Answer: b

82. The length of a rectangle is four times the width. Express the area, A, as a function of the length, l.

    (a) $A(l) = \dfrac{l}{4}$ (b) $A(l) = 4l^2$ (c) $A(l) = \dfrac{l^2}{4}$

    (d) $A(l) = \dfrac{9l}{4}$ (e) None of these

    1—Answer: c

83. The height of a rectangular solid is three times its width and the length is one-half the width. Express the volume, V, as a function of the width, w.

    (a) $V(w) = \tfrac{3}{2}w^3$ (b) $V(w) = \tfrac{9}{2}w$ (c) $V(w) = 6w^3$

    (d) $V(w) = \tfrac{7}{2}w$ (e) None of these

    1—Answer: a

84. The height of a rectangular solid is four inches more than the length and six times the width. Express the volume, V, as a function of the height, h.

    (a) $V(h) = \tfrac{1}{6}(h^2 - 4h)$ (b) $V(h) = \tfrac{1}{6}(h^3 - 4h^2)$ (c) $V(h) = \tfrac{1}{6}(3h - 4)$

    (d) $V(h) = 6h^3 - 24h^2$ (e) None of these

    2—Answer: b

85. Express the length of the diagonal, L, of a square as a function of the length, x, of a side.

    1—Answer: $L(x) = \sqrt{2}x$

86. Express the volume, V, of a cube as a function of the length, x, of a side.

    1—Answer: $V(x) = x^3$

87. Express the perimeter, P, of a square as a function of the length, x, of a side.

    1—Answer: $P(x) = 4x$

88. Express the area, A, of a square as a function of the length, x, of a side.

    1—Answer: $A(x) = x^2$

89. The expenditures for personal flowers, seeds, and potted plants, in billions of dollars, in the United States from 1980 to 1990 can be modeled by

    $$f(x) = \begin{cases} \frac{1}{10}(3x + 40), & 0 \leq t \leq 6 \\ \frac{1}{20}(11x + 63), & 7 \leq t \leq 10 \end{cases}$$

    where t is the year with $t = 0$ corresponding to 1980. Use this model to find the expenditures in 1985. (Source: *U.S. Bureau of Economic Analysis*)

    1—Answer: $5.5 billion

90. You invest $12,000 to start a business. Each unit costs $3.40 and is sold for $5.60. Let x be the number of units produced and sold. Write the profit P as a function of x. (P = Revenue − Cost)

    (a) $P = 9.00x + 12,000$
    (b) $P = 9.00x - 12,000$
    (c) $P = 2.20x + 12,000$
    (d) $P = 2.20x - 12,000$
    (e) None of these

    2—Answer: d

91. To produce x units of a product, there are fixed costs of $23,000 and a production cost of $4.78 per unit. Write the total cost C as a function of x.

    (a) $C = 4.78x + 23,000$
    (b) $C = 23,000 - 4.78x$
    (c) $C = \dfrac{4.78}{x} + 23,000$
    (d) $C = 23,004.78x$
    (e) None of these

    1—Answer: a

92. An open box is made from a rectangular piece of material by cutting equal squares from each corner and turning up the sides. Write the volume of the box as a function of x if the material is 24 inches by 16 inches.

    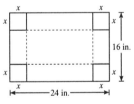

    (a) $V = (24 - x)(16 - x)$
    (b) $V = x(24 - x)(16 - x)$
    (c) $V = x(24 - 2x)(16 - 2x)$
    (d) $V = (24 - 2x)(16 - 2x)$
    (e) None of these

    1—Answer: c

93. An open box is made from a rectangular piece of material by cutting equal squares from each corner and turning up the sides. Write the volume of the box as a function of x if the material is 18 inches by 12 inches.

    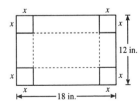

    1—Answer: $V = x(12 - 2x)(18 - 2x)$

94. Strips of width x are added to the four sides of a square, which are 16 inches on a side. Write the area A of the remaining figure as a function of x.

2—Answer: $A(x) = 64x + 256$

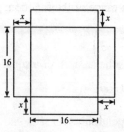

95. Strips of width x are cut from two sides of a square that is 16 inches on a side. Write the area A of the remaining square as a function of x.

2—Answer: $A(x) = (16 - x)^2$ or $A(x) = 256 - 32x + x^2$

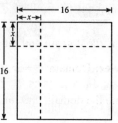

❑ 2.3 Analyzing Graphs of Functions

1. Find the range of the function: $y = \sqrt{9 - x^2}$.

    (a) $(-\infty, -3], [3, \infty)$ (b) $[-3, 3]$

    (c) $[0, 3]$ (d) $[3, \infty)$

    (e) None of these

1—Answer: c

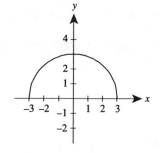

2. Find the domain and range of the function: $f(x) = |3 + x|$.

1—Answer: Domain: $(-\infty, \infty)$, Range: $[0, \infty)$

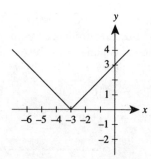

3. Find the range of the function: $f(x) = \sqrt{x^2 - 9}$.

    (a) $[-3, 3]$ (b) $(-\infty, -3], [3, \infty)$

    (c) $[0, \infty)$ (d) $(-\infty, \infty)$

    (e) None of these

1—Answer: c

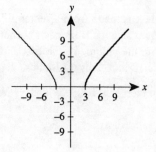

Section 2.3   Analyzing Graphs of Functions   149

4. Find the range of the function: $f(x) = x\sqrt{x+1}$.

   (a) $\left[-\dfrac{1}{2}, \infty\right)$

   (b) $\left[-\dfrac{\sqrt{2}}{4}, \infty\right)$

   (c) $[-1, \infty)$

   (d) $\left[-\dfrac{1}{2}, -\dfrac{\sqrt{2}}{4}\right]$

   (e) None of these

   1—Answer: b

5. Determine the domain and range of the function $f(x) = 3 - |x|$.

   1—Answer: Domain: $(-\infty, \infty)$, Range: $(-\infty, 3]$

6. Determine the domain and range of the function $f(x) = 3 - x^2$.

   1—Answer: Domain: $(-\infty, \infty)$, Range: $(-\infty, 3]$

7. Determine the domain and range of the function $f(x) = |3 - x|$.

   1—Answer: Domain: $(-\infty, \infty)$, Range: $[0, \infty)$

8. Determine the domain and range of the function $f(x) = x^3 + 3$.

   1—Answer: Domain: $(-\infty, \infty)$, Range: $(-\infty, \infty)$

9. Use a graphing utility to graph the function $f(x) = x^2 + 2$. Then determine the domain and range.

   2—T—Answer: 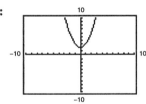 ; Domain: $(-\infty, \infty)$, Range: $[2, \infty)$

10. Use a graphing utility to graph the function $f(x) = |x^2 - 2|$. Then determine the domain and range.

    2—T—Answer: 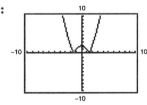 ; Domain: $(-\infty, \infty)$, Range: $[2, \infty)$

11. Use a graphing utility to graph the function $f(x) = -|x - 3|$. Then determine the domain and range.

    2—T—Answer: 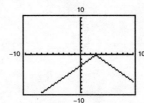 ; Domain: $(-\infty, \infty)$, Range: $(-\infty, 0]$

12. Use a graphing utility to graph the function $f(x) = \sqrt{x^2 - 4}$. Then determine the domain and range.

    2—T—Answer: 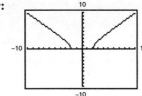 ; Domain: $(-\infty, -2], [2, \infty)$ Range: $[0, \infty)$

13. Use a graphing utility to graph the function $f(x) = 1 - \sqrt{x}$. Then determine the domain and range.

    2—T—Answer: 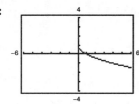 ; Domain: $[0, \infty)$ Range: $(-\infty, 1]$

14. Use the graph shown at the right to find $f(-1)$.

    (a) -1 (b) 0

    (c) 2   (d) 3

    (e) None of these

    1—Answer: d

    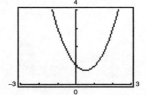

15. Use the graph shown at the right to find $f(-2)$.

    (a) 2   (b) -2

    (c) 3   (d) -6

    (e) None of these

    1—Answer: a

    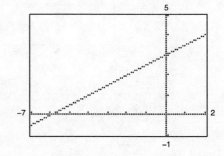

16. Use the graph shown at the right to find $f(3)$.

    (a) 3 (b) -3

    (c) -3 and 3 (d) 0

    (e) None of these

    1—Answer: d

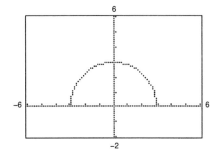

17. Which viewing rectangle shows the most complete graph of the function $f(x) = x^2 + x - 3$?

    (a)
    Xmin = -6
    Xmax = 6
    Xscl = 1
    Ymin = -1
    Ymax = 7
    Yscl = 1

    (b)
    Xmin = 0
    Xmax = 6
    Xscl = 1
    Ymin = -4
    Ymax = 4
    Yscl = 1

    (c)
    Xmin = -6
    Xmax = 6
    Xscl = 1
    Ymin = -4
    Ymax = 4
    Yscl = 1

    (d)
    Xmin = -9
    Xmax = 0
    Xscl = 1
    Ymin = -3
    Ymax = 3
    Yscl = 1

    (e) None of these

    1—T—Answer: c

18. Which viewing rectangle shows the most complete graph of the function $f(x) = x^2 - 6x + 1$?

    (a)
    Xmin = 0
    Xmax = 6
    Xscl = 1
    Ymin = -6
    Ymax = 0
    Yscl = 1

    (b)
    Xmin = -6
    Xmax = 0
    Xscl = 1
    Ymin = 0
    Ymax = 6
    Yscl = 1

    (c)
    Xmin = -6
    Xmax = 6
    Xscl = 1
    Ymin = -6
    Ymax = 6
    Yscl = 1

    (d)
    Xmin = -3
    Xmax = 9
    Xscl = 1
    Ymin = -9
    Ymax = 3
    Yscl = 1

    (e) None of these

    1—T—Answer: d

19. Which viewing rectangle shows the most complete graph of the function $f(x) = 5x\sqrt{100 - x^2}$?

    (a)
    Xmin = -20
    Xmax = 20
    Xscl = 4
    Ymin = -100
    Ymax = 100
    Yscl = 10

    (b)
    Xmin = -100
    Xmax = 100
    Xscl = 10
    Ymin = -100
    Ymax = 100
    Yscl = 100

    (c)
    Xmin = -100
    Xmax = 100
    Xscl = 10
    Ymin = -20
    Ymax = 20
    Yscl = 4

    (d)
    Xmin = -20
    Xmax = 20
    Xscl = 4
    Ymin = -300
    Ymax = 300
    Yscl = 50

    (e) None of these

    1—T—Answer: d

20. Use the vertical line test to determine in which case y is a function of x.

(a)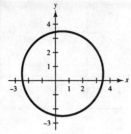

(b)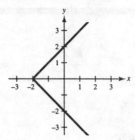

(c)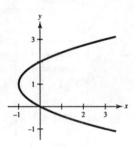

(d)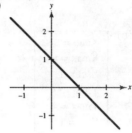

(e) None of these

1—Answer: d

21. Use the vertical line test to determine in which case y is a function of x.

(a)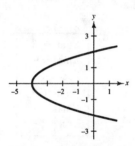

(b)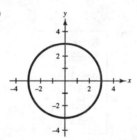

(c)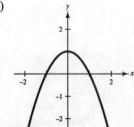

(d)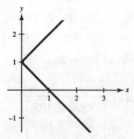

(e) None of these

1—Answer: c

22. Use the vertical line test to determine in which case y is a function of x.

(a)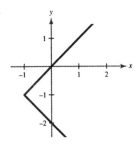

(b)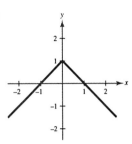

(c)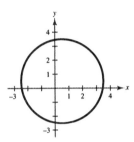

(d)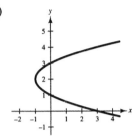

(e) None of these

1—Answer: b

23. Use the vertical line test to determine in which case y is a function of x.

(a)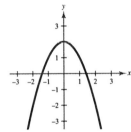

(b)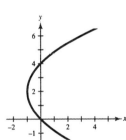

(c)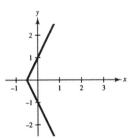

(d)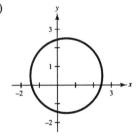

(e) None of these

1—Answer: a

24. Does the graph to the right depict y as a function of x?

(a) y is a function of x.

(b) y is not a function of x.

1—Answer: b

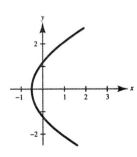

25. Does the graph at the right depict y as a function of x?

    (a) y is a function of x.

    (b) y is not a function of x.

    1—Answer: a

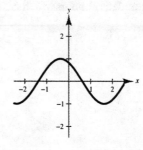

26. Which of the following graphs represent y as a function of x?

    (a)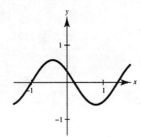

    (b)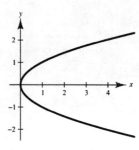

    (c)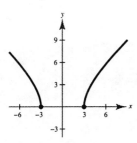

    (d)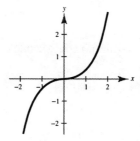

    (e) None of these

    1—Answer: a, c, and d

27. Which of the following graphs represent y as a function of x?

    (a)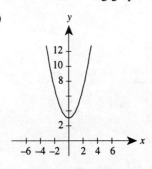

    (b)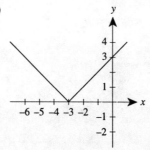

    (c)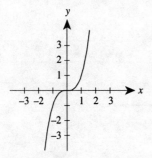

    (d) All of these are functions of x.

    (e) None of these are functions of x.

    1—Answer: d

28. Determine which of the following are graphs of odd functions.

(a)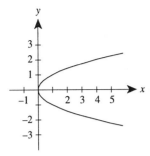

(b)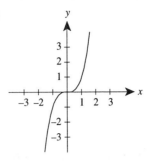

(c)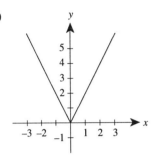

(d) All are odd functions.

(e) None of these are odd functions.

1—Answer: c

29. Determine which of the following are graphs of even functions.

(a)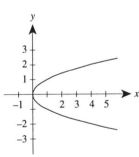

(b)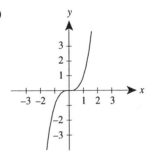

(c)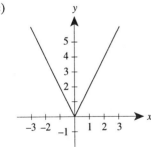

(d) All are even functions.

(e) None of these are even functions.

1—Answer: b

30. Determine which of the following are graphs of odd functions.

(a)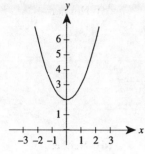

(b)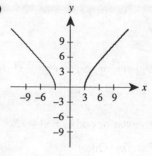

(c)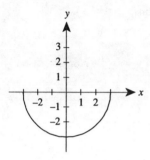

(d) All are odd functions.

(e) None of these are odd functions.

1—**Answer:** e

31. Determine which of the following are graphs of even functions.

(a)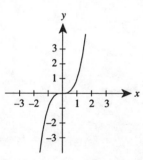

(b)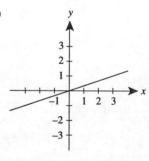

(c)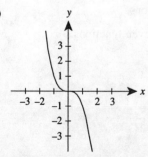

(d) All of these are even functions.

(e) None of these are even functions.

1—**Answer:** e

32. Is the following function even or odd? $f(x) = 3x^4 - x^2 + 2$

(a) Odd   (b) Even   (c) Both   (d) Neither

1—Answer: b

33. Is the following function even or odd? $y = -x^4 + 2x^2 - 1$

1—Answer: Even

34. Is the following function even or odd? $f(x) = 2x^3 + 3x^2$

(a) Even   (b) Odd   (c) Both   (d) Neither

1—Answer: d

35. Is the following function even or odd? $f(x) = 4x^3 + 3x$

(a) Even   (b) Odd   (c) Both   (d) Neither

1—Answer: b

36. Determine the interval(s) over which the function is increasing: $y = 2x^3 + 3x^2 - 12x$.

(a) $(-\infty, -2)(1, \infty)$ (b) $(-\infty, 20)(-7, \infty)$

(c) $(-2, 1)$ (d) $(20, -7)$

(e) None of these

2—Answer: a

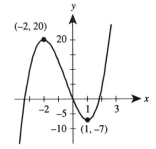

37. Determine the interval(s) over which the function is increasing: $y = x^4 - 2x^2$.

(a) $(-\infty, -1)$ (b) $(-\infty, -1), (0, 1)$

(c) $(-1, 0), (1, \infty)$ (d) $(-\infty, -1), (0, -1)$

(e) None of these

2—Answer: c

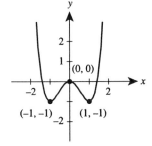

38. Determine the interval(s) over which the function is increasing: $y = \frac{2}{3}x^3 - x^2$.

2—Answer: $(-\infty, 0), (1, \infty)$

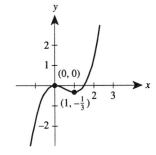

39. Determine the interval(s) over which the function is increasing: $y = \dfrac{1}{x^2 - 1}$.

2—**Answer:** $(-\infty, -1), (-1, 0)$

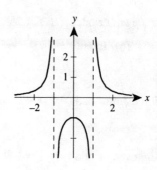

40. Use a graphing utility to determine the interval(s) over which the function is increasing: $f(x) = \frac{1}{3}x^3 - x + 1$.

2—T—**Answer:** $(-\infty, -1), (1, \infty)$

41. Use a graphing utility to determine the interval(s) over which the function is increasing: $f(x) = -\frac{1}{3}x^3 + \frac{1}{2}x^2 + 2x$.

2—T—**Answer:** $(-1, 2)$

42. Use a graphing utility to determine the interval(s) over which the function is increasing: $f(x) = \frac{1}{3}x^3 + \frac{1}{2}x^2 - 2x$.

2—T—**Answer:** $(-\infty, -2), (1, \infty)$

43. Use a graphing utility to determine the interval(s) over which the function is increasing: $f(x) = -\frac{1}{3}x^3 - \frac{1}{2}x^2 + 2x$.

2—T—**Answer:** $(-2, 1)$

44. Determine the open intervals in which the function is increasing, decreasing, or constant.

   (a) Increasing on $(-\infty, \infty)$
   (b) Increasing on $(-\infty, 0)$
       Decreasing on $(0, \infty)$
   (c) Increasing on $(-\infty, -2), (0, \infty)$
       Decreasing on $(-2, 0)$
   (d) Increasing on $(-\infty, 3)$
       Decreasing on $(3, \infty)$
   (e) None of these

1—**Answer:** c

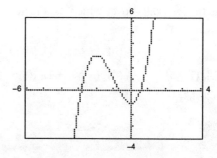

45. Determine the open intervals in which the function is increasing, decreasing, or constant.

   (a) Increasing on $(-5, 5)$
   (b) Increasing on $(-\infty, 0)$
       Decreasing on $(0, \infty)$
   (c) Increasing on $(0, 6)$
   (d) Increasing on $(-\infty, 6)$
       Decreasing on $(6, \infty)$
   (e) None of these

1—**Answer:** e

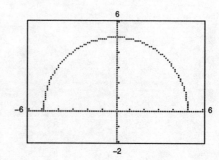

46. Sketch the graph of the function: $f(x) = \begin{cases} 3x - 1, & x < 1 \\ x^2 + 1, & x \geq 1 \end{cases}$.

2—Answer:

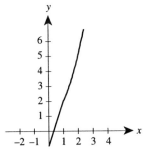

47. Identify the graph: $f(x) = \begin{cases} 2x + 1 & x \leq 1 \\ x^2 & x > 1 \end{cases}$.

(a)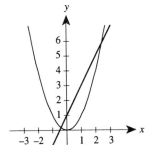

(b)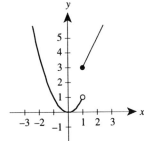

(c)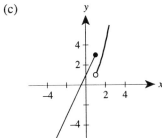

(d)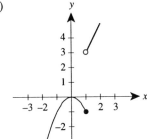

(e) None of these

2—Answer: c

48. Identify the graph: $f(x) = \begin{cases} -x^2 + 2, & x \leq 0 \\ x + 2, & x > 0 \end{cases}$.

(a)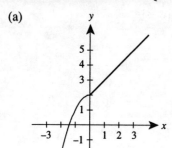

(b)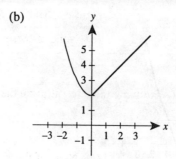

(c)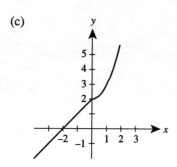

(d)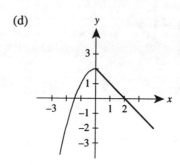

(e) None of these

1—Answer: a

49. Identify the graph: $f(x) = \begin{cases} 3x - 1, & x \leq 1 \\ 2 + x^2, & x > 1 \end{cases}$.

(a)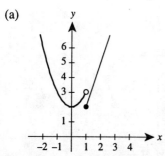

(b)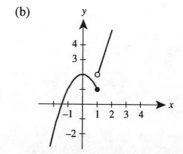

(c)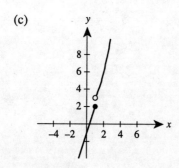

(d)

(e) None of these

2—Answer: c

50. Use a graphing utility to determine the interval(s) on the real axis for which $f(x) \geq 0$ for $f(x) = 9 - x^2$.
    (a) $(-\infty, \infty)$
    (b) $[-3, 3]$
    (c) $[0, 3]$
    (d) $(-\infty, -3), (3, \infty)$
    (e) None of these

    1—T—Answer: b

51. Use a graphing utility to determine the interval(s) on the real axis for which $f(x) \geq 0$ for $f(x) = \sqrt{x - 9}$.
    (a) $(-\infty, \infty)$
    (b) $[-9, 9]$
    (c) $[-3, 3]$
    (d) $[9, \infty)$
    (e) None of these

    1—T—Answer: d

52. Use a graphing utility to determine the interval(s) on the real axis for which $f(x) \geq 0$ for $f(x) = \sqrt{9 - x^2}$.
    (a) $(-\infty, \infty)$
    (b) $[-3, 3]$
    (c) $[0, 3]$
    (d) $(-\infty, -3], [3, \infty)$
    (e) None of these

    1—T—Answer: b

53. Use a graphing utility to determine the interval(s) on the real axis for which $f(x) \geq 0$ for $f(x) = \sqrt{4 - x^2}$.

    1—T—Answer: $[-2, 2]$

54. Use a graphing utility to determine the interval(s) on the real axis for which $f(x) \geq 0$ for $f(x) = \sqrt{x^2 - 4}$.

    1—T—Answer: $(-\infty, -2], [2, \infty)$

55. Write the height h of the rectangle as a function of x.
    (a) $h(x) = 2x + 3 - x^2$
    (b) $h(x) = x^2 - 2x + 3$
    (c) $h(x) = x^2 - 2x - 3$
    (d) $h(x) = 2x + 3 + x^2$
    (e) None of these

    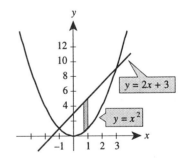

    2—Answer: a

56. Write the height h of the rectangle as a function of x.

(a) $h(x) = -x^2 + 4x + 1 - \frac{2}{3}x - \frac{2}{3}$
$= -x^2 + \frac{10}{3}x + \frac{1}{3}$

(b) $h(x) = -x^2 + 4x + 1 + \frac{2}{3}x + \frac{2}{3}$
$= -x^2 + \frac{14}{3}x + \frac{5}{3}$

(c) $h(x) = \frac{2}{3}x + \frac{2}{3} - x^2 - 4x - 1$
$= -x^2 - \frac{10}{3}x - \frac{1}{2}$

(d) $h(x) = \frac{2}{3}x + \frac{2}{3} + x^2 - 4x + 1$
$= x^2 - \frac{10}{3} + \frac{5}{3}$

(e) None of these

2—Answer: a

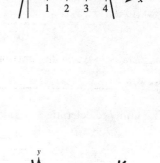

57. Write the height h of the rectangle as a function of x.

(a) $h(x) = x^2 - 5x + 3$

(b) $h(x) = -x^2 + 5x - 3$

(c) $h(x) = 3 - 5x + x^2$

(d) $h(x) = 5x - 3 + x^2$

(e) None of these

2—Answer: b

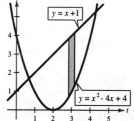

58. Write the height h of the rectangle as a function of x.

2—Answer: $h(x) = -x^2 + 5x - 4$

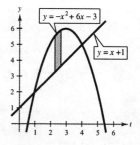

59. Write the height h of the rectangle as a function of x.

2—Answer: $h(x) = -x^2 + 3x - 1$

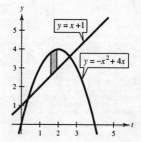

60. Use a graphing utility to graph the function $f(x) = [\![x + 2]\!]$.

1—T—Answer:
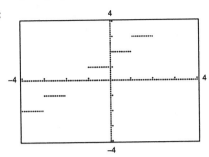

61. Use a graphing utility to graph the function $f(x) = 4 + [\![x]\!]$.

1—T—Answer:
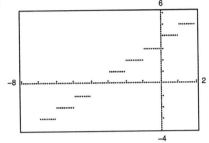

62. Use a graphing utility to graph: $f(x) = \begin{cases} 2x + 1, & x \leq 1 \\ x^2, & x > 1 \end{cases}$.

1—T—Answer:
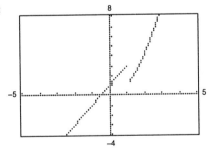

63. Use a graphing utility to graph: $f(x) = \begin{cases} -x^2 + 2, & x \leq 0 \\ x + 2, & x > 0 \end{cases}$.

1—T—Answer:
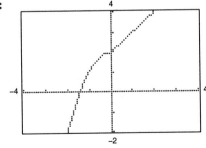

64. Sketch the graph of the function $f(x) = \begin{cases} -x, & \text{if } -2 \leq x < 0 \\ \frac{1}{4}x^2, & \text{if } 0 \leq x \leq 2 \end{cases}$.

2—Answer:

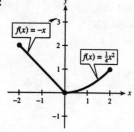

65. Sketch the graph of the function $f(x) = \begin{cases} 3 + 2x, & \text{if } -2 \leq x \leq 0 \\ 3 - x, & \text{if } 0 < x \leq 4 \end{cases}$.

2—Answer:

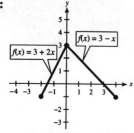

66. The height in feet of a ball thrown by a child is

$$y = -\tfrac{4}{5}x^2 + 6x + 4$$

where y is the horizontal distance (in feet) when the ball is thrown. Sketch the graph of the function using a graphing calculator. Use the graph to estimate the maximum height the ball reaches.

2—Answer:

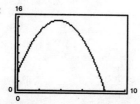

The maximum height ≈ 15 feet.

67. The height in feet of a ball thrown by a child is

$$y = -\tfrac{1}{4}x^2 + 4x + 5$$

where y is the horizontal distance (in feet) when the ball is thrown. Sketch the graph of the function using a graphing calculator. Use the graph to estimate the maximum height the ball reaches.

2—**Answer:**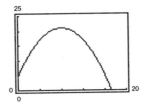

The maximum height ≈ 21 feet.

68. The height in feet of a ball thrown by a child is

$$y = -\tfrac{1}{2}x^2 + 2x + 3$$

where y is the horizontal distance (in feet) when the ball is thrown. Sketch the graph of the function using a graphing calculator. Use the graph to estimate the maximum height the ball reaches.

2—**Answer:**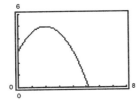

The maximum height ≈ 5 feet.

69. The height in feet of a ball thrown by a child is

$$y = -x^2 + 6x + 10$$

where y is the horizontal distance (in feet) when the ball is thrown. Sketch the graph of the function using a graphing calculator. Use the graph to estimate the maximum height the ball reaches.

2—**Answer:**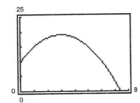

The maximum height ≈ 19 feet.

70. A ball is shot vertically upward at a speed of 64 feet per second. The equation that relates the position, s, of the ball as a function of time, t, is $s = 64t - 16t^2$, $0 \leq t \leq 4$. Sketch the graph over the interval.

1—Answer:

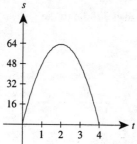

71. A ball is shot vertically upward at a speed of 32 feet per second. The equation that relates the position, s, of the ball as a function of time, t, is $s = 32t - 16t^2$, $0 \leq t \leq 2$. Sketch the graph over the interval.

1—Answer:

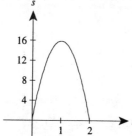

❏ 2.4 Translations and Combinations

1. Describe the transformation of the graph of $f(x) = x^2$ for the graph of $g(x) = (x + 9)^2$.

   (a) Vertical shift 9 units up
   (b) Vertical shift 9 units down
   (c) Horizontal shift 9 units to the right
   (d) Horizontal shift 9 units to the left
   (e) None of these

   1—Answer: d

2. Describe the transformation of the graph of $f(x) = |x|$ for the graph of $g(x) = |x| - 20$.

   (a) Vertical shift 20 units up
   (b) Vertical shift 20 units down
   (c) Horizontal shift 20 units to the right
   (d) Horizontal shift 20 units to the left
   (e) None of these

   1—Answer: b

3. Describe the transformation of the graph of $f(x) = \sqrt{x}$ for the graph of $g(x) = \sqrt{x - 5}$.

   (a) Vertical shift 5 units up
   (b) Vertical shift 5 units down
   (c) Horizontal shift 5 units to the right
   (d) Horizontal shift 5 units to the left
   (e) None of these

   1—Answer: c

4. Describe the transformation of the graph of $f(x) = x^3$ for the graph of $g(x) = 7 + x^3$.

  (a) Vertical shift 7 units up
  (b) Vertical shift 7 units down
  (c) Horizontal shift 7 units to the right
  (d) Horizontal shift 7 units to the left
  (e) None of these

  1—Answer: a

5. The graph at the right is a transformation of the graph of $f(x) = x^2$. Find an equation for the function.

  (a) $g(x) = (x - 3)^2$
  (b) $g(x) = x^2 + 3$
  (c) $g(x) = (x + 3)^2$
  (d) $g(x) = x^2 - 3$
  (e) None of these

  1—Answer: c

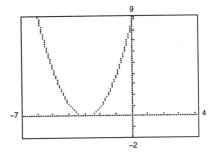

6. The graph at the right is a transformation of the graph of $f(x) = \sqrt{x}$. Find an equation for the function.

  (a) $g(x) = \sqrt{x} - 1$
  (b) $g(x) = \sqrt{x} + 1$
  (c) $g(x) = \sqrt{x - 1}$
  (d) $g(x) = \sqrt{x + 1}$
  (e) None of these

  1—Answer: a

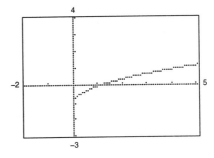

7. The graph at the right is a transformation of the graph of $f(x) = x^3$. Find an equation for the function.

  (a) $g(x) = x^3 + 3$
  (b) $g(x) = (x + 3)^3$
  (c) $g(x) = x^3 - 3$
  (d) $g(x) = (x - 3)^3$
  (e) None of these

  1—Answer: d

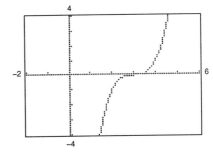

8. The graph at the right is a transformation of the graph of $f(x) = |x|$. Find an equation for the function.

  (a) $g(x) = |x + 4|$
  (b) $g(x) = |x| + 4$
  (c) $g(x) = |x| - 4$
  (d) $g(x) = |x - 4|$
  (e) None of these

  1—Answer: b

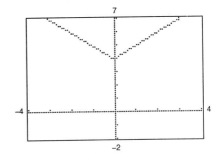

9. Graph $g(x) = |x + 2|$ using a transformation of the graph of $f(x) = |x|$.

(a)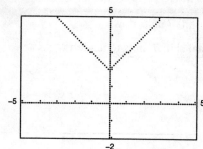

(b)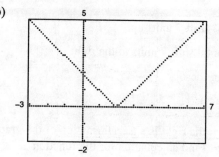

(c)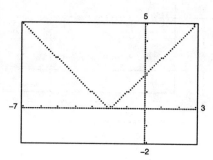

(d)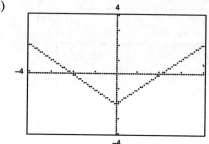

(e) None of these

1—T—Answer: c

10. Graph $g(x) = (x - 1)^2$ using a transformation of the graph of $f(x) = x^2$.

(a)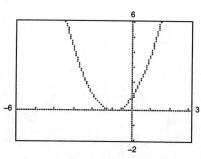

(b)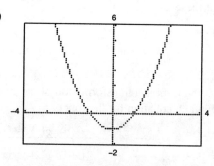

(c)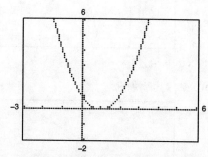

(d)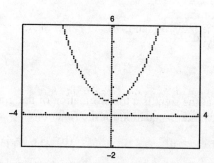

(e) None of these

1—T—Answer: c

11. Graph $g(x) = 3 + \sqrt{x}$ using a transformation of the graph of $f(x) = \sqrt{x}$.

    (a)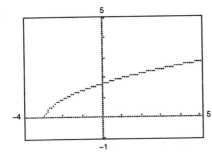

    (b)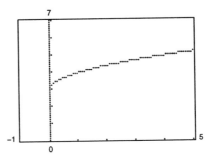

    (c)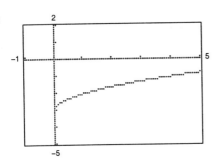

    (d)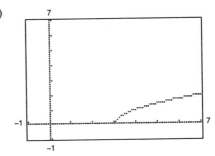

    (e) None of these

    1—T—Answer: b

12. Graph $g(x) = \sqrt[3]{x+4}$ using a transformation of the graph of $f(x) = \sqrt[3]{x}$.

    1—T—Answer: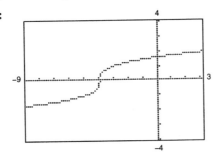

13. Describe the transformation of the graph of $f(x) = |x|$ for the graph of $g(x) = -|x|$.

    (a) Reflection in the x-axis
    (b) Reflection in the y-axis
    (c) Horizontal shift 1 unit to the right
    (d) Vertical shift 1 unit down
    (e) None of these

    1—Answer: a

14. Describe the transformation of the graph of $f(x) = x^6$ for the graph of $g(x) = (-x)^6$.

    (a) Reflection in the x-axis
    (b) Reflection in the y-axis
    (c) Horizontal shift 1 unit to the right
    (d) Vertical shift 1 unit down
    (e) None of these

    1—Answer: b

15. The graph at the right is a transformation of the graph of $f(x) = x^3$. Find the equation for the function.

1—Answer: $g(x) = -x^3$

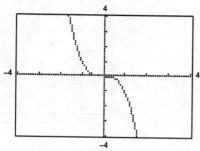

16. The graph at the right is a transformation of the graph of $f(x) = x^4$. Find the equation for the function.

1—Answer: $g(x) = (-x)^4$

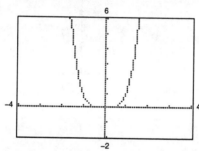

17. Graph $g(x) = (-x)^2$ using a transformation of the graph of $f(x) = x^2$.

(a)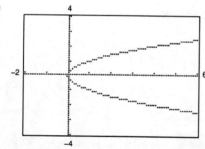

(b)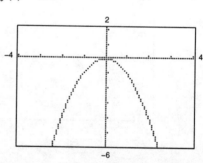

(c)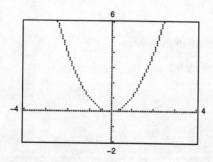

(d)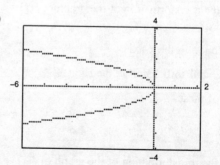

(e) None of these

1—T—Answer: c

18. Graph $g(x) = -\sqrt{x}$ using a transformation of the graph of $f(x) = \sqrt{x}$.

   (a)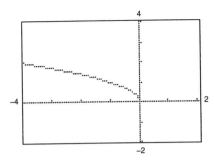

   (b)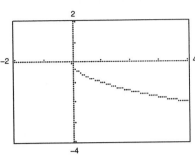

   (c)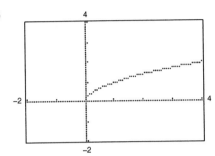

   (d)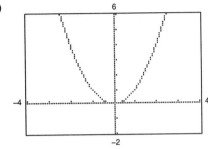

   (e) None of these

   1—T—Answer: b

19. Which sequence of transformations will yield the graph of $g(x) = (x + 1)^2 + 10$ from the graph of $f(x) = x^2$?

   (a) Horizontal shift 10 units to the right
   Vertical shift 1 unit up

   (b) Horizontal shift 1 unit to the left
   Vertical shift 10 units up

   (c) Horizontal shift 1 unit to the right
   Vertical shift 10 units up

   (d) Horizontal shift 10 units to the left
   Vertical shift 1 unit up

   1—Answer: b

20. Which sequence of transformations will yield the graph of $g(x) = -|x + 9|$ from the graph of $f(x) = |x|$?

   (a) Reflection in the x axis
   Horizontal shift 9 units to the left

   (b) Reflection in the y-axis
   Horizontal shift 9 units to the left

   (c) Reflection in the x-axis
   Horizontal shift 9 units to the right

   (d) Reflection in the y-axis
   Horizontal shift 9 units to the right

   1—Answer: a

21. What sequence of transformations will yield the graph of $g(x) = \sqrt[4]{x - 3} + 2$ from the graph of $f(x) = \sqrt[4]{x}$?

   1—Answer: Horizontal shift 3 units to the right
   Vertical shift 2 units up

22. A function is a reflection in the y-axis and a vertical shift 5 units up of the graph of $f(x) = |x|$. Write an equation for the function.

   (a) $g(x) = |x| + 5$
   (b) $g(x) = -|x| + 5$
   (c) $g(x) = |-x| + 5$
   (d) $g(x) = -|x| - 5$
   (e) None of these

   1—Answer: c

23. The graph at the right is a transformation of the graph of $f(x) = x^2$. Find an equation for the function.

   (a) $g(x) = (x + 3)^2 + 1$
   (b) $g(x) = (x + 1)^2 - 3$
   (c) $g(x) = (x - 3)^2 + 1$
   (d) $g(x) = (x + 1)^2 + 3$
   (e) None of these

   1—Answer: e

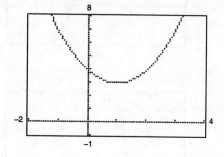

24. The graph at the right is a transformation of the graph of $f(x) = \sqrt{x}$. Find an equation for the function.

   (a) $g(x) = \sqrt{-x} + 2$
   (b) $g(x) = \sqrt{-x + 2}$
   (c) $g(x) = -\sqrt{x} + 2$
   (d) $g(x) = -\sqrt{x - 2}$
   (e) None of these

   1—Answer: c

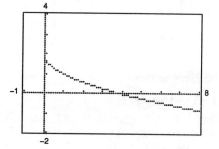

25. The graph at the right is a transformation of the graph of $f(x) = |x|$. Find an equation for the function.

   (a) $g(x) = |x - 1| - 4$
   (b) $g(x) = |x - 4| - 1$
   (c) $g(x) = |x - 1| + 4$
   (d) $g(x) = |x + 4| - 1$
   (e) None of these

   1—T—Answer: d

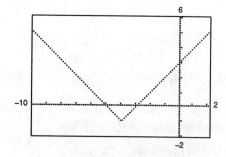

26. The graph at the right is a transformation of the graph of $f(x) = x^3$. Find an equation for the function.

   (a) $g(x) = -x^3 + 3$
   (b) $g(x) = -(x - 3)^3$
   (c) $g(x) = -(x + 3)^3$
   (d) $g(x) = -x^3 - 3$
   (e) None of these

   1—Answer: b

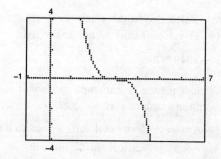

27. Graph $g(x) = |x + 2| - 3$ using a transformation of the graph of $f(x) = |x|$.

(a)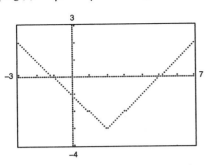

(b)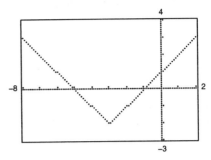

(c)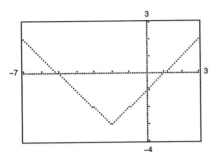

(d)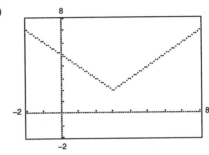

(e) None of these

1—T—**Answer:** c

28. Graph $g(x) = -x^2 + 2$ using a transformation of the graph of $f(x) = x^2$.

(a)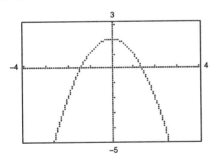

(b)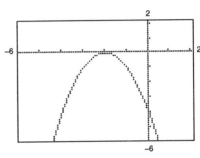

(c)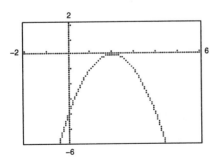

(d)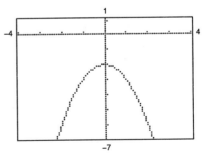

(e) None of these

1—T—**Answer:** a

174 Chapter 2 *Functions and Their Graphs*

29. Graph $g(x) = (x - 1)^3 + 1$ using a transformation of the graph of $f(x) = x^3$.

(a)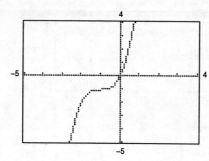

(b)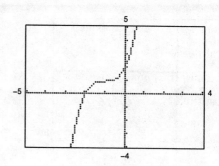

(c)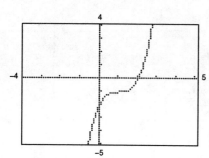

(d)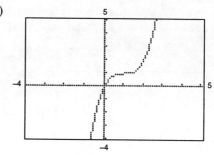

(e) None of these

1—T—**Answer:** d

30. Graph $g(x) = \sqrt{x - 4} - 3$ using a transformation of the graph of $f(x) = \sqrt{x}$.

1—T—**Answer:**

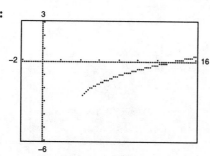

31. Describe the nonrigid transformation of the graph of $f(x) = x^2$ for the graph of $g(x) = 2x^2 + 1$.

(a) Vertical shift 1 unit down

Vertical stretch

(b) Vertical shift 1 unit up

Vertical shrink

(c) Horizontal shift 2 units to the left

Vertical shrink

(d) Vertical shift 1 unit up

Vertical stretch

(e) None of these

2—**Answer:** d

32. Describe the nonrigid transformation of the graph of $f(x) = \sqrt{x}$ for the graph of $g(x) = \frac{1}{3}\sqrt{x+4}$.

    (a) Horizontal shift 4 units to the left

    Vertical shrink

    (b) Horizontal shift 4 units to the left

    Vertical stretch

    (c) Horizontal shift 4 units to the right

    Vertical stretch

    (d) Horizontal shift 4 units to the right

    Vertical shrink

    (e) None of these

    2—Answer: a

33. Find an equation of the function whose graph is a horizontal shift 9 units to the right and a vertical stretch (by 4) of the graph of $f(x) = \sqrt[3]{x}$.

    (a) $g(x) = \frac{1}{4}\sqrt[3]{x+9}$
    (b) $g(x) = \frac{1}{4}\sqrt[3]{x-9}$
    (c) $g(x) = 4\sqrt[3]{x-9}$
    (d) $g(x) = 4\sqrt[3]{x+9}$
    (e) None of these

    2—T—Answer: c

34. Find an equation of the function whose graph is a vertical shift 5 units down and a vertical shrink (by 6) of the graph of $f(x) = |x|$.

    (a) $g(x) = \frac{1}{6}|x+5|$
    (b) $g(x) = \frac{1}{6}|x| - 5$
    (c) $g(x) = 6|x| - 5$
    (d) $g(x) = 6|x-5|$
    (e) None of these

    2—Answer: b

35. Given the graph of $y = x^4$ sketch the graph of $y = (x-2)^4 + 6$.

    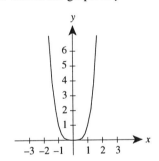

    2—Answer:

    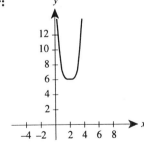

36. Given the graph of $y = x^2$ sketch the graph of $y = (x + 3)^2 - 1$.

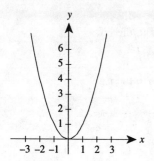

2—Answer:

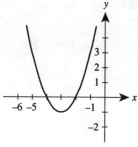

37. Given the graph of $y = -\sqrt{x}$, identify the graph of $y = 2 - \sqrt{x + 3}$.

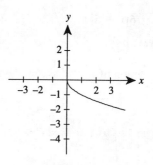

(a)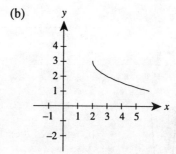

(b)

(c)

(d)

(e) None of these

1—Answer: d

38. Use the graph of $y = x^2$ to find a formula for the function $y = f(x)$.

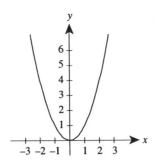
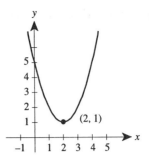

(a) $f(x) = (x - 2)^2 + 1$
(b) $f(x) = (x - 1)^2 + 2$
(c) $f(x) = (x + 2)^2 + 1$
(d) $f(x) = (x + 1)^2 - 2$
(e) None of these

2—Answer: a

39. Use the graph of $y = x^4$ to find a formula for the function $y = f(x)$.

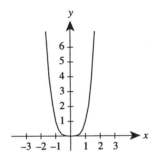
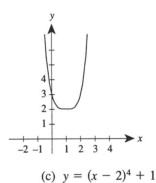

(a) $y = (x + 1)^4 + 2$
(b) $y = (x - 1)^4 + 2$
(c) $y = (x - 2)^4 + 1$
(d) $y = (x + 2)^4 + 1$
(e) None of these

1—Answer: b

40. Use the graph of $y = x^3$ to find a formula for the function $y = f(x)$.

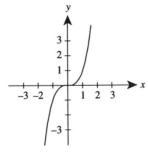
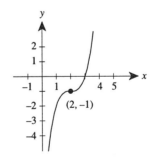

(a) $y = (x + 1)^3 - 2$
(b) $y = (x - 1)^3 - 2$
(c) $y = (x + 2)^3 - 1$
(d) $y = (x - 2)^3 - 1$
(e) None of these

1—Answer: d

41. Given $f(x) = 2x - 4$ and $g(x) = 1 + 3x$, find $(f + g)(x)$.

(a) $5x - 3$
(b) $x - 3$
(c) $-(x + 3)$
(d) 0
(e) None of these

1—Answer: a

42. Given $f(x) = 6$ and $g(x) = 2x^2 - 1$, find $(f - g)(x)$.

    (a) $2x^2 + 5$ (b) $2x^2 - 7$ (c) $-2x^2 + 7$

    (d) $-2x^2 + 5$ (e) None of these

    1—Answer: c

43. Given $f(x) = 2x$ and $g(x) = x - 1$, find $(fg)(x)$.

    (a) $x + 1$ (b) $2x^2 - 2x$ (c) $3x - 1$

    (d) $2x^2 - 1$ (e) None of these

    1—Answer: b

44. Given $f(x) = \dfrac{1}{x}$ and $g(x) = x^2 - 5$ find $(fg)x$.

    (a) $\dfrac{x}{x^2 - 5}$ (b) $x(x^2 - 5)$ (c) $\dfrac{x^2 - 5}{x}$

    (d) $x - 5$ (e) None of these

    1—T—Answer: c

45. Given $f(x) = x$ and $g(x) = 3x - 1$ find $(f/g)(x)$.

    (a) $3x^2 - x$ (b) $\dfrac{3x - 1}{x}$ (c) $\dfrac{x}{3x - 1}$

    (d) $4x - 1$ (e) None of these

    1—Answer: c

46. Given $f(x) = 1/x$ and $g(x) = x/4$, find $(f/g)(x)$.

    (a) $\dfrac{x^2}{4}$ (b) 4 (c) $\dfrac{x + 4}{4x}$

    (d) $\dfrac{4}{x^2}$ (e) None of these

    1—Answer: d

47. Given $f(x) = x - 2$ and $g(x) = 6 - 2x$, find $(f + g)(-2)$.

    (a) 6 (b) 2 (c) -2

    (d) -14 (e) None of these

    1—Answer: a

48. Given $f(x) = x$ and $g(x) = x^2 - 7$, find $(fg)(3)$.

    (a) -13 (b) 29 (c) 5

    (d) 6 (e) None of these

    1—Answer: d

49. Given $f(x) = 9x + 1$ and $g(x) = 4 - x$, find $(f - x)(5)$.

(a) 37   (b) 47   (c) 55

(d) -46 (e) None of these

1—Answer: b

50. Given $f(x) = x^2 + 3$ and $g(x) = x - 1$, find $(f/g)(-1)$.

(a) 2   (b) 0   (c) -2

(d) Undefined   (e) None of these

1—Answer: c

51. Using a graphing utility to graph $(f + g)(x)$ if $f(x) = 2x - 3$ and $g(x) = x + 5$.

(a)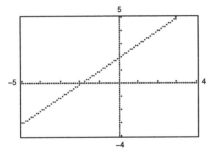

(b)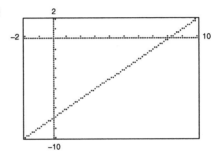

(c)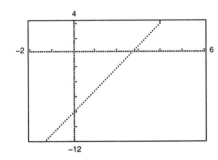

(d)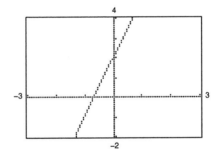

(e) None of these

1—T—Answer: d

52. Use a graphing utility to graph $(f - g)(x)$ if $f(x) = x^2$ and $g(x) = -x$.

(a)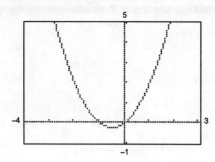

(b)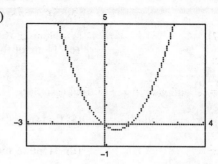

(c)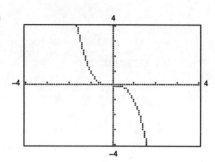

(d)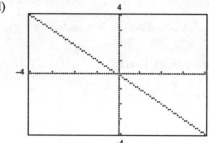

(e) None of these

1—T—Answer: a

53. Use a graphing utility to graph $(f/g)(x)$ if $f(x) = 2x^2 - x$ and $g(x) = x$.

1—T—Answer:

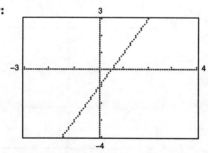

54. Use a graphing utility to graph $(fg)(x)$ if $f(x) = \dfrac{1}{x}$ and $g(x) = x + 2$.

1—T—Answer:

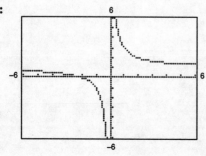

55. Given $f(x) = x - 7$ and $g(x) = 4x$, find $(f \circ g)(x)$.

(a) $3x - 7$ (b) $4x^2 - 7x$ (c) $4x - 7$

(d) $4(x - 7)$ (e) None of these

1—Answer: c

56. Given $f(x) = x^2$ and $g(x) = x + 5$, find $(g \circ f)(x)$.

(a) $(x + 5)^2$ (b) $x^2 + 5$ (c) $x^2 + 25$

(d) $x^2 + 5x^2$ (e) None of these

1—Answer: b

57. Given $f(x) = x^2 + 5$ and $g(x) = 6 - x$, find $(f \circ g)(x)$.

(a) $x^4 + 10$ (b) $x^4 + 25$ (c) $(x^2 + 5)^2 + 5$

(d) $x^2 + 10$ (e) None of these

1—Answer: c

58. Given $f(x) = x^2 - 2x$ and $g(x) = 3x + 2x$, find $(f \circ g)(x)$.

(a) $4x^2 + 8x + 3$ (b) $2x^2 - 4x + 3$ (c) $2x^3 - x^2 - 6x$

(d) $3x^2 + x$ (e) None of these

1—Answer: a

59. Given $f(x) = 4 - 2x^2$ and $g(x) = 2 - x$, find $(f \circ g)(x)$.

(a) $4x^2 - 16x + 20$ (b) $2x^2 - 4$ (c) $2x^2 - 2$

(d) $-2x^3 - 4x^2 - 4x + 8$ (e) None of these

1—Answer: e

60. Given $f(x) = \dfrac{1}{x^2}$ and $g(x) = \sqrt{x^2 + 4}$, find $(f \circ g)(x)$.

(a) $\dfrac{1}{x^2 + 4}$ (b) $\dfrac{1}{\sqrt{x^2 + 4}}$ (c) $x^2 + 4$

(d) $\dfrac{1}{x^2\sqrt{x^2 + 4}}$ (e) None of these

1—Answer: a

61. Given $f(x) = x + 4$ and $g(x) = 3x$, find $(f \circ g)(2)$.

(a) 36   (b) 10   (c) 32

(d) 16   (e) None of these

1—Answer: b

62 Given $f(x) = x^2$ and $g(x) = \sqrt{x - 6}$, find $(f \circ g)(-1)$.

(a) -7 (b) 7   (c) $\sqrt{7}$

(d) Undefined   (e) None of these

1—Answer: a

63. Given $f(x) = \dfrac{1}{x}$ and $g(x) = \dfrac{1}{x}$, find $(f \circ g)(9)$.

1—Answer: 9

64. Given $f(x) = x^3 + 4$ and $g(x) = \sqrt[3]{x}$, find $(f \circ g)(-3)$.

1—Answer: 1

65. Use a graphing utility to graph $(f \circ g)(x)$ if $f(x) = x - 1$ and $g(x) = x + 9$.

(a)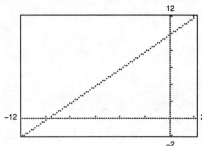

(b)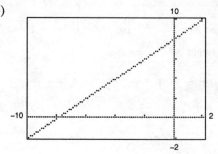

(c)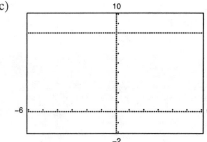

(d)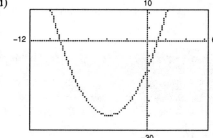

(e) None of these

1—T—Answer: b

66. Use a graphing utility to graph $(f \circ g)(x)$ if $f(x) = x^2 - 2$ and $g(x) = \sqrt{x + 6}$.

(a)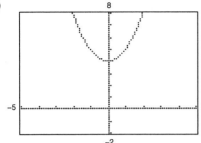

(b)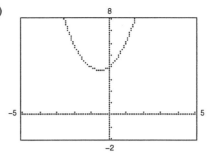

(c)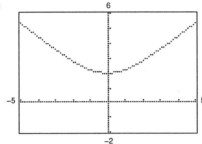

(d)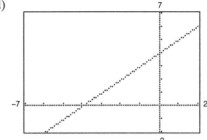

(e) None of these

1—T—Answer: d

67. Find functions f and g such that $(f \circ g)(x) = h(x)$: $h(x) = \sqrt{(x + 1)^2 - 3}$.
(a) $f(x) = x + 1, g(x) = \sqrt{x - 3}$
(b) $f(x) = \sqrt{(x + 1)^2}, g(x) = \sqrt{3}$
(c) $f(x) = (x + 1)^2, g(x) = -3$
(d) $f(x) = \sqrt{x^2 - 3}, g(x) = x + 1$
(e) None of these

2—M—Answer: d

68. Find functions f and g such that $(f \circ g)(x) = h(x)$: $h(x) = (x + 2)^4 - x - 2$.
(a) $f(x) = x^4 - x - 2, g(x) = x + 2$
(b) $f(x) = x^4 - x, g(x) = x + 2$
(c) $f(x) = x + 2, g(x) = x^4 - x - 2$
(d) $f(x) = x^4, g(x) = x - 2$
(e) None of these

2—M—Answer: b

69. Find functions f and g such that $(f \circ g)(x) = h(x)$: $h(x) = \dfrac{1}{x^2 - 2}$.

(a) $f(x) = \dfrac{1}{x - 2}, g(x) = x^2 - 2$
(b) $f(x) = x^2 - 2, g(x) = \dfrac{1}{x^2 - 2}$
(c) $f(x) = \dfrac{1}{x^2 - 2}, g(x) = \sqrt{x - 2}$
(d) $f(x) = \dfrac{1}{x^2}, g(x) = \sqrt{x^2 - 2}$
(e) None of these

2—Answer: d

70. Given $f(x) = \sqrt{x}$ and $g(x) = x^2 + 4$, find the domain of $(f \circ g)(x)$.

(a) $(-\infty, 0]$ (b) $[0, \infty)$ (c) $(-\infty, \infty)$

(d) $(-\infty, -2), (-2, 2)$ (e) None of these

2—Answer: c

71. Given $f(x) = \dfrac{1}{x^2 - 1}$ and $g(x) = x + 3$, find the domain of $(f \circ g)(x)$.

(a) $(-\infty, \infty)$ (b) $(-\infty, -1), (-1, 1), (1, \infty)$

(c) $(-\infty, -4), (-4, -2), (-2, \infty)$ (d) $[-3, \infty)$

(e) None of these

2—Answer: c

72. Given $f(x) = \dfrac{1}{\sqrt{x}}$ and $g(x) = x + 3$, find the domain of $(f \circ g)(x)$.

(a) $(0, \infty)$ (b) $(-3, \infty)$ (c) $(-\infty, -3), (-3, \infty)$

(d) $(-\infty, 0), (0, \infty)$ (e) None of these

2—Answer: b

73. Given $f(x) = \dfrac{1}{x^2 - 1}$ and $g(x) = 1 - x$, find the domain of $(f \circ g)(x)$.

(a) $(-\infty, 0), (0, 2), (2, \infty)$ (b) $(-\infty, -1), (-1, 1), (1, \infty)$

(c) $(1, \infty)$ (d) $(-1, 1)$ (e) None of these

2—Answer: a

74. The weekly cost of producing x units in a manufacturing process is given by the function $C(x) = 30x + 400$. If the number of units produced in t hours is given by $x(t) = 75t$, find $(C \circ x)(t)$.

2—Answer: $2250t + 400$

75. The cost of producing x units in a manufacturing process is given by the function $C(x) = 1.25x + 65$. The revenue obtained from selling x units is given by $R(x) = 2.75x - 0.0025x^2$. Determine the profit as a function of the number of units sold if $P = R - C$.

2—Answer: $P = -0.0025x^2 + 1.50x - 65$

76. An environmental study of a small town has shown that the average daily level of a certain pollutant in the air can be modeled by $P(n) = \sqrt{n^2 + 4}$ parts per million when the population is n hundred people. It is estimated that t years from now the population will be $n(t) = 4 + 0.25t^2$ hundred. Determine the level of pollutant as a function of time.

2—Answer: $(P \circ n)(t) = \sqrt{(4 + 0.25t^2)^2 + 4}$

$$= \sqrt{0.0625t^4 + 2t^2 + 20}$$

77. A company has determined that consumers will buy $L(p) = \dfrac{600}{p^2}$ liters of a soft drink when the price per liter is p dollars. It is also determined that the price per liter t months for now is $p(t) = 0.4t^2 + 2t + 1$. Determine the number of liters that consumers will buy as a function time in months.

2—Answer: $(L \circ p)(t) = \dfrac{600}{(0.4t^2 + 2t + 1)^2}$

$= \dfrac{600}{0.16t^4 + 1.6t^3 + 4.8t^2 + 4t + 1}$

❏ 2.5 Inverse Functions

1. Given $f(x) = x - 1$, identify the graph of $f^{-1}(x)$.

(a)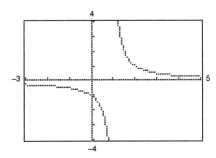

(b)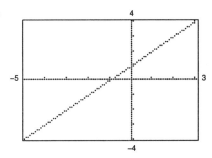

(c)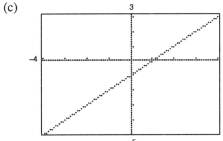

(d)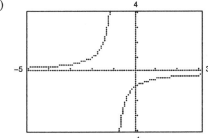

(e) None of these

1—Answer: b

2. Given $f(x) = \dfrac{x}{3}$, identify the graph of $f^{-1}(x)$.

(a)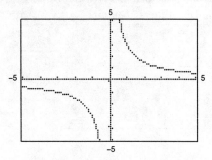

(b)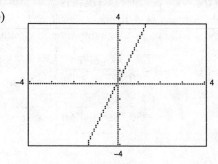

(c)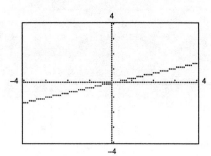

(d)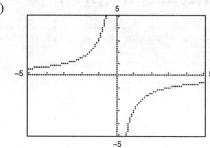

(e) None of these

1—Answer: b

3. Given $f(x) = 3 + x$, identify the graph of $f^{-1}(x)$.

(a)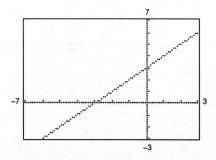

(b)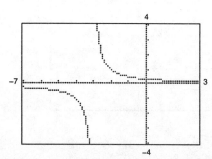

(c)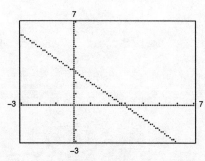

(d)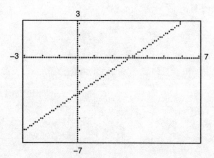

(e) None of these

1—Answer: d

4. Given $f(x) = 4x$, identify the graph of $f^{-1}(x)$.

(a)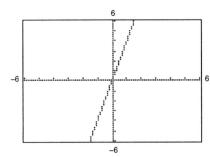

(b)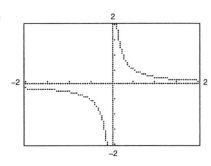

(c)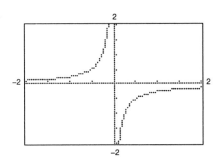

(d)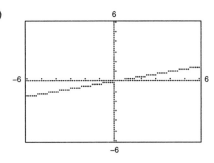

(e) None of these

1—**Answer:** d

5. Find the inverse of f informally: $f(x) = 2x$.

   1—**Answer:** $f^{-1}(x) = \dfrac{x}{2}$

6. Find the inverse of f informally: $f(x) = \dfrac{x}{3}$.

   1—**Answer:** $f^{-1}(x) = 3x$

7. Find the inverse of f informally $f(x) = x + 5$.

   1—**Answer:** $f^{-1}(x) = x - 5$

8. Find the inverse of f informally $f(x) = \sqrt[5]{x}$.

   1—**Answer:** $f^{-1}(x) = x^5$

9. Algebraically, determine which sets of functions are not inverses of each other.

   (a) $f(x) = x^2 + 1$
       $g(x) = \sqrt{x - 1}$

   (b) $f(x) = x^2 - 1$
       $g(x) = \sqrt{x + 1}$

   (c) $f(x) = 1 - x^2$
       $g(x) = \sqrt{1 + x^2}$

   (d) All of these are inverses of each other.

   (e) None of these are inverses of each other.

   1—**Answer:** c

10. Algebraically, determine which sets of functions are not inverses of each other.

   (a) $f(x) = x^3 + 2$

   $g(x) = \sqrt[3]{x - 2}$

   (b) $f(x) = 2 - x^3$

   $g(x) = \sqrt[3]{2 - x}$

   (c) $f(x) = x^3 - 2$

   $g(x) = \sqrt[3]{x + 2}$

   (d) All of these are inverses of each other.

   (e) None of these are inverses of each other.

   1—Answer: d

11. Algebraically, determine which sets of functions are not inverses of each other.

   (a) $f(x) = \dfrac{2}{x - 3}$

   $g(x) = \dfrac{x - 3}{2}$

   (b) $f(x) = \dfrac{5}{x}$

   $g(x) = \dfrac{5}{x}$

   (c) $f(x) = \dfrac{x}{2}$

   $g(x) = 2x$

   (d) All of these are inverses of each other.

   (e) None of these are inverses of each other.

   1—Answer: a

12. If f is a one-to-one function on its domain, the graph of $f^{-1}(x)$ is a reflection of the graph of $f(x)$ with respect to:

   (a) x-axis

   (b) y-axis

   (c) line $y = x$

   (d) line $y = -x$

   (e) origin

   1—Answer: c

13. Graphically, determine which sets of functions are not inverses of each other.

   (a) $f(x) = 9 + x$

   $g(x) = 9 - x$

   (b) $f(x) = x^2$

   $g(x) = -x^2$

   (c) $f(x) = \dfrac{x + 3}{3}$

   $g(x) = \dfrac{3}{x + 3}$

   (d) All of these are inverses of each other.

   (e) None of these are inverses of each other.

   2—T—Answer: e

14. Graphically determine which sets of functions are not inverses of each other.

   (a) $f(x) = x + 5$

   $g(x) = x - 5$

   (b) $f(x) = x^3$

   $g(x) = \sqrt[3]{x}$

   (c) $f(x) = \dfrac{x + 2}{4}$

   $g(x) = 4x - 2$

   (d) All of these are inverses of each other.

   (e) None of these are inverses of each other.

   2—T—Answer: d

15. Graphically determine which sets of functions are not inverses of each other.

   (a) $f(x) = \dfrac{1}{2}x - 1$

   $g(x) = 2x + 1$

   (b) $f(x) = \sqrt[5]{x}$

   $g(x) = \dfrac{1}{\sqrt[5]{x}}$

   (c) $f(x) = \dfrac{x - 1}{5}$

   $g(x) = x + \dfrac{1}{5}$

   (d) All of these are inverses of each other.

   (e) None of these are inverses of each other.

   2—T—Answer: e

16. Graphically, determine whether the functions $f(x) = (x + 1)^3$ and $g(x) = \sqrt[3]{x} - 1$ are inverses of each other.

1—Answer: Yes, they are inverses of each other.

17. Graphically, determine whether the functions $f(x) = \sqrt{x^2 - 5}$ and $g(x) = x^2 + 5$ are inverses of each other.

1—Answer: No, they are not inverses of each other.

18. In which graph does y not represent a one-to-one function of x?

(a)

(b)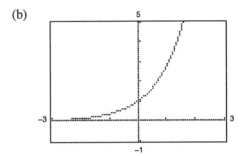

(c)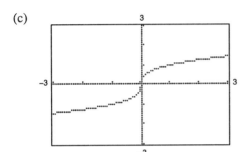

(d) All of these are one-to-one functions of x.

(e) None of these are one-to-one functions of x.

2—Answer: a

19. In which graph does y *not* represent a one-to-one function of x?

(a)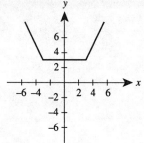

(b)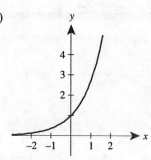

(c)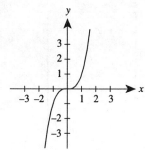

(d)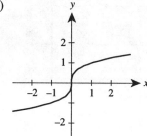

(e) None of these are one-to-one.

1—Answer: a

20. In which graph does y represent a one-to-one function of x?

(a)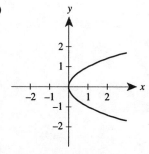

(b)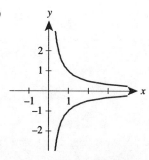

(c)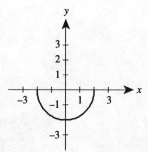

(d)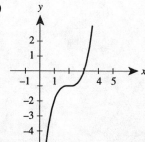

(e) None of these

1—Answer: d

21. In which graph does y *not* represent a one-to-one function of x?

(a)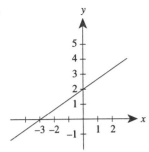

(b)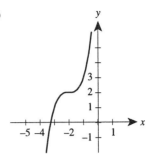

(c)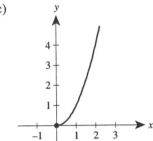

(d)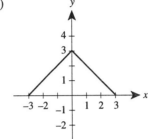

(e) All of these are one-to-one.

1—**Answer:** d

22. In which graph does y *not* represent a one-to-one function of x?

(a)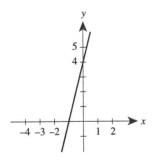

(b)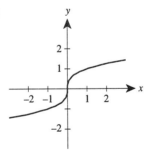

(c)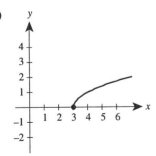

(d)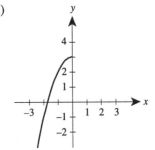

(e) All of these are one-to-one.

1—**Answer:** e

23. In which graph does y represent a one-to-one function of x?

(a)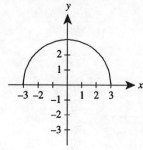

(b)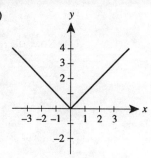

(c)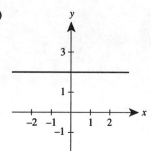

(d)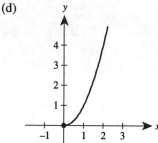

(e) None of these

1—**Answer:** d

24. In which graph does y *not* represent a one-to-one function of x?

(a)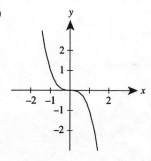

(b)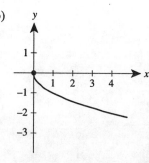

(c)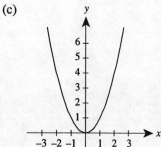

(d)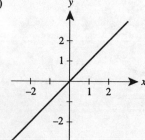

(e) None of these.

1—**Answer:** c

25. Determine which function is *not* one-to-one.

(a) $y = \sqrt[3]{x^2 + 1}$
(b) $y = \dfrac{2}{x}$
(c) $y = 7x - 2$
(d) $y = \sqrt{2 - x}$
(e) None of these

2—Answer: a

26. Determine which function is one-to-one.

(a) $y = |2 - x|$
(b) $y = x^2 + 2$
(c) $y = \sqrt{2 - x^2}$
(d) $y = \dfrac{1}{x + 2}$
(e) None of these

2—Answer: d

27. Determine which function is one-to-one.

(a) $y = |x + 1|$
(b) $y = \sqrt{5 + x}$
(c) $y = \sqrt{x^2 + 1}$
(d) $y = x^2 + 1$
(e) None of these

1—Answer: b

28. Determine which function is one-to-one.

(a) $y = \dfrac{9}{x}$
(b) $y = 3 + x^2$
(c) $y = |-4x|$
(d) $y = x^4$
(e) None of these

1—Answer: a

29. Find the inverse of the function $f(x) = 2 - x$.

(a) $\dfrac{1}{2 - x}$
(b) $2 + x$
(c) $-2 - x$
(d) $x - 2$
(e) None of these

1—Answer: e

30. Find the inverse of the function: $f(x) = \dfrac{x + 3}{2}$.

(a) $2x - \dfrac{2}{3}$
(b) $\dfrac{2}{x + 3}$
(c) $2x - 6$
(d) $2x - 3$
(e) None of these

1—Answer: d

31. Find the inverse of the function: $f(x) = \dfrac{4 + 5x}{7}$.

(a) $\dfrac{7}{5}(x - 4)$
(b) $\dfrac{1}{5}(7x - 4)$
(c) $-\dfrac{7}{4} - \dfrac{7}{5x}$
(d) $\dfrac{7}{4 + 5x}$
(e) None of these

1—Answer: b

194 Chapter 2 Functions and Their Graphs

32. Find the inverse of the function: $f(x) = \dfrac{2}{3x+1}$.

   (a) $\dfrac{3x-1}{2}$ (b) $\dfrac{2-x}{3x}$ (c) $\dfrac{3x+1}{2}$ (d) $\dfrac{1-x}{2}$ (e) None of these

   1—Answer: b

33. Find the inverse of the function: $f(x) = \dfrac{x}{6}$.

   (a) $6x$ (b) $\dfrac{x}{6}$ (c) $\dfrac{6}{x}$ (d) $x-6$ (e) None of these

   1—Answer: a

34. Determine whether the function $f(x) = \dfrac{6}{x}$ is one-to-one. If it is, find its inverse.

   1—Answer: f is one-to-one, $f^{-1}(x) = \dfrac{6}{x}$

35. Determine whether the function $f(x) = \dfrac{7}{x+2}$ is one-to-one. If it is, find its inverse.

   (a) Not one-to-one   (b) $f^{-1}(x) = \dfrac{x+2}{7}$ (c) $f^{-1}(x) = \dfrac{7-2x}{x}$

   (d) $f^{-1}(x) = -\dfrac{7}{x+2}$ (e) None of these

   2—Answer: c

36. Determine whether the function $f(x) = \dfrac{1}{x}$ is one-to-one. If it is, find its inverse.

   1—Answer: $f^{-1}(x) = \dfrac{1}{x}$

37. Given $f(x) = 7x + 2$, find $f^{-1}(x)$.

   (a) $7x + 2$ (b) $\dfrac{1}{7x+2}$ (c) $\dfrac{x-2}{7}$

   (d) $\dfrac{x}{7} - 2$ (e) None of these

   1—Answer: c

38. Given $f(x) = \sqrt{2x-1}$, find $f^{-1}(x)$.

   (a) $\sqrt{2y-1}, y \geq \tfrac{1}{2}$ (b) $x^2 + 1, x \geq 0$ (c) $\tfrac{1}{2}(x^2+1), x \geq 0$

   (d) $\dfrac{1}{\sqrt{2x-1}}, x \geq \tfrac{1}{2}$ (e) None of these

   2—Answer: c

39. Given $f(x) = 3x^3 - 1$, find $f^{-1}(x)$.

   (a) $\dfrac{1}{3x^3 - 1}$ (b) $3x^{-1} - 1$ (c) $3(x + 1)$

   (d) $\sqrt[3]{\dfrac{x + 1}{3}}$ (e) None of these

   1—Answer: d

40. Given $f(x) = 2x^2 + 1$ for $x \geq 0$, find $f^{-1}(x)$.

   2—Answer: $f^{-1}(x) = \sqrt{\dfrac{x - 1}{2}}$

41. Given $f(x) = \dfrac{2x + 1}{3}$, find $f^{-1}(x)$.

   1—Answer: $f^{-1}(x) = \dfrac{3x - 1}{2}$

42. Given $f(x) = 2x^5$, find $f^{-1}(x)$.

   (a) $\sqrt[5]{\dfrac{x}{2}}$ (b) $\dfrac{1}{2}\sqrt[5]{x}$ (c) $\dfrac{1}{2x^5}$

   (d) $\dfrac{2}{x^5}$ (e) None of these

   1—Answer: a

43. Restrict the domain of the function $f(x) = (x - 1)^2$ so that it is one-to-one. Then find the inverse and give its domain.

   2—Answer: Possible Answers: $f(x) = (x - 1)^2, x \geq 1$ or $f(x) = (x - 1)^2, x \leq 1$

   $f^{-1}(x) = 1 + \sqrt{x}, x \geq 0$ or $f^{-1}(x) = 1 - \sqrt{x}, x \geq 0$

44. Restrict the domain of the function $f(x) = |x + 5|$ so that it is one-to-one. Then find the inverse and give its domain.

   2—Answer: Possible Answers: $f(x) = |x + 5|, x \geq -5$ or $f(x) = |x + 5|, x \leq -5$

   $f^{-1}(x) = x - 5, x \geq 0$ or $f^{-1}(x) = -x - 5, x \geq 0$

45. Restrict the domain of the function $f(x) = (x + 2)^2$ so that it is one-to-one. Then find the inverse and give its domain.

   2—Answer: Possible Answers: $f(x) = (x + 2)^2, x \geq -2$ or $f(x) = (x + 2)^2, x \leq -2$

   $f^{-1}(x) = \sqrt{x} - 2, x \geq 0$ or $f^{-1}(x) = -\sqrt{x} - 2, x \geq 0$

46. Restrict the domain of the function $f(x) = |x - 1|$ so that it is one-to-one. Then find the inverse and give its domain.

   2—Answer: Possible Answers: $f(x) = |x - 1|, x \geq 1$ or $f(x) = |x - 1|, x \leq 1$

   $f^{-1}(x) = x + 1, x \geq 0$ or $f^{-1}(x) = -x + 1, x \geq 0$

47. Use a graphing utility to graph $f(x) = x + 1$ and its inverse on the same viewing rectangle.

(a)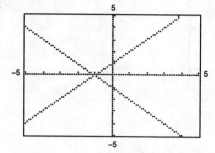

(b)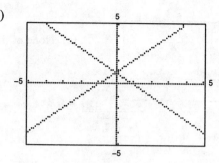

(c)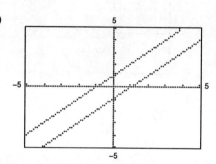

(d)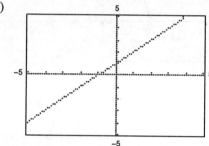

(e) None of these

2—T—Answer: c

48. Use a graphing utility to graph $f(x) = x^2 + 2$, $x \geq 0$ and its inverse on the same viewing rectangle.

(a)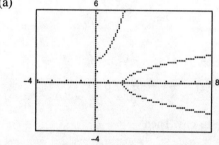

(b)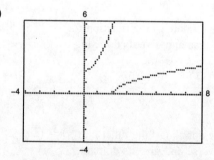

(c)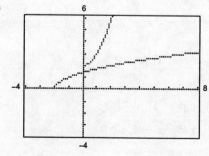

(d)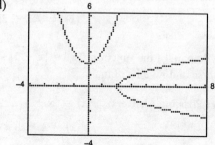

(e) None of these

2—T—Answer: b

49. Use a graphing utility to graph $f(x) = (x - 3)^2$, $x \geq 3$ and its inverse on the same viewing rectangle.

(a)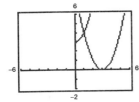

(b)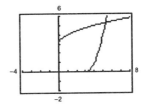

(c)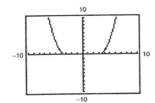

(d)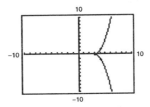

(e) None of these

2—T—Answer: b

50. Use a graphing utility to graph $f(x) = x^3 - 3$ and its inverse on the same viewing rectangle.

2—T—Answer: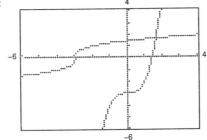

51. Use a graphing utility to graph $f(x) = \dfrac{x + 2}{3}$ and its inverse on the same viewing rectangle.

2—T—Answer: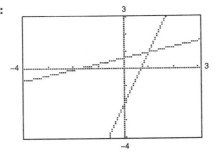

52. Given $f(x) = 1 - 3x$ and $g(x) = x + 2$, find $(f^{-1} \circ g^{-1})(1)$.

(a) $\frac{2}{3}$ (b) -6 (c) -11

(d) $-\frac{1}{3}$ (e) None of these

2—**Answer:** a

53. Given $f(x) = x^2$ and $g(x) = 2x - 3$, find $(g^{-1} \circ f^{-1})(9)$.

(a) 66   (b) -6 (c) -3

(d) 3   (e) None of these

2—**Answer:** d

54. Given $f(x) = x - 4$ and $g(x) = 3 - x$, find $(f^{-1} \circ g^{-1})(-1)$.

(a) 6   (b) 8   (c) -6

(d) -1 (e) None of these

2—**Answer:** b

55. Given $f(x) = \sqrt[3]{x}$ and $g(x) = x - 5$, find $(f^{-1} \circ g^{-1})(-3)$.

(a) -6 (b) -4 (c) 2

(d) -8 (e) None of these

2—**Answer:** e

56. Let $f(x) = 3 - x$ and $g(x) = x^3$. Find $(g^{-1} \circ f^{-1})(-5)$.

(a) $3 + \sqrt[3]{-5}$ (b) $\sqrt[3]{-2}$ (c) 2

(d) $\dfrac{1}{\sqrt[3]{-2}}$ (e) None of these

2—**Answer:** c

57. Let $f(x) = x + 2$ and $g(x) = \sqrt[3]{x}$. Find $(f^{-1} \circ g^{-1})(2)$.

2—**Answer:** 6

58. Let $f(x) = \sqrt[3]{x - 1}$ and $g(x) = x - 4$. Find $(f^{-1} \circ g^{-1})(-2)$.

2—**Answer:** 9

59. The function $y = 0.25x^2 + 10$, $0 \leq x \leq 20$ approximates the population of bacteria (in thousands) in terms of the hour x since the culture was exposed to the air. Find the inverse function. What does each variable represent in the inverse function?

2—**Answer:** $y = \sqrt{4(x - 10)} = 2\sqrt{x - 10}$, $x \geq 10$

y is the hour since exposure where x is the population of the culture in thousands.

60. The function $y = \sqrt{0.5x + 10}$, $0 \leq x \leq 10$ approximates the population of small town in thousands where x is the year with $x = 0$ representing 1980. Find the inverse function. What does each variable represent in the function?

2—Answer: $y = 2(x^2 - 10) = 2x^2 - 20$, $x \geq \sqrt{10}$

y is the year with $y = 0$ is representing 1980 in terms of x, the population in thousands of a small town

CHAPTER THREE
Zeros of Polynomial Functions

❑ 3.1 Quadratic Functions

1. Match the correct graph with the function: $f(x) = 2(x - 3)^2 - 1$.

   (a)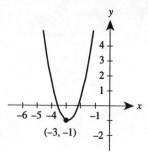

   (b)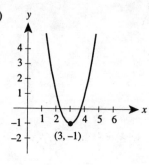

   (c)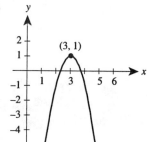

   (d)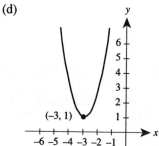

   (e) None of these

   1—Answer: b

2. Match the correct graph with the function: $f(x) = -\frac{1}{2}(x - 2)^2 + 1$.

   (a)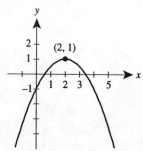

   (b)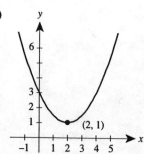

— CONTINUED ON NEXT PAGE —

2. — CONTINUED —

(c)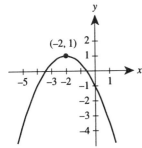

(d)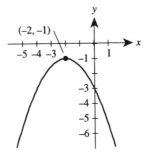

(e) None of these

1—Answer: a

3. Match the correct graph with the function: $f(x) = 3(x + 2)^2 - 1$

(a)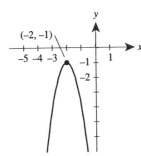

(b)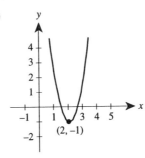

(c)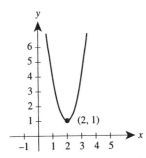

(d)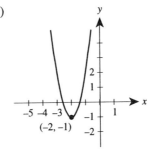

(e) None of these

1—Answer: d

4. Match the correct graph with the function: $f(x) = \frac{1}{3}(x + 3)^2 + 1$

(a)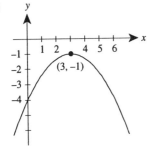

(b)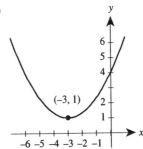

— CONTINUED ON NEXT PAGE —

202 Chapter 3 Zeros of Polynomial Functions

4. — CONTINUED —

(c)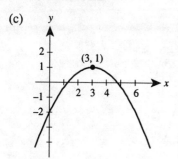

(d)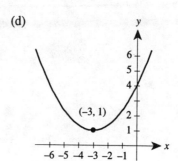

(e) None of these

1—Answer: b

5. Write the standard form of the equation of the parabola.

(a) $y = (x - 2)^2 + 3$
(b) $y = (x + 2)^2 - 3$
(c) $y = (x - 2)^2 - 3$
(d) $y = (x + 2)^2 + 3$
(e) None of these

1—Answer: c

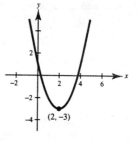

6. Write the standard form of the equation of the parabola.

(a) $y = (x - 3)^2 - 1$
(b) $y = -(x + 3)^2 - 1$
(c) $y = (x + 1)^2 + 3$
(d) $y = -(x - 1)^2 - 3$
(e) None of these

1—Answer: c

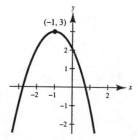

7. Write the standard form of the equation of the parabola.

1—Answer: $f(x) = -(x - 3)^2 - 1$

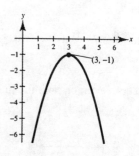

8. Match the correct equation with the parabola.

    (a) $y = 5(x - 1)^2 - 2$
    (b) $y = (x + 2)^2 - 1$
    (c) $y = (x - 2)^2 - 1$
    (d) $y = (x - 1)^2 + 2$
    (e) None of these

    1—Answer: c

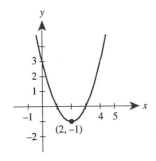

9. Match the correct equation with the parabola.

    (a) $y = -\frac{1}{4}(x - 4)^2 - 1$
    (b) $y = -\frac{1}{4}(x + 4)^2 - 1$
    (c) $y = -(x - 1) + 4$
    (d) $y = -(x - 1)^2 - 4$
    (e) None of these

    1—Answer: a

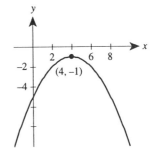

10. Write in the form $y = a(x - h)^2 + k$: $y = 2x^2 + 16x + 9$

    (a) $y = 2(x + 4)^2 - 7$
    (b) $y = 2(x + 2)^2 + 5$
    (c) $y = 2(x + 4)^2 - 23$
    (d) $y = 2(x + 8)^2 + 73$
    (e) None of these

    1—Answer: c

11. Write in the form $y = a(x - h)^2 + k$: $y = x^2 - 8x + 2$

    (a) $y = (x - 4)^2 - 18$
    (b) $y = (x - 4)^2 - 14$
    (c) $y = (x - 8)^2 + 66$
    (d) $y = (x - 4)^2 + 18$
    (e) None of these

    1—Answer: b

12. Write the form $y = a(x - h)^2 + k$: $y = -x^2 + 3x - 2$

    1—Answer: $y = -\left(x - \frac{3}{2}\right)^2 + \frac{1}{4}$

13. Write in the form $y = a(x - h)^2 + k$: $y = -2x^2 - 4x - 5$

    (a) $y = -2(x - 1)^2 - 2$
    (b) $y = (2x - 2)^2 - 1$
    (c) $y = -2(x + 2)^2 - 1$
    (d) $y = -2(x + 1)^2 - 3$
    (e) None of these

    2—M—Answer: d

14. Write in the form $y = a(x - h)^2 + k$: $y = 3x^2 + 12x + 17$

    (a) $y = (x + 2)^2 + \frac{13}{3}$
    (b) $y = 3(x + 2)^2 + 21$
    (c) $y = 3(x + 2)^2 + 5$
    (d) $y = (x + 2)^2 + \frac{29}{3}$
    (e) None of these

    2—Answer: c

15. Sketch the graph (and vertex) of the function: $f(x) = (x - 2)^2 + 6$

   1—Answer:

   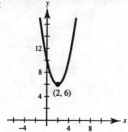

16. Sketch the graph (and vertex) of the function: $f(x) = (x + 5)^2 - 4$

   1—Answer:

   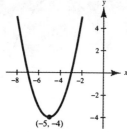

17. Sketch the graph (and vertex) of the function: $f(x) = -x^2 - 4x$

   2—Answer:

   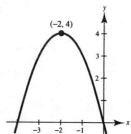

18. Sketch the graph of the function: $f(x) = x^2 + 4x + 3$. Identify the vertex and intercepts.

   (a)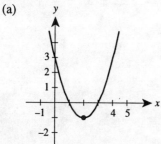

   (b)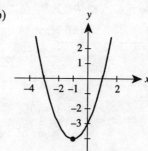

— CONTINUED ON NEXT PAGE —

18. — CONTINUED —

(c)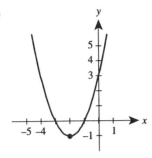

(d)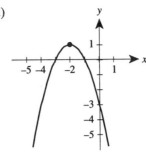

(e) None of these

2—**Answer:** c

19. Sketch the graph of the function: $f(x) = 4x^2 - 8x + 4$

(a)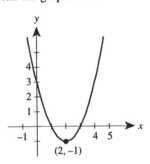

(b)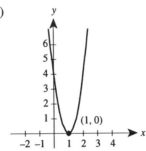

(c)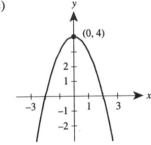

(d)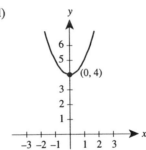

(e) None of these

2—**Answer:** b

20. Use a graphing utility to graph the quadratic function and identify the vertex and x-intercepts: $f(x) = x^2 - 6x + 8$

2—T—**Answer:** 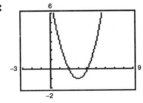 ; $(3,-1)$; $(2,0)$; $(4,0)$

21. Use a graphing utility to graph the quadratic function and identify the vertex and x-intercepts: $f(x) = x^2 - 2x + 3$

   2—T—Answer: 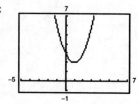 ; (1,2); no x-intercepts

22. Use a graphing utility to graph the quadratic function and identify the vertex and x-intercepts: $f(x) = -x^2 + 4x - 3$

   2—T—Answer: 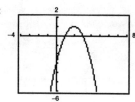 ; (2,1); (1,0); (3,0)

23. Use a graphing utility to graph the quadratic function and identify the vertex and x-intercepts: $f(x) = -2x^2 - 4x - 3$

   2—T—Answer: 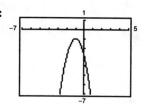 ; (-1,1); no x-intercepts

24. Use a graphing utility to graph the quadratic function and identify the vertex and x-intercepts: $f(x) = -x^2 - 4x + 5$

   2—T—Answer: 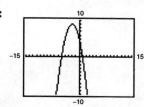 ; (-2,9); (-5,0); (1,0)

25. Use a graphing utility to graph the quadratic function and identify the vertex and x-intercepts: $f(x) = -x + 8x - 12$

   2—T—Answer: 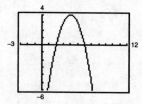 ; (4,4); (2,0); (6,0)

26. Find the vertex of the parabola: $y = 4x^2 + 8x + 1$
    (a) $(-2, 1)$
    (b) $(1, 13)$
    (c) $(0, 1)$
    (d) $(-1, -3)$
    (e) None of these

    2—Answer: d

27. Find the minimum point on the graph of $y = 2x^2 + 8x + 9$.
    (a) $(-2, 1)$
    (b) $(2, 33)$
    (c) $(2, 17)$
    (d) $(-2, -17)$
    (e) None of these

    2—Answer: a

28. Find the minimum point on the graph of $f(x) = x^2 - 4x + 14$.
    (a) $(2, 18)$
    (b) $(-2, 18)$
    (c) $(-2, 26)$
    (d) $(2, 10)$
    (e) None of these

    2—Answer: d

29. Find the maximum point on the graph of $f(x) = -3x^2 + 12x + 1$.
    (a) $(6, -5)$
    (b) $(-2, -19)$
    (c) $(2, 13)$
    (d) $(1, 14)$
    (e) None of these

    2—Answer: c

30. Find the x and y-intercepts: $y = x^2 - 5x + 4$.
    (a) $(0, -4), (0, 1), (4, 0)$
    (b) $(0, 4), (4, 0), (1, 0)$
    (c) $(0, -4), (-4, 0), (-1, 0)$
    (d) $(0, 4), (-4, 0), (-1, 0)$
    (e) None of these

    1—Answer: b

31. Find the x and y intercepts: $y = x^2 + 3x - 4$.
    (a) $(0, -4), (-4, 0), (1, 0)$
    (b) $(0, -4), (4, 0), (-1, 0)$
    (c) $(0, 4), (-4, 0), (0, 1)$
    (d) $(4, 0), (0, 4), (-1, 0)$
    (e) None of these

    1—Answer: a

32. Find the quadratic function that has a maximum point at $(-1, 17)$ and passes through $(7, 1)$.
    (a) $y = \frac{1}{4}(-x^2 - 2x + 16)$
    (b) $y = -\frac{1}{4}(x + 1)^2 + 17$
    (c) $y = (x - 7)^2 + 1$
    (d) $y = (x - 1)^2 + 17$
    (e) None of these

    2—Answer: b

33. Find the quadratic function that has a minimum at $(1, -2)$ and passes through $(0, 0)$.

   (a) $y = 2(x - 1)^2 - 2$
   (b) $y = 2(x + 1)^2 - 2$
   (c) $y = -2(x - 1)^2 + 2$
   (d) $y = -2(x + 1)^2 + 2$
   (e) None of these

   2—Answer: a

34. Find the quadratic function that has a maximum at $(-1, 2)$ and passes through $(0, 1)$.

   2—Answer: $f(x) = -(x + 1)^2 + 2$

35. Find the quadratic function whose graph opens upward and has x-intercepts at $(0, 0)$ and $(6, 0)$.

   2—Answer: $f(x) = x^2 - 6x$

36. Find the quadratic function whose graph opens upward and has x-intercepts at $(0, 0)$ and $(-6, 0)$.

   (a) $y = x^2 - 6x + 9$
   (b) $y = x^2 + 12x + 36$
   (c) $y = x^2 + 6x$
   (d) $y = x^3 + 12x^2 + 36x$
   (e) None of these

   1—Answer: c

37. Find the quadratic function whose graph opens downward and has x-intercepts at $(0, 0)$ and $(-6, 0)$.

   (a) $y = -x^2 - 6x + 9$
   (b) $y = -x^2 - 6x$
   (c) $y = -x^2 + 6x$
   (d) $y = x^2 + 6x$
   (e) None of these

   1—Answer: b

38. Find the quadratic function whose graph opens downward and has x-intercepts at $(2, 0)$ and $(-3, 0)$.

   (a) $y = 6 - x^2$
   (b) $y = 6 + x - x^2$
   (c) $y = 6 - x - x^2$
   (d) $y = 3x^2 - 2x$
   (e) None of these

   1—Answer: c

39. Find the quadratic function whose graph opens upward and has x-intercepts at $(-4, 0)$ and $(1, 0)$.

   (a) $y = x^2 + 3x - 4$
   (b) $y = 4 - 3x - x^2$
   (c) $y = x^2 + 5x + 4$
   (d) $y = x^2 - 3x - 4$
   (e) None of these

   1—Answer: a

40. Find the number of units that produce a maximum revenue, $R = 95x - 0.1x^2$, where R is the total revenue in dollars and x is the number of units sold.

   (a) 716
   (b) 475
   (c) 371
   (d) 550
   (e) None of these

   1—Answer: b

41. Find the number of units that produce a maximum revenue, $R = 400x - 0.01x^2$, where R is the total revenue in dollars and x is the number of units sold.

    (a) 15,000 (b) 32,000 (c) 4500
    (d) 20,000 (e) None of these

    1—Answer: d

42. The profit for a company is given by the equation

    $P = -0.0002x^2 + 140x - 250,000$

    where x is the number of units produced. What production level will yield a maximum profit?

    (a) 700,00 (b) 350,000 (c) 893
    (d) 350 (e) None of these

    2—Answer: b

43. The revenue R for a symphony concert is given by the equation

    $R = -\frac{1}{400}(x^2 - 4800x)$

    where x is the number of tickets sold. Determine the number of tickets that will yield maximum revenue.

    (a) 4800 (b) 12 (c) 48,000
    (d) 2400 (e) None of these

    2—Answer: d

44. The perimeter of a rectangle is 300 feet. What is the width of the rectangle of maximum area?

    (a) 100 feet (b) 50 feet (c) 75 feet
    (d) 60 feet (e) None of these

    2—Answer: c

45. A rancher wishes to enclose a rectangular corral with 320 feet of fencing. The fencing is only required on three sides because of an existing stone wall. What are the dimensions of the corral of maximum area?

    2—Answer: 80 feet by 160 feet

46. A rancher wishes to enclose a rectangular corral with 360 feet of fencing. The fencing is only required on three side because of an existing stone wall. What are the dimensions of the corral of maximum area?

    2—Answer: 90 feet by 180 feet

47. An object is thrown upward from a height of 48 feet with velocity 32 feet per second. Its height is given by $h(t) = -16t^2 + 32t + 48$ when t is time in seconds. Sketch the graph of the function. Use an appropriate window, estimate its maximum height and the time it strikes the ground. Then algebraically, identify its vertex and x-intercepts.

2—T—Answer: 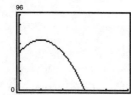 $(1, 64); (3, 0); h(t) = -16(t - 1)^2 + 64$; vertex: $(1, 64)$

x-intercept: $(3, 0); (-1, 0)$

❏ 3.2 Polynomial Functions of Higher Degree

1. Determine the left and right behavior of the graph: $y = 4x^2 - 2x + 1$.
   (a) Up to the left, down to the right
   (b) Down to the left, up to the right
   (c) Up to the left, up to the right
   (d) Down to the left, down to the right
   (e) None of these

   1—Answer: c

2. Determine the left and right behavior of the graph: $f(x) = -x^5 + 2x^2 - 1$.
   (a) Up to the left, down to the right
   (b) Down to the left, up to the right
   (c) Up to the left, up to the right
   (d) Down to the left, down to the right
   (e) None of these

   1—Answer: a

3. Determine the left and right behavior of the graph: $f(x) = 3x^5 - 7x^2 + 2$.
   (a) Down to the left, up to the right
   (b) Down to the left, up to the right
   (c) Up to the left, up to the right
   (d) Down to the left, down to the right
   (e) None of these

   1—Answer: a

4. Determine the left and right behavior of the graph: $f(x) = -4x^3 + 3x^2 - 1$.

   1—Answer: Up to the left, down to the right

5. Determine the left and right behavior of the graph: $f(x) = 3x^4 + 2x^3 + 7x^2 + x - 1$.

   1—Answer: Up to the left and right

6. Determine the left and right behavior of the graph: $f(x) = -2x^4 + 3x^3 + 5x^2$.
   (a) Up to the left, down to the right
   (b) Down to the left, up to the right
   (c) Up to the left, up to the right
   (d) Down to the left, down to the right
   (e) None of these

   1—Answer: d

7. Find all the real zeros of the polynomial function: $f(x) = x^3 - 3x^2 - 4x$.

    (a) $-1, 4$ (b) $-4, 1$ (c) $-1, 0, 4$ (d) $0, 4$ (e) None of these

    1—**Answer:** c

8. Find all the real zeros of the polynomial function: $f(x) = x^6 - x^2$.

    (a) 0 (b) $0, 1$ (c) 1 (d) $0, 1, -1$ (e) None of these

    1—**Answer:** d

9. Find all the real zeros of the polynomial function: $f(x) = x^3 + x$.

    (a) 0 (b) $0, 1$ (c) $0, 1, -1$ (d) $1, -1$ (e) None of these

    1—**Answer:** a

10. Find all the real zeros of the polynomial function: $f(x) = x^4 - 5x^2 - 36$.

    (a) $3, 2$ (b) ± 3 (c) $\pm 3, \pm 2$ (d) ± 2 (e) None of these

    2—**Answer:** b

11. Find all the real zeros of the polynomial function: $g(t) = t^3 + 3t^2 - 16t - 48$.

    (a) -3 (b) 3 (c) $-4, -3, 4$ (d) $-3, 4$ (e) None of these

    2—**Answer:** c

12. Find all the real zeros of the polynomial function: $f(x) = 9x^4 - 37x^2 + 4$.

    2—**Answer:** $\pm 2, \pm \frac{1}{3}$

13. Find a polynomial function with the given zeros: $0, -1, 2$.

    (a) $f(x) = x(x - 1)(x + 2)$ (b) $f(x) = x(x + 1)(x - 2)$
    (c) $f(x) = (x + 1)(x - 2)$ (d) $f(x) = (x + 1)^2(x - 2)$
    (e) None of these

    1—**Answer:** b

14. Find a polynomial function with the given zeros: $-2, -2, 1, 3$.

    (a) $f(x) = (x + 1)(x + 3)(x - 2)$ (b) $f(x) = (x - 2)^2(x - 1)(x - 3)$
    (c) $f(x) = (x + 2)(x - 1)(x - 3)$ (d) $f(x) = (x + 2)^2(x - 1)(x - 3)$
    (e) None of these

    1—**Answer:** d

15. Find a polynomial function with zeros: $1, 0, -3$.

    (a) $f(x) = x(x - 3)^3(x + 1)^2$ (b) $f(x) = x^2(x - 1)(x + 3)$
    (c) $f(x) = x(x - 3)(x - 1)$ (d) $f(x) = (x - 1)(x + 3)^2$

    1—**Answer:** b

212 Chapter 3 Zeros of Polynomial Functions

16. Find a polynomial function with zeros: 0, 1, −2.

   (a) $f(x) = x(x - 1)(x - 2)$ (b) $f(x) = x(x + 1)(x + 2)$

   (c) $f(x) = (x - 1)(x + 2)$ (d) $f(x) = x^2(x - 1)(x + 2)$

   (e) None of these

   1—Answer: d

17. Determine the correct function for the given graph.

   (a) $f(x) = x^5 + 2$ (b) $f(x) = -x^5 - 2$

   (c) $f(x) = x^4 + 2$ (d) $f(x) = x^4 + 2x^2$

   (e) None of these

   2—Answer: a

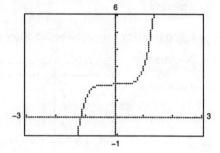

18. Determine the correct function for the given graph.

   (a) $f(x) = x^3 + x^2 - 6$ (b) $f(x) = -x^3 - x^2 + 6x$

   (c) $f(x) = x^3 + x^2 - 6x$ (d) $f(x) = x^4 + x^2 - 6x$

   (e) None of these

   2—Answer: c

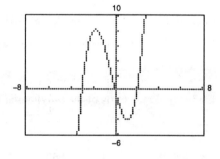

19. Determine the correct function for the given graph.

   (a) $f(x) = 2x^3 - 3x^2$ (b) $f(x) = 3x^4 - 2x^3$

   (c) $f(x) = 2x^3 + 3x^2$ (d) $f(x) = 3x^2 - 2x^3$

   (e) None of these

   2—Answer: d

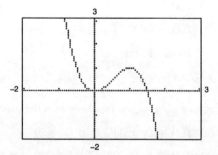

20. Determine the correct function for the given graph.

   (a) $f(x) = 2x^4 + 8x^2$ (b) $f(x) = -2x^4 - 8x^2$

   (c) $f(x) = 8x^2 - 2x^4$ (d) $f(x) = 2x^4 - 8x^2$

   (e) None of these

   2—Answer: d

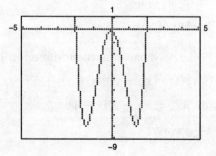

21. Use a graph utility to graph the function: $f(x) = 2x^3 - 3x^2$.

   1—T—Answer: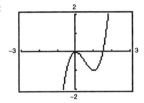

22. Use a graph utility to graph the function: $f(x) = -2x^4 + x$.

   1—T—Answer: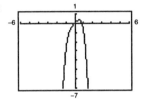

23. Use a graph utility to graph the function: $f(x) = -x^5 + 4$.

   1—T—Answer: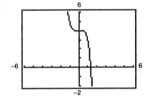

24. Use a graph utility to graph the function: $f(x) = x^4 + 2x^3 + 1$.

   1—T—Answer: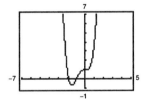

25. Which of the following viewing rectangles gives the best view of the basic characteristics of the graph of $f(x) = -\frac{1}{4}x^4 + x^3 - 1$?

(a)
```
Xmin = -3
Xmax = 3
Xscl = 1
Ymin = -3
Ymax = 3
Yscl = 1
```

(b)
```
Xmin = 1
Xmax = 4
Xscl = 1
Ymin = 0
Ymax = 6
Yscl = 1
```

(c)
```
Xmin = -4
Xmax = 6
Xscl = 1
Ymin = -3
Ymax = 7
Yscl = 1
```

(d)
```
Xmin = 0
Xmax = 6
Xscl = 1
Ymin = -1
Ymax = 7
Yscl = 1
```

(e) None of these

1—T—Answer: c

26. Which of the following viewing rectangles gives the best view of the basic characteristics of the graph of $f(x) = 2x^3 + 3$?

(a)
Xmin = 0
Xmax = 3
Xscl = 1
Ymin = 2
Ymax = 5
Yscl = 1

(b)
Xmin = -4
Xmax = 4
Xscl = 1
Ymin = -1
Ymax = 7
Yscl = 1

(c)
Xmin = -3
Xmax = 0
Xscl = 1
Ymin = -2
Ymax = 1
Yscl = 1

(d)
Xmin = -2
Xmax = 2
Xscl = 1
Ymin = 3
Ymax = 7
Yscl = 1

(e) None of these

1—T—Answer: b

27. Use the Intermediate Value Theorem to estimate the real zero in the internal $[0, 1]$: $f(x) = 3x^3 + 7x - 9$.

(a) Between 0.2 and 0.3
(b) Between 0.5 and 0.7
(c) Between 0.7 and 0.8
(d) Between 0.9 and 1.0
(e) None of these

2—T—Answer: d

28. Use the Intermediate Value Theorem to estimate the real zero in the interval $[1, 2]$: $f(x) = 3x^3 - 2x^2 - 2$.

(a) Between 1.0 and 1.1
(b) Between 1.1 and 1.2
(c) Between 1.3 and 1.4
(d) Between 0.7 and 0.8
(e) None of these

2—T—Answer: b

29. Use the Intermediate Value Theorem to estimate the real zero in the interval $[0, 1]$: $f(x) = 2x^3 + 7x^2 - 1$.

(a) Between 0.1 and 0.2
(b) Between 0.2 and 0.3
(c) Between 0.3 and 0.4
(d) Between 0.4 and 0.5
(e) None of these

2—T—Answer: c

30. Use the Intermediate Value Theorem to estimate the real zero in the interval $[1, 2]$: $f(x) = x^4 - 4x - 1$.

(a) Between 1.5 and 1.6
(b) Between 1.6 and 1.7
(c) Between 1.7 and 1.8
(d) Between 1.8 and 1.9
(e) None of these

2—T—Answer: b

31. The function $f(x)$ has a zero of 2 with multiplicity 3. We know

(a) since 3 is an odd number, the graph touches but does not cross the x-axis.

(b) since 3 is an odd number, the graph crosses the x-axis.

(c) since 2 is an even number, the graph touches but does not cross the x-axis.

(d) since 2 is an even number, the graph crosses the x-axis.

(e) None of these

1—Answer: b

32. The function $f(x)$ has a zero of -1 with multiplicity 1. We know
    (a) since the multiplicity is 1, the graph crosses the x-axis.
    (b) since the multiplicity is 1, the graph touches but does not cross the x-axis.
    (c) since the multiplicity is -1 the graph crosses the y-axis at -1.
    (d) since the zero is -1, the graph goes down to the left.
    (e) None of these

    1—Answer: a

33. The function $f(x)$ has a zero of 3 with multiplicity 2. We know
    (a) since the zero is 3, the graph crosses the y-axis at 3.
    (b) since the zero is 3, the graph goes up to the right.
    (c) since the multiplicity is 2, the graph crosses the x-axis.
    (d) since the multiplicity is 2, the graph touches but does not cross the x-axis.
    (e) None of these

    1—Answer: d

34. An open box is made from a 16-inch square piece of material by cutting equal squares with sides of length, x, from all corners and turning up the sides. The volume of the box is $V(x) = 4x(8 - x)^2$. Estimate the value of x for which the volume is maximum.

    (a) 2.7 inches   (b) 3.0 inches   (c) 1.9 inches   (d) 8 inches   (e) None of these

    2—Answer: a

35. An open box is made from a 24-inch square piece of material by cutting equal squares with sides of length, x, form all corners and turning up the sides. The volume of the box is $V(x) = 4x(12 - x)^2$. Estimate the value of x for which the volume is maximum.

    (a) 12 inches   (b) 3 inches   (c) 6 inches   (d) 4 inches   (e) None of these

    2—Answer: d

36. An open box is made from a 10-inch square piece of material by cutting equal squares with sides of length, x, from all corners and turning up the sides. The volume of the box is $V(x) = 4x(5 - x)^2$. Estimate the value of x for which the volume is maximum.

    (a) 2.0 inches   (b) 1.7 inches   (c) 3.4 inches   (d) 2.5 inches   (e) None of these

    2—Answer: b

37. An open box is to be made from a 16-inch square piece of material by cutting equal squares from each corner and turning up the sides. Verify the volume of the box is $V(x) = 4x(8 - x)^2$. Sketch the graph of the function using a graphing utility and use the graph to estimate the volume of x for which $V(x)$ is maximum.

   2—T—Answer: $V(x) = x \cdot (16 - 2x) \cdot (16 - 2x)$

   $= x \cdot 2(8 - x) \cdot 2(8 - x)$

   $= 4x(8 - x)^2$

   When $x \approx 2.67$ inches, $V(x)$ is a maximum ≈ 303.4 inches^3.

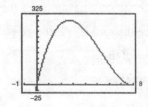

38. An open box is to be made from a 20-inch square piece of material by cutting equal squares from each corner and turning up the sides. Verify the volume of the box is $V(x) = 4x(10 - x)^2$. Sketch the graph of the function using a graphing utility and use the graph to estimate the volume of x for which $V(x)$ is maximum.

   2—T—Answer: $V(x) = x \cdot (20 - 2x) \cdot (20 - 2x)$

   $= x \cdot 2(10 - x) \cdot 2(10 - x)$

   $= 4x(10 - x)^2$

   When $x \approx 3.33$ inches, $V(x)$ is a maximum ≈ 592.6 inches^3.

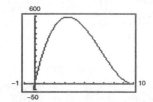

39. An open box is to be made from a 14 inch square piece of material by cutting equal squares from each corner and turning up the sides. Verify the volume of the box is $V(x) = 4x(7 - x)^2$. Sketch the graph of the function using a graphing utility and use the graph to estimate the volume of x for which $V(x)$ is maximum.

   2—T—Answer: $V(x) = x \cdot (14 - 2x) \cdot (14 - 2x)$

   $= x \cdot 2(7 - x) \cdot 2(7 - x)$

   $= 4x(7 - x)^2$

   When $x \approx 2.33$ inches, $V(x)$ is a maximum ≈ 203.3 inches^3.

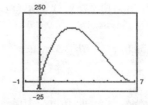

40. An open box is to be made from a 8-inch square piece of material by cutting equal squares from each corner and turning up the sides. Verify the volume of the box is $V(x) = 4x(4 - x)^2$. Sketch the graph of the function using a graphing utility and use the graph to estimate the volume of x for which $V(x)$ is maximum.

   2—T—Answer: $V(x) = x \cdot (8 - 2x) \cdot (8 - 2x)$

   $= x \cdot 2(4 - x) \cdot 2(4 - x)$

   $= 4x(4 - x)^2$

   When $x \approx 1.33$ inches, $V(x)$ is a maximum ≈ 37.9 inches^3.

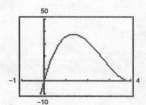

3.3 Polynomial and Synthetic Division

1. Divide $(9x^3 - 6x^2 - 8x - 3) \div (3x + 2)$

   (a) $3x^2 - \dfrac{8}{3}x - \dfrac{7/3}{3x+2}$

   (b) $3x^2 - 4x - 2 + \dfrac{7}{3x+2}$

   (c) $3x^2 - 4x - \dfrac{3}{3x+2}$

   (d) $3x^2 - 4x - \dfrac{16}{3} + \dfrac{23/3}{3x+2}$

   (e) None of these

   1—Answer: c

2. Divide: $(6x^3 + 7x^2 - 15x + 6) \div (2x - 1)$

   (a) $3x^2 + 2x - \dfrac{17}{2} - \dfrac{5}{2(2x-1)}$

   (b) $3x^2 + 5x - 5 + \dfrac{1}{2x-1}$

   (c) $3x^2 + 5x + 5 + \dfrac{11}{2x-1}$

   (d) $3x^2 + 4x - 17 + \dfrac{29/2}{2x-1}$

   (e) None of these

   1—Answer: b

3. Divide: $(3x^4 + 2x^3 - 3x + 1) \div (x^2 + 1)$

   (a) $3x^2 + 2x + 3 - \dfrac{5x+2}{x^2+1}$

   (b) $3x^2 + 2x - 3 + \dfrac{-5x+4}{x^2+1}$

   (c) $3x^2 - x^2 - 4 + \dfrac{5}{x^2+1}$

   (d) $3x^2 - x + 1 + \dfrac{-4x+5}{x^2+1}$

   (e) None of these

   1—Answer: b

4. Divide: $(6x^4 - 4x^3 + x^2 + 10x - 1) \div (3x + 1)$

   1—Answer: $2x^3 - 2x^2 + x + 3 - \dfrac{4}{3x+1}$

5. Divide: $(2x^4 + 7x - 2) \div (x^2 + 3)$

   2—Answer: $2x^2 - 6 + \dfrac{7x+16}{x^2+3}$

6. Divide by long division: $(2x^3 - x^2 - 3x + 4) \div (2x + 1)$

   (a) $x^2 - x - 1 + \dfrac{5}{2x+1}$

   (b) $x^2 - \dfrac{3}{2} + \dfrac{5}{2(2x+1)}$

   (c) $x^2 - x - 1 + \dfrac{3}{2x+1}$

   (d) $x^2 - x - 2 + \dfrac{6}{2x+1}$

   (e) None of these

   1—Answer: a

7. Divide by long division: $(x^4 + 3x^3 - 3x^2 - 12x - 4) \div (x^2 + 3x + 1)$

   (a) $x^2 - 2 - \dfrac{6x+2}{x^2+3x+1}$

   (b) $x^2 - 4$

   (c) $x^2 + 2 - \dfrac{18x-6}{x^2+3x+1}$

   (d) $x^2 - 4x$

   (e) None of these

   1—Answer: b

Chapter 3 Zeros of Polynomial Functions

8. Use synthetic division to divide: $(5x^4 - 2x^2 + 1) \div (x + 1)$

   (a) $5x^3 - 5x^2 + 3x - 3 + \dfrac{4}{x+1}$ (b) $5x^2 - 7x + 8$

   (c) $5x^2 + 3x + 4$ (d) $5x^3 + 5x^2 + 3x + 3 + \dfrac{4}{x+1}$

   (e) None of these

   1—Answer: a

9. Use synthetic division to divide: $(3x^4 + 4x^3 - 2x^2 + 1) \div (x + 2)$

   (a) $3x^3 + 10x^2 + 18x + 37$ (b) $3x^3 - 2x^2 + 2x - 3$

   (c) $3x^3 - 2x^2 + 2x - 4 + \dfrac{9}{x+2}$ (d) $3x^3 + 10x^2 + 18x + 36 + \dfrac{73}{x+2}$

   (e) None of these

   1—Answer: c

10. Use synthetic division to divide: $(x^4 + 2x^2 - x + 1) \div (x - 2)$

    1—Answer: $x^3 + 2x^2 + 6x + 11 + \dfrac{23}{x-2}$

11. Divide: $(x^3 + 4x^2 + 4x + 3) \div (x + 3)$

    (a) $x^2 + 7x - 17 + \dfrac{54}{x+3}$ (b) $x^2 - x + 7 - \dfrac{21}{x+3}$ (c) $x^2 + x + 1$

    (d) $x^2 + x + 1 + \dfrac{6}{x+3}$ (e) None of these

    1—Answer: c

12. Divide: $(x^3 + 8) \div (x + 2)$

    (a) $x^2 - 2x + 4$ (b) $x^2 + 4$ (c) $x^2 + 2x + 4$

    (d) $x^2 + 2x - 4$ (e) None of these

    1—Answer: a

13. Divide by synthetic division: $(x^3 - 6x^2 - 3x + 1) \div (x + 2)$

    (a) $x^2 - 4x - 11 - \dfrac{21}{x+2}$ (b) $x^2 - 8x + 13 - \dfrac{25}{x+2}$ (c) $x^2 - 4x + 5 - \dfrac{9}{x+2}$

    (d) $x^2 - 8x + 13 - \dfrac{27}{x+2}$ (e) None of these

    1—Answer: b

14. Divide by synthetic division: $(2x^3 + 3x^2 - 19x - 1) \div (x + 4)$

    (a) $2x^2 - 5x + 1 - \dfrac{3}{x+4}$ (b) $2x^2 - x - 15 + \dfrac{54}{x+4}$ (c) $2x^2 - 5x + 1 - \dfrac{5}{x+4}$

    (d) $2x^2 + 11x + 25 + \dfrac{99}{x+4}$ (e) None of these

    1—Answer: c

15. Divide using synthetic division: $(x^3 - x - 6) \div (x - 2)$.

(a) $x^2 - x - 3$ (b) $x^2 + 2x + 3$ (c) $x^2 + x + 2 + \dfrac{2}{x-2}$

(d) $x^2 + 2x + 5$ (e) None of these

1—Answer: b

16. Use synthetic division to determine which of the following is a solution of the equation: $3x^4 - 2x^3 + 26x^2 - 18x - 9 = 0$

(a) 3 (b) 1 (c) -3 (d) $\frac{1}{3}$ (e) None of these

2—Answer: b

17. Use synthetic division to determine which of the following are solutions of the equation: $3x^3 - 11x^2 - 6x + 8 = 0$

(a) $\frac{2}{3}$ (b) -1 (c) 4 (d) All of these (e) None of these

1—Answer: d

18. Use synthetic division to determine which of the following is a solution of the equation: $6x^4 - 11x^3 - 10x^2 + 19x - 6 = 0$

(a) 2 (b) 3 (c) -2 (d) -3 (e) None of these

1—Answer: a

19. Use synthetic division of factor completely: $x^3 - x^2 - 10x - 8$ [Hint: -2 is a zero.]

(a) $(x-2)(x-4)(x+1)$ (b) $-2, 4, -1$ (c) $(x+2)(x+1)(x-4)$

(d) $(x+2)(x+4)(x-1)$ (e) Does not factor

1—Answer: c

20. Use synthetic division to factor the polynomial $x^3 - 4x^2 - 7x + 10$ completely if -2 is a zero.

(a) $(x+2)(x+1)(x+5)$ (b) $-2, 1, 5$ (c) $(x+2)(x-1)(x-5)$

(d) $(x-2)(x-1)(x+5)$ (e) Does not factor

1—Answer: c

21. Use synthetic division to factor the polynomial $x^3 + 4x^2 + x - 6$ completely if 1 is a zero.

(a) $(x+1)(x+2)(x+3)$ (b) $(x-1)(x-2)(x-3)$ (c) $1, -2, -3$

(d) $(x-1)(x+2)(x+3)$ (e) Does not factor

1—Answer: d

220   Chapter 3   Zeros of Polynomial Functions

22. Use synthetic division to factor the polynomial $2x^3 - 7x^2 + 7x - 2$ completely if $\frac{1}{2}$ is a zero.

   (a) $2(x + \frac{1}{2})(x - 1)(x - 2)$ (b) $2(x - \frac{1}{2})(x + 1)(x + 2)$ (c) $2(x - \frac{1}{2})(x - 1)(x - 2)$
   (d) $2(x + \frac{1}{2})(x + 1)(x + 2)$ (e) Does not factor

   2—Answer: c

23. Factor the polynomial $x^3 + 3x^2 - 10x - 24$ completely knowing that $x - 3$ is a factor.

   (a) $(x - 3)(x + 2)(x + 4)$ (b) $(x - 3)(x - 2)(x - 4)$ (c) $(x - 3)(x + 1)(x + 7)$
   (d) $(x - 3)(x - 1)(x + 7)$ (e) None of these

   1—Answer: a

24. Express $f(x) = 3x^4 - 2x^2 + x - 1$ in the form $f(x) = (x - k)q(x) + r$ for $k = -1$.

   (a) $f(x) = (x - 1)(3x^3 + 3x^2 + x + 2) + 1$ (b) $f(x) = (x + 1)(3x^3 - 3x^2 + x) - 1$
   (c) $f(x) = (x - 1)(3x^3 + x^2 + 2x) + 1$ (d) $f(x) = (x + 1)(3x^3 - 5x^2 + 6x) - 7$
   (e) None of these

   1—Answer: b

25. Express $f(x) = 2x^3 - 3x + 2$ in the form $f(x) = (x - k)q(x) + r$ for $k = -2$.

   (a) $f(x) = (x - 2)(2x^2 - x) + 4$ (b) $f(x) = (x + 2)(2x^2 - 7x) + 16$
   (c) $f(x) = (x - 2)(2x^2 + 4x + 5)$ (d) $f(x) = (x + 2)(2x^2 - 4x + 5) - 8$
   (e) None of these

   1—Answer: d

26. Express $f(x) = 3x^4 - 7x^3 + x - 1$ in the form $f(x) = (x - k)q(x) + r$ for $k = 2$.

   (a) $f(x) = (x + 2)(3x^3 - 13x^2 + 26x - 51) + 101$ (b) $f(x) = (x - 2)(3x^3 - x^2 - 2x - 3) - 7$
   (c) $f(x) = (x + 2)(3x^3 - 13x^2 - 25x) + 49$ (d) $f(x) = (x - 2)(3x^3 - x^2 - x) - 3$
   (e) None of these

   1—Answer: b

27. Express $f(x) = 2x^3 - 5x^2 + 3$ in the form $f(x) = (x - k)q(x) + r$ for $k = -2$.

   (a) $f(x) = (x + 2)(2x^2 - 9x + 18) - 33$ (b) $f(x) = (x - 2)(2x^2 - x - 2) - 1$
   (c) $f(x) = (x + 2)(2x^2 - 9x) + 21$ (d) $f(x) = (x - 2)(2x^2 - x) + 1$
   (e) None of these

   1—Answer: a

28. Use synthetic division to find $f(-2)$: $f(x) = 4x^3 + 3x + 10$

(a) 20 (b) -20 (c) 36

(d) -28 (e) None of these

2—Answer: d

29. Use synthetic division to find $f(-3)$: $f(x) = 3x^3 + 2x^2 - 1$

(a) 98 (b) $3x^2 - 7x + 21 - \dfrac{64}{x+3}$ (c) -64

(d) 20 (e) None of these

1—Answer: c

30. Use synthetic division to find $f(3)$: $f(x) = x^4 + 2x^2 - x - 1$

(a) $x^3 + 3x^2 + 11x + 32 + \dfrac{95}{x-3}$ (b) 95

(c) 101 (d) 122 (e) None of these

1—Answer: b

31. Use synthetic division to find $f(-2)$: $f(x) = 3x^4 - 2x^2 + 1$

(a) 23 (b) -16 (c) 17 (d) 21 (e) None of these

1—Answer: e

32. Simplify the rational function: $f(x) = \dfrac{x^3 + 4x^2 - 3x + 10}{x+5}$

(a) 0 (b) $x^2 - x + 2$ (c) $5x^2 - 3x + 2$

(d) $x^2 + 9x + 42$ (e) None of these

1—Answer: b

33. Simplify the rational function: $f(x) = \dfrac{x^5 - 1}{x-1}$

(a) x^4 (b) $x^4 + x^3 + x^2 + x$ (c) $x^4 + x^3 + x^2 + x + 1$

(d) $x^4 - x^3 + x^2 - x + 1$ (e) None of these

1—Answer: c

34. Simplify the rational expression: $\dfrac{2x^4 - x^3 - 4x^2 + x - 3}{2x + 3}$

(a) $x^3 + x^2 - x + 1$ (b) $x^3 - 2x^2 - 5x - 1$ (c) $x^3 + x^2 - 5x - 1$

(d) $(x^3 - 2x^2 + x - 1)$ (e) None of these

1—Answer: d

Chapter 3 Zeros of Polynomial Functions

35. Simplify the rational expression: $\dfrac{4x^5 - 3x^4 + 8x^3 - 3x^2 + 6}{x^2 + 2}$

    (a) $4x^3 - 3x^2 + 6x + 3 + \dfrac{12x}{x^2 + 2}$ (b) $4x^3 - 3x^2 + 3$

    (c) $4x^3 - 3x^2 - 9 + \dfrac{24x}{x^2 + 2}$ (d) $4x^3 + 5x^2 + 6$

    (e) None of these

    2—Answer: b

36. A rectangular room has a volume of $3x^3 - 2x^2 - 11x + 10$ cubic feet. The height of the room is $x - 1$. Find the algebraic expression for the number of square feet of floor space in the room.

    (a) 20   (b) $3x^3 - 2x^2 - 10x + 9$ (c) $3x^2 + x - 10$

    (d) $3x^2 - 5x - 5 + \dfrac{15}{x - 1}$ (e) None of these

    2—Answer: c

37. A rectangular room has a volume of $3x^3 + 20x^2 + 27x + 10$ cubic feet. The height of the room is $x + 1$. Find the algebraic expression for the number of square feet of floor space in the room.

    (a) 126   (b) $3x^3 + 20x^2 + 26x + 9$ (c) $3x^2 + 23x + 50 + \dfrac{60}{x + 1}$

    (d) $3x^2 + 17x + 10$ (e) None of these

    2—Answer: d

38. A rectangular room has a volume of $3x^3 + 8x^2 + 5x + 2$ cubic feet. The height of the room is $x + 2$. Find the algebraic expression for the number of square feet of floor space in the room.

    (a) 86   (b) $3x^3 + 8x^2 + 4x$ (c) $3x^2 + 2x + 1$

    (d) $3x^2 + 14x + 33 + \dfrac{68}{x + 2}$ (e) None of these

    2—Answer: c

39. A rectangular room has a volume of $4x^3 - 7x^2 - 16x + 3$ cubic feet. The height of the room is $x - 3$. Find the algebraic expression for the number of square feet of floor space in the room.

    2—Answer: $4x^2 + 5x - 1$

40. A rectangular room has a volume of $2x^3 - 17x + 3$ cubic feet. The height of the room is $x + 3$. Find the algebraic expression for the number of square feet of floor space in the room.

    2—Answer: $2x^2 - 6x + 1$

41. Use the root-finding capabilities of a graphing utility to approximate the indicated zero. Use synthetic division to verify your result and then factor the polynomial completely: $f(x) = 6x^3 + 17x^2 - 4x - 3$

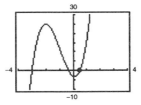

2—T—Answer: $x = \frac{1}{2}$; $f(x) = (2x - 1)(x + 3)(3x + 1)$

42. Use the root-finding capabilities of a graphing utility to approximate the indicated zero. Use synthetic division to verify your result and then factor the polynomial completely: $f(x) = 10x^3 + 3x^2 - 16x + 3$

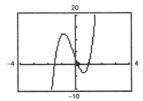

2—T—Answer: $x = \frac{1}{5}$; $f(x) = (5x - 1)(2x + 3)(x - 1)$

43. Use the root-finding capabilities of a graphing utility to approximate the indicated zero. Use synthetic division to verify your result and then factor the polynomial completely: $f(x) = 3x^3 + 2x^2 - 7x + 2$

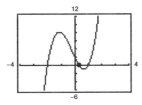

2—T—Answer: $x = \frac{1}{3}$; $f(x) = (3x - 1)(x + 2)(x - 1)$

44. Use the root-finding capabilities of a graphing utility to approximate the indicated zero. Use synthetic division to verify your result and then factor the polynomial completely: $f(x) = 6x^3 + 13x^2 + 4$

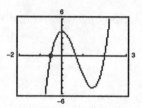

2—T—Answer: $x = -\frac{1}{2}$; $f(x) = (2x + 1)(x - 2)(3x - 2)$

❑ 3.4 Real Zeros of Polynomial Functions

1. Use Descarte's Rule of Signs to determine the possible number of positive and negative zeros: $f(x) = 5x^4 - 3x^3 - 4x + 2$

    (a) 2 positive, 2 negative
    (b) 2 or 0 positive, 0 negative
    (c) 4 positive, 0 negative
    (d) 0 positive, 4 negative
    (e) None of these

    1—Answer: b

2. Use Descarte's Rule of Signs to determine the possible number of positive and negative zeros: $f(x) = x^3 + 2x - 1$

    (a) 1 positive, 0 negative
    (b) 0 positive, 1 negative
    (c) 3 or 1 positive, 0 or 2 negative
    (d) 1 positive, 2 negative
    (e) None of these

    1—Answer: a

3. Use Descarte's Rule of Signs to determine the possible number of positive and negative zeros: $f(x) = 6x^5 - 6x^3 + 10x + 5$

    (a) 4 or 2 or 0 positive, 1 negative
    (b) 3 or 1 positive, 2 or 4 negative
    (c) 2 or 0 positive, 3 or 1 negative
    (d) 1 positive, 0 negative
    (e) None of these

    1—Answer: c

Section 3.4   Real Zeros of Polynomial Functions   225

4. Use Descarte's Rule of Signs to determine the possible number of positive and negative zeros: $f(x) = x^3 + 1$

   (a) 3 positive, 0 negative
   (b) 0 positive, 1 negative
   (c) 1 positive, 2 negative
   (d) 1 positive, 0 negative
   (e) None of these

   1—Answer: b

5. Use Descarte's Rule of Signs to determine which of the following polynomial functions has a possible 0 or 2 positive zeros and 0 negative zeros.

   (a) $f(x) = 5x^4 - 3x^3 - 4x + 2$
   (b) $f(x) = 4x^4 - 3x^3 - x^2 + 1$
   (c) $f(x) = 3x^4 - 4x^2 - 4x + 2$
   (d) Both a and b
   (e) Both b and c

   1—Answer: a

6. Use Descarte's Rule of Signs to determine which of the following polynomial functions has a possible 0 or 2 negative zeros and 0 positive zeros.

   (a) $f(x) = 3x^4 + x^3 + 4x^2 + x$
   (b) $f(x) = 5x^4 + 3x^3 + x^2 + 1$
   (c) $f(x) = 2x^4 - 4x^2 + x$
   (d) Both a and b
   (e) Both b and c

   1—Answer: b

7. Given $f(x) = x^4 - 2x^3 + x^2 - x - 5$, determine the possible number of negative zeros.

   (a) None
   (b) Either 3 or 1
   (c) Exactly one
   (d) Either 2 or 0
   (e) None of these

   1—Answer: c

8. Given $f(x) = x^4 - 3x^3 + x^2 - 6x - 5$, determine the possible number of negative zeros.

   (a) None
   (b) Either 3 or 1
   (c) Exactly one
   (d) Either 2 or 0
   (e) None of these

   1—Answer: c

9. List the possible rational zeros of the function: $f(x) = 3x^5 + 2x^2 - 3x + 2$

   (a) $\pm 3, \pm 2, \pm \frac{3}{2}, \pm 1, \pm \frac{2}{3}$
   (b) $\pm 3, \pm \frac{1}{3}, \pm 2, \pm \frac{1}{2}, \pm 1$
   (c) $\pm 2, \pm 1, \pm \frac{2}{3}, \pm \frac{1}{3}$
   (d) $\pm 3, \pm 1, \pm \frac{3}{2}, \pm \frac{1}{2}$
   (e) None of these

   1—Answer: c

10. List the possible rational zeros of the function: $f(x) = 3x^5 - 2x^3 + 3x - 5$

    (a) $\pm\frac{5}{3}, \pm 3, \pm\frac{1}{3}, \pm 5, \pm 1, \pm\frac{1}{5}, \pm\frac{3}{5}$
    (b) $\pm\frac{3}{5}, \pm 1, \pm\frac{1}{5}, \pm 3$
    (c) $\pm\frac{1}{3}, \pm 1, \pm\frac{5}{3}, \pm 5$
    (d) $\pm 1, \pm 3, \pm\frac{3}{5}, \pm\frac{5}{3}$
    (e) None of these

    1—Answer: c

11. List the possible rational zeros of the function: $f(x) = 3x^5 + 7x^3 - 3x^2 + 2$

    (a) $\pm\frac{2}{3}, \pm\frac{3}{2}, \pm 2, \pm 3$
    (b) $\pm\frac{1}{3}, \pm\frac{2}{3}, \pm 1, \pm 2$
    (c) $\pm\frac{3}{2}, \pm\frac{1}{2}, \pm 3, \pm 1$
    (d) $\pm\frac{3}{2}, \pm\frac{2}{3}, \pm\frac{1}{2}, \pm\frac{1}{3}$
    (e) None of these

    1—Answer: b

12. Which of the following is *not* a possible rational zero of: $f(x) = 2x^3 + 5x^2 + 3$

    (a) $\frac{1}{2}$
    (b) $-\frac{3}{2}$
    (c) $\frac{2}{3}$
    (d) 3
    (e) All of these are possible rational zeros.

    1—Answer: c

13. Which of the following is *not* a possible rational zero of: $f(x) = 5x^3 + 3x^2 + 2$

    (a) $-\frac{5}{2}$
    (b) 2
    (c) $\frac{2}{5}$
    (d) $\frac{1}{5}$
    (e) All of these are possible.

    1—Answer: a

14. List the possible rational zeros of the function: $f(x) = 3x^5 + 2x^2 - 3x + 2$

    (a) $\pm\frac{2}{3}, \pm 1, \pm\frac{3}{4}, \pm 2, \pm 3$
    (b) $\pm\frac{1}{3}, \pm\frac{1}{2}, \pm 1, \pm 2, \pm 3$
    (c) $\pm\frac{1}{3}, \pm\frac{2}{3}, \pm 1, \pm 2$
    (d) $\pm\frac{1}{2}, \pm 1, \pm\frac{3}{2}, \pm 3$
    (e) None of these

    1—Answer: c

15. List the possible rational zeros of the function: $f(x) = 2x^4 - 3x^2 + 15$

    1—Answer: $\pm\frac{1}{2}, \pm 1, \pm\frac{3}{2}, \pm\frac{5}{2}, \pm 3, \pm 5, \pm\frac{15}{2}, \pm 15$

16. Given $f(x) = 3x^3 + 4x - 1$, determine whether $x = -2$ is an upper bound for the zeros of f, a lower bound for the zeros of f, or neither.

    2—Answer: Lower bound

17. Which of the following are upper bounds for the zeros of f:
    $f(x) = x^5 + 2x^4 - x^3 - 2x^2 - 30x - 60$

    (a) 1
    (b) 2
    (c) 3
    (d) Both b and c
    (e) None of these

    2—Answer: c

18. Which of the following are lower bounds for the zeros of f:
    $f(x) = 6x^4 + 3x^3 + 5x - 10$

    (a) 0   (b) -1 (c) -2

    (d) Both -1 and -2 (e) None of these

    2—Answer: c

19. Which of the following are upper bounds for the zeros of f:
    $f(x) = 3x^3 + x^2 - 7x + 2$

    (a) 1   (b) 2   (c) 3   (d) Both b and c   (e) None of these

    2—Answer: d

20. Which of the following are lower bounds for the zeros of f:
    $f(x) = 3x^3 + x^2 - 11x - 5$

    (a) -3 (b) -2 (c) -1 (d) All of these   (e) None of these

    2—Answer: a

21. Which of the following are lower bounds for the zeros of f:
    $f(x) = 3x^4 + x^2 + 3x - 1$

    (a) -3 (b) -2 (c) -1 (d) All of these   (e) None of these

    2—Answer: d

22. Find all of the real zeros of the function: $f(x) = 2x^3 + 14x^2 + 24x$

    (a) 0, 3, 4   (b) 3, 4   (c) $-4, -3, 0$

    (d) 0, 1, 6   (e) None of these

    1—Answer: c

23. Find all of the real zeros of the function: $f(x) = 3x^4 - 27x^3 + 54x^2$

    (a) 0, 3, 9, 2   (b) 0, 6, 3   (c) 0, 9, 2

    (d) 0, 6   (e) None of these

    1—Answer: b

24. Find all of the real zeros of the function: $f(x) = 6x^4 + 32x^3 - 70x^2$

    (a) 0, -1, 5   (b) 0, $-7, \frac{5}{3}$ (c) $\frac{7}{3}, 5$

    (d) 0, $-1, -7, \frac{5}{3}$ (e) None of these

    1—Answer: b

Chapter 3 Zeros of Polynomial Functions

25. Find all of the real zeros of the function: $f(x) = x^3 + 6x^2 + 12x + 7$

(a) $\dfrac{-5 \pm \sqrt{3}}{2}$ (b) $-1, \dfrac{-5 \pm \sqrt{3}}{2}$ (c) -1

(d) $-1, 7$ (e) None of these

2—Answer: c

26. Find all of the real zeros of the function: $f(x) = 4x^3 - 3x - 1$

(a) $1, -\frac{1}{2}$ (b) $1, \frac{1}{2}, -\frac{1}{2}$ (c) $\frac{1}{2}, 1$

(d) 1 (e) None of these

2—Answer: a

27. Find all of the real roots: $x^3 - 7x + 6 = 0$

(a) $-3, 1, 2$ (b) $-2, -1, 3$ (c) $-6, -1, 1$

(d) $-1, 1, 6$ (e) None of these

1—Answer: a

28. Find all of the real roots: $2x^3 + 5x^2 - x - 6 = 0$

(a) $-3, -1, 1$ (b) $-1, \frac{3}{2}, 2$ (c) $-2, -\frac{3}{2}, 1$

(d) $-6, 2, 5$ (e) None of these

1—Answer: c

29. Find all of the real roots: $x^3 - 5x^2 + 5x - 1 = 0$

(a) $1, 2, \pm 2\sqrt{3}$ (b) $1, 2 \pm \sqrt{3}$ (c) 1

(d) $-1, 2 \pm 2\sqrt{3}$ (e) None of these

2—Answer: b

30. Find all of the real roots: $3x^4 - 4x^3 + 4x^2 - 4x + 1 = 0$

(a) $-1, \frac{1}{3}, 1$ (b) $-1, -\frac{1}{3}, 1$ (c) $\frac{1}{3}, 1$ (d) 1 (e) None of these

1—Answer: c

31. Find all of the real roots: $x^3 + 8x^2 + 17x + 6 = 0$

(a) -3 (b) $-3, \dfrac{-5 \pm \sqrt{17}}{2}$ (c) $-3, -1, 2$

(d) $-3, \dfrac{-5 \pm \sqrt{33}}{2}$ (e) None of these

2—Answer: b

32. Find the real zeros of the function: $f(x) = x^3 - \frac{4}{3}x^2 - \frac{5}{3}x + \frac{2}{3}$

(a) $-1, \frac{1}{3}, 2$ (b) $1, \frac{2}{3}, -2$ (c) $-2, -\frac{1}{3}, 1$

(d) $-1, \frac{2}{3}, 1$ (e) None of these

2—Answer: a

33. Find all of the real zeros of the function: $f(x) = x^3 - \frac{9}{2}x^2 + \frac{11}{2}x - \frac{3}{2}$

(a) $1, 4 \pm \sqrt{13}$ (b) $\frac{3}{2}, \frac{3 \pm \sqrt{13}}{2}$ (c) $\frac{3}{2}, \frac{3 \pm \sqrt{5}}{2}$

(d) $1, 4 \pm \sqrt{19}$ (e) None of these

2—Answer: c

34. Find all of the real zeros of the function: $f(x) = x^3 - \frac{11}{3}x^2 + \frac{5}{3}x + 1$

(a) $-\frac{1}{3}, 1, 3$ (b) $3, \pm 1$ (c) $1, \pm 3$

(d) $3, 1 \pm \sqrt{2}$ (e) None of these

2—Answer: a

35. List the possible rational zeros of the function. Then use a graphing utility to graph the function to eliminate some of the possible zeros. Finally determine all real zeros of the function: $f(x) = 3x^3 - x^2 - 12x + 4$

2—T—Answer: $\pm\frac{1}{3}, \pm\frac{2}{3}, \pm 1, \pm\frac{4}{3}, \pm 2, \pm 4$; 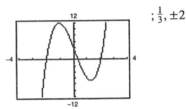 ; $\frac{1}{3}, \pm 2$

36. List the possible rational zeros of the function. Then use a graphing utility to graph the function to eliminate some of the possible zeros. Finally determine all real zeros of the function: $f(x) = 2x^3 - x^2 - 18x + 9$

2—T—Answer: $\pm\frac{1}{2}, \pm 1, \pm\frac{3}{2}, \pm 3, \pm\frac{9}{2}, \pm 9$; 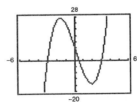 ; $\frac{1}{2}, \pm 3$

37. List the possible rational zeros of the function. Then use a graphing utility to graph the function to eliminate some of the possible zeros. Finally determine all real zeros of the function: $f(x) = 2x^3 - 7x^2 + x + 10$

2—T—Answer: $\pm\frac{1}{2}, \pm 1, \pm\frac{5}{2}, \pm 5, \pm 10$; 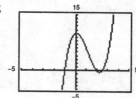 ; $-1, 2, \frac{5}{2}$

38. List the possible rational zeros of the function. Then use a graphing utility to graph the function to eliminate some of the possible zeros. Finally determine all real zeros of the function: $f(x) = 5x^3 - 12x^2 - 11x + 6$

2—T—Answer: $\pm\frac{1}{5}, \pm\frac{2}{5}, \pm\frac{3}{5}, \pm 1, \pm\frac{6}{5}, \pm 2, \pm 3, \pm 6$; 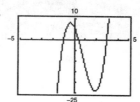 ; $-1, \frac{2}{5}, 3$

❑ 3.5 The Fundamental Theorem of Algebra

1. Write as a product of linear factors: $x^4 + 25x^2 + 144$
   - (a) $(x^2 + 9)(x^2 + 16)$
   - (b) $(x + 3i)(x + 3i)(x + 4i)(x + 4i)$
   - (c) $(x + 3i)(x - 3i)(x + 4i)(x - 4i)$
   - (d) $(x - 3i)(x - 3i)(x - 4i)(x - 4i)$
   - (e) None of these

   1—Answer: c

2. Write as a product of linear factors: $f(x) = x^4 - 3x^2 - 28$
   - (a) $(x^2 + 4)(x^2 - 7)$
   - (b) $(x - 2i)(x + 2i)(x - \sqrt{7})(x + \sqrt{7})$
   - (c) $(x + 2i)(x + 2i)(x + \sqrt{7})(x - \sqrt{7})$
   - (d) $(x - 2i)(x - 2i)(x - \sqrt{7})(x + \sqrt{7})$
   - (e) None of these

   1—Answer: b

3. Write as a product of linear factors: $f(x) = x^4 - 5x^3 + 8x^2 - 20x + 16$
   - (a) $(x + 2)(x - 2)(x - 4)(x - 1)$
   - (b) $(x + 4)(x + 1)(x - 2i)(x + 2i)$
   - (c) $(x - 4)(x - 1)(x + 2i)(x - 2i)$
   - (d) $(x + 4)(x + 1)(x + 2i)(x + 2i)$
   - (e) None of these

   2—Answer: c

4. Write as a product of linear factors: $x^2 - 16$

   1—**Answer:** $(x + 2)(x - 2)(x + 2i)(x - 2i)$

5. Write as a product of linear factors: $f(x) = x^4 - 100$

   1—**Answer:** $f(x) = (x + \sqrt{10})(x - \sqrt{10})(x + \sqrt{10}i)(x - \sqrt{10}i)$

6. Write as a product of linear factors: $f(x) = x^4 - 6x^3 - 4x^2 + 40x + 32$

   (a) $(x - 4)(x + 2)(x + 2 + \sqrt{8})(x + 2 - \sqrt{8})$ (b) $(x + 4)(x - 2)(x - 2 + \sqrt{8})(x - 2 - \sqrt{8})$
   (c) $(x - 4)(x - 2)(x - 2 + \sqrt{8})(x - 2 - \sqrt{8})$ (d) $(x + 4)(x + 2)(x + 2 + \sqrt{8})(x + 2 - \sqrt{8})$
   (e) None of these

   2—**Answer:** a

7. Write as a product of linear factors: $f(x) = x^4 + 2x^3 - 5x^2 - 18x - 36$

   (a) $(x + 3)(x + 3)(x + 1 + \sqrt{3})(x + 1 - \sqrt{3})$ (b) $(x - 3)(x - 3)(x - 1 + \sqrt{6})(x - 1 - \sqrt{6})$
   (c) $(x - 3)(x + 3)(x + 1 + \sqrt{3}i)(x + 1 - \sqrt{3}i)$ (d) $(x - 3)(x + 3)(x - 1 + \sqrt{3}i)(x - 1 - \sqrt{3}i)$
   (e) None of these

   2—**Answer:** c

8. Write as a product of linear factors: $f(x) = x^4 - 49$

   (a) $(x - \sqrt{7}i)(x + \sqrt{7}i)$ (b) $(x - 7i)^2(x + 7)^2$
   (c) $(x - \sqrt{7})^2(x + \sqrt{7})^2$ (d) $(x - \sqrt{7})(x + \sqrt{7})(x - \sqrt{7}i)(x + \sqrt{7}i)$
   (e) None of these

   1—**Answer:** d

9. Write as a product of linear factors: $f(x) = x^4 = 13x^2 + 36$

   (a) $(x - 2i)^2(x - 3i)^2$ (b) $(x - 2i)(x + 2i)(x - 3i)(x - 3i)$
   (c) $(x - \sqrt{2}i)(x + \sqrt{2}i)(x - \sqrt{3}i)(x + \sqrt{3}i)$ (d) $(x - \sqrt{2}i)^2(x - \sqrt{3}i)^2$
   (e) None of these

   1—**Answer:** b

10. Find a polynomial with real coefficients that has zeros: $0, 3, -3, i$, and $-i$

    (a) $f(x) = x^5 - 8x^3 - 9x$ (b) $f(x) = x^5 - 10x^3 + 9x$ (c) $f(x) = x^3 - 4x^2 + 3$
    (d) $f(x) = x^5 - 9x$ (e) None of these

    1—**Answer:** a

11. Find a fourth degree polynomial function that has zeros: $1, -1, 0$, and 2

    1—**Answer:** $f(x) = x^4 - 2x^3 - x^2 + 2x$

12. Find a fourth degree polynomial with real coefficients that has zeros: $1, -1, i, -i$

    (a) $x^4 + 1$ (b) $x^4 - 1$ (c) $x^4 + 2x^2 + 1$
    (d) $x^4 - 2x^2 + 1$ (e) None of these

    1—**Answer:** b

13. Find a third degree polynomial with real coefficients that has zeros: $0, 2, -i, 2+i$

    (a) $x^3 - 5x^2 + 4x$
    (b) $x^3 + 4x^2 - 5x$
    (c) $x^3 - 4x^2 + 5x$
    (d) $x^3 + 5x$
    (e) None of these

    1—Answer: c

14. Find a fourth degree polynomial with real coefficients that has zeros: $2, 3, \sqrt{2}i$

    (a) $x^4 - 5x^3 + 8x^2 - 10x + 12$
    (b) $x^4 - 5x^2 + 6$
    (c) $x^4 - 5x^3 + 6x^2$
    (d) $x^4 - 5x^3 + 8x^2 - 10x - 12$
    (e) None of these

    2—Answer: a

15. Find a fourth degree polynomial that has zeros: $1, -3, 2i$

    (a) $x^4 - 2x^3 + x^2 - 8x - 12$
    (b) $x^4 + 2x^3 - 7x^2 - 8x + 12$
    (c) $x^4 + 2x^3 + x^2 + 8x - 12$
    (d) $x^4 - 2x^3 - 7x^2 + 8x - 12$
    (e) None of these

    1—Answer: c

16. Find a fourth degree polynomial that has zeros: $3, -2, i$

    (a) $x^4 - x^3 - 5x^2 - x - 6$
    (b) $x^4 + x^3 + 5x^2 - x - 6$
    (c) $x^4 - x^3 + 5x^2 + x - 6$
    (d) $x^4 + x^3 - 5x^2 + x - 6$
    (e) None of these

    1—Answer: a

17. Find a third degree polynomial that has zeros: $-2, -4i$

    (a) $x^3 - 4x^2 + 4x - 32$
    (b) $x^3 + 4x^2 - 4x + 16$
    (c) $x^3 - 2x^2 - 14x + 32$
    (d) $x^3 + 2x^2 + 16x + 32$
    (e) None of these

    1—Answer: d

18. Find a third degree polynomial with zeros: 6 and $-2i$

    (a) $x^3 + 6x^2 + 2x + 12$
    (b) $x^3 - 6x^2 + 4x - 24$
    (c) $x^3 - 6x^2 + 2x - 12$
    (d) $x^3 + 6x^2 + 4x + 24$
    (e) None of these

    1—Answer: b

19. Write the polynomial as a product of factors irreducible over the rational numbers: $f(x) = x^4 - 3x^2 - 28$

    (a) $(x^2 + 4)(x^2 - 7)$
    (b) $(x - 2i)(x + 2i)(x - \sqrt{7})(x + \sqrt{7})$
    (c) $(x^2 + 4)(x - \sqrt{7})(x + \sqrt{7})$
    (d) $(x - 2i)(x + 2i)(x^2 - 7)$
    (e) None of these

    1—Answer: a

20. Write the polynomial as a product of factors irreducible over the rational numbers: $f(x) = x^4 + 4x^2 - 45$

(a) $(x^2 - 5)(x - 3i)(x + 3i)$
(b) $(x^2 - 5)(x^2 + 9)$
(c) $(x - \sqrt{5})(x + \sqrt{5})(x - 3i)(x + 3i)$
(d) $(x - \sqrt{5})(x + \sqrt{5})(x^2 + 9)$
(e) None of these

1—Answer: b

21. Write the polynomial as a product of factors irreducible over the rational numbers: $f(x) = x^4 - 1$

(a) $(x^2 - 1)(x^2 + 1)$
(b) $(x^2 - 1)(x - i)(x + i)$
(c) $(x - 1)(x + 1)(x - i)(x + i)$
(d) $(x - 1)(x + 1)(x^2 + 1)$
(e) None of these

1—Answer: d

22. Write the polynomial as a product of factors irreducible over the real numbers: $f(x) = x^4 - 3x^2 - 28$

(a) $(x^2 + 4)(x^2 - 7)$
(b) $(x - 2i)(x + 2i)(x - \sqrt{7})(x + \sqrt{7})$
(c) $(x^2 + 4)(x - \sqrt{7})(x + \sqrt{7})$
(d) $(x - 2i)(x + 2i)(x^2 - 7)$
(e) None of these

1—Answer: c

23. Write the polynomial as a product of factors irreducible over the real numbers: $f(x) = x^4 + 4x^2 - 45$

(a) $(x^2 - 5)(x - 3i)(x + 3i)$
(b) $(x^2 - 5)(x^2 + 9)$
(c) $(x - \sqrt{5})(x + \sqrt{5})(x - 3i)(x + 3i)$
(d) $(x - \sqrt{5})(x + \sqrt{5})(x^2 + 9)$
(e) None of these

1—Answer: d

24. Write the polynomial as a product of factors irreducible over the real numbers: $f(x) = x^4 + 23x^2 - 50$

(a) $(x^2 + 25)(x^2 - 2)$
(b) $(x^2 - 2)(x - 5i)(x + 5i)$
(c) $(x - \sqrt{2})(x + \sqrt{2})(x^2 + 25)$
(d) $(x - \sqrt{2})(x + \sqrt{2})(x - 5i)(x + 5i)$
(e) None of these

1—Answer: c

25. Write the polynomial in completely factored form: $f(x) = x^4 - 16$

(a) $(x^2 - 4)(x^2 + 4)$
(b) $(x - 2)(x + 2)(x^2 + 4)$
(c) $(x - 2)(x + 2)(x - 2i)(x + 2i)$
(d) $(x^2 - 4)(x - 2i)(x + 2i)$
(e) None of these

1—Answer: c

26. Write the polynomial f in completely factored form:
$f(x) = 3x^4 - 4x^3 + 4x^2 - 4x + 1$

   2—Answer: $f(x) = (x - 1)(3x - 1)(x + i)(x - i)$

27. Write the polynomial f in completely factored form: $f(x) = x^4 - 2x^3 - x^2 + 2x$

   1—Answer: $f(x) = x(x - 1)(x + 1)(x - 2)$

28. Write the polynomial in completely factored form: $f(x) = x^3 - x + 6$
   - (a) $(x + 2)(x^2 - 2x + 3)$
   - (b) $(x + 2)(x - 1 + \sqrt{2}i)(x + 1 - \sqrt{2}i)$
   - (c) $(x + 2)(x - 1 - \sqrt{2}i)(x - 1 + \sqrt{2}i)$
   - (d) $(x + 2)(x + 1 - \sqrt{2}i)(x + 1 + \sqrt{2}i)$
   - (e) None of these

   2—Answer: c

29. Use the fact that $3i$ is a zero of f to find the remaining zeros:
$f(x) = x^4 - 6x^3 + 14x^2 - 54x + 45$
   - (a) $0, \pm 3i$
   - (b) $-1, -5, \pm 3i$
   - (c) $2, 3, \pm 3i$
   - (d) $1, 5, \pm 3i$
   - (e) None of these

   1—Answer: d

30. Use the fact that i is a zero of f to find the remaining zeros:
$f(x) = x^4 - 5x^3 + 7x^2 - 5x + 6$

   2—Answer: $2, 3, \pm i$

31. Use the fact that $1 - 2i$ is zero of f to find the remaining zeros:
$f(x) = x^3 - 3x^2 + 7x - 5$

   1—Answer: $1, 1 \pm 2i$

32. Use the fact that $2 \pm \sqrt{3}i$ is a zero of f to find the remaining zeros:
$f(x) = x^4 - 6x^3 + 12x^2 - 2x - 21$

   2—Answer: $-1, 3, 2 \pm \sqrt{3}i$

33. Find all of the zeros of the function: $f(x) = x^4 + 25x^2 + 144$
   - (a) $\pm 2\sqrt{3}, \pm 5$
   - (b) $-3i, -3i, -4i, -4i$
   - (c) $\pm 3i, \pm 4i$
   - (d) $3i, 3i, 4i, 4i$
   - (e) None of these

   1—Answer: c

34. Find all of the zeros of the function: $f(x) = x^3 - \frac{9}{2}x^2 + \frac{11}{2}x - \frac{3}{2}$
   - (a) $\frac{3}{2}, \frac{3 \pm \sqrt{5}}{2}$
   - (b) $1, -2, 3$
   - (c) $\frac{3}{2}, \frac{3 \pm \sqrt{5}i}{2}$
   - (d) $\frac{3}{2}, 1 \pm i$
   - (e) None of these

   1—Answer: a

35. Find all of the zeros of the function: $f(x) = x^3 + 6x^2 + 12x + 7$

   (a) $-1, 7$
   (b) $-1, \dfrac{-5 \pm \sqrt{3}}{2}$
   (c) -1
   (d) $-1, \dfrac{-5\sqrt{3}i}{2}$
   (e) None of these

   1—Answer: d

36. Find all of the zeros of the function: $f(x) = x^4 - 5x^3 + 8x^2 - 20x + 16$

   (a) $1, 4, \pm 2$
   (b) $-4, -1, \pm 2i$
   (c) $1, 4, \pm 2i$
   (d) $-4, -1, -2i, -2i$
   (e) None of these

   2—Answer: c

37. Find all of the zeros of the function: $f(x) = 3x^4 - 7x^3 + 21x^2 - 63x - 54$

   2—Answer: $-\frac{2}{3}, 3, \pm 3i$

38. Find all the zeros of the function: $f(x) = x^4 + 2x^3 + 3x^2 - 2x - 4$. Use a graphing utility to graph the function to discard any rational zeros that are obviously not zeros of the function.

   2—T—Answer: 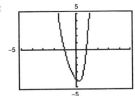 $; \pm 1; -1 \pm \sqrt{3}i$

39. Find all the zeros of the function: $f(x) = x^4 + 2x^3 + x^2 - 8x - 20$. Use a graphing utility to graph the function to discard any rational zeros that are obviously not zeros of the function.

   2—T—Answer: 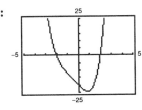 $; \pm 2, -1 \pm 2i$

40. Find all the zeros of the function: $f(x) = x^4 + 2x^3 + 2x^2 - 8x - 24$. Use a graphing utility to graph the function to discard any rational zeros that are obviously not zeros of the function.

   2—T—Answer: 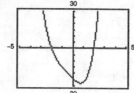 ; $\pm 2, -1 \pm \sqrt{5}i$

41. Find all the zeros of the function: $f(x) = x^4 - 2x^3 + 5x^2 - 8x + 4$. Use a graphing utility to graph the function to discard any rational zeros that are obviously not zeros of the function.

   2—T—Answer: 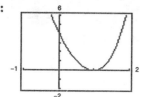 ; $1, \pm 2i$

❑ 3.6 Mathematical Modeling

1. Find a mathematical model for the statement "y varies directly as the cube root of x."

   (a) $y = \dfrac{k}{\sqrt[3]{x}}$ (b) $y = k\sqrt[3]{x}$ (c) $y\sqrt[3]{x} = 1$

   (d) $\dfrac{y}{x^3} = 1$ (e) None of these

   1—Answer: b

2. Find a mathematical model for the statement "y varies directly as the cube of x."

   (a) $y = kx^3$ (b) $y = \dfrac{k}{x^3}$ (c) $x^3y = 1$

   (d) $\dfrac{\sqrt[3]{x}}{y} = 1$ (e) None of these

   1—Answer: a

3. Find a mathematical model for the statement "y is inversely proportional to $x + 3$."

   (a) $y = k(x + 3)$ (b) $y = \dfrac{k}{(x + 3)}$ (c) $\dfrac{y}{x + 3} = 1$

   (d) $xy + 3 = k$ (e) None of these

   1—Answer: b

Section 3.6 Mathematical Modeling 237

4. Find a mathematical model for the statement "y is inversely proportional to $7 - x$."

   (a) $\dfrac{y}{7-x} = 1$

   (b) $7y - x = k$

   (c) $y = k(7 - x)$

   (d) $y = \dfrac{k}{7-x}$

   (e) None of these

   1—Answer: d

5. Find a mathematical model for the statement "y varies jointly with the square of x and the square root of z."

   (a) $\dfrac{y}{x^2} = \dfrac{k}{\sqrt{z}}$

   (b) $y = \dfrac{k}{x^2\sqrt{z}}$

   (c) $y = kx^2\sqrt{z}$

   (d) $y = \dfrac{kx^2}{\sqrt{z}}$

   (e) None of these

   1—Answer: c

6. Find a mathematical model for the statement "y varies jointly with x and the cube of z and inversely with w."

   1—Answer: $y = \dfrac{kxz^3}{w}$

7. Which of the following sentences describes the formula to find the volume of a right circular cone, $V = \dfrac{\pi r^2 h}{3}$?

   (a) The volume of a right circular cone is directly proportional to the height and jointly proportional to the square of the radius.

   (b) The volume of a right circular cone is indirectly proportional to the height and radius.

   (c) The volume of a right circular cone varies inversely with the height and the square of the radius.

   (d) The volume of a right circular cone is jointly proportional to the height and the square of the radius.

   (e) None of these

   1—Answer: d

8. Which of the following sentences describes the formula to find the lateral surface area of a right circular cylinder, $S = 2\pi rh$?

   (a) The lateral surface area of a right circular cylinder is inversely proportional to the radius and the height.

   (b) The lateral surface area of a right circular cylinder is jointly proportional to the radius and the height.

   (c) The lateral surface area of a right circular cylinder is directly proportional to the radius and the height.

   (d) The lateral surface area of a right circular cylinder varies indirectly with the radius and height.

   (e) None of these

   1—Answer: b

9. Write a sentence using variation terminology to describe the formula to find the volume of a right circular cylinder, $V = \pi r^2 h$.

   (a) The volume is directly proportional to radius and height.

   (b) The volume is inversely proportional to the height and the square of the radius.

   (c) The volume is jointly proportional to the height and the square of the radius.

   (d) The volume varies inversely with the height and the square of the radius.

   (e) None of these

   1—Answer: c

10. Write a sentence using variation terminology to describe the equation relating the distance, d, a spring is stretched to the force, F, applied, $d = kF$.

    (a) The distance a spring is stretched is jointly proportional to the force applied.

    (b) The distance a spring is stretched varies inversely with the force applied.

    (c) The distance a spring is stretched varies directly with the force applied.

    (d) The distance a spring is stretched is indirectly proportional to the force applied.

    (e) None of these

    1—Answer: c

11. y is directly proportional to x and $y = 35$ when $x = 5$. Find the constant of proportionality.

    (a) $\frac{1}{7}$ (b) 7   (c) 175

    (d) $\frac{1}{175}$ (e) None of these

    1—Answer: b

12. x varies inversely with the square of y. If $x = 1$ when $y = 5$, find the constant of proportionality.

    1—Answer: 25

13. h varies jointly with the square of x and the cube root of y. If $h = \frac{1}{4}$ when $x = 2$ and $y = 8$, find the constant of proportionality.

    (a) $\frac{1}{4}$ (b) 4   (c) $\frac{1}{32}$

    (d) 64   (e) None of these

    1—Answer: c

14. W varies directly with the square of x and inversely with the cube root of y. $W = \frac{1}{2}$ when $x = 6$ and $y = 8$. Find the constant of proportionality.

    (a) $\frac{1}{9}$ (b) $\frac{1}{36}$ (c) $\frac{1}{6}$

    (d) $\frac{9}{2}$ (e) None of these

    1—Answer: b

15. V varies jointly with g and the square of t. Find the constant of proportionality if $V = -144$ when $t = 3$ and $g = -32$.

    (a) 5   (b) $\frac{1}{36}$ (c) $\frac{1}{3}$

    (d) $\frac{1}{2}$ (e) None of these

    1—Answer: d

16. T varies directly with the square root of L and inversely with the square root of g. If $t = \pi/2$ when $L = 2$ and $g = 32$, find the constant of proportionality.

    (a) 2π (b) $\frac{\pi}{8}$ (c) π

    (d) $\frac{\pi}{2}$ (e) None of these

    1—Answer: a

17. A is jointly proportional to P and $(1 + N)$. Find the constant of proportionality if $A = 180$ when $P = 300$ and $N = 2$.

    (a) 1   (b) $\frac{1}{2}$ (c) $\frac{1}{5}$

    (d) 10   (e) None of these

    1—Answer: c

18. x varies directly as y and $x = 3$ when $y = 10$. Find the equation that relates x and y.

    (a) $3x = 10y$ (b) $10x = 3y$ (c) $xy = 30$

    (d) $\frac{x}{y} = 30$ (e) None of these

    1—Answer: b

19. P varies jointly with x and y and inversely with the square of z. If $P = -17/50$ when $x = 3, y = 17$, and $z = 10$, find the equation relating the variables.

    (a) $P = \dfrac{-17(x + y)}{10z^2}$ (b) $P = \dfrac{-867x^2}{5000xy}$ (c) $P = \dfrac{-2xy}{3z^2}$

    (d) $P = \dfrac{-17x^2}{10(x + y)}$ (e) None of these

    2—Answer: c

20. z varies jointly with x and the square of y and inversely with w. If $z = 5$ when $x = 5, y = 3$, and $w = 6$, find the equation that relates the variables.

    (a) $z = \dfrac{35w}{3(x + y^2)}$ (b) $z = \dfrac{75w}{2xy^2}$ (c) $z = \dfrac{15(x + y^2)}{7w}$

    (d) $z = \dfrac{2xy^2}{3w}$ (e) None of these

    2—Answer: d

Chapter 3 Zeros of Polynomial Functions

21. z varies jointly with x and the square of y and inversely with w. If $z = 3$ when $x = 2$, $y = 3$ and $w = 4$, find the mathematical model that relates the variables.

(a) $z = \dfrac{xy^2w}{24}$ (b) $z = \dfrac{2xy^2}{w}$ (c) $z = \dfrac{2xy^2}{3w}$

(d) $z = \dfrac{12(x + y^2)}{11}$ (e) None of these

2—Answer: c

22. P varies jointly with x and w and inversely with the square root of y. If $P = 30$ when $x = 2$, $w = 5$ and $y = 9$, find the mathematical model relating the variables.

(a) $P = \dfrac{9xw}{\sqrt{y}}$ (b) $P = \dfrac{240xw}{y^2}$ (c) $P = \dfrac{90(x + y)}{\sqrt{y}}$

(d) $P = \dfrac{100\sqrt{y}}{xy}$ (e) None of these

2—Answer: a

23. x varies jointly with x and $y + 1$ and inversely with z. If $x = -\dfrac{1}{6}$ when $y = 2$ and $z = 3$, find the mathematical model relating the variables.

(a) $x = -\dfrac{z}{3y(y + 1)}$ (b) $x = -\dfrac{y(y + 1)}{12z}$ (c) $x = -\dfrac{1}{16}\left(2y + 1 + \dfrac{1}{z}\right)$

(d) $x = -\dfrac{2y + 1}{10z}$ (e) None of these

2—Answer: b

24. V varies directly as the square of P and inversely as Q. If $V = 2$ when $P = 2$ and $Q = 4$, find a mathematical model that relates these variables.

(a) $V = \dfrac{2P^2}{Q}$ (b) $V = \dfrac{Q}{P^2}$ (c) $V = \dfrac{8}{17}\left(P^2 + \dfrac{1}{Q}\right)$

(d) $V = \dfrac{8}{17}\left(Q + \dfrac{1}{P^2}\right)$ (e) None of these

2—Answer: a

25. x varies jointly with y and the square of z and inversely with w. If $x = 4/3$ when $y = 1$, $z = -2$, and $w = 7$, find x when $y = 2$, $z = -1$, and $w = 6$.

2—Answer: $\dfrac{7}{9}$

26. V varies directly with the cube of P and inversely with Q. If $V = 2$ when $P = 2$ and $Q = 8$, find V when $P = 1$ and $Q = 2$.

(a) 4 (b) 2 (c) 1 (d) $\dfrac{1}{2}$ (e) None of these

2—Answer: c

27. V varies jointly with P and the square of Q and inversely with the cube of S. $V = 8$ when $P = 12$, $Q = 4$ and $S = 2$. Find V when $P = 6$, $Q = 1$ and $S = 1$.

(a) 16 (b) 8 (c) 4 (d) 2 (e) None of these

2—Answer: d

Section 3.6   Mathematical Modeling   241

28. x varies jointly with y and the cube of z and inversely with w. $x = 64$ when $y = 2, z = 4$ and $w = 6$. Find x when $y = 1, z = 3$ and $w = 15$.

   (a) $\frac{84}{165}$ (b) $\frac{3}{7}$ (c) 16   (d) $\frac{27}{5}$ (e) None of these

   2—Answer: d

29. z varies directly as the square of x and inversely as y. $z = \frac{3}{2}$ when $x = 3$ and $y = 4$. Find z when $x = 12$ and $y = 6$.

   (a) $\frac{207}{5}$ (b) 16   (c) 3   (d) $\frac{21}{2}$ (e) None of these

   2—Answer: d

30. x is jointly proportional to y and $w - y$. $x = 30$ when $w = 20$ and $y = 5$. Find x when $y = 10$ and $w = 30$.

   (a) 10   (b) 20   (c) 40   (d) 80   (e) None of these

   2—Answer: d

31. x is jointly proportional to y and $y + 1$ and inversely proportional to w. If $x = 20$ when $y = 5$ and $w = 3$, find x when $y = 7$ and $w = 4$.

   (a) 16   (b) 28   (c) $\frac{45}{22}$ (d) $\frac{9}{2}$ (e) None of these

   2—Answer: b

32. z is jointly proportional to x and the square of y. If x and y are both tripled what happens to z?

   (a) z is tripled.   (b) z is multiplied by $3\sqrt{3}$.   (c) z is multiplied by 27.
   (d) z is multiplied by $\frac{1}{27}$.   (e) None of these

   2—Answer: c

33. Ohms Law states that the voltage E across a given resistor is proportional to the current I. If the voltage is 12 volts when there is a current of 2 amps, what is the voltage when the current is 1.5 amps?

   (a) 8 volts   (b) 16 volts   (c) 9 volts
   (d) 15 volts   (e) None of these

   2—Answer: c

34. Boyle's Law states that for a fixed amount of gas, the volume of the gas at a constant temperature is inversely proportional to the pressure. If a certain gas occupies 9.84 liters at a pressure of 50 cm.Hg., what is the pressure when the volume is increased to 10 liters?

   (a) 0.02 cm.Hg.   (b) 4920 cm.Hg.   (c) 49.2 cm.Hg.
   (d) 50.8 cm.Hg.   (e) None of these

   2—T—Answer: c

35. Ohms Law states that the voltage E across a given resistor is proportional to the current I. If the voltage is 12 volts when there is a current of $\frac{3}{2}$ amps, what is the voltage when the current is 1 amp.

    (a) 8 volts (b) 9 volts (c) 11 volts
    (d) 6 volts (e) None of these

    2—T—Answer: a

36. Boyle's Law states that for a fixed amount of gas, the volume of the gas at a constant temperature is inversely proportional to the pressure. If a certain gas occupies 9.84 liters at a pressure of 50 cm.Hg., what is the pressure when the volume is increased to 12 liters?

    (a) 43.2 cm.Hg. (b) 45.1 cm.Hg. (c) 39.8 cm.Hg.
    (d) 41.0 cm.Hg. (e) None of these

    2—T—Answer: d

37. The table shows the per capita consumption of broccoli, y (in pounds), for the years 1980 through 1989.

Year, x	1980	1981	1982	1983	1984	1985	1986	1987	1988	1989
Pounds, y	1.6	1.8	2.2	2.3	2.7	2.9	3.5	3.6	4.2	4.5

    Construct a scatter plot for the data, let $t = 0$ represent 1980. Find the least squares regression line that fits this data, and graph the linear model on the same set of axes as the scatter plot.

    2—T—Answer: 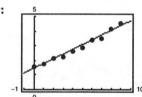 ; $y = 0.325x + 1.465$

38. The table shows the total amount, y in millions of dollars, spent by the federal government on mathematical research from 1980 through 1990.

Year, x	1980	1981	1982	1983	1984	1985	1986	1987	1988	1989	1990
Amount, y	91	118	128	134	151	184	185	205	212	230	245

    Construct a scatter plot for the data, let $t = 0$ represent 1980. Find the least squares regression line that fits the data and graph the model on the same set of axes as the scatter plot.

    2—T—Answer: 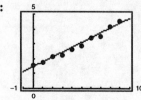 ; $y = 14.963x + 96.363$

39. The table shows the number of households, y in millions, in the U.S. that owned computers between 1984 and 1991.

Year, x	1984	1985	1986	1987	1988	1989	1990	1991
Households, y	6.0	11.3	14.2	16.2	19.2	21.3	25.3	26.6

Construct a scatter plot for the data. Let $t = 4$ represent 1984. Find the least squares regression line that fits the data and graph the model on the same set of axes as the scatter plot.

2—T—Answer: 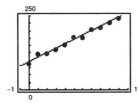 ; $y = 2.839x - 3.782$

CHAPTER FOUR
Rational Functions and Conics

❑ 4.1 Rational Functions and Asymptotes

1. Determine the horizontal asymptote of the graph of $f(x) = \dfrac{x}{x+1}$ by using its graph at the right.

   (a) $y = -1$ (b) $y = 1$

   (c) $x = -1$ (d) $x = 1$

   (e) None of these

   1—Answer: b

   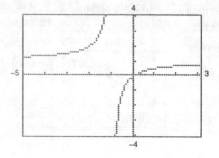

2. Determine the horizontal asymptote of the graph of $f(x) = \dfrac{2x^2 - 1}{x^2 + 3}$ by using its graph at the right.

   (a) $x = 2$ (b) $y = -\dfrac{1}{2}$

   (c) $y = 2$ (d) $x = -\dfrac{1}{2}$

   (e) None of these

   1—Answer: c

   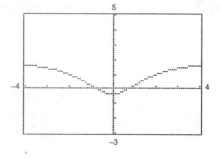

3. Find the vertical asymptote(s): $f(x) = \dfrac{7}{x+2}$

   (a) $x = -2$ (b) $y = -2$ (c) $(0, -2)$

   (d) $y = 0$ (e) None of these

   1—Answer: a

4. Find the vertical asymptote(s): $f(x) = \dfrac{1}{(x+2)(x-5)}$

   (a) $x = -2, x = 5$ (b) $y = 1$ (c) $y = 0$

   (d) $y = 1, y = 0$ (e) None of these

   1—Answer: a

5. Find the vertical asymptote(s): $f(x) = \dfrac{x+3}{(x-2)(x+5)}$

   (a) $y = 2, y = -5, y = -3$ (b) $x = 2, x = -5, x = -3, x = 1$

   (c) $x = 1$ (d) $x = 2, x = -5$ (e) None of these

   1—Answer: d

6. Find the vertical asymptote(s): $f(x) = \dfrac{x+5}{x^2+4}$

   (a) $x = -2, x = 2$ (b) $x = -5$ (c) $x = 0$

   (d) $y = -2, y = 2$ (e) None of these

   1—Answer: e

Section 4.1 Rational Functions and Asymptotes

7. Find the vertical asymptote(s): $f(x) = x + 2 - \dfrac{3}{x}$

   (a) $x = -2, x = 0$ (b) $y = 0$ (c) $y = -2$

   (d) $x = 0$ (e) None of these

   1—Answer: d

8. Find the vertical asymptote(s): $f(x) = \dfrac{x + 2}{x^2 - 9}$

   (a) $x = 3$ (b) $x = -2, x = -3, x = 3$ (c) $y = 0, x = -2$

   (d) $x = -3, x = 3$ (e) None of these

   1—Answer: d

9. Find the horizontal asymptote(s): $f(x) = \dfrac{x^2 - 1}{x^2 + 9}$

   (a) $y = 1$ (b) $y = 0$ (c) $x = 1$

   (d) $x = \pm 1$ (e) None of these

   1—Answer: a

10. Find the horizontal asymptote(s): $f(x) = \dfrac{3x - 2}{x + 2}$

    (a) $y = 0$ (b) $x = -2$ (c) $x = \tfrac{1}{3}$

    (d) $y = 3$ (e) None of these

    1—Answer: d

11. Find the horizontal asymptote(s): $f(x) = \dfrac{x^2 - 4}{x^2 - 9}$

    (a) $x = \pm 3$ (b) $x = \pm 3$ (c) $y = 1$

    (d) $y = 0$ (e) None of these

    1—Answer: c

12. Find the horizontal asymptote(s): $f(x) = \dfrac{7}{x - 4}$

    (a) $x = 4$ (b) $y = 0$ (c) $y = 7$

    (d) $x = 0$ (e) None of these

    1—Answer: b

13. Find the horizontal asymptote(s): $f(x) = \dfrac{3x^2 + 2x - 16}{x^2 - 7}$

    1—Answer: $y = 3$

14. Find the horizontal asymptote(s): $f(x) = \dfrac{x^2 - 3}{(x - 2)(x + 1)}$

    1—Answer: $y = 1$

15. Find the domain: $f(x) = \dfrac{x + 2}{x^2 - 3x + 2}$

    (a) All reals except $x = -2, 1, 2$ (b) All reals except $x = -2$

    (c) All reals except $x = 1, 2$ (d) All reals

    (e) None of these

    1—Answer: c

16. Find the domain: $f(x) = \dfrac{3x-2}{x^2+9}$

(a) All reals (b) All reals except $x = \pm 3$ (c) All reals except $x = \tfrac{1}{3}$

(d) All reals except $x = \tfrac{1}{3}, \pm 3$ (e) None of these

1—Answer: a

17. Find the domain: $f(x) = \dfrac{x^2}{x+1}$

(a) All reals (b) All reals except $x = -1$ (c) All reals except $x = 0$

(d) All reals except $x = -1, 0$ (e) None of these

1—Answer: b

18. Find the domain: $f(x) = \dfrac{x^3-1}{x^2-4}$

(a) All reals (b) All reals except $x = 2$ (c) All reals except $x = 1$

(d) All reals except $x = 1, 2$ (e) None of these

1—Answer: e

19. Find the domain: $f(x) = \dfrac{x^3+1}{x^2+4}$

(a) All reals (b) All reals except $x = -2$ (c) All reals except $x = -1$

(d) All reals except $x = -2, -1$ (e) None of these

1—Answer: a

20. Find the domain: $f(x) = \dfrac{3x^2 - 4x + 1}{x+2}$

(a) All reals (b) All reals except $x = \tfrac{1}{3}, 1$

(c) All reals except $x = -2, \tfrac{1}{3}, 1$ (d) All reals except $x = -2$

(e) None of these

1—Answer: d

21. Find the domain: $f(x) = \dfrac{x}{x^2 + 3x - 4}$

1—Answer: All reals except $x = -4, 1$

22. Find the domain: $f(x) = \dfrac{4+x}{x^2 - 10}$

1—Answer: All reals except $x = \pm\sqrt{10}$

23. Use a graphing utility with the given range setting to graph $f(x) = \dfrac{x+1}{x-3}$ and its horizontal asymptote(s).

2—T—Answer:

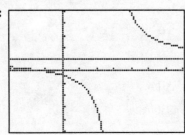

```
Xmin = -3
Xmax = 7
Xscl = 1
Ymin = -5
Ymax = 5
Yscl = 1
```

24. Use a graphing utility with the given range setting to graph $f(x) = \dfrac{x^2}{x^2 - 6x + 9}$ and its horizontal asymptote(s).

   Xmin = -12
   Xmax = 16
   Xscl = 2
   Ymin = -2
   Ymax = 6
   Yscl = 1

   2—T—Answer:
   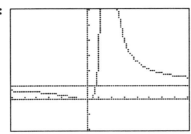

25. Find the real zeros of f: $f(x) = \dfrac{2x - 1}{x^2 + 2}$

    (a) $(-\sqrt{2}, 0)$ (b) $(\frac{1}{2}, 0)$ (c) $(-\sqrt{2}, 0), (\frac{1}{2}, 0)$
    (d) $(\sqrt{2}, 0)$ (e) None of these

    1—Answer: b

26. Find the real zeros of f: $f(x) = \dfrac{x + 2}{x - 1}$

    (a) $(1, 0)$ (b) $(-2, 0)(1, 0)$ (c) $(1, -2)$
    (d) $(-2, 0)$ (e) None of these

    1—Answer: d

27. Use a graphing utility to find the horizontal asymptote(s) of $f(x) = \dfrac{x - 3}{|x| - 2}$.

    (a) $y = -1, y = 1$ (b) $y = 2$ (c) $x = 2$
    (d) $x = -2, x = 2$ (e) None of these

    2—T—Answer: a

28. Use a graphing utility to find the horizontal asymptote(s) of $f(x) = \dfrac{3x + 1}{2 + |x|}$.

    (a) $y = -\frac{3}{2}, y = \frac{3}{2}$ (b) $y = -3, y = 3$ (c) $x = -2, x = 2$
    (d) $x = -3, x = 3$ (e) None of these

    2—T—Answer: b

29. Find the real zeros of f: $f(x) = \dfrac{6x + 3}{x^2 - 1}$

    (a) $-\frac{1}{2}$ (b) $\frac{1}{2}$ (c) ± 1
    (d) $-\frac{1}{2}, \pm 1$ (e) None of these

    1—Answer: a

30. Find the real zeros of f: $f(x) = \dfrac{x^2 - 9}{2x - 1}$

    (a) $\frac{1}{2}$ (b) $-\frac{1}{2}$ (c) ± 3
    (d) $\frac{1}{2}, \pm 3$ (e) None of these

    1—Answer: c

31. Suppose the cost C of removing $p\%$ of pollutants is

$$C = \frac{25{,}000p}{100 - p}, \quad 0 \le p < 100.$$

Find the cost of removing 60%.

(a) $167   (b) $25,000   (c) $37,500

(d) $375   (e) None of these

1—Answer: c

32. Suppose the cost C of removing $p\%$ of pollutants is

$$C = \frac{25{,}000p}{100 - p}, \quad 0 \le p < 100.$$

Find the cost of removing 90%.

(a) $225,000   (b) $2778   (c) $2

(d) $2250   (e) None of these

1—Answer: a

33. A herd of 40 elk is introduced into state game lands. The population of the herd is expected to follow the model

$$P(t) = \frac{40(1 + 2t)}{1 + 0.2t}, \quad t \ge 0.$$

Determine the limiting size of the herd.

2—Answer: 400 elk

34. A herd of 40 elk is introduced into state game lands. The population of the herd is expected to follow the model

$$P(t) = \frac{40(1 + 3t)}{1 + 0.2t}, \quad t \ge 0.$$

Determine the limiting size of the herd.

2—Answer: 600 elk

35. Find the vertical asymptote(s): $f(x) = \dfrac{x^2 - 9}{x^2 - 6x + 8}$

1—Answer: $x = 2, x = 4$

36. Find the vertical asymptote(s): $f(x) = \dfrac{x^2 - 4}{x^2 - 6x + 5}$

1—Answer: $x = 1, x = 5$

4.2 Graphs of Rational Functions

1. Find the intercepts: $f(x) = \dfrac{x^2 - 16}{x^2 - 9}$

   (a) $(-4, 0), (4, 0), (0, -3), (0, 3)$
   (b) $(-4, 0), (4, 0)$
   (c) $(0, 1), (-4, 0), (4, 0)$
   (d) $\left(0, \tfrac{16}{9}\right), (-4, 0), (4, 0)$
   (e) None of these

   1—**Answer:** d

2. Find the intercepts: $f(x) = \dfrac{x - 14}{2x + 7}$

   (a) $(0, -2), (14, 0)$
   (b) $(-14, 0), \left(\tfrac{1}{2}, 0\right)$
   (c) $(14, 0), \left(0, \tfrac{1}{2}\right)$
   (d) $(14, 0), \left(0, -\tfrac{7}{2}\right)$
   (e) None of these

   1—**Answer:** a

3. Find any horizontal and vertical asymptotes: $f(x) = \dfrac{x^2}{x^2 + 9}$

   1—**Answer:** $y = 1$

4. Find any horizontal and vertical asymptotes: $f(x) = \dfrac{x}{x^3 - 1}$

   1—**Answer:** $y = 0, x = 1$

5. Match the rational function with the correct graph: $f(x) = \dfrac{3 + x}{x - 1}$

   (a)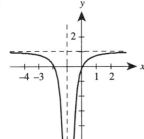

   (b)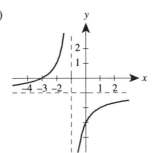

   (c)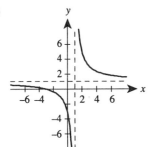

   (d)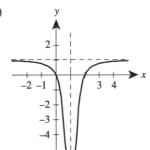

   (e) None of these

   1—**Answer:** c

6. Match the rational function with the correct graph: $f(x) = \dfrac{6}{x+2}$

(a)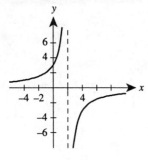

(b)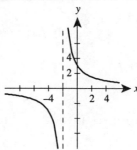

(c)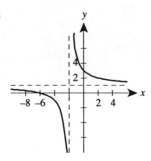

(d)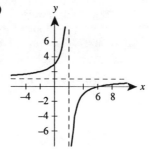

(e) None of these

1—**Answer:** b

7. Match the rational function with the correct graph: $f(x) = \dfrac{x^2}{x+2}$

(a)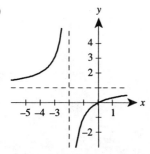

(b)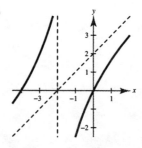

(c)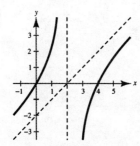

(d)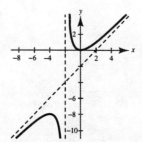

(e) None of these

2—**Answer:** d

8. Match the graph with the correct function.

   (a) $f(x) = \dfrac{1}{2x + 1}$
   (b) $f(x) = \dfrac{x - 1}{2x + 1}$
   (c) $f(x) = \dfrac{x^2 + 2x + 2}{2x - 1}$
   (d) $f(x) = \dfrac{x^3 + 2x^2 + x - 2}{2x + 1}$
   (e) None of these

   1—Answer: c

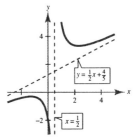

9. Match the graph with the correct function.

   (a) $f(x) = \dfrac{x + 3}{x - 1}$
   (b) $f(x) = x + 3$
   (c) $f(x) = \dfrac{x - 1}{x^2 + 2x - 3}$
   (d) $f(x) = \dfrac{x^2 + 2x - 3}{x - 1}$
   (e) None of these

   1—Answer: d

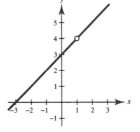

10. Match the graph with the correct function.

    (a) $f(x) = \dfrac{x - 5}{x + 3}$
    (b) $f(x) = \dfrac{5 - x}{x + 3}$
    (c) $f(x) = \dfrac{x + 5}{x + 3}$
    (d) $f(x) = \dfrac{x + 5}{x + 3}$
    (e) None of these

    1—Answer: c

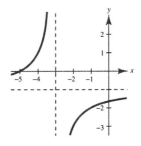

11. Use a graphing utility to graph $f(x) = \dfrac{x - 2}{x + 2}$.

    (a)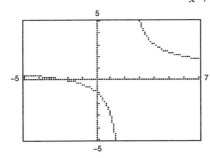
    (b)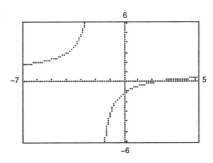
    (c)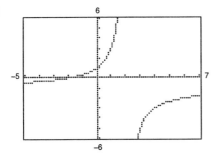
    (d)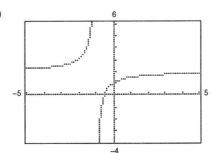
    (e) None of these

    1—T—Answer: b

12. Use a graphing utility to graph $f(x) = \dfrac{3x+1}{x}$.

(a)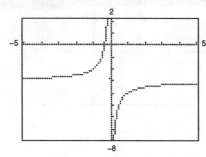
(b)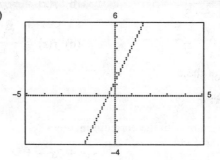
(c)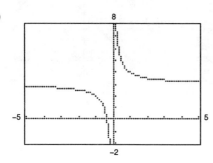
(d)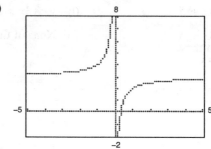

(e) None of these

1—T—Answer: c

13. Use a graphing utility to graph
$$f(x) = \dfrac{2}{x-1}.$$
1—T—Answer:

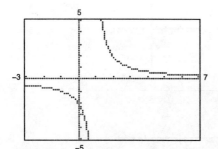

14. Use a graphing utility to graph
$$f(x) = \dfrac{x}{x^2-1}.$$
1—T—Answer:

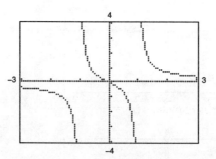

15. Match the graph with the correct function.

(a) $f(x) = \dfrac{3}{x-2}$ (b) $f(x) = \dfrac{3x}{x^2-4}$

(c) $f(x) = \dfrac{3}{x^2-4}$ (d) $f(x) = \dfrac{3}{x+2}$

(e) None of these

1—Answer: c

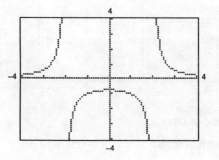

16. Match the graph with the correct function.

    (a) $f(x) = \dfrac{1}{x-3}$ (b) $f(x) = \dfrac{1}{x+3}$

    (c) $f(x) = \dfrac{x}{x-3}$ (d) $f(x) = \dfrac{x}{x+3}$

    (e) None of these

    1—Answer: d

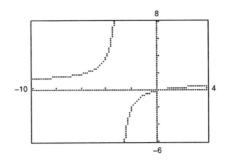

17. Find the slant asymptote: $f(x) = \dfrac{3x^2 + 2x - 1}{x - 1}$

    (a) $y = -3x + 5$ (b) $y = 3x + 5$ (c) $y = 3x - 5$

    (d) $y = -3x - 5$ (e) None of these

    1—Answer: b

18. Find the slant asymptote: $f(x) = x - 2 + \dfrac{3}{x + 4}$

    (a) $y = 0$ (b) $y = x + 4$ (c) $y = x - 2$

    (d) $y = x + 3$ (e) None of these

    1—Answer: c

19. Find the slant asymptote: $f(x) = \dfrac{x^3 + 7x^2 - 1}{x^2 + 1}$

    (a) $y = 1$ (b) $y = x + 7$ (c) $y = x - 8$

    (d) $y = x + 1$ (e) None of these

    1—Answer: b

20. Find the slant asymptote: $f(x) = \dfrac{x^2 + 2x - 1}{x - 1}$

    (a) $y = 1$ (b) $y = x - 1$ (c) $y = x + 1$

    (d) $y = x + 3$ (e) None of these

    1—Answer: d

21. Find the slant asymptote: $f(x) = \dfrac{x^3 + 2x^2 - 3}{x + 1}$

    (a) $y = -1$ (b) $y = x^2 + x - 1$ (c) $y = 0$

    (d) $y = x^2 + 2x$ (e) None of these

    1—Answer: e

22. Find the horizontal or slant asymptote: $f(x) = \dfrac{x^2 + 3x + 1}{x + 1}$

    (a) $x = -1$ (b) $y = x + 3$ (c) $y = x + 2$

    (d) $y = 0$ (e) None of these

    2—Answer: c

23. Find the vertical, horizontal, or slant asymptotes: $f(x) = \dfrac{x^3 - 2x^2 + 5}{x^2}$

    1—Answer: $x = 0,\ y = x - 2$

24. Find the vertical, horizontal, or slant asymptotes: $f(x) = \dfrac{x-2}{x^2 - 2x - 3}$

    1—Answer: $x = -1, y = 3, y = 0$

25. Find the vertical, horizontal, or slant asymptotes: $f(x) = \dfrac{3x^2 - 2x + 4}{x - 3}$

    1—Answer: $x = 3, y = 3x + 7$

26. Label all intercepts and asymptotes, and sketch the graph: $f(x) = \dfrac{x+2}{x+1}$

    2—Answer:

    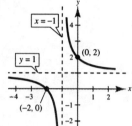

27. Label all intercepts and asymptotes, and sketch the graph: $f(x) = \dfrac{3x+2}{x-5}$

    2—Answer:

    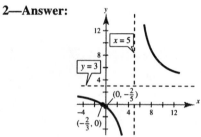

28. Label all intercepts and asymptotes, and sketch the graph: $f(x) = \dfrac{x^2 + x - 2}{x - 3}$

    2—Answer:

    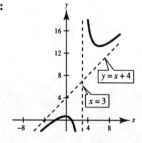

29. The concentration of a mixture is given by

    $$C = \dfrac{2x + 9}{3(x + 12)}.$$

    Use a graphing utility using the indicated range setting to determine what the concentration approaches.

    (a) 33%   (b) 67%

    (c) 50%   (d) 75%

    (e) None of these

    2—Answer: b

    ```
    Xmin = 0
    Xmax = 280
    Xscl = 50
    Ymin = 0
    Ymax = 1
    Yscl = .1
    ```

30. The concentration of a mixture is given by

$$C = \frac{3x + 8}{4(x + 8)}.$$

Use a graphing utility using the indicated range setting to determine what the concentration approaches.

(a) 75% (b) 25%

(c) 50% (d) 33%

(e) None of these

```
Xmin = 0
Xmax = 200
Xscl = 50
Ymin = 0
Ymax = 1
Yscl = .1
```

2—**Answer:** a

31. The population of a bacteria culture is given by

$$P = \frac{20t + 2}{5 + 0.2t}, \quad t \geq 0$$

where t is time in hours. Use a graphing utility using the indicated range setting, graph the function and determine the value which the population approaches.

```
Xmin = 0
Xmax = 200
Xscl = 25
Ymin = 0
Ymax = 125
Yscl = 25
```

2—T—**Answer:** 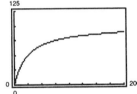 ;100

32. The population of a bacteria culture is given by

$$P = \frac{20t + 2}{10 + 0.02t}, \quad t \geq 0$$

where t is time in hours. Use a graphing utility using the indicated range setting, graph the function and determine the value which the population approaches.

```
Xmin = 0
Xmax = 15000
Xscl = 1000
Ymin = 0
Ymax = 1250
Yscl = 250
```

2—T—**Answer:** 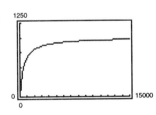 ;1000

4.3 Partial Fractions

1. Find the partial fraction decomposition: $\dfrac{1-x}{2x^2+x}$

   1—Answer: $\dfrac{1}{x} - \dfrac{3}{2x+1}$

2. Find the partial fraction decomposition: $\dfrac{7x-2}{3x^2-x}$

   1—Answer: $\dfrac{2}{x} - \dfrac{3}{3x+1}$

3. Find the partial fraction decomposition: $\dfrac{4x+23}{x^2-x-6}$

   (a) $\dfrac{4x}{x+2} + \dfrac{23}{x-3}$ (b) $\dfrac{2}{x+2} - \dfrac{13}{x-3}$ (c) $\dfrac{5}{x-3} - \dfrac{2}{x+2}$

   (d) $\dfrac{7}{x-3} - \dfrac{3}{x+2}$ (e) None of these

   1—Answer: d

4. Find the partial fraction decomposition: $\dfrac{7}{3x^2+5x-2}$

   (a) $\dfrac{6}{(3x-2)} + \dfrac{1}{x-1}$ (b) $\dfrac{3}{3x-1} - \dfrac{1}{x+2}$ (c) $\dfrac{6}{x+2} - \dfrac{18}{3x-1}$

   (d) $\dfrac{4}{3x-1} + \dfrac{2}{x+2}$ (e) None of these

   1—Answer: b

5. Find the partial fraction decomposition: $\dfrac{5x+3}{x^2-3x-10}$

   (a) $\dfrac{2}{x+5} - \dfrac{7}{x-2}$ (b) $\dfrac{7}{x-5} - \dfrac{2}{x+2}$ (c) $\dfrac{2}{x-5} + \dfrac{3}{x+2}$

   (d) $\dfrac{4}{x-5} + \dfrac{1}{x+2}$ (e) None of these

   1—Answer: d

6. Find the partial fraction decomposition: $\dfrac{-5x-3}{x^2-9}$

   (a) $\dfrac{-7}{x-3} + \dfrac{2x-3}{(x-3)^2}$ (b) $\dfrac{-3}{x-3} - \dfrac{2}{x+3}$ (c) $\dfrac{1}{x+3} - \dfrac{6}{x-3}$

   (d) $\dfrac{4}{x-3} - \dfrac{9}{x+3}$ (e) None of these

   1—Answer: b

7. Find the partial fraction decomposition: $\dfrac{x-7}{x^2-1}$

    (a) $\dfrac{3}{x+1} - \dfrac{4}{x-1}$
    (b) $-\dfrac{2}{x+1} + \dfrac{1}{x-1}$
    (c) $\dfrac{3}{x+1} - \dfrac{2}{x-1}$
    (d) $\dfrac{4}{x+1} - \dfrac{3}{x-1}$
    (e) None of these

    1—Answer: d

8. Find the partial fraction decomposition: $\dfrac{-3}{2x^2+3x}$

    (a) $-\dfrac{1}{x} + \dfrac{2}{2x+3}$
    (b) $\dfrac{3}{x} - \dfrac{9}{2x+3}$
    (c) $\dfrac{-1}{x} - \dfrac{2}{2x+3}$
    (d) $\dfrac{6}{2x+3} - \dfrac{9}{x}$
    (e) None of these

    1—Answer: a

9. Find the partial fraction decomposition: $\dfrac{3x^2-7x+1}{(x-1)^3}$

    2—Answer: $\dfrac{3}{x-1} - \dfrac{1}{(x-1)^2} - \dfrac{3}{(x-1)^3}$

10. Find the partial fraction decomposition: $\dfrac{2x+1}{(x+1)^2}$

    (a) $\dfrac{-2}{x+1} + \dfrac{1}{(x+1)^2}$
    (b) $\dfrac{1}{x+1} - \dfrac{2x}{(x+1)^2}$
    (c) $\dfrac{-6}{x+1} - \dfrac{1}{(x+1)^2}$
    (d) $\dfrac{-2}{x+1} + \dfrac{-2x+3}{(x+1)^2}$
    (e) None of these

    1—Answer: a

11. Find the partial fraction decomposition: $\dfrac{5x-2}{(x-1)^2}$

    (a) $\dfrac{5}{x-1} + \dfrac{-2}{(x-1)^2}$
    (b) $\dfrac{-2}{x-1} + \dfrac{5x}{(x-1)^2}$
    (c) $\dfrac{5}{x-1} + \dfrac{3}{(x-1)^2}$
    (d) $\dfrac{3}{(x-1)} + \dfrac{-2x+1}{(x-1)^2}$
    (e) None of these

    1—Answer: c

12. Find the partial fraction decomposition: $\dfrac{6x-13}{x^2-2x+1}$

    (a) $\dfrac{6x}{(x-1)} + \dfrac{13}{(x-1)^2}$
    (b) $\dfrac{6}{x-1} - \dfrac{7}{(x-1)^2}$
    (c) $\dfrac{6}{x-1} - \dfrac{6x+7}{(x-1)^2}$
    (d) $\dfrac{4}{x-1} + \dfrac{2}{(x-1)^2}$
    (e) None of these

    1—Answer: b

13. Find the partial fraction decomposition: $\dfrac{2x^2 - 9x + 11}{(x - 2)^3}$

    (a) $\dfrac{3}{x - 2} + \dfrac{4}{(x - 2)^2} - \dfrac{1}{(x - 2)^3}$

    (b) $\dfrac{2}{x - 2} + \dfrac{4x + 1}{(x - 2)^2} + \dfrac{x^2 - 3x}{(x - 2)^3}$

    (c) $\dfrac{2}{x - 2} + \dfrac{-1}{(x - 2)^2} + \dfrac{1}{(x - 2)^3}$

    (d) $\dfrac{2x^2}{(x - 2)^3} - \dfrac{9x}{(x - 2)^2} - \dfrac{11}{(x - 2)}$

    (e) None of these

    2—Answer: c

14. Find the partial fraction decomposition: $\dfrac{5x^2 + 12x + 10}{(x + 1)^3}$

    2—Answer: $\dfrac{5}{x + 1} + \dfrac{2}{(x + 1)^2} + \dfrac{3}{(x + 1)^3}$

15. Find the partial fraction decomposition: $\dfrac{9x^2 + x - 1}{x^2(x + 1)}$

    (a) $\dfrac{2}{x} - \dfrac{1}{x^2} + \dfrac{7}{x + 1}$

    (b) $\dfrac{20}{x} - \dfrac{1}{x^2} - \dfrac{11}{x + 1}$

    (c) $\dfrac{9}{x} + \dfrac{1}{x^2} - \dfrac{1}{x + 1}$

    (d) $\dfrac{-1}{x^2} + \dfrac{9}{x + 1}$

    (e) None of these

    2—Answer: a

16. Find the partial fraction decomposition: $\dfrac{-9}{(x + 1)^2(x - 2)}$

    (a) $\dfrac{1}{x + 1} + \dfrac{3}{(x + 1)^2} - \dfrac{1}{x - 2}$

    (b) $\dfrac{10}{x + 1} - \dfrac{4}{(x + 1)^2} - \dfrac{15}{x - 2}$

    (c) $\dfrac{2}{x + 1} - \dfrac{1}{(x + 1)^2} - \dfrac{1}{x - 2}$

    (d) $\dfrac{1}{x + 1} - \dfrac{3}{(x + 1)^2} + \dfrac{5}{x - 2}$

    (e) None of these

    2—Answer: a

17. Find the partial fraction decomposition: $\dfrac{-5x^2 - 19x - 28}{x^3 + 4x^2 + 4x}$

    (a) $-\dfrac{5x^2}{x^3} - \dfrac{19x}{4x^2} - \dfrac{28}{4x}$

    (b) $-\dfrac{5x^2}{x} - \dfrac{19x}{x + 2} - \dfrac{28}{(x + 2)^2}$

    (c) $\dfrac{2}{x} - \dfrac{5}{x + 2} + \dfrac{16}{(x + 2)^2}$

    (d) $-\dfrac{7}{x} + \dfrac{2}{x + 2} + \dfrac{5}{(x + 2)^2}$

    (e) None of these

    2—Answer: d

18. Find the partial fraction decomposition: $\dfrac{12x^2 - 13x - 3}{(x - 1)^2(x + 3)}$

    2—Answer: $\dfrac{3}{x - 1} - \dfrac{1}{(x - 1)^2} + \dfrac{9}{x + 3}$

19. Find the partial fraction decomposition: $\dfrac{2x^2 + 6x - 11}{(x - 3)(x + 2)^2}$

    (a) $\dfrac{6}{x + 2} + \dfrac{1}{(x + 2)^2} - \dfrac{1}{x - 3}$ (b) $\dfrac{-3}{x + 2} + \dfrac{5}{(x + 2)^2} - \dfrac{2}{x - 3}$

    (c) $\dfrac{2}{x + 2} + \dfrac{7}{(x + 2)^2} - \dfrac{11}{x - 3}$ (d) $\dfrac{1}{x + 2} + \dfrac{3}{(x + 2)^2} + \dfrac{1}{x - 3}$

    (e) None of these

    2—Answer: d

20. Find the partial fraction decomposition: $\dfrac{17x^2 - 14x + 3}{x^3 - x^2}$

    (a) $\dfrac{5}{x} + \dfrac{2}{x^2} + \dfrac{5}{x - 1}$ (b) $\dfrac{1}{x} + \dfrac{1}{x^2} + \dfrac{1}{x - 1}$ (c) $\dfrac{11}{x} - \dfrac{3}{x^2} + \dfrac{6}{x - 1}$

    (d) $\dfrac{7}{x} + \dfrac{2}{x^2} - \dfrac{8}{x - 1}$ (e) None of these

    2—Answer: c

21. Find the partial fraction decomposition: $\dfrac{3x^2 - 31x - 25}{(x + 1)(x^2 - 7x - 8)}$

    2—Answer: $\dfrac{4}{x + 1} - \dfrac{1}{(x + 1)^2} - \dfrac{1}{x - 8}$

22. Find the partial fraction decomposition: $\dfrac{-x^2 - 7x + 27}{x(x^2 + 9)}$

    (a) $\dfrac{3}{x} + \dfrac{4}{x + 3} - \dfrac{7}{x^2 + 9}$ (b) $\dfrac{3}{x} - \dfrac{4x + 7}{x^2 + 9}$ (c) $\dfrac{2}{x} + \dfrac{3x + 5}{x^2 + 9}$

    (d) $\dfrac{2}{x} + \dfrac{3}{x + 3} - \dfrac{5}{x^2 + 9}$ (e) None of these

    2—Answer: b

23. Find the partial fraction decomposition: $\dfrac{5x^2 - 9x + 12}{(x - 2)(x^2 + x + 1)}$

    (a) $\dfrac{5x^2}{x - 2} - \dfrac{9x + 12}{x^2 + x + 1}$ (b) $-\dfrac{4/9}{x - 2} + \dfrac{22/9}{x + 1} - \dfrac{26/3}{(x + 1)^2}$

    (c) $\dfrac{2}{x - 2} + \dfrac{3x - 5}{x^2 + x + 1}$ (d) $\dfrac{5}{x - 2} + \dfrac{2x - 7}{x^2 + x + 1}$

    (e) None of these

    2—Answer: c

24. Find the partial fraction decomposition: $\dfrac{x^2 - x - 4}{x(x^2 + 2)}$

    2—Answer: $-\dfrac{2}{x} + \dfrac{3x - 1}{x^2 + 2}$

25. Find the partial fraction decomposition: $\dfrac{4-x}{x(x^2+4)}$

   2—Answer: $\dfrac{1}{x} - \dfrac{x+1}{x^2+4}$

26. Find the partial fraction decomposition: $\dfrac{x^2-4x+1}{(x-3)(x^2+1)}$

   (a) $\dfrac{-1/5}{x-3} + \dfrac{2}{x+1} + \dfrac{4/5}{x+1}$
   (b) $\dfrac{1}{x-3} + \dfrac{-4x+1}{x^2+1}$
   (c) $\dfrac{1}{x-3} - \dfrac{4}{x+1} + \dfrac{1}{x+1}$
   (d) $\dfrac{-1/5}{x-3} + \dfrac{(6/5)x - (2/5)}{x^2+1}$
   (e) None of these

   2—Answer: d

27. Find the partial fraction decomposition: $\dfrac{7x^2+24x-1}{(x^2+2)(x+5)}$

   (a) $\dfrac{2}{x+5} + \dfrac{5x-1}{x^2+2}$
   (b) $\dfrac{3}{x+5} + \dfrac{5x+2}{x^2+2}$
   (c) $\dfrac{3}{x+5} - \dfrac{1}{x^2+2}$
   (d) $\dfrac{1}{x+5} - \dfrac{3x}{x^2+2}$
   (e) None of these

   2—Answer: a

28. Find the partial fraction decomposition: $\dfrac{x^2+11x+2}{(x-3)(x^2+2)}$

   (a) $\dfrac{x-6}{x^2+2} + \dfrac{4}{x-3}$
   (b) $\dfrac{7x-5}{x^2+2} - \dfrac{1}{x-3}$
   (c) $\dfrac{-3x+2}{x^2+2} + \dfrac{4}{x-3}$
   (d) $\dfrac{2x+1}{x^2+2} + \dfrac{3}{x-3}$
   (e) None of these

   2—Answer: c

29. Find the partial fraction decomposition: $\dfrac{x^2-x+2}{(x^2+2)^2}$

   2—Answer: $\dfrac{1}{x^2+2} - \dfrac{x}{(x^2+2)^2}$

30. Find the partial fraction decomposition: $\dfrac{2x^3-x^2+2x+2}{(x^2+1)^2}$

   (a) $\dfrac{x+1}{x^2+1} + \dfrac{x-1}{(x^2+1)^2}$
   (b) $\dfrac{2x-1}{x^2+1} + \dfrac{3}{(x^2+1)^2}$
   (c) $\dfrac{x-3}{x^2+1} + \dfrac{2x+5}{(x^2+1)^2}$
   (d) $\dfrac{7}{x^2+1} + \dfrac{-3x+2}{(x^2+1)^2}$
   (e) None of these

   2—Answer: b

31. Find the partial fraction decomposition: $\dfrac{2x^2 - 6x + 4}{(x^2 + 1)^2}$

(a) $\dfrac{3x + 1}{x^2 + 1} - \dfrac{x + 1}{(x^2 + 1)^2}$ (b) $\dfrac{4}{x^2 + 1} - \dfrac{6x}{(x^2 + 1)^2}$ (c) $\dfrac{2x - 6}{x^2 + 1} + \dfrac{4}{(x^2 + 1)^2}$

(d) $\dfrac{2}{x^2 + 1} - \dfrac{6x - 2}{(x^2 + 1)^2}$ (e) None of these

1—Answer: d

32. Find the partial fraction decomposition: $\dfrac{3x^2 + 9x + 4}{(x^2 + x + 1)^2}$

(a) $\dfrac{3}{x^2 + x + 1} + \dfrac{6x + 1}{(x^2 + x + 1)^2}$ (b) $\dfrac{2x - 1}{x^2 + x + 1} + \dfrac{4x + 7}{(x^2 + x + 1)}$

(c) $\dfrac{9x - 1}{x^2 + x + 1} + \dfrac{4x}{(x^2 + x + 1)^2}$ (d) $\dfrac{3x + 6}{x^2 + x + 1} + \dfrac{3x - 6}{(x^2 + x + 1)^2}$

(e) None of these

2—Answer: a

33. Find the partial fraction decomposition: $\dfrac{6x^3 + 24x - 7}{(x^2 + 4)^2}$

(a) $\dfrac{-2x + 1}{x^2 + 4} + \dfrac{16x - 5}{(x^2 + 4)^2}$ (b) $\dfrac{-4x + 3}{x^2 + 4} + \dfrac{x + 1}{(x^2 + 4)^2}$ (c) $\dfrac{6x}{x^2 + 4} - \dfrac{7}{(x^2 + 4)^2}$

(d) $\dfrac{2 - 3x}{x^2 + 4} + \dfrac{4x - 1}{(x^2 + 4)^2}$ (e) None of these

1—Answer: c

34. Find the partial fraction decomposition: $\dfrac{x^3 + x}{(x^2 + x + 1)^2}$

(a) $\dfrac{x + 1}{x^2 + x + 1} - \dfrac{x + 1}{(x^2 + x + 1)^2}$ (b) $\dfrac{x - 1}{x^2 + x + 1} + \dfrac{x + 1}{(x^2 + x + 1)^2}$

(c) $\dfrac{x + 1}{x^2 + x + 1} + \dfrac{x - 1}{(x^2 + x + 1)^2}$ (d) $\dfrac{x - 1}{x^2 + x + 1} + \dfrac{x - 1}{(x^2 + x + 1)^2}$

(e) None of these

2—Answer: b

35. Find the partial fraction decomposition: $\dfrac{5x^3 + 4x^2 + 7x + 3}{(x^2 + 2)(x^2 + 1)}$

2—Answer: $\dfrac{2x - 1}{x^2 + 1} + \dfrac{3x + 5}{x^2 + 2}$

36. Find the partial fraction decomposition: $\dfrac{5x + 5}{(x^2 + 1)(x^2 + 5)}$

2—Answer: $\dfrac{x + 1}{x^2 + 1} - \dfrac{x}{x^2 + 5}$

37. Find the partial fraction decomposition: $\dfrac{x^3 + x^2 + 2x - 2}{x^2 - 1}$

   2—Answer: $x + 1 + \dfrac{1}{x - 1} + \dfrac{2}{x + 1}$

38. Find the partial fraction decomposition: $\dfrac{x^3 - x^2 + 4}{x^2 - 1}$

   (a) $x - 1 + \dfrac{x}{x - 1} + \dfrac{1}{x + 1}$
   (b) $x + 1 + \dfrac{4}{x + 1} + \dfrac{-2}{x - 1}$
   (c) $x - 1 + \dfrac{2}{x - 1} - \dfrac{1}{x + 1}$
   (d) $x + 1 - \dfrac{3}{x + 1} + \dfrac{1}{x - 1}$
   (e) None of these

   2—Answer: c

39. Find the partial fraction decomposition: $\dfrac{3x^4 + x^2 - 2}{x^2 - 1}$

   (a) $3x^2 - 2 - \dfrac{2}{x - 1} + \dfrac{1}{x + 1}$
   (b) $3x^2 - 2$
   (c) $3x^2 - 4 + \dfrac{2}{x + 1} - \dfrac{1}{x - 1}$
   (d) $3x^2 + 4 + \dfrac{1}{x - 1} - \dfrac{1}{x + 1}$
   (e) None of these

   2—Answer: d

❏ 4.4 Conics

1. Match the graph with the correct equation.

   (a) $\dfrac{x^2}{1} + \dfrac{y^2}{3} = 1$
   (b) $\dfrac{x^2}{3} + \dfrac{y^2}{1} = 1$
   (c) $\dfrac{x^2}{9} + \dfrac{y^2}{1} = 1$
   (d) $\dfrac{x^2}{2} + \dfrac{y^2}{9} = 1$
   (e) None of these

   1—Answer: e

2. Match the graph with the correct equation.

   (a) $\dfrac{x^2}{16} - \dfrac{y^2}{4} = 1$
   (b) $\dfrac{x^2}{4} - \dfrac{y^2}{16} = 1$
   (c) $\dfrac{y^2}{16} - \dfrac{x^2}{4} = 1$
   (d) $\dfrac{y^2}{4} - \dfrac{x^2}{16} = 1$
   (e) None of these

   1—Answer: a

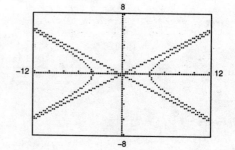

3. Match the graph with the correct equation.

   (a) $\dfrac{x^2}{4} + \dfrac{y^2}{2} = 1$
   (b) $\dfrac{y^2}{4} + \dfrac{x^2}{2} = 1$
   (c) $\dfrac{x^2}{16} + \dfrac{y^2}{4} = 1$
   (d) $\dfrac{y^2}{16} - \dfrac{x^2}{4} = 1$
   (e) None of these

   1—Answer: c

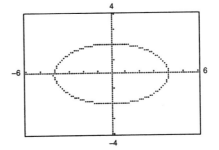

4. Match the graph with the correct equation.

   (a) $y = 2x^2$
   (b) $y = -4x^2$
   (c) $x = 3y^2$
   (d) $x = -2y^2$
   (e) None of these

   1—Answer: d

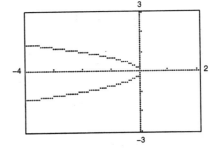

5. Match the graph with the correct equation.

   (a) $\dfrac{x^2}{9} - \dfrac{y^2}{4} = 1$
   (b) $\dfrac{x^2}{4} - \dfrac{y^2}{9} = 1$
   (c) $\dfrac{y^2}{9} - \dfrac{x^2}{4} = 1$
   (d) $\dfrac{y^2}{4} - \dfrac{x^2}{9} = 1$
   (e) None of these

   1—Answer: c

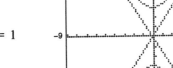

6. Use a graphing utility to graph:

   $x^2 = 24(y - 2)$

   1—T—Answer:

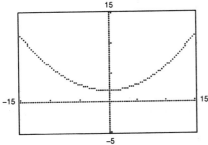

7. Use a graphing utility to graph:

   $x^2 + 5y^2 = 5$

   1—T—Answer:

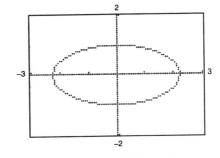

8. Use a graphing utility to graph: $\dfrac{x^2}{9} - \dfrac{y^2}{4} = 1$

   1—T—Answer:

   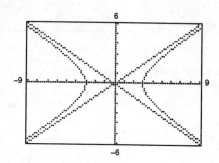

9. Find the focus of the parabola: $y^2 = -32x$

   (a) $(8, 0)$ (b) $(-8, 0)$ (c) $(0, 8)$

   (d) $(0, -8)$ (e) None of these

   1—Answer: b

10. Find the directrix of the parabola: $y^2 = x$

    (a) $y = \frac{1}{4}$ (b) $x = \frac{1}{4}$ (c) $y = -\frac{1}{4}$

    (d) $x = -\frac{1}{4}$ (e) None of these

    1—Answer: d

11. Find the focus of the parabola: $x = -16y^2$

    (a) $(-4, 0)$ (b) $\left(-\frac{1}{64}, 0\right)$ (c) $(0, -4)$

    (d) $\left(0, -\frac{1}{64}\right)$ (e) None of these

    1—Answer: b

12. Find the directrix of the parabola: $x = -4y^2$

    (a) $x = 1$ (b) $y = 1$ (c) $x = \frac{1}{16}$

    (d) $y = \frac{1}{16}$ (e) None of these

    1—Answer: c

13. Find the focus of the parabola: $y = -16x^2$

    (a) $\left(-\frac{1}{64}, 0\right)$ (b) $\left(0, -\frac{1}{64}\right)$ (c) $(-4, 0)$

    (d) $(0, -4)$ (e) None of these

    1—Answer: b

14. Find an equation of the parabola with vertex at $(0, 0)$ and focus at $(-3, 0)$.

    (a) $x^2 = -12y$ (b) $y^2 = -12x$ (c) $x^2 = 12y$

    (d) $y^2 = 12x$ (e) None of these

    1—Answer: b

15. Find the standard equation of the parabola with vertex at $(0, 0)$ and directrix $x = 7$.

    (a) $x^2 = -28y$ (b) $x^2 = \frac{7}{4}y$ (c) $y^2 = -\frac{4}{7}x$

    (d) $y^2 = -28x$ (e) None of these

    1—Answer: d

16. Find the standard equation of the parabola with vertex at (0, 0) and directrix $x = 5/2$.

    (a) $x^2 = 10y$ (b) $y^2 = 10x$ (c) $x^2 = -\frac{5}{8}y$

    (d) $y^2 = \frac{8}{5}x$ (e) None of these

    1—Answer: b

17. Find the standard equation of the parabola that has a horizontal axis and a vertex at (0, 0), and passes through the point (2, −4).

    2—Answer: $y^2 = 8x$

18. Find an equation of the parabola with vertex at (0, 0) and focus (0, 1).

    (a) $y = \frac{1}{4}x^2$ (b) $y = 4x^2$ (c) $x = -\frac{1}{4}y^2$

    (d) $x = -4y^2$ (e) None of these

    1—Answer: a

19. Find the foci of the ellipse: $\dfrac{x^2}{81} + \dfrac{y^2}{225} = 1$

    (a) $(0, 12), (0, -12)$ (b) $(12, 0), (-12, 0)$ (c) $(0, 3\sqrt{34}), (0, -3\sqrt{34})$

    (d) $(3\sqrt{34}, 0), (-3\sqrt{34}, 0)$ (e) None of these

    1—Answer: a

20. Find the length of the minor axis: $\dfrac{x^2}{9} + \dfrac{y^2}{16} = 1$

    (a) 3   (b) 4   (c) 8

    (d) $\sqrt{7}$ (e) None of these

    1—Answer: e

21. If the length of each latus rectum of an ellipse is $\dfrac{2b^2}{a}$, determine the combined length of the 2 latus recta of the ellipse $\dfrac{x^2}{4} + y^2 = 1$.

    (a) 1   (b) 2   (c) 4

    (d) 64   (e) None of these

    2—Answer: b

22. Find the vertices: $9x^2 + 4y^2 = 36$

    (a) $(3, 0), (-3, 0)$ (b) $(0, 2), (0, -2)$ (c) $(0, 3), (0, -3)$

    (d) $(0, \sqrt{5}), (0, -\sqrt{5})$ (e) None of these

    1—Answer: c

23. Find the vertices: $\dfrac{x^2}{9} + \dfrac{y^2}{5} = 1$

    (a) $(-2, 0), (2, 0)$ (b) $(0, -2), (0, 2)$ (c) $(4, 0), (-4, 0)$

    (d) $(0, 4), (0, -4)$ (e) None of these

    1—Answer: a

24. Find the standard equation of the ellipse with center at (0, 0), one focus at (3, 0), and a major axis of length 12.

   1—Answer: $\dfrac{x^2}{36} + \dfrac{y^2}{27} = 1$

25. Find the standard equation of the ellipse with vertices at (± 4, 0) and foci at (± 3, 0).

   (a) $\dfrac{x^2}{7} + \dfrac{y^2}{16} = 1$ (b) $\dfrac{x^2}{16} + \dfrac{y^2}{9} = 1$ (c) $\dfrac{x^2}{16} + \dfrac{y^2}{7} = 1$

   (d) $\dfrac{x^2}{16} - \dfrac{y^2}{9} = 1$ (e) None of these

   1—Answer: c

26. Find the standard equation of the ellipse with vertices at (0, ± 7) and foci at $(0, \pm\sqrt{13})$.

   (a) $\dfrac{x^2}{13} + \dfrac{y^2}{49} = 1$ (b) $\dfrac{x^2}{49} + \dfrac{y^2}{36} = 1$ (c) $\dfrac{x^2}{36} + \dfrac{y^2}{13} = 1$

   (d) $\dfrac{x^2}{36} + \dfrac{y^2}{49} = 1$ (e) None of these

   2—Answer: d

27. Find the standard equation of the ellipse that passes through the point $(2, 6\sqrt{2})$ and has endpoints of (± 6, 0) on the minor axis.

   2—Answer: $\dfrac{x^2}{36} + \dfrac{y^2}{81} = 1$

28. Find the standard equation of the ellipse with center at (0, 0), a focus at $(2\sqrt{35}, 0)$ and minor axis of length 4.

   (a) $\dfrac{x^2}{70} + \dfrac{y^2}{16} = 1$ (b) $\dfrac{x^2}{70} + \dfrac{y^2}{144} = 1$ (c) $\dfrac{x^2}{4} + \dfrac{y^2}{70} = 1$

   (d) $\dfrac{x^2}{144} + \dfrac{y^2}{4} = 1$ (e) None of these

   1—Answer: d

29. Find the foci of the hyperbola: $2y^2 - 9x^2 - 18 = 0$

   (a) $(\pm\sqrt{11}, 3)$ (b) $(0, \pm\sqrt{7})$ (c) $(0, \pm\sqrt{11})$

   (d) $(\pm\sqrt{7}, 0)$ (e) None of these

   1—Answer: c

30. Find the foci of the hyperbola: $\dfrac{x^2}{144} - \dfrac{y^2}{36} = 1$

   (a) $(\pm 6\sqrt{5}, 0)$ (b) $(0, \pm 6\sqrt{5})$ (c) $(\pm 6\sqrt{3}, 0)$

   (d) $(0, \pm 6\sqrt{3})$ (e) None of these

   1—Answer: a

31. Find the vertices: $\dfrac{y^2}{81} - \dfrac{x^2}{144} = 1$

   (a) $(\pm 9, 0)$ (b) $(0, \pm 9)$ (c) $(\pm 12, 0)$
   (d) $(0, \pm 12)$ (e) None of these

   1—Answer: b

32. Find the vertices: $\dfrac{x^2}{36} - \dfrac{y^2}{25} = 1$

   (a) $(\pm 5, 0)$ (b) $(0, \pm 5)$ (c) $(0, \pm 6)$
   (d) $(\pm 6, 0)$ (e) None of these

   1—Answer: d

33. Find the equations of the asymptotes of the hyperbola: $\dfrac{x^2}{25} - \dfrac{y^2}{81} = 1$

   (a) $y = \pm \dfrac{5}{9} x$ (b) $y = \pm \dfrac{\sqrt{106}}{5} x$ (c) $y = \pm \dfrac{9}{5} x$
   (d) $y = \pm \dfrac{5\sqrt{106}}{106} x$ (e) None of these

   1—Answer: c

34. Find the equations of the asymptotes of the hyperbola: $\dfrac{y^2}{36} - \dfrac{x^2}{25} = 1$

   (a) $y = \pm \dfrac{5}{6} x$ (b) $y = \pm \dfrac{6}{5} x$ (c) $y = \pm \dfrac{\sqrt{11}}{6} x$
   (d) $y = \dfrac{\sqrt{11}}{5} x$ (e) None of these

   1—Answer: b

35. Find an equation of the hyperbola with center at $(0, 0)$, vertices at $(\pm 3, 0)$, and foci at $(\pm 3\sqrt{5}, 0)$.

   (a) $\dfrac{x^2}{9} - \dfrac{y^2}{45} = 1$ (b) $\dfrac{y^2}{9} - \dfrac{x^2}{45} = 1$ (c) $\dfrac{x^2}{9} - \dfrac{y^2}{36} = 1$
   (d) $\dfrac{x^2}{9} - \dfrac{y^2}{54} = 1$ (e) None of these

   1—Answer: c

36. Find an equation of the hyperbola with center at $(0, 0)$, vertices at $(0, \pm 9)$, and asymptotes $y = \pm \tfrac{9}{2} x$.

   (a) $\dfrac{x^2}{2} - \dfrac{y^2}{9} = 1$ (b) $\dfrac{x^2}{81} - \dfrac{y^2}{4} = 1$ (c) $\dfrac{y^2}{81} - \dfrac{x^2}{4} = 1$
   (d) $\dfrac{x^2}{77} - \dfrac{y^2}{4} = 1$ (e) None of these

   2—Answer: c

37. Find an equation of the hyperbola with vertices (±12, 0) and foci (±13, 0).

   (a) $\dfrac{x^2}{144} - \dfrac{y^2}{169} = 1$ (b) $\dfrac{x^2}{169} - \dfrac{y^2}{144} = 1$ (c) $\dfrac{y^2}{169} - \dfrac{x^2}{25} = 1$

   (d) $\dfrac{x^2}{144} - \dfrac{y^2}{25} = 1$ (e) None of these

   1—Answer: d

38. The width of an elliptical window is 6 feet. The height is 4 feet. Find an equation for the elliptical shape of the window.

   (a) $\dfrac{x^2}{4} + \dfrac{y^2}{36} = 1$ (b) $\dfrac{x^2}{9} + \dfrac{y^2}{4} = 1$ (c) $\dfrac{x^2}{36} + \dfrac{y^2}{9} = 1$

   (d) $\dfrac{x^2}{4} + \dfrac{y^2}{9} = 1$ (e) None of these

   2—Answer: b

39. The width of an elliptical window is 4 feet. The width of the window is 8 feet. Find an equation for the elliptical shape of the window.

   (a) $\dfrac{x^2}{4} + \dfrac{y^2}{16} = 1$ (b) $\dfrac{x^2}{4} + \dfrac{y^2}{8} = 1$ (c) $\dfrac{x^2}{64} + \dfrac{y^2}{16} = 1$

   (d) $\dfrac{x^2}{16} + \dfrac{y^2}{64} = 1$ (e) None of these

   2—Answer: a

❑ 4.5 Translations of Conics

1. Write in standard form: $x^2 + 4x - 8y + 4 = 0$

   (a) $y = \dfrac{1}{8}(x + 2)^2$ (b) $x(x + 4) = 8\left(y - \dfrac{1}{2}\right)$ (c) $(x + 2)^2 = 4(2)y$

   (d) $\dfrac{(x + 2)^2}{8y} = 1$ (e) None of these

   1—Answer: c

2. Write in standard form: $4x^2 + 9y^2 - 8x + 72y + 4 = 0$

   (a) $\dfrac{(x - 1)^2}{36} + \dfrac{(y + 4)^2}{16} = 1$ (b) $\dfrac{(x - 1)^2}{144} + \dfrac{(y + 8)^2}{64} = 1$

   (c) $\dfrac{(x - 1)^2}{13/4} + \dfrac{(y + 4)^2}{13/9} = 1$ (d) $\dfrac{(x - 4)^2}{327} + \dfrac{(y + 36)^2}{436/3} = 1$

   (e) None of these

   2—Answer: a

3. Write in standard form: $9x^2 - 4y^2 - 54x + 8y + 41 = 0$

   (a) $\dfrac{(y+1)^2}{41/4} - \dfrac{(x-3)^2}{41/9} = 1$

   (b) $\dfrac{(x-3)^2}{31/9} - \dfrac{(y-1)^2}{31/4} = 1$

   (c) $\dfrac{(x-3)^2}{4} - \dfrac{(y-1)^2}{9} = 1$

   (d) $\dfrac{(y+1)^2}{14} + \dfrac{(x-3)^2}{63/2} = 1$

   (e) None of these

   2—Answer: c

4. Write in standard form: $4x^2 - 5y^2 - 16x - 30y - 9 = 0$

   (a) $\dfrac{(x-4)^2}{11} - \dfrac{(y-3)^2}{4} = 1$

   (b) $\dfrac{(y+3)^2}{4} - \dfrac{(x-2)^2}{5} = 1$

   (c) $\dfrac{(y-3)^2}{6} - \dfrac{(x+2)^2}{9} = 1$

   (d) $\dfrac{(x+2)^2}{4} - \dfrac{(y+3)^2}{6} = 1$

   (e) None of these

   2—Answer: b

5. Find the vertex of the parabola: $(x+3)^2 - 8(y+6) = 0$

   (a) $(3, 6)$ (b) $(-3, -6)$ (c) $(-3, -4)$

   (d) $(-1, -6)$ (e) None of these

   1—Answer: b

6. Find the directrix: $y = \tfrac{1}{2}(x-2)^2$

   (a) $y = -\tfrac{1}{2}$ (b) $y = \tfrac{1}{8}$ (c) $x = 2$

   (d) $x = -8$ (e) None of these

   1—Answer: a

7. Find the focus of a parabola with directrix $x = 2$ and vertex $(6, 2)$.

   (a) $(8, 2)$ (b) $(6, 4)$ (c) $(10, 2)$

   (d) $(6, 6)$ (e) None of these

   1—Answer: c

8. Find the vertex: $(y-2)^2 - 8(x+1) = 0$

   (a) $(-2, -1)$ (b) $(-1, -2)$ (c) $(2, -1)$

   (d) $(-1, 2)$ (e) None of these

   1—Answer: d

9. Find an equation of the parabola with vertex $(2, -3)$ and focus $(2, 0)$.

   (a) $y^2 + 6y - 12x + 33 = 0$ (b) $x^2 - 4x + 12y + 40 = 0$ (c) $x^2 - 4x - 12y - 32 = 0$

   (d) $y^2 + 6y + 12x - 15 = 0$ (e) None of these

   2—Answer: c

10. Find an equation of the parabola with vertex $(1, -1)$ and focus $(1, 0)$.

    (a) $x^2 - 2x - 4y - 3 = 0$ (b) $x^2 + 2x - 4y + 5 = 0$ (c) $y^2 - 4x - 2y - 2 = 0$

    (d) $y^2 - 4x - 2y - 3 = 0$ (e) None of these

    2—Answer: a

11. Find an equation of the parabola with focus $(-8, 1)$ and directrix $x = 0$.
    (a) $(y - 1)^2 = 4(x - 4)$
    (b) $(x - 4)^2 = -16(y + 1)$
    (c) $(y - 1)^2 = -16(x + 4)$
    (d) $(x + 4)^2 = 2(y - 1)$
    (e) None of these

    2—Answer: c

12. Find an equation of the parabola with vertex $(2, -3)$ and directrix $y = 1$.
    (a) $(x - 2)^2 = -16(y + 3)$
    (b) $(x - 2)^2 = -8(y + 3)$
    (c) $(x + 2)^2 = 8(y - 3)$
    (d) $(y + 3)^2 = 4(x - 2)$
    (e) None of these

    1—Answer: a

13. Find the center of the ellipse: $9x^2 + 4y^2 - 36x - 24y - 36 = 0$
    (a) $(2, 3)$
    (b) $(3, -2)$
    (c) $(2\sqrt{3}, 3\sqrt{3})$
    (d) $(6, 48)$
    (e) None of these

    2—Answer: a

14. Find the center of the ellipse: $5x^2 + 2y^2 - 20x + 24y + 82 = 0$

    2—Answer: $(2, -6)$

15. Find the vertices of the ellipse: $\dfrac{(x - 1)^2}{4} + \dfrac{(y + 3)^2}{9} = 1$
    (a) $(-1, 0), (-1, 6)$
    (b) $(1, 0), (1, -6)$
    (c) $(-1, -1), (-1, 5)$
    (d) $(1, -5), (1, -1)$
    (e) None of these

    2—Answer: b

16. Find the vertices of the ellipse: $\dfrac{(x + 5)^2}{25} + \dfrac{(y - 2)^2}{9} = 1$
    (a) $(5, 0), (5, 10)$
    (b) $(-5, 0), (-5, 10)$
    (c) $(-10, 2), (0, 2)$
    (d) $(-8, 2), (-2, 2)$
    (e) None of these

    1—Answer: c

17. Find an equation of the ellipse with center at $(-1, 3)$, vertex at $(3, 3)$, and a minor axis of length 2.
    (a) $\dfrac{x^2}{16} + \dfrac{y^2}{4} = 1$
    (b) $\dfrac{x^2}{4} + \dfrac{y^2}{16} = 1$
    (c) $\dfrac{(x + 1)^2}{1} + \dfrac{(y - 3)^2}{16} = 1$
    (d) $\dfrac{(x + 1)^2}{16} + \dfrac{(y - 3)^2}{1} = 1$
    (e) None of these

    2—Answer: d

18. Find an equation of the ellipse with foci at $(0, 2)$, and vertices at $(0, 0)$, and $(0, 10)$.

    2—Answer: $25x^2 + 16y^2 - 160y = 0$

19. Determine an equation of the ellipse with foci at $(-4, -1)$, and $(-4, -3)$, and vertices $(-4, 0)$ and $(-4, -4)$.

    (a) $\dfrac{(x-4)^2}{3} + \dfrac{(y-4)^2}{4} = 1$

    (b) $\dfrac{(x+4)^2}{3} + \dfrac{(y+2)^2}{4} = 1$

    (c) $\dfrac{(x+4)^2}{9} + \dfrac{(y+2)^2}{4} = 1$

    (d) $\dfrac{(x-4)^2}{4} + \dfrac{(y-4)^2}{1} = 1$

    (e) None of these

    2—Answer: b

20. Determine an equation of the ellipse with vertices $(-1, 10)$ and $(-1, 2)$ and minor axis with length 6.

    (a) $\dfrac{(x-1)^2}{16} + \dfrac{(y-6)^2}{9} = 1$

    (b) $\dfrac{(x+1)^2}{64} + \dfrac{(y-6)^2}{36} = 1$

    (c) $\dfrac{(x+1)^2}{16} + \dfrac{(y-4)^2}{9} = 1$

    (d) $\dfrac{(x+1)^2}{9} + \dfrac{(y-6)^2}{16} = 1$

    (e) None of these

    1—Answer: d

21. Find the center of the hyperbola: $3x^2 - 4y^2 - 6x - 16y + 7 = 0$

    (a) $(1, -2)$
    (b) $(4, 3)$
    (c) $(1, -8)$
    (d) $(3, -8)$
    (e) None of these

    2—Answer: a

22. Find the vertices of the hyperbola: $\dfrac{(x-2)^2}{9} - \dfrac{(y+7)^2}{12} = 1$

    (a) $(5, -7)(-1, -7)$
    (b) $(2, -4)(2, -10)$
    (c) $(-2, 10)(-2, 4)$
    (d) $(1, 7), (-5, 7)$
    (e) None of these

    1—Answer: a

23. Determine the vertices: $\dfrac{(x+1)^2}{16} - \dfrac{(y-6)^2}{25} = 1$

    (a) $(-5, 6), (3, 6)$
    (b) $(-6, 6)(4, 6)$
    (c) $(-1, 2)(-1, 10)$
    (d) $(-1, 1)(-1, 11)$
    (e) None of these

    1—Answer: a

24. Determine the foci: $\dfrac{(x+1)^2}{16} - \dfrac{(y-6)^2}{25} = 1$

    (a) $(-1, 6)(1, -6)$
    (b) $(-1, 9)(-1, 3)$
    (c) $(-1 - \sqrt{41}, 6)(-1 + \sqrt{41}, 6)$
    (d) $(-1, 6 - \sqrt{41})(-1, 6 + \sqrt{41})$
    (e) None of these

    2—Answer: c

25. Find the equation of the hyperbola with vertices at $(0, -1)$ and $(4, -1)$, and foci at $(-2, -1)$ and $(6, -1)$.

    (a) $\dfrac{(x-2)^2}{16} - \dfrac{(y+1)^2}{4} = 1$

    (b) $\dfrac{(x-2)^2}{4} - \dfrac{(y+1)^2}{12} = 1$

    (c) $\dfrac{(y+1)^2}{12} - \dfrac{(x-2)^2}{4} = 1$

    (d) $\dfrac{(x+2)^2}{4} - \dfrac{(y-1)^2}{12} = 1$

    (e) None of these

    2—Answer: b

26. Find the standard equation of the hyperbola with center at $(2, 5)$, one focus at $(2, 15)$, and transverse axis of length 12.

    2—Answer: $\dfrac{(y-5)^2}{36} - \dfrac{(x-2)^2}{64} = 1$

27. Find the equation of the hyperbola with foci $(-3, -4)$, $(-3, 4)$ and transverse axis of length 6.

    (a) $\dfrac{(x+3)^2}{16} - \dfrac{y^2}{9} = 1$

    (b) $\dfrac{y^2}{16} - \dfrac{(x+3)^2}{9} = 1$

    (c) $\dfrac{y^2}{9} - \dfrac{(x+3)^2}{7} = 1$

    (d) $\dfrac{(x+3)^2}{9} - \dfrac{y^2}{7} = 1$

    (e) None of these

    2—Answer: c

28. Find an equation of the hyperbola with foci $(-7, -7)(-1, -7)$ and vertices $(-6, -7)(-2, -7)$.

    (a) $\dfrac{(x-4)^2}{9} - \dfrac{(y-7)^2}{4} = 1$

    (b) $\dfrac{(y+7)^2}{4} - \dfrac{(x+4)^2}{9} = 1$

    (c) $\dfrac{(y-7)^2}{4} - \dfrac{(x-4)^2}{5} = 1$

    (d) $\dfrac{(x+4)^2}{4} - \dfrac{(y+7)^2}{5} = 1$

    (e) None of these

    2—Answer: d

29. Classify the graph of $3x^2 + 6x - 4y + 12 = 0$.

    (a) Circle (b) Hyperbola (c) Ellipse

    (d) Parabola (e) None of these

    1—Answer: d

30. Classify the graph of each of the following equations.

    (a) $3x^2 + 3y^2 - 6x + 18y + 10 = 0$

    (b) $3x^2 - 2y^2 + 6x - 8y + 1 = 0$

    (c) $x^2 + 2x + 4y^2 + 1 = 0$

    1—Answer: (a) Circle (b) Hyperbola (c) Ellipse

31. Classify the graph of $2x^2 - 5y^2 + 4x - 6 = 0$.

    (a) Circle (b) Hyperbola (c) Ellipse

    (d) Parabola (e) None of these

    1—Answer: d

32. Classify the graph of $3x^2 + 3y^2 - 4x + 5y - 16 = 0$.

    (a) Circle (b) Parabola (c) Ellipse
    (d) Hyperbola (e) None of these

    1—Answer: a

33. For a science project you plan to demonstrate the elliptical orbit of a planet with the sun at one of the foci. If the center of the ellipse is (0, 0), and the length of the major axis is 120 cm, what is the smallest distance from the planet to the sun, if the sun is located at the point (1, 0)?

    (a) 61 cm (b) 59 cm (c) 119 cm
    (d) 59.99 cm (e) None of these

    2—Answer: b

34. For a science project you plan to demonstrate the elliptical orbit of a planet with the sun at one of the foci. If the center of the ellipse is (0, 0), the length of the major axis is 120 cm, and the sun is located at the point (1, 0), what is the equation in standard form of the ellipse?

    2—Answer: $\dfrac{x^2}{3600} + \dfrac{y^2}{3599} = 1$

35. Use a graphing utility to graph the conics $4x^2 + y^2 = 36$ and $y = x^2 - 4$ on the same viewing rectangle and approximate the coordinates of any points of intersection.

    2—T—Answer: 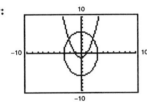 $(2.63, 2.90), (-2.63, 2.90)$

36. Use a graphing utility to graph the conics $y^2 - x^2 = 2$ and $x^2 + y^2 = 16$ on the same viewing rectangle and approximate the coordinates of any points of intersection.

    2—T—Answer: 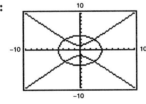 $(2.65, 3), (-2.65, 3),$
    $(-2.65, -3), (2.65, -3)$

37. Given the equation $x^2 - 6xy + y^2 + 1 = 0$, use the quadratic formula to solve for y, and use a graphing utility to graph the resulting equations. Identify the conic.

    2—T—Answer: $y = 3x \pm \sqrt{8x^2 - 1}$; 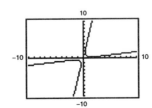 ; hyperbola

38. Given the equation $x^2 - 8xy + y^2 + 1 = 0$, use the quadratic formula to solve for y, and use a graphing utility to graph the resulting equations. Identify the conic.

   2—T—Answer: $y = -4x \pm \sqrt{15x^2 - 1}$; 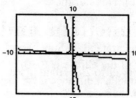 ; hyperbola

39. Given the equation $2x^2 + 2xy + y^2 - 1 = 0$, use the quadratic formula to solve for y, and use a graphing utility to graph the resulting equations. Identify the conic.

   2—T—Answer: $y = -x \pm \sqrt{1 - x^2}$; 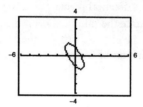 ; ellipse

40. Given the equation $5x^2 - 4xy + y^2 - 1 = 0$, use the quadratic formula to solve for y, and use a graphing utility to graph the resulting equations. Identify the conic.

   2—T—Answer: $y = 2x \pm \sqrt{1 - x^2}$; 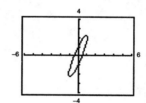 ; ellipse

CHAPTER FIVE
Exponential and Logarithmic Functions

❏ 5.1 Exponential Functions and Their Graphs

1. Evaluate: $5.1(1.32)^{\sqrt{2}}$ Round your answer to 2 decimal places.

   (a) 14.83   (b) 27.69   (c) 9.52

   (d) 7.55   (e) None of these

   1—T—Answer: d

2. Evaluate: $4.7e^{\sqrt{3}}$ Round your answer to 2 decimal places.

   (a) 82.48   (b) 74.10   (c) 26.57

   (d) 22.13   (e) None of these

   1—T—Answer: c

3. Evaluate: $\sqrt[3]{e}$ Round your answer to 2 decimal places.

   (a) 0.91   (b) 1.40   (c) 0.05

   (d) 20.09   (e) None of these

   1—T—Answer: b

4. Evaluate: $(2)(4^{2e})$ Round your answer to 2 decimal places.

   (a) 86.99   (b) 81,228.08   (c) 12,343.03

   (d) 3751.18   (e) None of these

   1—T—Answer: d

5. Evaluate when $t = 15$: $300e^{-0.076t}$

   (a) 95.95   (b) 39.31   (c) 0.000479718

   (d) -1906.12 (e) None of these

   1—T—Answer: a

6. Evaluate: $\dfrac{3e^{(0.0721)(52)}}{(1 - 0.0721)}$

   (a) 4.2727   (b) 180.6908   (c) 137.3653

   (d) -410.3055 (e) None of these

   1—T—Answer: c

7. Evaluate when $t = 3$: $y = \dfrac{300}{1 + e^{-2t}}$

   (a) 299.2582   (b) 213.3704   (c) 300.0025

   (d) 107.4591   (e) None of these

   1—T—Answer: a

8. Evaluate when $x = 65$: $200 - 5e^{0.002x}$

   1—T—Answer: 194.3059

9. Evaluate when $x = -20$: $16e^{-0.015x}$

   1—T—Answer: 21.5977

10. Match the graph with the correct function.

    (a) $f(x) = 4^x - 5$ (b) $f(x) = 4^x + 5$

    (c) $f(x) = 4^{-x} + 5$ (d) $f(x) = 4^{-x} - 5$

    1—Answer: a

    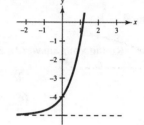

11. Match the graph with the correct function.

    (a) $y = 3^{x-1}$ (b) $y = 3^x - 1$

    (c) $y = 3^{1-x}$ (d) $y = 3^{-x} - 1$

    1—Answer: b

    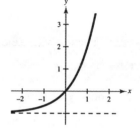

12. Match the graph with the correct function.

    (a) $f(x) = \left(\dfrac{1}{2}\right)^x - 1$ (b) $f(x) = 3^{-x^2} - 1$

    (c) $f(x) = 3^{x+1}$ (d) $f(x) = 4^{-x}$

    2—Answer: d

13. Without using a graphing utility, sketch the graph of $f(x) = 3^x - 5$.

    1—Answer:

    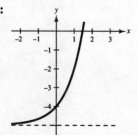

14. Without using a graphing utility, sketch the graph of $f(x) = 3^x - 2$.

    1—Answer:

    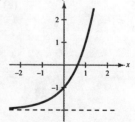

15. Identify the graph of $f(x) = e^x$.

(a)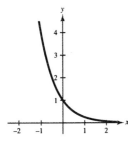

(b)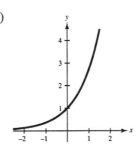

(c)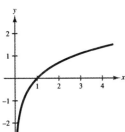

(d)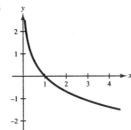

(e) None of these

1—Answer: b

16. Match the exponential function with the correct graph: $y = 2 - e^x$.

(a)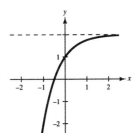

(b)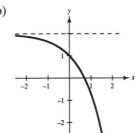

(c)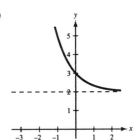

(d)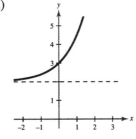

(e) None of these

1—Answer: b

17. Match the exponential function with the correct graph: $y = \left(\frac{1}{5}\right)^x + 1$

(a)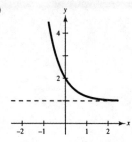

(b)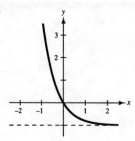

(c)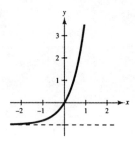

(d)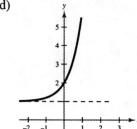

(e) None of these

1—Answer: a

18. The domain of $f(x) = 1 + e^{-x}$ is:

   (a) $(-\infty, \infty)$ (b) $(0, \infty)$ (c) $(-1, \infty)$

   (d) $(1, \infty)$ (e) None of these

   1—Answer: a

19. The domain of $f(x) = 3 - e^x$ is:

   (a) $(3, \infty)$ (b) $[0, \infty)$ (c) $(-\infty, \infty)$

   (d) $(-\infty, 3)$ (e) None of these

   1—Answer: c

20. The range of $f(x) = 1 + e^{-x}$ is:

   (a) $(-\infty, \infty)$ (b) $(0, \infty)$ (c) $(-1, \infty)$

   (d) $(1, \infty)$ (e) None of these

   2—Answer: d

21. The range of $f(x) = 3 - e^x$ is:

   (a) $(3, \infty)$ (b) $[0, \infty)$ (c) $(-\infty, \infty)$

   (d) $(-\infty, 3)$ (e) None of these

   2—Answer: d

22. Use a graphing utility to graph $f(x) = 2^{x-1}$.

1—T—Answer: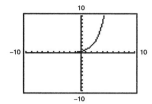

23. Use a graphing utility to graph $y = 3^{x+1} - 1$.

1—T—Answer: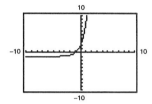

24. Use a graphing utility to graph $y = \left(\frac{2}{5}\right)^x$.

1—T—Answer: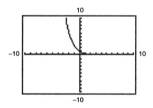

25. Use a graphing utility to graph $s(t) = 2e^{0.5t}$.

1—T—Answer: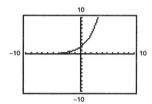

26. Use a graphing utility to graph the function $f(x) = \dfrac{2}{1 + e^{-0.2/x}}$. Use the graph to find any asymptotes of the function.

2—T—Answer: 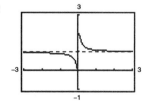 ; Vertical asymptote: $x = 0$
Horizontal asymptote: $y = 1$

27. Use a graphing utility to graph the function $f(x) = \dfrac{4}{2 + e^{-0.2/x}}$. Use the graph to find any asymptotes of the function.

2—T—Answer: 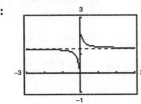 ; Vertical asymptote: $x = 0$

Horizontal asymptote: $y = \dfrac{4}{3}$

28. Use a graphing utility to graph the function $f(x) = x^2 e^x$. Use the graph to determine the intervals for which the function is increasing and decreasing and any relative extrema.

2—T—Answer: 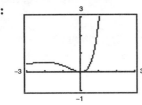 ; Increasing on $(-\infty, -2)$ and $(0, \infty)$

Decreasing on $(-2, 0)$

Relative maximum: $(-2, 0.541)$ or $\left(-2, \dfrac{4}{e^2}\right)$

Relative minimum: $(0, 0)$

29. Use a graphing utility to graph the function $f(x) = xe^{-x}$. Use the graph to determine the intervals for which the function is increasing and decreasing and any relative extrema.

2—T—Answer: 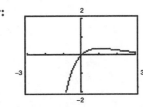 ; Increasing on $(-\infty, 1)$

Decreasing on $(1, \infty)$

Relative maximum: $(1, 0.368)$ or $\left(1, \dfrac{1}{e}\right)$

30. $1500 is invested at a rate of 8% compounded quarterly. What is the balance at the end of 5 years?

(a) $1624.67 (b) $2237.74 (c) $2228.92

(d) $2226.04 (e) None of these

1—Answer: c

31. $1500 is invested at a rate of 10% compounded quarterly. What is the balance at the end of 12 years?

(a) $1657.70 (b) $3512.55 (c) $4955.47

(d) $4980.18 (e) None of these

1—Answer: c

32. $2100 is invested at a rate of 7% compounded monthly. What is the balance at the end of 10 years?

1—Answer: $4220.29

33. $3500 is invested at a rate of 9% compounded continuously. What is the balance at the end of 18 years?

    (a) $68,932.98 (b) $17,685.82 (c) $17,493.53
    (d) $8608.61   (e) None of these

    1—Answer: b

34. $3500 is invested at a rate of $4\frac{1}{2}$% compounded continuously. What is the balance at the end of 10 years?

    (a) $315,059.96 (b) $5472.45 (c) $5221.39
    (d) $5489.09   (e) None of these

    1—Answer: d

35. $2000 is invested at a rate of $7\frac{1}{2}$% compounded continuously. What is the balance at the end of 20 years?

    1—Answer: $8963.38

36. Determine the amount of money that should be invested at a rate of 8% compounded quarterly to produce a final balance of $20,000 in 10 years.

    (a) $16,406.97 (b) $9057.81 (c) $18,463.80
    (d) $9081.26   (e) None of these

    2—Answer: b

37. Determine the amount of money that should be invested at a rate of $6\frac{1}{2}$% compounded monthly to produce a final balance of $15,000 in 20 years.

    (a) $4102.34   (b) $5216.07 (c) $2458.83
    (d) $14,056.14 (e) None of these

    2—Answer: a

38. Determine the amount of money that should be invested at a rate of 7% compounded continuously to produce a final balance of $15,000 in 20 years.

    2—Answer: $3698.95

39. A certain population decreases according to the equation $y = 300 - 5e^{0.2t}$. Find the initial population and the population (to the nearest integer) when $t = 10$.

    2—Answer: 295, 263

40. A certain population grows according to the equation $y = 40e^{0.025t}$. Find the initial population and the population (to the nearest integer) when $t = 50$.

    2—Answer: 40, 140

41. A certain population increases according to the model $P(t) = 250e^{0.47t}$. Use the model to determine the population when $t = 5$. Round your answer to the nearest integer.

    (a) 40   (b) 1597 (c) 1998
    (d) 2621 (e) None of these

    1—Answer: d

42. A certain population increases according to the model $P(t) = 250e^{0.47t}$. Use the model to determine the population when $t = 10$. Round your answer to the nearest integer.

(a) 400 (b) 4091 (c) 27,487

(d) 23,716 (e) None of these

1—Answer: c

43. A certain population increases according to the model $P(t) = 250e^{0.47t}$. Use the model to determine the population when $t = 8$. Round your answer to the nearest integer.

(a) 400 (b) 2621 (c) 10,737

(d) 27,487 (e) None of these

1—Answer: c

❑ 5.2 Logarithmic Functions and Their Graphs

1. Evaluate: $\log_7 7$

(a) 1 (b) 0 (c) 2

(d) 49 (e) None of these

1—Answer: a

2. Evaluate: $\log_a a^3$

(a) a^3 (b) a (c) 3

(d) $3a$ (e) None of these

1—Answer: c

3. Evaluate: $\log_a \dfrac{1}{a}$

(a) 1 (b) -1 (c) a

(d) $\dfrac{1}{a}$ (e) None of these

1—Answer: b

4. Evaluate: $\ln e^{1-x}$

(a) e^{1-x} (b) e (c) $1-x$

(d) $\ln(1-x)$ (e) None of these

1—Answer: c

5. Evaluate: $\ln 3.76$

(a) 1.3244 (b) 0.5752 (c) 42.9484

(d) 5754.3994 (e) None of these

1—Answer: a

Section 5.2 Logarithmic Functions and Their Graphs   283

6. Evaluate: $\log \sqrt{18}$
   - (a) $\sqrt{18}$
   - (b) 4.2426
   - (c) 1.4452
   - (d) 0.6276
   - (e) None of these

   1—Answer: d

7. Evaluate: $\ln(1 + \sqrt{2})$
   - (a) 0.3828
   - (b) 0.8814
   - (c) 0.3466
   - (d) 0.1505
   - (e) None of these

   1—Answer: b

8. Evaluate: $\log(1 + \sqrt{2})$
   - (a) 0.3828
   - (b) 0.8814
   - (c) 0.3466
   - (d) 0.1505
   - (e) None of these

   1—Answer: a

9. Write the logarithmic form: $4^3 = 64$
   - (a) $4 \log 3 = 64$
   - (b) $\log_4 64 = 3$
   - (c) $\log_3 4 = 64$
   - (d) $\log_3 64 = 4$

   1—Answer: b

10. Write the logarithmic form: $5^2 = 25$

    1—Answer: $\log_5 25 = 2$

11. Write the logarithmic form: $3^5 = 243$

    1—Answer: $\log_3 243 = 5$

12. Write the exponential form: $\log_b 37 = 2$
    - (a) $37^2 = b$
    - (b) $2^b = 37$
    - (c) $b = 10$
    - (d) $b^2 = 37$
    - (e) None of these

    1—Answer: d

13. Write the exponential form: $\log_b 7 = 13$
    - (a) $7^{13} = b$
    - (b) $b^{13} = 7$
    - (c) $b^7 = 13$
    - (d) $7^b = 13$
    - (e) None of these

    1—Answer: b

14. Write the exponential form: $\log_7 b = 12$
    - (a) $7^{12} = b$
    - (b) $b^7 = 12$
    - (c) $7^b = 12$
    - (d) $b^{12} = 7$
    - (e) None of these

    1—Answer: a

15. Evaluate: $\dfrac{15 \ln 23}{\ln 7 - \ln 2}$
    - (a) 37.5429
    - (b) 23.4767
    - (c) 34.8698
    - (d) 22,218,828.26
    - (e) None of these

    1—Answer: a

16. Evaluate: $\dfrac{3 \ln 5}{7 \ln 6 - 2 \ln 7}$

    (a) -3.8222 (b) -2.6559 (c) 0.5582

    (d) -11.6058 (e) None of these

    1—Answer: c

17. Evaluate: $\dfrac{16 \ln 5}{1 + 2 \ln 3}$

    (a) 918.3228 (b) 27.9482 (c) 8.0542

    (d) 22.5538 (e) None of these

    1—Answer: c

18. Evaluate: $\dfrac{16 \ln (1/2)}{3 \ln 10}$

    1—Answer: -1.6055

19. Find the domain of the function: $f(x) = \ln(3x + 1)$

    (a) $(-\infty, \infty)$ (b) $\left(-\tfrac{1}{3}, \infty\right)$ (c) $(0, \infty)$

    (d) $\left(\tfrac{1}{3}, \infty\right)$ (e) None of these

    1—Answer: b

20. Find the domain of the function: $f(x) = 3 \log(5x - 2)$

    (a) $(-\infty, \infty)$ (b) $(0, \infty)$ (c) $\left(\tfrac{2}{5}, \infty\right)$

    (d) $(0.064, \infty)$ (e) None of these

    1—Answer: c

21. Find the domain of the function: $f(x) = 3 + \ln(x - 1)$

    (a) $(-\infty, \infty)$ (b) $(0, \infty)$ (c) $(1, \infty)$

    (d) $(3, \infty)$ (e) None of these

    1—Answer: c

22. Find the domain of the function: $f(x) = 3 - \log(x^2 - 1)$

    2—Answer: $(-\infty, -1), (1, \infty)$

23. Find the domain of the function: $f(x) = \log_3 (x^2 - 4)$

    2—Answer: $(-\infty, -2), (2, \infty)$

24. Find the vertical asymptote: $f(x) = \ln(x + 2)$

    (a) $x = 2$ (b) $x = 0$ (c) $y = 2$

    (d) $x = -2$ (e) None of these

    2—Answer: d

25. Find the vertical asymptote: $f(x) = 2 + \ln x$

    (a) $x = 2$ (b) $y = 2$ (c) $x = 0$

    (d) $x = -2$ (e) None of these

    2—Answer: c

26. Match the graph with the correct function.

(a) $f(x) = -3 + \ln x$ (b) $f(x) = 3 + \ln x$

(c) $f(x) = \ln(x - 3)$ (d) $f(x) = \ln(x + 3)$

1—Answer: d

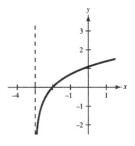

27. Match the graph with the correct function.

(a) $f(x) = 3 + \log x$ (b) $f(x) = \log(x + 3)$

(c) $f(x) = \frac{1}{3} \log x$ (d) $f(x) = 3 \log x$

1—Answer: a

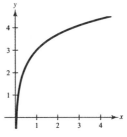

28. Match the graph with the correct function.

(a) $f(x) = e^x$ (b) $f(x) = e^{x-1}$

(c) $f(x) = \ln x$ (d) $f(x) = \ln(x - 1)$

1—Answer: d

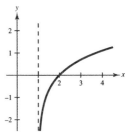

29. Sketch the graph: $f(x) = 1 + \log_5 x$

1—Answer:

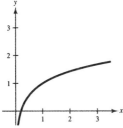

30. Sketch the graph: $y = \ln(1 - x)$

2—Answer:

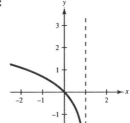

31. Use a graphing utility to graph $f(x) = \ln(x + 1)$. Determine the domain, x-intercept, and vertical asymptote of the function.

2—T—Answer:

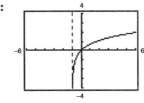

Domain: $(-1, \infty)$

x-intercept: $(0, 0)$

Vertical asymptote: $x = -1$

32. Use a graphing utility to graph $f(x) = \ln(2 - x)$. Determine the domain, x-intercept, and vertical asymptote of the function.

   2—T—Answer: 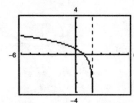 Domain: $(-\infty, 2)$
   x-intercept: $(0, 0)$
   Vertical asymptote: $x = 2$

33. Use a graphing utility to graph $f(x) = \dfrac{\ln x}{x}$. Use the graph to determine the intervals in which the function is increasing and decreasing and approximate any relative maximum or minimum.

   2—T—Answer: 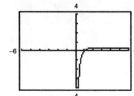 Increasing on $(0, 2.718)$ or $(0, e)$
   Decreasing on $(2.718, \infty)$ or (e, ∞)
   Relative maximum: $(2.718, 0.368)$ or $\left(e, \dfrac{1}{e}\right)$

34. Use a graphing utility to graph $f(x) = x \ln x$. Use the graph to determine the intervals in which the function is increasing and decreasing and approximate any relative maximum or minimum.

   2—T—Answer: 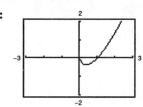 Increasing on $(0.368, \infty)$ or $\left(\dfrac{1}{e}, \infty\right)$
   Decreasing on $(0, 0.368)$ or $\left(0, \dfrac{1}{e}\right)$
   Relative minimum: $(0.368, -0.368)$ or $\left(\dfrac{1}{e}, -\dfrac{1}{e}\right)$

35. Students in an algebra class were given an exam and then tested monthly with an equivalent exam. The average score for the class was given by the human memory model

   $$f(t) = 85 - 16 \log_{10}(t + 1), \quad 0 \le t \le 12$$

   where t is the time in months. What is the average score after 3 months?

   (a) 77 (b) 67 (c) 75
   (d) 63 (e) None of these

   2—Answer: a

36. Students in an algebra class were given an exam and then tested monthly with an equivalent exam. The average score for the class was given by the human memory model

   $$f(t) = 85 - 16 \log_{10}(t + 1), \quad 0 \le t \le 12$$

   where t is the time in months. What is the average score after 5 months?

   (a) 73 (b) 74 (c) 59
   (d) 56 (e) None of these

   2—Answer: a

37. Students in an algebra class were given an exam and then tested monthly with an equivalent exam. The average score for the class was given by the human memory model

$$f(t) = 85 - 16 \log_{10}(t+1), \quad 0 \le t \le 12$$

where t is the time in months. What is the average score after 10 months?

(a) 69   (b) 48   (c) 47
(d) 68   (e) None of these

2—Answer: d

38. A principal P invested at $7\frac{1}{2}\%$ interest compounded continuously increases to an amount K times the original principal after t years, where t is given by $t = \dfrac{\ln K}{0.075}$. Determine the number of years necessary to triple the investment

(Hint: $K = 3$).

(a) 6.4   (b) 14.6   (c) 12.8
(d) 8.2   (e) None of these

2—Answer: b

39. A principal P invested at $6\frac{1}{2}\%$ interest compounded continuously increases to an amount K times the original principal after t years, where t is given by $t = \dfrac{\ln K}{0.065}$. Determine the number of years necessary to triple the investment

(Hint: $K = 3$).

(a) 7.3   (b) 9.2   (c) 14.8
(d) 16.9   (e) None of these

2—Answer: d

40. A principal P invested at $7\frac{1}{2}\%$ interest compounded continuously increases to an amount K times the original principal after t years, where t is given by $t = \dfrac{\ln K}{0.085}$. Determine the number of years necessary to triple the investment

(Hint: $K = 3$).

(a) 5.6   (b) 8.2   (c) 12.9
(d) 15.1   (e) None of these

2—Answer: c

❑ 5.3 Properties of Logarithms

1. Use the change of base formula to identify the expression that is equivalent to $\log_2 7$.

(a) $\dfrac{\log 2}{\log 7}$ (b) $\dfrac{\ln 2}{\ln 7}$ (c) $\dfrac{\ln 7}{\ln 2}$
(d) $2 \log 7$ (e) None of these

1—Answer: c

2. Use the change of base formula to identify the expression that is equivalent to $\log_3 5$.

   (a) $\dfrac{\log 5}{\log 3}$
   (b) $\dfrac{\ln 3}{\ln 5}$
   (c) $5 \ln 3$
   (d) $\log \dfrac{5}{3}$
   (e) None of these

   1—Answer: a

3. Use the change of base formula to identify the expression that is equivalent to $\log_3 10$.

   (a) $\dfrac{\ln 3}{\ln 10}$
   (b) $10 \log 3$
   (c) $\ln \dfrac{10}{3}$
   (d) $\dfrac{1}{\log 3}$
   (e) None of these

   2—Answer: d

4. Evaluate $\log_4 7$ using the change of base formula.

   (a) 0.2430
   (b) 0.5596
   (c) 0.7124
   (d) 1.4037
   (e) None of these

   2—Answer: d

5. Evaluate $\log_{1/2} 13$ using the change of base formula.

   (a) 2.5649
   (b) 1.1139
   (c) −0.2702
   (d) −3.7004
   (e) None of these

   2—Answer: d

6. Evaluate $\log_7 15$ using the change of base formula.

   (a) 1.3917
   (b) 12.6765
   (c) 2.1429
   (d) 0.7186
   (e) None of these

   2—Answer: a

7. Evaluate $\log_5 22$ using the change of base formula.

   2—Answer: 1.9206

8. Evaluate $\log_5 17$ using the change of base formula.

   2—Answer: 1.7604

9. Write as a sum, difference, or multiple of logarithms: $\log \sqrt[3]{\dfrac{a^2 b}{c}}$

   (a) $\sqrt[3]{\dfrac{2 \log a + \log b}{\log c}}$
   (b) $\dfrac{1}{3}\left(\dfrac{2 \log a + \log b}{\log c}\right)$
   (c) $\dfrac{1}{3}(2 \log a + \log b - \log c)$
   (d) $\sqrt[3]{2 \log a^2 + \log b - \log c}$
   (e) None of these

   1—Answer: c

10. Write as a sum, difference, or multiple of logarithms: $\log_b \left(\dfrac{x^3 y^2}{\sqrt{w}} \right)$

(a) $x^3 + y^3 - \sqrt{w}$

(b) $\dfrac{1}{3} \log_b x + \dfrac{1}{2} \log_b y - 2 \log_b w$

(c) $3 \log_b x + 2 \log_b y - \dfrac{1}{2} \log_b w$

(d) $\dfrac{3 \log x + 2 \log y}{(1/2) \log w}$

(e) None of these

1—Answer: c

11. Write as a sum, difference, or multiple of logarithms: $\ln \dfrac{5x}{\sqrt[3]{x^2 + 1}}$

1—Answer: $\ln 5 + \ln x - \dfrac{1}{3} \ln(x^2 + 1)$

12. The expression $\log_2 \sqrt{\dfrac{x^2}{y}}$ is equivalent to:

(a) $\dfrac{1}{2} [\log_2 x - \log_2 y]$

(b) $\log_2 x - \dfrac{1}{2} \log_2 y$

(c) $\dfrac{1}{2} [\log_2 x + \log_2 y]$

(d) $\log_2 x + \dfrac{1}{2} \log_2 y$

(e) None of these

1—Answer: b

13. Write as the logarithm of a single quantity: $\tfrac{1}{4} \log_b 16 - 2 \log_b 5 + \log_b 7$

(a) $\dfrac{14}{25}$

(b) $\log_b \dfrac{2}{175}$

(c) 1

(d) $\log_b \dfrac{14}{25}$

(e) None of these

1—Answer: d

14. Write as the logarithm of a single quantity: $\dfrac{1}{5} [3 \log(x + 1) + 2 \log(x - 1) - \log 7]$

2—Answer: $\log \sqrt[5]{\dfrac{(x + 1)^3 (x - 1)^2}{7}}$

15. Write as the logarithm of a single quantity: $\tfrac{1}{2} [\ln(x + 1) + 2 \ln(x - 1)] + \tfrac{1}{3} \ln x$

(a) $\ln \sqrt[3]{x} \sqrt{(x + 1)(x^2 - 1)}$

(b) $\ln \sqrt[3]{x} \sqrt{x^2 - 1}$

(c) $\ln \sqrt{x(x^2 - 1)}$

(d) $\ln \sqrt[3]{x(x + 1)(x - 1)^2}$

(e) None of these

2—Answer: e

16. Write as the logarithm of a single quantity: $\log_2(x - 2) + \log_2(x + 2)$

(a) $-2 + 2 \log_2 x$

(b) $\log_2(x^2 - 4)$

(c) $2 \log_2 x$

(d) $\log_2 2x$

(e) None of these

1—Answer: b

17. Evaluate $\log_a 24$, given that $\log_a 2 = 0.4307$ and $\log_a 3 = 0.6826$.

(a) 0.8820

(b) 1.9747

(c) 0.2940

(d) 1.1133

(e) None of these

2—Answer: b

18. Evaluate $\log_b\left(\dfrac{14}{3b}\right)$, given that $\log_b 2 = 0.2789$, $\log_b 3 = 0.4421$, and $\log_b 7 = 0.7831$.

    2—Answer: -0.3801

19. Evaluate $\log_b \sqrt{10b}$, given that $\log_b 2 = 0.3562$ and $\log_b 5 = 0.8271$.

    2—Answer: 1.09165

20. Simplify: $\ln 5e^3$

    (a) $3 + \ln 5$ (b) $3 \ln 5$ (c) $3 + 3 \ln 5$
    (d) $5e^3$ (e) None of these

    2—Answer: a

21. Simplify the following:

    (a) $\log_b b$ (b) $\log_b\left(\dfrac{M}{N}\right)$ (c) $b^{\log_b x}$

    1—Answer: (a) 1   (b) $\log_b M - \log_b N$ (c) x

22. Simplify: $\log_a \sqrt[3]{a}$

    (a) 1 (b) -3 (c) 0
    (d) $\tfrac{1}{3}$ (e) None of these

    1—Answer: d

23. Simplify: $\ln \sqrt[3]{e^2 x}$

    (a) $\dfrac{2e}{3} + \dfrac{1}{3}\ln x$ (b) $\dfrac{2}{3} + \ln \dfrac{x}{3}$ (c) $\dfrac{2}{3} + \dfrac{1}{3}\ln x$
    (d) $\dfrac{2e}{3} + \ln \dfrac{x}{3}$ (e) None of these

    2—Answer: c

24. Simplify: $\ln \sqrt[4]{e^3 x}$

    (a) $\dfrac{3}{4} + \dfrac{1}{4}\ln x$ (b) $\dfrac{3}{4} + \ln \dfrac{x}{4}$ (c) $\dfrac{3e}{4} + \dfrac{1}{4}\ln x$
    (d) $\dfrac{3e}{4} + \ln \dfrac{x}{4}$ (e) None of these

    2—Answer: a

25. Simplify: $\ln \sqrt[5]{e^3 x}$

    (a) $\dfrac{3e}{5} + \dfrac{1}{5}\ln x$ (b) $\dfrac{3e}{5} + \ln \dfrac{x}{5}$ (c) $\dfrac{3}{5} + \ln \dfrac{x}{5}$
    (d) $\dfrac{3}{5} + \dfrac{1}{5}\ln x$ (e) None of these

    2—Answer: d

26. Evaluate $\log_a 16$, given that $\log_a 2 = 0.4307$.

    (a) 0.0344 (b) 1.7228 (c) 4.4307
    (d) 1.8168 (e) None of these

    1—Answer: b

27. Evaluate $\log_a 18$, given that $\log_a 2 = 0.2789$, $\log_a 3 = 0.4421$.

  (a) 1.1631   (b) 0.2466   (c) 0.0349
  (d) 1.4420   (e) None of these

  1—Answer: a

28. Evaluate $\log_a \frac{9}{2}$, given that $\log_a 2 = 0.2789$, $\log_a 3 = 0.4421$.

  (a) -0.0834 (b) 1.1631   (c) -0.3264
  (d) 0.6053   (e) None of these

  1—Answer: d

29. Simplify: $\log_6 \sqrt{6}$

  (a) 2.4495   (b) -2 (c) 1
  (d) $\frac{1}{2}$ (e) None of these

  2—Answer: d

30. Simplify: $\log_2 \frac{1}{16}$

  (a) 4   (b) -4 (c) 8
  (d) $\frac{1}{2}$ (e) None of these

  1—Answer: b

31. Simplify: $\ln \sqrt{e^3}$

  (a) $\ln \frac{3}{2}$ (b) $\ln \frac{2}{3}$ (c) $\frac{3}{2}$
  (d) $\frac{2}{3}$ (e) None of these

  1—Answer: c

32. Simplify: $\log_b 3b^4$

  (a) $4 \log_b 3 + 1$ (b) $4 + 4 \log_b 3$ (c) $4 + \log_b 3$
  (d) 12   (e) None of these

  1—Answer: c

33. Simplify: $\log_b \sqrt{4b^3}$

  (a) $\log_b 2 + \sqrt{b^3}$ (b) $\frac{3}{2} + \frac{3}{2} \log_b 4$ (c) $\frac{3}{2} + 3 \log_b 2$
  (d) $\frac{3}{2} + \log_b 2$ (e) None of these

  1—Answer: d

34. Use a graphing utility to graph both equations $y_1 = \ln \sqrt{x}$ and $y_2 = \dfrac{\ln x}{2}$ on the same viewing rectangle. Are the expressions equivalent? Explain.

  2—T—Answer: 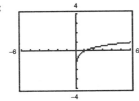 ; Yes, $\ln \sqrt{x} = \ln x^{1/2} = \dfrac{1}{2} \ln x = \dfrac{\ln x}{2}$.

35. Use a graphing utility to graph both equations $y_1 = \ln\left(\dfrac{x+1}{x-1}\right)$ and $y_2 = \ln(x+1) - \ln(x-1)$ on the same viewing rectangle. Are the expressions equivalent? Explain.

2—T—Answer: 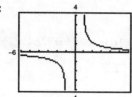 Yes; $\ln\left(\dfrac{x+1}{x-1}\right) = \ln(x+1) - \ln(x-1)$

❑ 5.4 Exponential and Logarithmic Equations

1. Solve for x: $3^{2x} = 81$

(a) $x = 13.5$ (b) $x = \frac{1}{4}$ (c) $x = 4$

(d) $x = 2$ (e) None of these

1—Answer: d

2. Solve for x: $16 = 2^{7x-5}$

(a) 0.1143 (b) -0.3010 (c) $\frac{13}{7}$

(d) $\frac{9}{7}$ (e) None of these

1—Answer: d

3. Solve for x: $27^x = 81$

(a) $\frac{3}{4}$ (b) $-\frac{1}{3}$ (c) $\frac{4}{3}$

(d) $\frac{2}{3}$ (e) None of these

1—Answer: c

4. Solve for x: $\log_x 8 = -3$

(a) 2 (b) 512 (c) $\frac{1}{2}$

(d) -2 (e) None of these

1—Answer: c

5. Solve for x: $\ln e^{4x} = 60$

(a) 2.7832 (b) 15 (c) 1.0236

(d) 2.7081 (e) None of these

1—Answer: b

Section 5.4   Exponential and Logarithmic Equations   293

6. Solve for x: $\ln e^{2x+1} = 9$

   (a) $\dfrac{-1 + \ln 9}{2}$
   (b) $\dfrac{9}{2 \ln e} - \dfrac{1}{2}$
   (c) 23
   (d) 4
   (e) None of these

   1—Answer: d

7. Solve for x: $25^{x-2} = 5^{3x}$

   1—Answer: -4

8. Solve for x: $2x + \ln e^{4x} = 12$

   1—Answer: 2

9. Which of the following equations is not true?

   (a) $b^{\log_b c} = c$
   (b) $\log_1 b = b$
   (c) $\log_b b = 1$
   (d) All of these equations are false.
   (e) All of these equations are true.

   1—Answer: b

10. Simplify: $e^{3 \ln 2}$

    (a) 6
    (b) 8
    (c) 9
    (d) 5
    (e) None of these

    1—Answer: b

11. Simplify: $e^{2 \ln(x+1)}$

    (a) $(x + 1)^2$
    (b) $2(x + 1)$
    (c) $e^2 \ln(x + 1)$
    (d) $x + 1$
    (e) None of these

    1—Answer: a

12. Simplify: $3e^{2 \ln x}$

    (a) $3x$
    (b) $3xe^2$
    (c) $3x^2$
    (d) $\ln x^3$
    (e) None of these

    1—Answer: c

13. Simplify: $2e^{3 \ln(x+1)}$

    (a) $2(x + 1)e^3$
    (b) $6(x + 1)$
    (c) $3(x + 1) \ln 2$
    (d) $2(x + 1)^3$
    (e) None of these

    1—Answer: d

14. Simplify: $3 + \ln e^{5x}$

    (a) $\dfrac{\ln 3}{5x}$
    (b) $\ln 3 + 5x$
    (c) $3 + 5x$
    (d) $5x \ln 3$
    (e) None of these

    1—Answer: c

15. Simplify: $7 + 2 \ln e^{5x}$

(a) $7 + 10x$ (b) $7 + 2^{5x}$ (c) $45x$

(d) $7 + 2e^{5x}$ (e) None of these

1—Answer: a

16. Simplify: $7 + \ln e^{5x}$

(a) $5x + \ln 7$ (b) $7 + 5x$ (c) $\dfrac{\ln 7}{5x}$

(d) $35x$ (e) None of these

1—Answer: b

17. Solve for x: $3^{5x+1} = 5$

(a) 0.1022   (b) 0.0930   (c) 0.1333

(d) 0.2218   (e) None of these

1—Answer: b

18. Solve for t: $e^{-0.0097t} = 12$

(a) -256.1759 (b) -1237.1134 (c) $16{,}778{,}844.47$

(d) -2.5886 (e) None of these

1—Answer: a

19. Solve for x: $3^{2x} = 5^{x-1}$

(a) -0.5563 (b) -1 (c) -2.7381

(d) 15.2755 (e) None of these

2—Answer: c

20. Solve for x: $2^{x-1} = 5^{2x+6}$

2—Answer: -4.0977

21. Solve for x: $16^x = 8^{2x-1}$

2—Answer: $\dfrac{3}{2}$

22. Solve for x: $3^{1-x} = 5^x$

(a) $\ln \dfrac{1}{5}$ (b) $\ln \dfrac{3}{5}$ (c) $\dfrac{\ln 3}{\ln 15}$

(d) $(\ln 3)\ln(15)$ (e) None of these

2—Answer: c

23. Solve for x: $2^{1-x} = 3^x$

(a) $\dfrac{\ln 2}{\ln 6}$ (b) $\ln \dfrac{1}{3}$ (c) $\ln \dfrac{2}{3}$

(d) $\ln 3 + \ln 2$ (e) None of these

2—Answer: a

24. Solve for x: $\ln x = 5.3670$

1—Answer: 214.2192

25. Solve for x: $\log_x 16 = 5$

2—Answer: 1.7411

26. Solve for x: $\log(3x + 7) + \log(x - 2) = 1$

   (a) $\frac{8}{3}$ (b) $3, -\frac{8}{3}$ (c) 2

   (d) $2, -\frac{5}{3}$ (e) None of these

   2—Answer: a

27. Solve for x: $\ln(7 - x) + \ln(3x + 5) = \ln(24x)$

   (a) $\frac{6}{11}$ (b) $\frac{7}{3}$ (c) $\frac{7}{3}, -5$

   (d) $\frac{6}{11}, 5$ (e) None of these

   2—Answer: b

28. Solve for x: $\log(7 - x) - \log(3x + 2) = 1$

   (a) $\frac{19}{31}$ (b) $-\frac{13}{31}$ (c) $-\frac{27}{29}$

   (d) $\frac{9}{4}$ (e) None of these

   2—Answer: b

29. Solve for x: $\log x + \log(x + 3) = 1$

   2—Answer: 2

30. Solve for x: $x^2 - 4x = \log_2 32$

   2—Answer: $-1, 5$

31. Solve for x: $\log_3(x^2 + 5) = \log_3(4x^2 - 2x)$

   1—Answer: $-1, \frac{5}{3}$

32. Use a graphing utility to graph $f(x) = 5e^{x+1} - 10$ and approximate its zero accurate to three decimal places.

   (a) 1.693 (b) -0.307 (c) 0.588

   (d) -1.693 (e) None of these

   2—T—Answer: b

33. Use a graphing utility to graph $f(x) = 2e^{x/2} - 14$ and approximate its zero accurate to three decimal places.

   (a) 2.639 (b) 0.973 (c) 1.946

   (d) 3.892 (e) None of these

   2—T—Answer: d

34. Use a graphing utility to graph $f(x) = 2e^{x-5} - 10$ and approximate its zero accurate to three decimal places.

   (a) 0.322 (b) -3.391 (c) 6.609

   (d) 8.047 (e) None of these

   2—T—Answer: c

35. Use a graphing utility to graph $f(x) = 10e^{2x+1} - 5$ and approximate its zero accurate to three decimal places.

   2—T—Answer: -0.847

36. Use a graphing utility to graph $f(x) = 4e^{x-5/2} - 2$ and approximate its zero accurate to three decimal places.

    2—T—**Answer:** 3.614

37. Use a graphing utility to approximate the point of intersection of the graphs of $y_1 = 10$ and $y_2 = e^{x+1}$. Round the result to three decimal places.

    2—T—**Answer:** (1.303, 10)

38. Use a graphing utility to approximate the point of intersection of the graphs of $y_1 = 5$ and $y_2 = 2\ln(x + 1)$. Round the result to three decimal places.

    2—T—**Answer:** (11.064, 5)

39. Use a graphing utility to approximate the point of intersection of the graphs of $y_1 = 200$ and $y_2 = 1250e^{-x/2}$. Round the result to three decimal places.

    2—T—**Answer:** (3.665, 200)

40. Use a graphing utility to approximate the point of intersection of the graphs of $y_1 = \frac{4}{5}$ and $y_2 = 2\ln(x - 2)$ Round the result to three decimal places.

    2—T—**Answer:** (3.492, 0.8)

41. Find the number of years required for a $3000 investment to double at a 7% interest rate compounded continuously.

    2—T—**Answer:** 9.9 years

42. Find the number of years required for a $2000 investment to triple at an 8% interest rate compounded continuously.

    (a) 12.6   (b) 13.7   (c) 11.2

    (d) 15.1   (e) None of these

    2—T—**Answer:** b

43. Find the number of years required for a $2000 investment to triple at an $9\frac{1}{2}$% interest rate compounded continuously.

    (a) 12.6   (b) 13.7   (c) 11.6

    (d) 15.1   (e) None of these

    2—T—**Answer:** c

44. The yield V (in millions of cubic feet per acre) for the forest at age t years is given by $V = 6.7e^{-48.1/t}$. Find the time necessary to have a yield of 1.7 million cubic feet.

    2—T—**Answer:** 35 years

45. The yield V (in millions of cubic feet per acre) for the forest at age t years is given by $V = 6.7e^{-48.1/t}$. Find the time necessary to have a yield of 2.1 million cubic feet.

    (a) 22.1   (b) 25.2   (c) 39.8

    (d) 41.5   (e) None of these

    2—T—**Answer:** d

5.5 Exponential and Logarithmic Models

1. If $3700 is invested at $11\frac{1}{2}\%$ interest compounded continuously, find the balance, B, in the account after 5 years.

    (a) $3918.99   (b) $20,754.65   (c) $6575.38
    (d) $7376.75   (e) None of these

    1—Answer: c

2. If $9200 is invested at $9\frac{1}{2}\%$ interest compounded continuously, find the balance, B, in the account after 10 years.

    (a) $22,628.35   (b) $25,040.56   (c) $17,940.00
    (d) $23,788.53   (e) None of these

    1—Answer: d

3. Find the balance after 15 years if $1500 is invested in an account that pays $8\frac{1}{2}\%$ compounded quarterly.

    (a) $5273.72   (b) $5296.82   (c) $1978.13
    (d) $1632.98   (e) None of these

    1—Answer: b

4. Find the balance after 10 years if $1500 is invested in an account that pays $7\frac{1}{2}\%$ compounded quarterly.

    (a) $3153.52   (b) $4151.16   (c) $2625.00
    (d) $2997.10   (e) None of these

    1—Answer: a

5. Determine the principal P that must be invested at a rate of 8% compounded quarterly so that the balance B in 40 years will be $200,000. $\left[B = P\left(1 + \frac{r}{n}\right)^{nt} \right]$

    (a) $90,578.10   (b) $47,539.00   (c) $12,416.00
    (d) $8414.00   (e) None of these

    1—Answer: d

6. Determine the principal that must be invested at a rate of $7\frac{1}{2}\%$ compounded quarterly so that the balance in 20 years will be $35,000.

    (a) $2333.33   (b) $14,000.00   (c) $9635.17
    (d) $7918.78   (e) None of these

    1—Answer: d

7. Determine the principal that must be invested at a rate of 9% compounded monthly so that the balance in 20 years will be $35,000.

    (a) $12,500.00   (b) $9470.02   (c) $6914.23
    (d) $5824.45   (e) None of these

    1—Answer: d

8. Determine the principal that must be invested at a rate of $9\frac{1}{2}\%$ compounded quarterly so that the balance in 15 years will be $40,000.

   1—Answer: $9781.94

9. An initial deposit of $2000 is compounded continuously at an annual percentage rate of 9%. Find the effective yield.

   (a) 9.4%   (b) 9.2%   (c) $188.00
   (d) $180.00   (e) None of these

   2—Answer: a

10. An initial deposit of $3000 is compounded continuously at an annual percentage rate of $7\frac{1}{2}\%$. Find the effective yield.

    (a) $225.00   (b) $3233.65   (c) 7.8%
    (d) 8.0%   (e) None of these

    2—Answer: c

11. An initial deposit of $3000 is compounded continuously at an annual percentage rate of $6\frac{1}{2}\%$. Find the effective yield.

    (a) $3201.48   (b) $195.00   (c) 6.9%
    (d) 6.7%   (e) None of these

    2—Answer: d

12. An initial deposit of $2500 is compounded continuously at 7%. Find the effective yield.

    2—Answer: 7.25%

13. An initial deposit of $3000 is made in a savings account for which the interest is compounded continuously. The balance will double in seven years. What is the annual rate of interest for this account?

    (a) 4.3%   (b) 6.2%   (c) 8.1%
    (d) 9.9%   (e) None of these

    2—Answer: d

14. An initial deposit of $4000 is made in a savings account for which the interest is compounded continuously. The balance will triple in 15 years. What is the annual rate of interest for this account?

    (a) 6.2%   (b) 7.3%   (c) 7.9%
    (d) 8.2%   (e) None of these

    2—Answer: b

15. Determine the annual rate of interest compounded continuously for the sum of money in an account to double in 10 years.

    (a) 6.9%   (b) 7.4%   (c) 8.2%
    (d) 9.9%   (e) None of these

    2—Answer: a

16. Determine the annual rate of interest compounded continuously for the sum of money in an account to become four times the original amount in 15 years.

2—Answer: 9.2%

17. The ice trays in a freezer are filled with water at 68° F. The freezer maintains a temperature of 20° F. According to Newton's Law of Cooling, the water temperature T is related to the time t (in hours) by the equation

$$kt = \ln\frac{T-20}{68-20}.$$

After 1 hour, the water temperature in the ice trays is 49° F. Use the fact that $T = 49$ when $t = 1$ to find how long it takes the water to freeze (water freezes at 32° F).

(a) 3.27 hours (b) 2.75 hours (c) 5.10 hours

(d) 1.17 hours (e) None of these

2—Answer: b

18. The ice trays in a freezer are filled with water at 60° F. The freezer maintains a temperature of 20° F. According to Newton's Law of Cooling, the water temperature T is related to the time t (in hours) by the equation

$$kt = \ln\frac{T-20}{60-20}.$$

After 1 hour, the water temperature in the ice trays is 44° F. Use the fact that $T = 44$ when $t = 1$ to find how long it takes the water to freeze (water freezes at 32° F).

(a) 2.4 hours (b) 3.2 hours (c) 1.7 hours

(d) 5.1 hours (e) None of these

2—Answer: a

19. The ice trays in a freezer are filled with water at 50° F. The freezer maintains a temperature of 0° F. According to Newton's Law of Cooling, the water temperature T is related to the time t (in hours) by the equation

$$kt = \ln\frac{T}{50}.$$

After 1 hour, the water temperature in the ice trays is 43° F. Use the fact that $T = 43$ when $t = 1$ to find how long it takes the water to freeze (water freezes at 32° F).

(a) 2.4 hours (b) 3.0 hours (c) 3.6 hours

(d) 2.1 hours (e) None of these

2—Answer: b

20. The ice trays in a freezer are filled with water at 60° F. The freezer maintains a temperature of 0° F. According to Newton's Law of Cooling, the water temperature T is related to the time t (in hours) by the equation

$$kt = \ln \frac{T}{60}.$$

After 1 hour, the water temperature in the ice trays is 51° F. Use the fact that $T = 51$ when $t = 1$ to find how long it takes the water to freeze (water freezes at 32° F).

2—**Answer:** 3.9 hours

21. The spread of a flu virus through a certain population is modeled by

$$y = \frac{1000}{1 + 990e^{-0.7t}},$$

where y is the total number infected after t days. In how many days will 820 people be infected with the virus?

(a) 10 days (b) 11 days (c) 12 days
(d) 13 days (e) None of these

2—**Answer:** c

22. The spread of a flu virus through a certain population is modeled by

$$y = \frac{1000}{1 + 990e^{-0.7t}},$$

where y is the total number infected after t days. In how many days will 690 people be infected with the virus?

(a) 10 days (b) 11 days (c) 12 days
(d) 13 days (e) None of these

2—**Answer:** b

23. The spread of a flu virus through a certain population is modeled by

$$y = \frac{1000}{1 + 990e^{-0.7t}},$$

where y is the total number infected after t days. In how many days will 900 people be infected with the virus?

(a) 11 days (b) 13 days (c) 15 days
(d) 17 days (e) None of these

2—**Answer:** b

Section 5.5 Exponential and Logarithmic Models 301

24. The spread of a flu virus through a certain population is modeled by

$$y = \frac{1000}{1 + 990e^{-0.7t}},$$

where y is the total number infected after t days. In how many days will 530 people be infected with the virus?

(a) 13 days (b) 12 days (c) 11 days

(d) 10 days (e) None of these

2—Answer: d

25. The relationship between the level of sound β, in decibels, and the intensity of sound, I, in watts per centimeter squared is given by

$$\beta = 10 \log_{10}\left(\frac{I}{10^{-16}}\right).$$

Determine the level of sound when $I = 10^{-12}$.

(a) 93 (b) 74 (c) 56

(d) 40 (e) None of these

1—Answer: d

26. The relationship between the level of sound β, in decibels, and the intensity of sound, I, in watts per centimeter squared is given by

$$\beta = 10 \log_{10}\left(\frac{I}{10^{-16}}\right).$$

Determine the level of sound when $I = 10^{-10}$.

(a) 92 (b) 60 (c) 51

(d) 100 (e) None of these

1—Answer: b

27. The relationship between the level of sound β, in decibels, and the intensity of sound, I, in watts per centimeter squared is given by

$$\beta = 10 \log_{10}\left(\frac{I}{10^{-16}}\right).$$

Determine the level of sound when $I = 10^{-8}$.

(a) 50 (b) 60 (c) 70

(d) 80 (e) None of these

1—Answer: d

28. The relationship between the level of sound β, in decibels, and the intensity of sound, I, in watts per centimeter squared is given by

$$\beta = 10 \log_{10}\left(\frac{I}{10^{-16}}\right).$$

Determine the level of sound when $I = 10^{-6}$.

(a) 100
(b) 90
(c) 80
(d) 70
(e) None of these

1—Answer: a

29. The relationship between the level of sound β, in decibels, and the intensity of sound, I, in watts per centimeter squared is given by

$$\beta = 10 \log_{10}\left(\frac{I}{10^{-16}}\right).$$

Determine the level of sound when $I = 10^{-5}$.

(a) 100
(b) 110
(c) 120
(d) 125
(e) None of these

1—Answer: b

30. The pH of a solution is determined by $pH = -\log_{10}[H^+]$ where pH is a measure of the hydrogen ion concentration $[H^+]$, measured in moles per liter. Find the pH of a solution for which $[H^+] = 7.61 \times 10^{-6}$.

(a) -5.12
(b) 5.12
(c) -11.79
(d) 11.79
(e) None of these

2—Answer: b

31. The pH of a solution is determined by $pH = -\log_{10}[H^+]$ where pH is a measure of the hydrogen ion concentration $[H^+]$, measured in moles per liter.

Find the pH of a solution for which $[H^+] = 5.93 \times 10^{-7}$.

2—Answer: 6.23

32. The demand equation for a certain product is given by $p = 450 - 0.4e^{0.007x}$. Find the demand x if the price charged is $300.

1—Answer: 847

33. The demand equation for a certain product is given by $p = 450 - 0.4e^{0.007x}$. Find the demand x if the price charged is $250.

1—Answer: 888

34. Find the constant k so that the exponential function $y = 3e^{kt}$ passes through the points (0, 3) and (3, 5).

2—Answer: $k = \frac{1}{3}\ln\frac{5}{3}$

35. Find the constant k so that the exponential function $y = 2e^{kt}$ passes through the points (0, 2) and (2, 5).

2—Answer: $k = \frac{1}{2}\ln\frac{5}{2}$

36. The number N of bacteria in a culture is given by

$$N = 200e^{kt}.$$

If $N = 300$ when $t = 4$ hours, find k (to the nearest tenth) and then determine approximately how long it will take for the number of bacteria to triple in size.

2—Answer: $k = 0.1$, $t \approx 11$ hours

37. Write an equation for the amount Q of a radioactive substance with a half-life of 30 days, if 10 grams are present when $t = 0$.

2—Answer: $Q(t) = 10e^{-0.0231t}$

38. A state game commission releases 200 deer into state game lands. The commission believes that the growth of the herd will be modeled by

$$p(t) = \frac{400}{1 + 2e^{-0.025t}}$$

where t is measured in years. Use a graphing utility to graph the function, using an appropriate window. Determine the following.

(a) The population after 10 years, 50 years, and 100 years

(b) The limiting size of the herd

2—T—Answer: 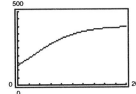 (a) ≈ 156 deer, ≈ 254 deer, ≈ 343 deer

(b) 400 deer

39. A state game commission releases 200 deer into state game lands. The commission believes that the growth of the herd will be modeled by

$$p(t) = \frac{600}{1 + 3e^{-0.025t}}$$

where t is measured in years. Use a graphing utility to graph the function, using an appropriate window. Determine the following.

(a) The population after 10 years, 50 years, and 100 years

(b) The limiting size of the herd

2—T—Answer: 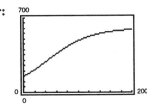 (a) ≈ 180 deer, ≈ 323 deer, ≈ 481 deer

(b) 600 deer

CHAPTER SIX
Systems of Equations and Inequalities

❏ 6.1 Solving Systems of Equations

1. Solve the system by the method of substitution:

    $x + y = 1$
    $x^2 + 3y^2 = 21$

    (a) $\left(\frac{3}{2}, -3\right)$ (b) $\left(3, -\frac{3}{2}\right)$ (c) $\left(-\frac{3}{2}, \frac{5}{2}\right), (3, -2)$
    (d) $\left(\frac{3}{2}, -\frac{1}{2}\right), (-3, 4)$ (e) No solution

    1—Answer: c

2. Solve the system by the method of substitution:

    $2x^2 + 2y^2 = 7$
    $x + y^2 = 7$

    (a) $(2.8, 2.0), (-0.5, 7.3)$ (b) $(4.6, 1.5), (-2.6, 3.1)$ (c) $(2.8, -0.5)$
    (d) $(4.6), (-2.6)$ (e) No solution

    2—Answer: e

3. Solve the system graphically:

    $x^2 + y^2 = 25$
    $x - y = 1$

    1—T—Answer:

    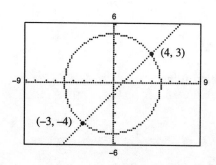

4. Solve the system graphically:

    $x^2 + 2y - 5 = 0$
    $3x^2 - y - 1 = 0$

    1—T—Answer:

    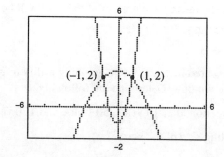

5. Solve the system by the method of substitution:

$$y = \frac{1}{x}$$
$$x + 5y = 6$$

1—Answer: $(1, 1), \left(5, \frac{1}{5}\right)$

6. Solve the system graphically:

$$x^2 - 4y = 17$$
$$x - 2y = 1$$

1—T—Answer:

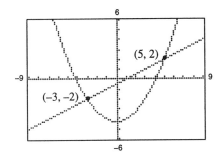

7. Solve the system graphically:

$$2x + y = 1$$
$$-x + 2y = 7$$

1—T—Answer:

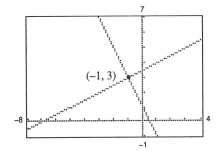

8. Solve the system by the method of substitution:

$$5x + y = 11$$
$$3x - 2y = 4$$

(a) $\left(\frac{15}{13}, \frac{68}{13}\right)$ (b) $(2, 21)$ (c) $(2, 1)$
(d) $\left(\frac{15}{8}, -15\right)$ (e) None of these

1—Answer: c

9. Solve the system graphically:

$$3x + 4y = 2$$
$$2x + y = 3$$

1—T—Answer:

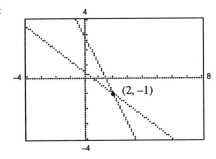

10. Solve the system by the method of substitution:

$$0.1x - 0.3y = 1.2$$
$$3x - 2y = 71$$

(a) $(5, 27)$ (b) $(a, 5a)$ (c) $(27, 5)$
(d) $\left(\frac{61}{3}, 5\right)$ (e) None of these

1—Answer: c

11. Solve the system by the method of substitution:

    $\frac{1}{3}x - \frac{3}{5}y = -2$

    $2x - y = 14$

    (a) $\left(\frac{136}{23}, \frac{50}{23}\right)$ (b) $(12, 10)$ (c) $(12, -38)$

    (d) No solution   (e) None of these

    1—Answer: b

12. Solve the system by the method of substitution:

    $x^2 + 2y = 6$

    $2x + y = 3$

    (a) $(4, -5)$ (b) $(2, 1)$ (c) $(0, 3)$

    (d) $(0, 3)$ and $(4, -5)$ (e) None of these

    1—Answer: d

13. Solve the system graphically:

    $x^2 + 2y = -6$

    $x - y = 3$

    1—T—Answer:

    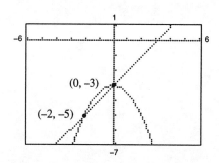

14. Solve the system by the method of substitution:

    $2x^2 - y = -2$

    $x - y = -2$

    1—T—Answer: $(0, 2)$ and $\left(\frac{1}{2}, \frac{5}{2}\right)$

15. Solve the system by the method of substitution:

    $3x + 2y = 12$

    $5x - y = 23$

    (a) $\left(-\frac{14}{13}, -\frac{344}{13}\right)$ (b) $\left(\frac{58}{13}, -\frac{9}{13}\right)$ (c) $\left(\frac{35}{13}, -\frac{124}{13}\right)$

    (d) $(-17, 108)$ (e) None of these

    1—Answer: b

16. Solve the system by the method of substitution:

    $6x + 2y = 7$

    $4x - 7y = -37$

    (a) $\left(4, -\frac{53}{7}\right)$ (b) $\left(6, \frac{61}{7}\right)$ (c) $\left(6, -\frac{29}{2}\right)$

    (d) $\left(-\frac{1}{2}, 5\right)$ (e) None of these

    1—Answer: d

17. Solve the system by the method of substitution:

    $x - 2y = 0$

    $4y - 3x = 10$

    (a) $(-10, -5)$ (b) $(6, 3)$ (c) $\left(\frac{1}{2}, \frac{1}{4}\right)$

    (d) $\left(1, \frac{1}{2}\right)$ (e) None of these

    1—Answer: a

18. Solve the system graphically:

    $x - y = 4$

    $3x - 2y = 14$

    1—Answer:

    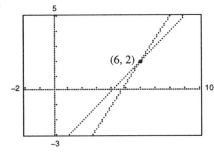

19. Solve the system:

    $x + y = 16$

    $\frac{1}{2}x + \frac{1}{6}y = 2$

    (a) $(4, 12)$ (b) $(-2, 18)$ (c) $(-4, 20)$

    (d) $(2, 14)$ (e) None of these

    1—Answer: b

20. Find all points of intersection of the graphs:

    $x^2 + y^2 = 3$

    $2x^2 - y = 0$

    (a) $\left(\frac{3}{2}, -2\right)$ (b) $\left(\pm\frac{\sqrt{3}}{2}, \frac{3}{2}\right)$ (c) $\left(\pm\frac{\sqrt{3}}{2}, \frac{3}{2}\right), (\pm 1, -2)$

    (d) $(2, 14)$ (e) None of these

    1—Answer: b

21. Use a graphing utility to find all points of intersection of the graphs:

    $(x - 3)^2 + y^2 = 4$

    $-2x + y^2 = 0$

    2—T—Answer:

    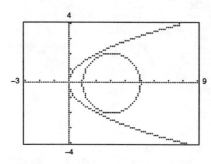

    No points of intersection

22. Use a graphing utility to find all points of intersection of the graphs:

    $x^2 - 4x + y = 0$

    $x - y = 0$

    1—T—Answer:

    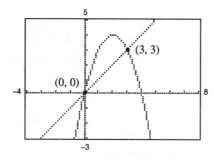

23. Use a graphing utility to find all points of intersection of the graphs:

    $2x^2 - y - 1 = 0$

    $2x^2 + y - 3 = 0$

    1—T—Answer:

    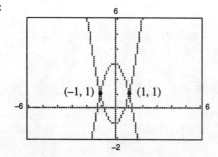

24. Find all points of intersection of the graphs:

    $3x - y = -2$

    $x^3 - y = 0$

    (a) $(2, 8), (1, 1)$ (b) $(-2, -8), (-1, -1)$ (c) $(-2, -8), (1, 1)$

    (d) $(2, 8), (-1, -1)$ (e) None of these

    1—Answer: d

25. Find the number of points of intersection of the graphs:

    $x^2 + y = 3$

    $x^2 + y^2 = 1$

    (a) 4 (b) 3 (c) 2

    (d) 1 (e) 0

    1—Answer: e

26. Find the number of points of intersection of the graphs:

    $x^2 + y^2 = 2$

    $2x + y = 10$

    (a) 4 (b) 3 (c) 2

    (d) 1 (e) 0

    1—Answer: e

27. Find the number of points of intersection of the graphs:

    $x^2 + y = 1$

    $x^2 - y = 3$

    (a) 4 (b) 3 (c) 2

    (d) 1 (e) 0

    1—Answer: c

28. Find the number of points of intersection:

    $x^2 + y^2 = 5$

    $x + 2y - 5 = 0$

    (a) 4 (b) 3 (c) 2

    (d) 1 (e) 0

    1—Answer: d

29. A total of $11,000 is invested in two funds paying 7% and 8% simple interest. If the yearly interest for both funds totals $865, determine the amount invested at 8%.

    (a) $9500 (b) $6500 (c) $1500

    (d) $4500 (e) None of these

    1—Answer: a

30. A total of $50,000 is invested in two funds paying 8% and 10% simple interest. If the yearly interest for both funds totals $4660, determine the amount invested at 8%.

    (a) $33,000 (b) $24,000 (c) $26,000

    (d) $17,000 (e) None of these

    1—Answer: d

31. A total of $12,000 is invested in two funds paying 7% and $9\frac{1}{2}$% simple interest. If the annual interest totals $913.75, determine the amount invested at 7%.

   (a) $7325   (b) $840   (c) $9050

   (d) $10,000   (e) None of these

   1—Answer: c

32. A total of $6000 is invested in two funds paying $5\frac{1}{4}$% and 6% simple interest. If the annual interest totals $327, determine the amount invested at 6%.

   (a) $1600   (b) $3750   (c) $6000

   (d) $270   (e) None of these

   1—Answer: a

33. If the total cost of running a business is given by the equation $C = 450x + 1000$ and the revenue is given by the equation $R = 500x$, find the sales necessary to break even.

   (a) 220   (b) 11   (c) 20

   (d) 2000   (e) None of these

   1—Answer: c

34. If the total cost of running a business is given by the equation $C = 4.16x + 75,000$ and the revenue is given by the equation $R = 7.91x$, find the sales necessary to break even.

   (a) 6214   (b) 20,000   (c) 200

   (d) 9482   (e) None of these

   1—Answer: b

35. Suppose you are setting up for a small business and have invested $5000 to produce an item that will sell for $9. If each unit can be produced for $7, how many units must you sell to break even?

   (a) 25   (b) 2500   (c) 556

   (d) 714   (e) None of these

   1—Answer: b

36. Suppose you are setting up for a small business and have invested $18,000 to produce an item that will sell for $20.65. If each unit can be produced for $13.45, determine the number of units that you must sell in order to break even.

   (a) 2500   (b) 872   (c) 1338

   (d) 250   (e) None of these

   1—Answer: a

6.2 Two-Variable Linear Systems

1. Solve the linear system by the method of elimination:

    $7x - 3y = 26$

    $2x + 5y = 25$

    (a) $\left(-5, -\frac{61}{3}\right)$
    (b) $(5, 3)$
    (c) Infinitely many solutions
    (d) No solution
    (e) None of these

    1—Answer: b

2. Solve the linear system by the method of elimination:

    $2x + 4y = 7$

    $3x + 6y = 5$

    (a) $\left(1, \frac{5}{4}\right)$
    (b) $(0, 0)$
    (c) Infinitely many solutions
    (d) No solution
    (e) None of these

    1—Answer: d

3. Solve the linear system by the method of elimination:

    $6x - 5y = 4$

    $3x + 2y = 1$

    (a) $\left(\frac{13}{27}, -\frac{2}{9}\right)$
    (b) $\left(-\frac{2}{9}, -\frac{8}{5}\right)$
    (c) $\left(-\frac{8}{5}, -\frac{68}{25}\right)$
    (d) $\left(2, \frac{8}{5}\right)$
    (e) None of these

    1—Answer: a

4. Solve the linear system by the method of elimination:

    $7x + y = 3$

    $21x + 5y = 11$

    1—Answer: $\left(\frac{2}{7}, 1\right)$

5. Use the method of elimination to find the value of y in the solution of the system of equations:

    $2x - 3y = 5$

    $2x + 3y = -3$

    (a) $\frac{1}{2}$
    (b) $-\frac{3}{4}$
    (c) $-\frac{4}{3}$
    (d) $\frac{4}{3}$
    (e) None of these

    1—Answer: c

6. Use the method of elimination to find the value of *x* in the solution of the system of equations:

   $2x - y = 5$

   $2x + 2y = -9$

   (a) 2   (b) $\frac{1}{2}$ (c) -5

   (d) $\frac{21}{2}$ (e) None of these

   1—Answer: b

7. Use the method of elimination to find the value of *y* in the solution of the system of equations:

   $5x + 2y = -1$

   $-15x + 8y = 10$

   (a) $\frac{1}{2}$ (b) $\frac{9}{10}$ (c) $\frac{9}{14}$

   (d) 0   (e) None of these

   1—Answer: a

8. Find the value of *y* in the solution of the system of equations:

   $-2x + 3y = 5$

   $3x - 2y = 0$

   (a) 0   (b) 1   (c) 3

   (d) -1 (e) None of these

   1—Answer: c

9. Find the value of *y* in the solution of the system of equations:

   $3x + 7y = 15$

   $-5x + 2y = 16$

   (a) 3   (b) -2 (c) 2

   (d) $\frac{6}{7}$ (e) None of these

   1—Answer: a

10. Solve this system by the method of elimination and verify the solution with a graphing utility:

    $2x - 5y = -4$

    $4x + 3y = 5$

    2—T—Answer: $\left(\frac{1}{2}, 1\right)$

    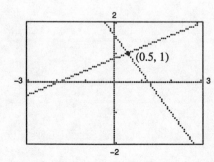

11. Solve this system by the method of elimination and verify the solution with a graphing utility:

    $6x + y = -2$

    $4x - 3y = 17$

    2—T—Answer: $\left(\frac{1}{2}, -5\right)$

    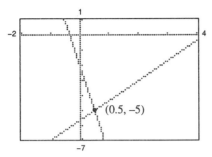

12. Solve this system by the method of elimination and verify the solution with a graphing utility:

    $\dfrac{6}{x} - \dfrac{8}{y} = 2$

    $\dfrac{9}{2x} - \dfrac{6}{y} = \dfrac{3}{2}$

    2—T—Answer: Infinitely many solutions

    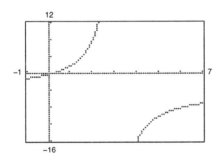

13. Solve the system by the method of elimination:

    $\dfrac{2}{x} - \dfrac{3}{y} = 8$

    $\dfrac{3}{x} + \dfrac{3}{y} = 2$

    (a) $\left(2, -\frac{4}{3}\right)$ (b) $\left(\frac{1}{2}, -\frac{3}{4}\right)$ (c) Infinitely many solutions

    (d) No solution (e) None of these

    1—Answer: b

14. Solve the system by the method of elimination:

$$\frac{6}{x} + \frac{3}{y} = 8$$

$$\frac{9}{x} + \frac{5}{y} = 16$$

(a) $\left(\frac{2}{3}, 2\right)$ (b) $\left(\frac{3}{2}, \frac{1}{2}\right)$ (c) $\left(2, \frac{3}{2}\right)$

(d) $\left(\frac{1}{2}, \frac{2}{3}\right)$ (e) None of these

2—Answer: b

15. Solve the system of equations:

$$\frac{6}{x} + \frac{1}{y} = -2$$

$$\frac{4}{x} - \frac{3}{y} = 17$$

1—Answer: $\left(2, -\frac{1}{5}\right)$

16. Solve the following system of equations for x:

$$\frac{3}{x} - \frac{2}{y} = 5$$

$$\frac{1}{x} + \frac{4}{y} = 4$$

(a) $\frac{1}{2}$ (b) 2 (c) 5

(d) $\frac{1}{5}$ (e) None of these

1—Answer: a

17. Solve the following system of equations for x:

$$\frac{5}{2x} - \frac{3}{y} = \frac{1}{3}$$

$$\frac{1}{x} + \frac{2}{y} = \frac{2}{3}$$

(a) $\frac{1}{2}$ (b) 6 (c) $\frac{1}{3}$

(d) 3 (e) None of these

1—Answer: d

18. Solve the following system of equations for y:

$$\frac{2}{x} + \frac{3}{y} = 7$$

$$\frac{3}{x} - \frac{1}{y} = 16$$

(a) 5 (b) -1 (c) $\frac{1}{5}$

(d) 2 (e) None of these

1—Answer: b

19. Solve the following system of equations for x:

$$\frac{5}{x} - \frac{3}{y} = 2$$

$$\frac{2}{x} + \frac{5}{y} = -24$$

(a) $-\frac{1}{2}$ (b) $-\frac{1}{4}$ (c) 5

(d) $-\frac{1}{3}$ (e) None of these

1—Answer: a

20. Solve the following system of equations for x:

$$\frac{5}{x} - \frac{7}{y} = 11$$

$$\frac{3}{x} + \frac{2}{y} = -12$$

(a) $-\frac{1}{2}$ (b) -2 (c) -1

(d) 1 (e) None of these

1—Answer: e

21. Solve the system of linear equations graphically.

$$\tfrac{1}{3}x - \tfrac{3}{5}y = -2$$

$$2x - y = 14$$

2—T—Answer:

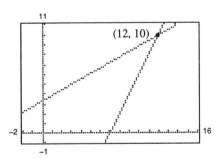

22. Solve the system of linear equations graphically.

$$3x + 2y = 8$$

$$6x + 4y = 10$$

2—T—Answer: No solutions

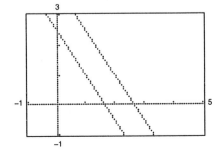

23. Solve the system of linear equations:

    $6x - 8y = 2$

    $\frac{9}{2}x - 6y = \frac{3}{2}$

    (a) $\left(\frac{3}{2}, 4\right)$ (b) $\left(\frac{2}{3}, \frac{1}{4}\right)$ (c) Infinitely many solutions

    (d) No solutions (e) None of these

    2—Answer: c

24. Solve the system of linear equations:

    $0.06x + 0.02y = 0.08$

    $0.09x + 0.05y = 0.16$

    (a) $\left(\frac{3}{2}, \frac{1}{2}\right)$ (b) $\left(\frac{3}{2}, 2\right)$ (c) $\left(\frac{1}{2}, \frac{2}{3}\right)$

    (d) $\left(2, \frac{3}{2}\right)$ (e) None of these

    2—Answer: b

25. Solve the following system of linear equations for x:

    $x + y = 1000$

    $0.03x + 0.04y = 31.50$

    (a) 325 (b) 540 (c) 675

    (d) 850 (e) None of these

    1—Answer: d

26. Solve the following system of linear equations for x:

    $x + 2.5y = 900$

    $5x - 2y = 150$

    (a) 300 (b) 150 (c) 900

    (d) 0 (e) None of these

    1—Answer: b

27. Solve the following system of linear equations for y:

    $3x + 4.5y = 825$

    $0.2x + 0.5y = 89$

    (a) 100 (b) 150 (c) 170

    (d) 190 (e) None of these

    1—Answer: c

28. A twenty-pound mixture of two kinds of candy sells for $30.52. One kind of candy in the mixture sells for $1.35 per pound. The other kind sells for $1.79 per pound. How much of the cheaper-priced candy is in the mixture?

    (a) 8 pounds (b) 10 pounds (c) 12 pounds

    (d) 14 pounds (e) None of these

    2—Answer: c

29. How many liters of a 40% solution of acid must be combined with a 15% solution to obtain 30 liters of a 20% solution?

   2—Answer: 6

30. The perimeter of a rectangle is 91 feet and the length is 8 feet more than twice the width. Find the dimensions of the rectangle.

   2—Answer: $L = 33$ feet, $W = 12.5$ feet

31. A total of $15,000 is invested in two corporate bonds that pay $7\frac{1}{4}$% and 9% simple interest. The annual income from both bonds is $1280. Determine how much is invested at 9%.

   (a) $9000 (b) $4000 (c) $11,000
   (d) $6000 (e) None of these

   1—Answer: c

32. Suppose the demand and supply equations for a certain product are given by

   $p = 220 - 0.0002x$ Demand equation
   $p = 90 + 0.0003x$ Supply equation

   where p is the price in dollars and x represents the number of units. Find the point of equilibrium.

   1—Answer: $x = 260,000$ and $p = \$168$

33. Suppose the demand and supply equations for a certain product are given by

   $p = 860 - 0.05x$ Demand equation
   $p = 420 + 0.15x$ Supply equation

   Find the point (x, p) of equilibrium.

   (a) (12,800, 220) (b) (1700, 775) (c) (420, 839)
   (d) (2200, 750) (e) None of these

   1—Answer: d

34. Find the least squares regression line $y = ax + b$ for the points $(0.6, 9.3)$, $(1.2, 12.0)$ and $(1.8, 15.2)$ if

   $$nb + \left(\sum_{i=1}^{n} x_i\right)a = \sum_{i=1}^{n} y_i \quad \text{and} \quad \left(\sum_{i=1}^{n} x_i\right)b + \left(\sum_{i=1}^{n} x_i^2\right)a = \sum_{i=1}^{n} x_i y_i.$$

   2—T—Answer: $y = 4.92x + 6.27$

35. Find the least squares regression line $y = ax + b$ for the points $(1, 2)$, $(2, 4)$, $(3, 5)$ and $(4, 7)$ if

   $$nb + \left(\sum_{i=1}^{n} x_i\right)a = \sum_{i=1}^{n} y_i \quad \text{and} \quad \left(\sum_{i=1}^{n} x_i\right)b + \left(\sum_{i=1}^{n} x_i^2\right)a = \sum_{i=1}^{n} x_i y_i.$$

   2—T—Answer: $y = 1.6x + 0.5$

36. Find the least squares regression line $y = ax + b$ by solving the following system for a and b.

$$5b + 10a = 12.3$$
$$10b + 30a = 29.1$$

(a) $y = 0.45x + 1.56$ (b) $y = 0.26x + 1.94$ (c) $y = 0.6x + 1.26$
(d) $y = 3.2x - 3.94$ (e) None of these

1—Answer: a

37. Find the least squares regression line $y = ax + b$ by solving the following system for a and b.

$$5b + 10a = 9.2$$
$$10b + 30a = 21.1$$

1—Answer: $y = 0.27x + 1.3$

❑ 6.3 Multivariable Linear Systems

1. Use the method of back-substitution to find the value of x for the solution of the system of equations.

$$x + 2y + z = 15$$
$$5y - 2z = -16$$
$$z = 3$$

(a) 22 (b) 16 (c) $\frac{15}{7}$
(d) 8 (e) None of these

1—Answer: b

2. Use the method of back-substitution to find the value of x for the solution of the system of equations.

$$x + 2y - z = 26$$
$$y + 3z = 5$$
$$z = -2$$

(a) 4 (b) 26 (c) 6
(d) 2 (e) None of these

1—Answer: d

3. Use the method of back-substitution to find the solution of the system of equations.

$$x + y + z = 2$$
$$y - z = 5$$
$$z = -2$$

1—Answer: $(1, 3, -2)$

4. Use Gaussian elimination to solve the system of equations:

$$x - 6y + z = 1$$
$$-x + 2y - 4z = 3$$
$$7x - 10y + 3z = -25$$

(a) $(5, 1, 2)$ (b) $(-5, -1, 0)$ (c) $(-1, 3, 1)$

(d) No solution (e) None of these

1—Answer: b

5. Use Gaussian elimination to solve the system of equations:

$$x + 2y + z = 6$$
$$2x - y + 3z = -2$$
$$x + y - 2z = 0$$

1—Answer: $(-1, 3, 1)$

6. Solve the system of linear equations:

$$x + 3y + z = 0$$
$$5x - y + z + w = 0$$
$$2x + 2z + w = 2$$
$$3x + 2z - w = 10$$

2—Answer: $x = 0, y = -1, z = 3, w = -4$

7. Solve the system of linear equations:

$$x - y + z = 5$$
$$3x + 2y - z = -2$$
$$2x + y + 3z = 10$$

(a) $(1, -1, 3)$ (b) $(2, -5, -2)$ (c) $(-1, 7, 13)$

(d) $(3, -9, -7)$ (e) No solution

1—Answer: a

8. Solve the system of linear equations:

$$x + y + 3z = 0$$
$$2x - y - 3z = -9$$
$$x + 2y + 3z = 1$$

(a) $\left(-3a, a, \dfrac{2a}{3}\right)$ (b) $\left(-1, 2, -\dfrac{1}{3}\right)$ (c) $\left(-3, 1, \dfrac{2}{3}\right)$

(d) No solution   (e) None of these

1—Answer: c

9. Solve the system of linear equations:

$$6x - 9y + 4z = -7$$
$$2x + 6y - z = 6$$
$$4x - 3y + 2z = -2$$

(a) $\left(\dfrac{1}{2}, \dfrac{2}{3}, -1\right)$ (b) $\left(\dfrac{11}{21}, 1, -\dfrac{2}{7}\right)$ (c) $\left(a, \dfrac{31a}{15}, \dfrac{44a}{5}\right)$

(d) No solution   (e) None of these

2—Answer: a

10. Solve the system of linear equations:

$$x + y - z = -1$$
$$2x + 3y - z = -2$$
$$-3x - 2y + 2z = -3$$

1—Answer: $(5, -3, 3)$

11. Solve the system of linear equations.

$$x - 3y + 2z = -11$$
$$x + 4y - 5z = 17$$
$$-2x + y - z = 6$$

1—Answer: $(-1, 2, -2)$

12. Solve the system of linear equations:

$$2x + y - z = 3$$
$$x - 3y + z = 7$$
$$3x + 5y - 3z = 0$$

(a) $\left(a, \dfrac{3a - 10}{2}, \dfrac{7a - 16}{2}\right)$ (b) $\left(\dfrac{3a + 10}{3}, a, 6a - 21\right)$

(c) $(2, -2, -1)$ (d) No solution

(e) None of these

2—Answer: d

13. Solve the system of linear equations:

$$x + y - 2z = 1$$
$$3x + y + z = 4$$
$$-x - 3y + 9z = 10$$

1—Answer: No solution

14. Solve the system of linear equations.

$$2x - 4y + z = 7$$
$$x + 3y - z = 2$$
$$-5x + 15y - 4z = 10$$

2—Answer: No solution

15. Solve the system of linear equations:

$$x - y + z = 2$$
$$2x + 3y + z = 7$$
$$3x + 2y + 2z = -8$$

(a) $(1, 0, 1)$ (b) $(6, 4, 4)$ (c) $(1, 2, 3)$

(d) No solution (e) None of these

1—Answer: d

16. Solve the system of linear equations:

$$x - 2y - z = 7$$
$$-3x + 6y + 3z = 0$$

(a) $(7 + 5a, 2a, a)$ (b) $(1, 1, -8)$ (c) $(1, 0, 1)$

(d) No solution (e) None of these

2—Answer: d

17. Solve the system of linear equations:

$$x + y + z = 4$$
$$x - 3y - z = 1$$
$$2x - 2y = 9$$

(a) $\left(-1, \dfrac{4}{7}, \dfrac{31}{7}\right)$ (b) $\left(a, \dfrac{2a - 9}{2}, 17 - 4a\right)$

(c) $\left(\dfrac{9 + 2a}{2}, a, -\dfrac{4a + 1}{2}\right)$ (d) No solution

(e) None of these

1—Answer: d

18. Solve the system of linear equations:

$$2x - 4y + z = 5$$
$$x + y + z = 3$$
$$6x + 5z = 17$$

(a) $\left(\dfrac{5}{6}, \dfrac{1}{6}, 0\right)$ (b) $\left(a, \dfrac{a - 2}{5}, \dfrac{17 - 6a}{5}\right)$

(c) $\left(a, \dfrac{2a - 3}{5}, \dfrac{a + 7}{5}\right)$ (d) No solution

(e) None of these

1—Answer: b

19. Solve the system of linear equations:

    $3x + 4y - 2z = 6$
    $x + y + z = 2$
    $x + 2y - 4z = 2$

    (a) $(20, -15, -3)$ (b) $\left(\dfrac{a}{2}, \dfrac{5a}{4}, \dfrac{a}{4}\right)$ (c) $(2 - 6a, 5a, a)$

    (d) No solution   (e) None of these

    2—Answer: c

20. Solve the system of linear equations:

    $x + y + z = 2$
    $3x - 2y + z = 7$
    $5y + 2z = -1$

    (a) $\left(\dfrac{2}{5}, -\dfrac{7}{5}, 3\right)$ (b) $\left(a, \dfrac{2a}{3}, \dfrac{a}{3}\right)$

    (c) $\left(\dfrac{11 - 3a}{5}, \dfrac{-1 - 2a}{5}, a\right)$ (d) No solution

    (e) None of these

    2—Answer: c

21. Solve the system of linear equations:

    $2x + y + z = 1$
    $x + 4y + 2z = 7$
    $-5x + y + 5z = 4$

    2—Answer: $\left(\dfrac{-3 + 6a}{7}, \dfrac{13 - 5a}{7}, a\right)$

22. Solve the system of linear equations:

    $x - y - z = 0$
    $2x + 4y + z = 0$
    $3x + y - z = 0$

    1—Answer: $(a, -a, 2a)$ where a is any real number.

23. Solve the system of linear equations:

    $2x + 3y + 3z = 6$
    $-x + y + z = 2$

    (a) $(2a - 4, 2 - a, a)$ (b) $(0, 2 - a, a)$ (c) $(2a, 2 + a, a)$

    (d) No solutions   (e) None of these

    2—Answer: b

24. Solve the system of linear equations:

    $2x + 3y - z = 6$
    $x - y + 2z = 1$

    (a) $\left(a, \dfrac{13 - 5a}{5}, \dfrac{9 - 5a}{5}\right)$
    (b) $\left(\dfrac{10 - 2a}{3}, a, \dfrac{5 - 2a}{3}\right)$
    (c) $\left(3, -\dfrac{2}{5}, -\dfrac{6}{5}\right)$
    (d) No solutions
    (e) None of these

    1—Answer: a

25. Solve the system of linear equations:

    $2x + 3y - 3z = 7$
    $-3x + y + z = 2$

    1—Answer: $\left(\dfrac{1 + 6a}{11}, \dfrac{25 + 7a}{11}, a\right)$

26. Find the value of b that makes the system inconsistent:

    $6x + by = 14$
    $-2x + 3y = 2$

    (a) -9
    (b) -3
    (c) $\frac{1}{7}$
    (d) $\frac{1}{3}$
    (e) None of these

    1—Answer: a

27. Find the value of a that makes the system inconsistent:

    $-3x + 5y = 2$
    $ax - 10y = 0$

    (a) 3
    (b) 6
    (c) 2
    (d) -6
    (e) None of these

    1—Answer: b

28. Find an equation of the parabola, $y = ax^2 + bx + c$, that passes through $(1, 4)$, $(-1, 0)$, and $(2, -3)$. Verify your result with a graphing utility.

    (a) $y = 4x^2 + 2x - 2$
    (b) $y = 3x^2 + 2x - 7$
    (c) $y = -3x^2 + 2x + 5$
    (d) $y = 4x^2$
    (e) None of these

    2—T—Answer: c

    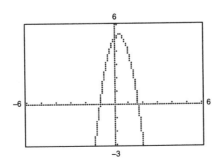

29. Find an equation of the parabola, $y = ax^2 + bx + c$, that passes through $(0, 5)$, $(2, -5)$, and $(-3, -40)$.

    (a) $y = 3x^2 - 2x - 7$ (b) $y = -4x^2 + 3x + 5$ (c) $y = 4x^2 + 3x + 5$

    (d) $y = 9x^2 - 121$ (e) None of these

    2—Answer: b

30. Find an equation of the parabola, $y = ax^2 + bx + c$, that passes through $(0, -5)$, $(2, 1)$, and $(-1, -14)$. Verify your result with a graphing utility.

    2—T—Answer: $y = -2x^2 + 7x - 5$

    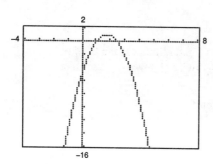

31. Find an equation of the parabola, $y = ax^2 + bx + c$, that passes through $(1, 1)$, $(-1, 11)$, and $(3, 23)$. Verify your result with a graphing utility.

    2—T—Answer: $y = 4x^2 - 5x + 2$

    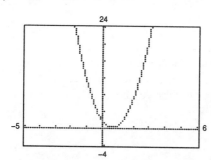

32. Find the value of c in the quadratic equation, $y = ax^2 + bx + c$, if its graph passes through the points $(1, 0)$, $(-1, -6)$, and $(2, 9)$.

    (a) -5 (b) -4 (c) 3

    (d) 11   (e) None of these

    2—Answer: a

33. Find the value of b in the quadratic equation, $y = ax^2 + bx + c$, if its graph passes through the points $(-1, 4)$, $(1, -2)$, and $(2, -2)$.

    (a) -3 (b) 2   (c) -2

    (d) -1 (e) None of these

    2—Answer: a

34. Find an equation of the parabola, $y = ax^2 + bx + c$, that passes through the points $(1, -2), (-2, 19)$, and $(3, 4)$.

    2—Answer: $y = 2x^2 - 5x + 1$

35. Find an equation of the circle, $x^2 + y^2 + Dx + Ey + F = 0$, that passes through $(9, -3), (2, 4)$, and $(-5, -3)$.

    (a) $x^2 + y^2 + 3x - 2y + 10 = 0$ (b) $x^2 + y^2 - 4x + 6y - 36 = 0$

    (c) $x^2 + y^2 - 8x + 2y - 12 = 0$ (d) $x^2 + y^2 + 2x - 7y + 1 = 0$

    (e) None of these

    2—Answer: b

36. The sum of three positive numbers is 19. Find the second number if the third is three times the first and the second is one more than twice the first.

    (a) 7   (b) 13   (c) 1

    (d) 9   (e) None of these

    1—Answer: a

37. The sum of three positive numbers is 180. Find the first number if the third is four times the first and the second is thirty-six less than twice the third.

    (a) 12   (b) 36   (c) 24

    (d) 60   (e) None of these

    2—Answer: c

38. A total of $7000 is invested in three separate accounts. Some of the money was invested at 6%, some at 8%, and the remaining at 9%. Find the amount invested at each rate if the total interest for one year was $555 and the amount invested at 8% was three times the amount invested at 9%. (Assume simple interest.)

    2—Answer: $1000 at 6%, $4500 at 8%, and $1500 at 9%

39. Write the partial fraction decomposition: $\dfrac{8x + 6}{x(x + 1)(x + 2)}$

    (a) $\dfrac{2}{x+1} - \dfrac{5}{x+2} + \dfrac{3}{x}$ (b) $\dfrac{7}{x+1} + \dfrac{1}{x+2} - \dfrac{6}{x}$

    (c) $\dfrac{1}{x+1} + \dfrac{3}{x+2} - \dfrac{1}{x}$ (d) $\dfrac{6}{x+1} - \dfrac{1}{x+2} - \dfrac{1}{x}$

    (e) None of these

    1—Answer: a

40. Write the partial fraction decomposition: $\dfrac{5}{x^2 - 7x + 12}$

    1—Answer: $\dfrac{5}{x-4} - \dfrac{5}{x-3}$

41. Find the position equation $s = \frac{1}{2}at^2 + v_0t + s_0$ for an object at the given heights moving vertically at the specified times. Use a graphing utility to plot the points and graph the parabola.

At $t = 1$ second, $s = 64$.

At $t = 2$ seconds, $s = 96$.

At $t = 3$ seconds, $s = 96$.

2—T—Answer: $a = -32, b = 80, c = 0; s = -16t^2 + 80t$

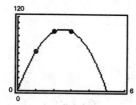

42. Find the position equation $s = \frac{1}{2}at^2 + v_0t + s_0$ for an object at the given heights moving vertically at the specified times. Use a graphing utility to plot the points and graph the parabola.

At $t = 1$ second, $s = 36$.

At $t = 2$ seconds, $s = 36$.

At $t = 3$ seconds, $s = 4$.

2—T—Answer: $a = -32, b = 48, c = 4; s = -16t^2 + 48t + 4$

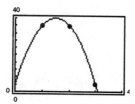

43. Find the least squares regression parabola $y = ax^2 + bx + c$ by solving the following system of linear equations for a, b, and c. Then use the least squares regression capabilities of a graphing utility to confirm the result. Then graph the data points and the parabola.

$$nc + \left(\sum_{i=1}^{n} x_i\right)b + \left(\sum_{i=1}^{n} x_i^2\right)a = \sum_{i=1}^{n} y_i$$

$$\left(\sum_{i=1}^{n} x_i\right)c + \left(\sum_{i=1}^{n} x_i^2\right)b + \left(\sum_{i=1}^{n} x_i^3\right)a = \sum_{i=1}^{n} x_i y_i$$

$$\left(\sum_{i=1}^{n} x_i^2\right)c + \left(\sum_{i=1}^{n} x_i^3\right)b + \left(\sum_{i=1}^{n} x_i^4\right)a = \sum_{i=1}^{n} x_i^2 y_i$$

Points: $(-2, -4)$, $(0, 4)$, $(2, 4)$, $(4, -4)$

2—T—Answer: $a = -1, b = 2, c = 4; y = -x^2 + 2x + 4$

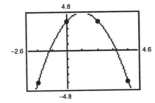

44. Find the least squares regression parabola $y = ax^2 + bx + c$ by solving the following system of linear equations for a, b, and c. Then use the least squares regression capabilities of a graphing utility to confirm the result. Then graph the data points and the parabola.

$$nc + \left(\sum_{i=1}^{n} x_i\right)b + \left(\sum_{i=1}^{n} x_i^2\right)a = \sum_{i=1}^{n} y_i$$

$$\left(\sum_{i=1}^{n} x_i\right)c + \left(\sum_{i=1}^{n} x_i^2\right)b + \left(\sum_{i=1}^{n} x_i^3\right)a = \sum_{i=1}^{n} x_i y_i$$

$$\left(\sum_{i=1}^{n} x_i^2\right)c + \left(\sum_{i=1}^{n} x_i^3\right)b + \left(\sum_{i=1}^{n} x_i^4\right)a = \sum_{i=1}^{n} x_i^2 y_i$$

Points: $(-1, 11)$, $(0, 1)$, $(2, -7)$, $(4, 1)$

2—T—Answer: $a = 2, b = -8, c = 1; y = 2x^2 - 8x + 1$

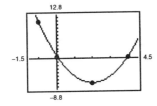

6.4 Systems of Inequalities

1. Match the graph with the correct inequality.

   (a) $y < x^2 + 3x - 1$
   (b) $y > x^2 + 3x - 1$
   (c) $y \leq x^2 + 3x - 1$
   (d) $y \geq x^2 + 3x - 1$

   1—Answer: c

   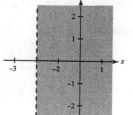

2. Match the graph with the correct inequality.

   (a) $y > -2$
   (b) $y < -2$
   (c) $x > -2$
   (d) $x \geq -2$

   1—Answer: c

   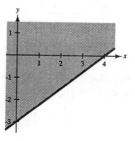

3. Match the graph with the correct inequality.

   (a) $3x - 4y < 12$
   (b) $3x - 4y \leq 12$
   (c) $3x - 4y > 12$
   (d) $3x - 4y \geq 12$

   1—Answer: b

   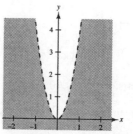

4. Match the graph with the correct inequality.

   (a) $4x^2 - y \leq 0$
   (b) $4x^2 - y > 0$
   (c) $4x^2 - y < 0$
   (d) $4x^2 - y^2 \geq 0$

   1—Answer: b

5. Use a graphing utility to graph the inequality
   $x^2 + (y - 1)^2 \leq 25$.

   1—T—Answer:

   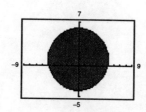

6. Use a graphing utility to graph the inequality
   $3x^2 + y \geq 6$.

   1—T—Answer: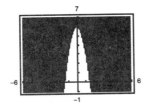

7. Use a graphing utility to graph the inequality
   $y > e^x$.

   1—T—Answer: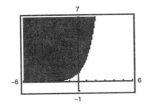

8. Sketch the graph of the inequality
   $x + y \leq 2$.

   (a)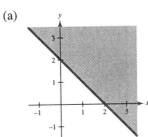

   (b)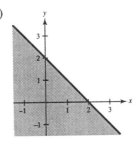

   (c)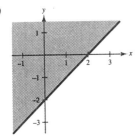

   (d)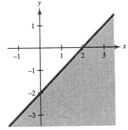

   (e) None of these

   1—Answer: b

9. Sketch the graph of the inequality
   $x - y > 1$.

   (a)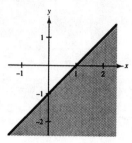

   (b)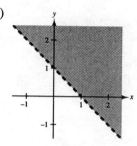

   (c)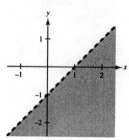

   (d)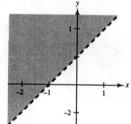

   (e) None of these

   1—Answer: c

10. Use a graphing utility to graph the inequality
    $3x^2 + y \geq 6$.

    1—T—Answer: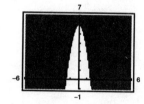

11. Identify the sketch of the inequality

$$\frac{x^2}{9} - \frac{y^2}{16} \le 1.$$

(a)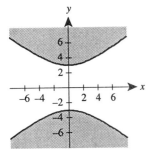

(b)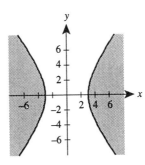

(c)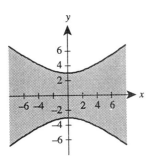

(d)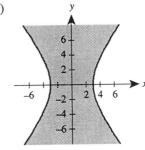

(e) None of these

1—Answer: d

12. Identify the sketch of the inequality $(x - 1)^2 + (y + 2)^2 > 4$.

(a)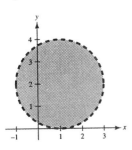

(b)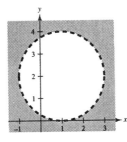

(c)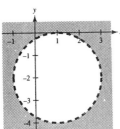

(d)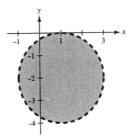

(e) None of these

1—Answer: c

13. Sketch the graph of the system of inequalities.

$$y \geq -|x + 2|$$
$$x \leq 0$$
$$y \leq 0$$

2—Answer:

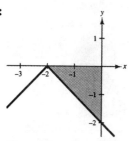

14. Sketch the graph of the system of inequalities.

$$x + y \geq 2$$
$$y \leq x$$
$$y > 0$$

1—Answer:

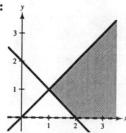

15. Sketch the graph of the system of inequalities.

$$2x + 3y \leq 6$$
$$x - 2y \geq -2$$

1—Answer:

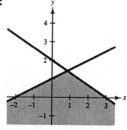

16. Sketch the graph of the system of inequalities.

$$2y - 3x \leq 10$$
$$2y \geq x^2$$

2—Answer:

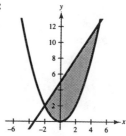

17. Identify the sketch of the graph of the solution for the system:

$$x + y \geq 2$$
$$y \geq 2$$
$$x \geq 2$$

(a)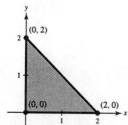

(b)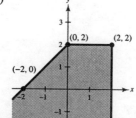

(c)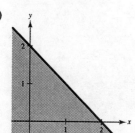

(d)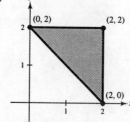

(e) None of these

1—Answer: d

18. Identify the sketch of the graph of the solution for the system:

$x + y > 1$

$x < 2$

$y \geq 0$

(a)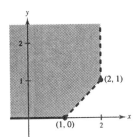

(b)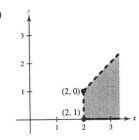

(c)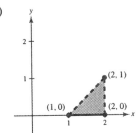

(d)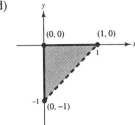

(e) None of these

1—Answer: c

19. Match the graph with the correct system of inequalities.

(a) $3x + y \leq 5$
$\quad y \geq 1$
$\quad x \geq 1$

(b) $3x + y \leq 5$
$\quad y \leq 1$
$\quad x \geq 1$

(c) $y \geq 5 + 3x$
$\quad y \leq 1$
$\quad x \leq 1$

(d) $3x + y \leq 5$
$\quad y \geq 1$
$\quad x \leq 1$

(e) None of these

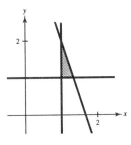

2—Answer: a

20. Match the graph with the correct system of inequalities.

(a) $x + 2y \leq 6$
$\quad x - y \leq 2$
$\quad y \geq 0$

(b) $x + 2y \geq 6$
$\quad x - y \geq 2$
$\quad x \geq 0$

(c) $x + 2y \leq 6$
$\quad x - y \geq 2$
$\quad x \geq 0$

(d) $x + 2y \geq 6$
$\quad x - y \geq 2$
$\quad y \geq 0$

(e) None of these

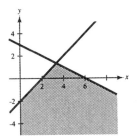

1—Answer: c

21. Match the graph with the correct system of inequalities.

    (a) $x + 2y \leq 4$
    $x \leq y$
    $x \geq 0$

    (b) $x + 2y \geq 4$
    $x \leq y$
    $y \geq 0$

    (c) $x + 2y \leq 4$
    $x \leq y$
    $y \geq 0$

    (d) $x + 2y \leq 4$
    $y \leq x$
    $y \geq 0$

    (e) None of these

    1—Answer: a

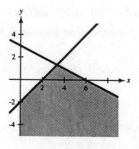

22. Match the graph with the correct system of inequalities.

    (a) $x + 2y \leq 4$
    $x \leq y$
    $x \geq 0$

    (b) $x + 2y \geq 4$
    $y \leq x$
    $y \geq 0$

    (c) $x + 2y \geq 4$
    $y \leq x$
    $y \geq 0$

    (d) $x + 2y \geq 4$
    $x \leq y$
    $y \geq 0$

    (e) None of these

    2—Answer: b

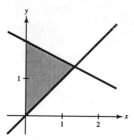

23. Match the graph with the correct system of inequalities.

    (a) $x^2 + y^2 \geq 16$
    $3x + 2y \leq 6$

    (b) $x^2 + y^2 \leq 16$
    $3x + 2y \leq 6$

    (c) $x^2 + y^2 \geq 16$
    $3x + 2y \geq 6$

    (d) $x^2 + y^2 \leq 16$
    $3x + 2y \geq 6$

    (e) None of these

    2—Answer: d

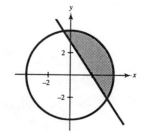

24. Match the graph with the correct system of inequalities.

    (a) $x^2 + y^2 \geq 9$
    $y \geq x^2$

    (b) $x^2 + y^2 \geq 9$
    $y \leq x^2$

    (c) $x^2 + y^2 \leq 9$
    $y \leq x^2$

    (d) $x^2 + y^2 \leq 9$
    $y \geq x^2$

    (e) None of these

    2—Answer: c

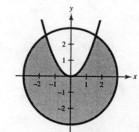

25. Find a set of inequalities that describe the triangular region with vertices at (0, 0), (3, 3), and (5, 0).

    2—Answer: $y \geq 0$, $y \leq x$, and $3x + 2y \leq 15$

26. For a circle of radius 3 and center at the origin, find a set of inequalities that describes the first quadrant sector bounded by radial lines that pass through $(3/\sqrt{2}, 3/\sqrt{2})$ and $(0, 3)$.

2—Answer: $y \geq 0$, $y \geq x$, and $x^2 + y^2 \leq 9$

27. Find the vertices of the region described by the system of inequalities:

$$3x + y \leq 4$$
$$2x - y \geq 1$$
$$x + 2y \geq -2$$

1—Answer: $(0, -1)$, $(1, 1)$, and $(2, -2)$

28. Find a set of inequalities that describe the triangular region with vertices $(0, 0)$, $(5, 5)$, and $(3, 0)$.

(a) $5x - 2y \geq 15$
$y \leq x$
$y \geq 0$

(b) $5x - 2y \geq 15$
$y \geq x$
$y \geq 0$

(c) $5x - 2y \leq 15$
$y \leq x$
$y \geq 0$

(d) $5x - 2y \leq 15$
$y \geq x$
$y \geq 0$

(e) None of these

2—Answer: c

29. Find a set of inequalities that describe the triangular region with vertices $(0, 0)$, $(8, 8)$, and $(10, 0)$.

(a) $4x + y \leq 40$
$y \leq x$
$y \geq 0$

(b) $4x + y \leq 40$
$y \geq x$

(c) $4x + y \geq 40$
$y \leq x$

(d) $4x + y \geq 40$
$y \leq x$
$y \geq 0$

(e) None of these

2—Answer: a

30. Find the vertices of the region described by the system of inequalities.

$$2x - y \leq 5$$
$$3x + y \geq 0$$
$$y \leq 0$$

(a) $(1, -3)$, $(0, 0)$, $(0, -5)$
(b) $(2, -1)$, $(0, 0)$, $(\frac{5}{2}, 0)$
(c) $(1, -3)$, $(0, 0)$, $(\frac{5}{2}, 0)$
(d) $(2, -1)$, $(0, 0)$, $(0, -5)$
(e) None of these

2—Answer: c

Chapter 6   Systems of Equations and Inequalities

31. Find the consumer surplus if the demand equation is $p = 110 - 20x$ and the supply equation is $p = 50 + 10x$.

    (a) $20   (b) $30   (c) $40
    (d) $50   (e) None of these

    2—Answer: c

32. Find the producer surplus if the demand equation is $p = 110 - 20x$ and the supply equation is $p = 50 + 10x$.

    (a) $20   (b) $30   (c) $40
    (d) $50   (e) None of these

    2—Answer: a

33. Find the producer surplus if the demand equation is $p = 90 - 10x$ and the supply equation is $p = 30 + 20x$.

    (a) $20   (b) $30   (c) $40
    (d) $50   (e) None of these

    2—Answer: c

34. Find the consumer surplus if the demand equation is $p = 100 - 0.1x$ and the supply equation is $p = 10 + 0.4x$.

    (a) $1250   (b) $1590   (c) $1620
    (d) $6480   (e) None of these

    2—Answer: c

35. Find the producer surplus if the demand equation is $p = 50 - 0.1x$ and the supply equation is $p = 10 + 0.4x$.

    (a) $1280   (b) $1200   (c) $560
    (d) $320   (e) None of these

    2—Answer: a

36. Find the consumer surplus if the demand equation is $p = 80 - \frac{1}{2}x$ and the supply equation is $p = 20 + \frac{5}{2}x$.

    (a) $500   (b) $300   (c) $100
    (d) $60   (e) None of these

    2—Answer: c

37. A small electronics company produces two types of CD players, standard and deluxe. The standard model requires 2 hours to manufacture and 2 hours to assemble. While the deluxe model requires 2 hours to manufacture and 3 hours to assemble. If the company has 40 hours per week of manufacturing time and 42 hours per week of assembly time. Find a system of inequalities describing the different numbers of standard and deluxe models that can be produced.

2—**Answer:** $2x + 2y \leq 40$

$2x + 3y \leq 42$

$x \geq 0$

$y \geq 0$

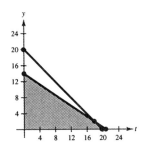

where x represents the number of standard models produced and y represents the number of deluxe models produced.

38. A company produces two types of four-pound gift boxes of candy, regular and chewy. The regular box has 2 pounds of chocolates and 2 pounds of caramels. The chewy box has 1 pound of chocolates and 3 pounds of caramels. The company has at most 60 pounds of chocolate available and at most 96 pounds of caramels available. Find a system of inequalities describing the number of regular boxes and chewy boxes that the company can produce.

2—**Answer:** $2x + y \leq 60$

$2x + 3y \leq 96$

$x \geq 0$

$y \geq 0$

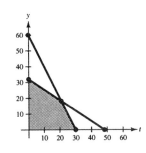

where x represents the number of regular boxes produced and y represents the number of chewy boxes produced.

6.5 Linear Programming

1. Find the maximum value of the objective function $z = 5x + 6y$ subject to the constraints:

$$x \geq 0$$
$$y \geq 0$$
$$x + 2y \leq 8$$
$$3x + 3y \leq 15$$

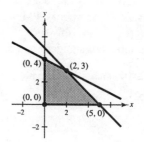

(a) 30　　　　　　　　(b) 28

(c) 25　　　　　　　　(d) 24

(e) None of these

1—Answer: b

2. Find the maximum value of the objective function $z = 10x + 8y$ subject to the constraints:

$$x \geq 0$$
$$y \geq 0$$
$$x + y \leq 5$$
$$3x + y \leq 12$$
$$-2x + y \leq 2$$

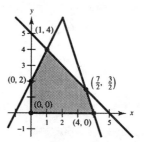

(a) 40　　　　　　　　(b) 50

(c) 42　　　　　　　　(d) 47

(e) None of these

1—Answer: d

3. Find the maximum value of the objective function $z = 3x + 2y$ subject to the constraints:

$$x \geq 0$$
$$y \geq 0$$
$$x + y \leq 4$$
$$x + 3y \leq 6$$

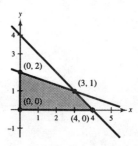

(a) 12　　　　　　　　(b) 11

(c) 10　　　　　　　　(d) 9

(e) None of these

1—Answer: a

4. Find the maximum value of the objective function $z = 9x + 6y$ subject to the constraints:

$$x \geq 0$$
$$y \geq 0$$
$$x + y \leq 3$$
$$2x + 5y \leq 10$$
$$x - y \leq 1$$

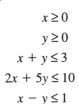

(a) 9   (b) 12

(c) 24   (d) 36

(e) None of these

1—**Answer:** c

5. Find the maximum value of the objective function $z = 4x + 2y$ subject to the constraints:

$$x \geq 0$$
$$y \geq 2$$
$$y \leq 3 - x$$

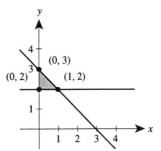

(a) 2   (b) 4

(c) 6   (d) 8

(e) None of these

1—**Answer:** b

6. Find the maximum value of the objective function $z = 5x + 4y$ subject to the constraints:

$$3x + 2y \leq 6$$
$$2y - x \geq 2$$
$$x \geq 0$$

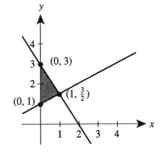

(a) 4   (b) 9

(c) 12   (d) 15

(e) None of these

1—**Answer:** a

7. Find the maximum value of the objective function $z = 40x + 35y$ subject to the constraints:

$$x \geq 0$$
$$y \geq 0$$
$$5x + 4y \leq 300$$
$$x + y \leq 70$$

(a) 2400   (b) 2625   (c) 2800

(d) 2550   (e) None of these

1—**Answer:** d

8. Find the maximum value of the objective function $z = 40x + 35y$ subject to the constraints:

$$x \geq 0$$
$$y \geq 0$$
$$4x + 6y \leq 300$$
$$x + y \leq 70$$

(a) 1750   (b) 2750   (c) 2800

(d) 2450   (e) None of these

1—**Answer:** c

9. Find the maximum value of the objective function $z = 40x + 60y$ subject to the constraints:

$$x \geq 0$$
$$y \geq 0$$
$$3x + 7y \leq 298$$
$$x + y \leq 50$$

(a) 2000   (b) 3000   (c) 3973

(d) 2740   (e) None of these

1—**Answer:** d

10. Find the maximum value of the objective function $z = 30x + 10y$ subject to the constraints:

$$x \geq 0$$
$$y \geq 0$$
$$2x + 4y \leq 60$$
$$x + y \leq 25$$

(a) 750   (b) 150   (c) 900

(d) 650   (e) None of these

1—**Answer:** a

11. Find the maximum value of the objective function $P = 8x + 10y$ subject to the constraints:

$$x \geq 0$$
$$y \geq 0$$
$$x + y \leq 160$$
$$x - 3y \geq 0$$

(a) 1280 (b) 1360 (c) 1500

(d) 0 (e) None of these

2—Answer: b

12. Find the maximum value of the function $P = 6x + 5y$ subject to the following constraints:

$$x \geq 0$$
$$y \geq 0$$
$$x + y \leq 76$$
$$x - 3y \geq 0$$

(a) 380 (b) 437 (c) 456

(d) 519 (e) None of these

2—Answer: c

13. Find the maximum value of the objective function $C = 3x + 2y$ subject to the constraints:

$$x \geq 0$$
$$y \geq 0$$
$$3x + 4y \leq 25$$
$$3x - y \leq 5$$

2—Answer: 17

14. Find the maximum value of the objective function $z = 6x + 5y$ subject to the constraints:

$$x \geq 0$$
$$y \geq 0$$
$$x + y \leq 76$$
$$x - 3y \geq 0$$

(a) 380 (b) 437 (c) 456

(d) 519 (e) None of these

1—Answer: c

15. Find the maximum value of the objective function $z = 3x + 2y$ subject to the constraints:

$$x \geq 0$$
$$y \geq 0$$
$$3x + 4y \leq 25$$
$$3x - y \leq 5$$

2—Answer: 17

16. Find the maximum value of the objective function $z = 3x - 4y$ subject to the constraints:

$$x \geq 0$$
$$y \geq 0$$
$$-2x + 3y \leq 5$$
$$5x + y \leq 30$$
$$3x + 2y \leq 12$$

(a) 10   (b) −5   (c) 12

(d) 24   (e) None of these

2—Answer: a

17. Find the maximum value of the objective function $z = 2x + 5y$ subject to the constraints:

$$x \geq 0$$
$$y \geq 0$$
$$2x + y \leq 12$$
$$x - 4y \leq -3$$

(a) 7   (b) 28   (c) 20

(d) No solution   (e) None of these

1—Answer: b

18. Find the maximum value of the objective function $z = 6x - 2y$ subject to the constraints:

$$x \geq 0$$
$$y \geq 0$$
$$x + y \leq 10$$
$$4x + y \geq 12$$

(a) 60   (b) 12   (c) 18

(d) 44   (e) None of these

1—Answer: a

19. Find the minimum and maximum values of the objective function $z = 4x + 16y$ subject to the constraints:

$$x \geq 0$$
$$y \geq 0$$
$$3x + y \leq 23$$
$$x + 2y \leq 16$$
$$3x \geq 2y$$

2—Answer: $z = 0$, minimum; $z = 112$, maximum

20. Find the maximum value of the objective function $C = 10x + 12y$ subject to the following constraints:

$$x \geq 0$$
$$y \geq 0$$
$$x + y \leq 36$$
$$x - 2y \geq 0$$

(a) 0   (b) 360   (c) 384

(d) 432   (e) None of these

2—Answer: a

21. Find the minimum value of the objective function $C = 2x + y$ subject to the following constraints:

$$x \geq 0$$
$$y \geq 0$$
$$3x + 2y \geq 90$$
$$2x + 3y \leq 105$$

2—Answer: 51

22. Find the minimum value of the objective function $z = 6x - 2y$ subject to the constraints:

$$x \geq 0$$
$$y \geq 0$$
$$x + y \leq 10$$
$$4x + y \geq 13$$

(a) 60   (b) -12 (c) -4

(d) -18 (e) None of these

1—Answer: b

23. Find the minimum value of the objective function $z = 3x + 2y$ subject to the constraints:

$$x \geq 0$$
$$y \geq 0$$
$$x + y \geq 4$$
$$3x + y \geq 6$$

(a) 0   (b) 8   (c) 9

(d) 12   (e) None of these

1—Answer: c

24. Find the minimum value of the objective function $z = 7x + 6y$ subject to the constraints:

    $x \geq 0$
    $y \geq 0$
    $5x + 2y \geq 16$
    $3x + 7y \geq 27$

    (a) $23\frac{1}{7}$ (b) 32   (c) $22\frac{2}{5}$
    (d) 0   (e) None of these

    1—Answer: b

25. Find the minimum value of the objective function $z = 3x + 4y$ subject to the constraints:

    $x \geq 0$
    $y \geq 0$
    $x + y \geq 6$
    $x + 2y \geq 8$

    (a) 0   (b) 20   (c) 16
    (d) 18   (e) None of these

    1—Answer: b

26. Find the minimum value of the objective function $z = 4x + y$ subject to the constraints:

    $x \geq 0$
    $y \geq 0$
    $5x + 3y \geq 15$
    $3x + 6y \geq 18$

    (a) 0   (b) 3   (c) 5
    (d) 9   (e) None of these

    1—Answer: c

27. A merchant plans to sell two models of an item at costs of $350 and $400. The $350 model yields a profit of $85 and the $400 model yields a profit of $90. The total demand per month for the two models will not exceed 150. Find the number of units of each model that should be stocked each month in order to maximize the profit. Assume the merchant can invest no more than $56,000 for inventory of these items.

    2—Answer: 80 of the $350 model; 70 of the $400 model

28. A merchant plans to sell two models of an item at costs of $350 and $500. The $350 model yields a profit of $45 and the $500 model yields a profit of $60. The total demand per month for the two models will not exceed 145. Find the number of units of each model that should be stocked each month in order to maximize the profit. Assume the merchant can invest no more than $56,000 for inventory of these items.

    2—Answer: 110 of the $350 model; 35 of the $500 model

29. A merchant plans to sell two models of an item at costs of $350 and $500. The $350 model yields a profit of $80 and the $500 model yields a profit of $100. The total demand per month for the two models will not exceed 150. Find the maximum monthly profit. Assume the merchant can invest no more than $63,000.

(a) $P = \$33,600$ (b) $P = \$13,400$ (c) $P = \$12,600$

(d) $P = \$12,000$ (e) None of these

2—**Answer:** b

30. A merchant plans to sell two models of an item at costs of $350 and $500. The $350 model yields a profit of $60 and the $500 model yields a profit of $75. The total demand per month for the two models will not exceed 100. Find the maximum monthly profit. Assume the merchant can invest no more than $41,000.

(a) $P = \$6600$ (b) $P = \$6150$ (c) $P = \$7020$

(d) $P = \$7500$ (e) None of these

2—**Answer:** a

31. A company produces two models of calculators at two different plants. In one day Plant A can produce 140 of Model I and 35 of Model II. In one day Plant B can produce 60 of Model I and 90 of Model II. The company needs to produce at least 460 Model I and 340 of Model II. Find the minimum cost. Assume it costs $1200 per day to operate Plant A and $900 per day for Plant B.

(a) $C = \$11,640$ (b) $C = \$8730$ (c) $C = \$5100$

(d) $C = \$3948$ (e) None of these

2—**Answer:** c

32. A company produces two models of calculators at two different plants. In one day Plant A can produce 60 of Model I and 70 of Model II. In one day Plant B can produce 80 of Model I and 40 of Model II. The company needs to produce at least 460 Model I and 340 of Model II. Find the minimum cost. Assume it costs $1200 per day to operate Plant A and $900 per day for Plant B.

(a) $C = \$7800$ (b) $C = \$9200$ (c) $C = \$7371$

(d) $C = \$5175$ (e) None of these

2—**Answer:** a

33. A company produces two models of calculators at two different plants. In one day Plant A can produce 70 of Model I and 40 of Model II. In one day Plant B can produce 80 of Model I and 90 of Model II. The company needs to produce at least 1370 Model I and 1270 of Model II. Find the minimum cost. Assume it costs $900 per day to operate Plant A and $1200 per day for Plant B.

(a) $C = \$32,570$ (b) $C = \$28,575$ (c) $C = \$20,550$

(d) $C = \$19,500$ (e) None of these

2—**Answer:** d

34. A company produces two models of calculators at two different plants. In one day Plant A can produce 150 of Model I and 250 of Model II. In one day Plant B can produce 175 of Model I and 140 of Model II. The company needs to produce at least 3075 Model I and 3760 of Model II. Find the minimum cost. Assume it costs $3500 per day to operate Plant A and $3100 per day for Plant B.

(a) $C = \$83,266$ (b) $C = \$62,900$ (c) $C = \$71,750$

(d) $C = \$59,200$ (e) None of these

2—**Answer:** b

35. A manufacturer wants to maximize the profit on two products. The first product yields a profit of $2.30 per unit, and the second product yields a profit of $1.50 per unit. Market tests and available resources have indicated the following constraints:

   1. The combined production level should not exceed 900 units per month.

   2. The demand for the first product is less than or equal to half the demand for the second product.

   Find the maximum profit.

   2—**Answer:** $1590

36. Use a graphing utility to sketch the region determined by the constraints. Then find the minimum and maximum values of the objective function, subject to the constraints.

   Objective function: $z = 2x + 5y$

   Constraints: $x \geq 0$
   $y \geq 0$
   $x + 2y \leq 20$
   $2x + 3y \geq 36$

   2—**Answer:** 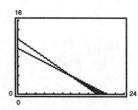 $z = 32$, minimum; $z = 40$, maximum

37. Use a graphing utility to sketch the region determined by the constraints. Then find the minimum and maximum values of the objective function, subject to the constraints.

Objective function: $z = 2x + y$

Constraints:
$$x \geq 0$$
$$y \geq 0$$
$$x + 6y \geq 48$$
$$7x + 2y \leq 56$$

2—Answer: 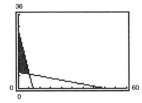 $z = 8$, minimum; $z = 28$, maximum

38. Use a graphing utility to sketch the region determined by the constraints. Then find the minimum and maximum values of the objective function, subject to the constraints.

Objective function: $z = 3x + 4y$

Constraints:
$$x \geq 0$$
$$y \geq 0$$
$$2x + 3y \geq 18$$
$$8x + 2y \leq 32$$

2—Answer: 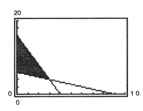 $z = 24$, minimum; $z = 64$, maximum

39. Use a graphing utility to sketch the region determined by the constraints. Then find the minimum and maximum values of the objective function, subject to the constraints.

Objective function: $z = 5x + 3y$

Constraints:
$$x \geq 0$$
$$y \geq 0$$
$$5x + y \geq 14$$
$$-5x + 6y \leq 49$$
$$7x + 2y \leq 77$$
$$4x + 9y \geq 44$$

2—Answer: $z = 22$, minimum; $z = 77$, maximum

CHAPTER SEVEN
Matrices and Determinants

❑ 7.1 Matrices and Systems of Equations

1. Determine the order of the matrix: $\begin{bmatrix} 2 & 7 & 9 \\ 3 & 5 & -1 \end{bmatrix}$

   (a) 2×3 (b) 3×2 (c) 3

   (d) 2   (e) None of these

   1—Answer: a

2. Determine the order of the matrix: $\begin{bmatrix} 1 & 3 \\ 0 & 6 \\ 2 & 1 \\ 4 & 7 \end{bmatrix}$

   (a) 2×4 (b) 4×2 (c) 4

   (d) 2   (e) None of these

   1—Answer: b

3. Determine the order of the matrix: $\begin{bmatrix} 2 & 1 \\ -1 & 5 \\ 4 & -3 \end{bmatrix}$

   (a) 6   (b) 2×3 (c) 5

   (d) 3×2 (e) None of these

   1—Answer: d

4. Determine which of the following matrices is in row-echelon form.

   (a) $\begin{bmatrix} 1 & 3 & -1 & 4 \\ 0 & 2 & 1 & 1 \\ 0 & 0 & 1 & 4 \end{bmatrix}$ (b) $\begin{bmatrix} 1 & 2 & 3 & 4 \\ 0 & 1 & 4 & 5 \\ 0 & 0 & 0 & 0 \end{bmatrix}$ (c) $\begin{bmatrix} 2 & -1 & 6 & 3 \\ 4 & 1 & 2 & 0 \\ 0 & 1 & 2 & 0 \end{bmatrix}$

   (d) All of these   (e) None of these

   1—Answer: b

5. Determine which of the following matrices is in row-echelon form.

   (a) $\begin{bmatrix} 0 & 3 & 2 & -1 \\ 4 & 0 & 1 & 0 \\ 0 & 0 & 2 & 5 \end{bmatrix}$ (b) $\begin{bmatrix} 3 & 1 & 2 & 5 \\ 0 & 2 & 0 & 4 \\ 0 & 0 & 3 & 7 \end{bmatrix}$ (c) $\begin{bmatrix} 1 & 2 & 3 & 4 \\ 0 & 2 & 4 & 7 \\ 0 & 0 & 1 & 0 \end{bmatrix}$

   (d) All of these   (e) None of these

   1—Answer: e

6. Determine which of the following matrices is in row-echelon form.

(a) $\begin{bmatrix} 1 & 2 & 4 & 6 \\ 0 & 1 & 3 & 2 \\ 0 & 0 & 1 & 0 \end{bmatrix}$
(b) $\begin{bmatrix} 3 & 4 & 7 & 0 \\ 6 & 2 & 1 & 4 \\ 3 & 2 & 1 & 3 \end{bmatrix}$
(c) $\begin{bmatrix} 1 & 6 & 4 & 2 \\ 0 & 2 & 3 & 1 \\ 0 & 0 & 1 & 0 \end{bmatrix}$

(d) All of these
(e) None of these

1—Answer: a

7. Determine which of the following matrices is in row-echelon form.

(a) $\begin{bmatrix} 1 & 2 & 3 & 4 \\ 0 & 1 & 7 & 2 \\ 0 & 0 & 1 & 5 \end{bmatrix}$
(b) $\begin{bmatrix} 1 & 0 & 0 & 3 \\ 0 & 1 & 0 & 2 \\ 0 & 0 & 1 & 5 \end{bmatrix}$
(c) $\begin{bmatrix} 1 & 0 & 4 & 7 \\ 0 & 1 & 0 & 2 \\ 0 & 0 & 1 & 2 \end{bmatrix}$

(d) All of these
(e) None of these

1—Answer: d

8. Determine which matrix is in row-echelon form.

(a) $\begin{bmatrix} 1 & 5 \\ 0 & 1 \\ 0 & 0 \end{bmatrix}$
(b) $\begin{bmatrix} 0 & 0 & 0 \\ 0 & 1 & 2 \end{bmatrix}$
(c) $\begin{bmatrix} 1 & -4 & 3 & 7 \\ 0 & 1 & 2 & -1 \\ 0 & 0 & 3 & 5 \end{bmatrix}$

(d) $[3]$
(e) None of these

1—Answer: a

9. Determine which matrix is in row-echelon form.

(a) $\begin{bmatrix} 1 & -2 \\ 0 & 1 \\ 0 & 0 \end{bmatrix}$
(b) $\begin{bmatrix} 1 & 0 & 4 & -2 \\ 0 & 1 & 7 & 5 \\ 0 & 0 & 0 & 0 \end{bmatrix}$
(c) $\begin{bmatrix} 0 & 0 & 0 \\ 0 & 1 & 2 \end{bmatrix}$

(d) $\begin{bmatrix} 1 & 1 \\ 0 & 1 \end{bmatrix}$
(e) None of these

1—Answer: b

10. Write the matrix in reduced row-echelon form: $\begin{bmatrix} 3 & 1 & 1 & 7 \\ 1 & -2 & 0 & 5 \\ 1 & 1 & 2 & 6 \end{bmatrix}$

(a) $\begin{bmatrix} 1 & 1 & 2 & 6 \\ 0 & 3 & 2 & 1 \\ 0 & 0 & 11 & 31 \end{bmatrix}$
(b) $\begin{bmatrix} 1 & 0 & 0 & \frac{21}{11} \\ 0 & 1 & 0 & -\frac{17}{11} \\ 0 & 0 & 1 & \frac{31}{11} \end{bmatrix}$

(c) $\begin{bmatrix} 1 & -2 & 0 & 5 \\ 0 & 1 & \frac{1}{7} & -\frac{8}{7} \\ 0 & 0 & 11 & 31 \end{bmatrix}$
(d) $\begin{bmatrix} 1 & 0 & 0 & \frac{3}{11} \\ 0 & 1 & 0 & \frac{7}{11} \\ 0 & 0 & 1 & -\frac{14}{11} \end{bmatrix}$

(e) None of these

1—Answer: b

11. Write the matrix in reduced row-echelon form: $\begin{bmatrix} 3 & 6 & -2 & 28 \\ -2 & -4 & 5 & -37 \\ 1 & 2 & 9 & -39 \end{bmatrix}$

(a) $\begin{bmatrix} 1 & 2 & 1 & 1 \\ 0 & 0 & 1 & -5 \\ 0 & 0 & 0 & 0 \end{bmatrix}$ (b) $\begin{bmatrix} 0 & 0 & 0 & 0 \\ 1 & 2 & 0 & 6 \\ 0 & 0 & 1 & -5 \end{bmatrix}$ (c) $\begin{bmatrix} 1 & 2 & 0 & 6 \\ 0 & 0 & 1 & -5 \\ 0 & 0 & 0 & 0 \end{bmatrix}$

(d) $\begin{bmatrix} 1 & 2 & 1 & 1 \\ 0 & 0 & 1 & -5 \\ 0 & 0 & 0 & 3 \end{bmatrix}$ (e) None of these

1—Answer: c

12. Write the matrix in reduced row-echelon form: $\begin{bmatrix} 1 & 3 & -8 & 13 \\ 2 & -1 & 6 & -19 \\ -5 & 1 & 2 & 44 \end{bmatrix}$

(a) $\begin{bmatrix} 1 & 0 & 0 & -7 \\ 0 & 1 & 0 & 8 \\ 0 & 0 & 1 & \frac{1}{2} \end{bmatrix}$ (b) $\begin{bmatrix} 1 & 0 & 6 & -4 \\ 0 & 1 & 2 & 9 \\ 0 & 0 & 2 & 1 \end{bmatrix}$ (c) $\begin{bmatrix} 1 & 0 & 6 & -4 \\ 0 & 1 & -4 & 6 \\ 0 & 0 & 0 & 0 \end{bmatrix}$

(d) $\begin{bmatrix} 1 & 1 & 8 & 5 \\ 0 & 1 & 2 & 9 \\ 0 & 0 & 2 & 1 \end{bmatrix}$ (e) None of these

1—Answer: a

13. Write the matrix in reduced row-echelon form: $\begin{bmatrix} 1 & 2 & -1 & 3 \\ 7 & -1 & 0 & 2 \\ 3 & 2 & 1 & -1 \end{bmatrix}$

1—Answer: $\begin{bmatrix} 1 & 0 & 0 & \frac{5}{16} \\ 0 & 1 & 0 & \frac{3}{16} \\ 0 & 0 & 1 & -\frac{37}{16} \end{bmatrix}$

14. Write the matrix in reduced row-echelon form: $\begin{bmatrix} 21 & 14 & -7 & 10 \\ 7 & 7 & 7 & -1 \\ 3 & -14 & 28 & 23 \end{bmatrix}$

1—Answer: $\begin{bmatrix} 1 & 0 & 0 & \frac{9}{7} \\ 0 & 1 & 0 & -\frac{9}{7} \\ 0 & 0 & 1 & -\frac{1}{7} \end{bmatrix}$

15. Use variables x, y, and z to write the system of linear equations represented by the augmented matrix:

$$\begin{bmatrix} 2 & -1 & 0 & \vdots & 4 \\ 0 & 3 & 1 & \vdots & -2 \\ 1 & -3 & 1 & \vdots & 1 \end{bmatrix}$$

1—Answer:
$$2x - y = 4$$
$$3y + z = -2$$
$$x - 3y + z = 1$$

16. Find the solution to the system of linear equations with the augmented matrix:

$$\begin{bmatrix} 2 & -1 & \vdots & 3 \\ 0 & 1 & \vdots & 2 \end{bmatrix}$$

(a) $(3, 2)$ (b) $(2, 3)$ (c) $(-1, 3)$

(d) $\left(\frac{5}{2}, 2\right)$ (e) None of these

1—Answer: d

17. Find the solution to the system of linear equations with the augmented matrix:

$$\begin{bmatrix} 1 & 2 & -1 & \vdots & 4 \\ 0 & 2 & 1 & \vdots & -3 \\ 0 & 0 & 2 & \vdots & -4 \end{bmatrix}$$

(a) $\left(3, -\frac{1}{2}, -2\right)$ (b) $(4, -3, -4)$ (c) $(1, 2, -1)$

(d) $\left(10, -\frac{5}{2}, -2\right)$ (e) None of these

1—Answer: a

18. Find the solution to the system of linear equations with the augmented matrix:

$$\begin{bmatrix} 1 & 0 & 1 & \vdots & 0 \\ 0 & 1 & -2 & \vdots & 1 \end{bmatrix}$$

(a) $(a, 1 + 2a, -a)$ (b) $(-a, 2a + 1, a)$ (c) $(a, 1 - 2a, a)$

(d) $(-a, 1 - 2a, a)$ (e) None of these

1—Answer: b

19. Find the solution to the system of linear equations with the augmented matrix:

$$\begin{bmatrix} 2 & 1 & \vdots & 7 \\ 5 & -3 & \vdots & 1 \end{bmatrix}$$

(a) $(3, 1)$ (b) $(1, 1)$ (c) $(4, -1)$

(d) $(2, 3)$ (e) None of these

1—Answer: d

Chapter 7   Matrices and Determinants

20. Find the solution to the system of linear equations with the augmented matrix:

$$\begin{bmatrix} 1 & 0 & 1 & \vdots & 1 \\ 2 & 1 & -1 & \vdots & 2 \end{bmatrix}$$

(a) $(2a, -3a, a)$ (b) $(1 - a, 3a, a)$ (c) $(a + 2, 4a, -a)$

(d) $(3a, 0, 2a)$ (e) None of these

1—Answer: b

21. Find the solution to the system of linear equations with the augmented matrix:

$$\begin{bmatrix} 3 & 1 & 1 & \vdots & 3 \\ 2 & -1 & 3 & \vdots & 2 \end{bmatrix}$$

(a) $(a, 3a, -7a)$ (b) $(3 + a, 2a, a)$ (c) $(2a, 1 + a, a)$

(d) $(1 - a, a, a)$ (e) None of these

1—Answer: d

22. Write the augmented matrix for the given system of linear equations.

$$2x - 3y = 12$$
$$x + y = 16$$

1—Answer: $\begin{bmatrix} 2 & -3 & \vdots & 12 \\ 1 & 1 & \vdots & 16 \end{bmatrix}$

23. Write the augmented matrix for the given system of linear equations.

$$3x - y = 2$$
$$x + 2y = 7$$

(a) $\begin{bmatrix} 3 & 1 & \vdots & -2 \\ 1 & 2 & \vdots & -7 \end{bmatrix}$ (b) $\begin{bmatrix} 3 & 1 & \vdots & 2 \\ -1 & 2 & \vdots & 7 \end{bmatrix}$ (c) $\begin{bmatrix} 3 & -1 & \vdots & 2 \\ 1 & 2 & \vdots & 7 \end{bmatrix}$

(d) $\begin{bmatrix} -3 & 1 & \vdots & -2 \\ 1 & 2 & \vdots & -7 \end{bmatrix}$ (e) None of these

1—Answer: c

24. Write the augmented matrix for the given system of linear equations.

$$7x + 2y = 12$$
$$x - y = 16$$

(a) $\begin{bmatrix} 7 & 2 & \vdots & 12 \\ 1 & 1 & \vdots & 16 \end{bmatrix}$ (b) $\begin{bmatrix} 7 & 2 & \vdots & -12 \\ 1 & 1 & \vdots & -16 \end{bmatrix}$ (c) $\begin{bmatrix} 7 & 2 & \vdots & 12 \\ 1 & -1 & \vdots & 16 \end{bmatrix}$

(d) $\begin{bmatrix} 7 & 1 & \vdots & 12 \\ 2 & -1 & \vdots & 16 \end{bmatrix}$ (e) None of these

1—Answer: c

25. Write the augmented matrix for the given system of linear equations.

$$3x + 2y = 5$$
$$7x - 6y = 12$$

(a) $\begin{bmatrix} 3 & 2 & \vdots & 5 \\ 7 & -6 & \vdots & 12 \end{bmatrix}$
(b) $\begin{bmatrix} 3 & 2 & \vdots & -5 \\ 7 & 6 & \vdots & -12 \end{bmatrix}$
(c) $\begin{bmatrix} 3 & 2 & \vdots & 5 \\ 7 & 6 & \vdots & 12 \end{bmatrix}$

(d) $\begin{bmatrix} 3 & 7 & \vdots & 5 \\ 2 & -6 & \vdots & 12 \end{bmatrix}$
(e) None of these

1—Answer: a

26. Form the augmented matrix for the system of equations.

$$y - 3z = 5$$
$$2x + z = -1$$
$$4x - y = 0$$

(a) $\begin{bmatrix} 1 & -3 \\ 2 & 1 \\ 4 & -1 \end{bmatrix}$
(b) $\begin{bmatrix} 1 & -3 & \vdots & 5 \\ 2 & 1 & \vdots & -1 \\ 4 & -1 & \vdots & 0 \end{bmatrix}$
(c) $\begin{bmatrix} 0 & 1 & -3 \\ 2 & 0 & 1 \\ 4 & -1 & 0 \end{bmatrix}$

(d) $\begin{bmatrix} 0 & 1 & -3 & \vdots & 5 \\ 2 & 0 & 1 & \vdots & -1 \\ 4 & -1 & 0 & \vdots & 0 \end{bmatrix}$
(e) None of these

1—Answer: d

27. Use Gauss-Jordan elimination to solve the system of linear equations.

$$x + 2y + z = 7$$
$$3x + z = 2$$
$$x - y - z = 1$$

(a) $\left(\frac{2}{3}, \frac{20}{3}, -7\right)$
(b) $\left(-\frac{7}{3}, \frac{1}{6}, 9\right)$
(c) $\left(\frac{11}{6}, \frac{13}{3}, -\frac{7}{2}\right)$

(d) $\left(-\frac{19}{3}, -\frac{2}{3}, \frac{44}{3}\right)$

1—Answer: c

28. Use the matrix capabilities of a graphing utility to reduce the augmented matrix and solve the system of equations.

$$x + y + z = 5$$
$$2x - y - 3z = 5$$
$$-2x + 2y + z = -11$$

1—T—Answer: $(5, -1, 2)$

354 Chapter 7 Matrices and Determinants

29. Use the matrix capabilities of a graphing utility to reduce the augmented matrix and solve the system of equations.

$$3x + 2y - 5z = -10$$
$$2x + 4y + z = 0$$
$$x - 6y - 4z = -3$$

1—T—Answer: $\left(\frac{1}{2}, -\frac{3}{4}, 2\right)$

30. Use Gaussian elimination with back-substitution or Gauss-Jordan elimination to solve the following system of linear equations.

$$3x + 2y + z = 7$$
$$x - y + z = 6$$
$$x + z = 5$$

(a) $(2, -1, 3)$ (b) $\left(1, -\frac{1}{2}, 5\right)$ (c) $(-1, 1, 2)$
(d) $(0, 4, -1)$ (e) None of these

1—Answer: a

31. Use Gaussian elimination with back-substitution or Gauss-Jordan elimination to solve the following system of linear equations.

$$2x + y - z = -3$$
$$4x - y + z = 6$$
$$2x + 3y + 2z = 9$$

(a) $(1, -1, 4)$ (b) $\left(\frac{1}{2}, 0, 4\right)$ (c) $\left(\frac{1}{2}, 2, 0\right)$
(d) $\left(\frac{3}{2}, -9, 0\right)$ (e) None of these

1—Answer: b

32. Use Gaussian elimination with back-substitution or Gauss-Jordan elimination to solve the following system of linear equations.

$$2x + 3y - 4z = 4$$
$$x - y - 5z = 0$$
$$-2x + 4y + 5z = 9$$

(a) $(-3, 2, -1)$ (b) $(3, -2, -1)$ (c) $(2, 1, 1)$
(d) $(0, 4, 2)$ (e) None of these

1—Answer: a

33. Find an equation of the parabola that passes through the points $(1, 4)$, $(2, 5)$, and $(-1, -4)$.

(a) $y = -3x^2 + 10x + 5$ (b) $y = x^2 - x - 1$ (c) $y = -2x^2 + 12x - 11$
(d) $y = 2x^2 - 4x + 1$ (e) None of these

1—Answer: d

34. Find an equation of the parabola that passes through the points $(1, -1)$, $(2, 1)$, and $(3, 7)$.

    (a) $y = x^2 + 5x - 5$
    (b) $y = x^2 - x - 1$
    (c) $y = -2x^2 + 12x - 11$
    (d) $y = 2x^2 - 4x + 1$
    (e) None of these

    1—Answer: d

35. Find an equation of the parabola that passes through the points $(-1, 0)$, $(1, 4)$, and $(2, 15)$.

    (a) $y = x^2 + 2x + 1$
    (b) $y = -x^2 + 2x + 3$
    (c) $y = -2x^2 + 2x + 4$
    (d) $y = 2x^2 + 3x + 1$
    (e) None of these

    1—Answer: e

36. Find the equation of the parabola that passes through the given points. Use a graphing utility to verify your result.

    1—T—Answer: $y = 2x^2 + 5x - 3$

    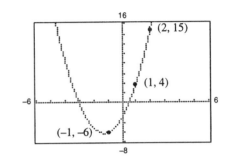

37. Find the equation of the parabola that passes through the given points. Use a graphing utility to verify your result.

    1—T—Answer: $y = x^2 - x + 4$

    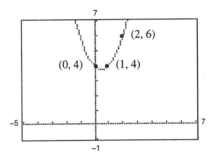

38. A small corporation borrowed $900,000; some at 7%, some at 8%, and some at 11%. How much was borrowed at 8% if the annual interest was $72,250 and the amount borrowed at 8% was $50,000 more than the amount borrowed at 11%?

    (a) $225,000
    (b) $175,000
    (c) $650,000
    (d) $450,000
    (e) None of these

    2—Answer: a

39. A small corporation borrowed $900,000; some at 7%, some at 8%, and some at 11%. How much was borrowed at 11% if the annual interest was $71,500 and the amount borrowed at 7% was twice the amount borrowed at 8%?

    (a) $400,000
    (b) $800,000
    (c) $150,000
    (d) $300,000
    (e) None of these

    2—Answer: c

7.2 Operations with Matrices

1. Find x: $\begin{bmatrix} 3x+2 & 5 & 2 \\ 7 & 2y & x \\ 4 & -1 & y+1 \end{bmatrix} = \begin{bmatrix} x-4 & 5 & y/2 \\ 7 & 8 & -3 \\ 4 & -1 & 5 \end{bmatrix}$

   (a) 4   (b) -3 (c) -1
   (d) 6   (e) None of these

   1—Answer: b

2. Find x: $\begin{bmatrix} 9-2x & y & 0 \\ 4 & 2-y & 3 \\ 6 & \frac{2}{5}x & 5 \end{bmatrix} = \begin{bmatrix} -1 & 3 & 0 \\ 4 & -1 & x-2 \\ 6 & 2 & x \end{bmatrix}$

   (a) -2 (b) -3 (c) 4
   (d) 5   (e) None of these

   1—Answer: d

3. Find x: $\begin{bmatrix} x-1 & 4 & 5 \\ 2 & -1 & 5 \\ -6 & x & 7 \end{bmatrix} = \begin{bmatrix} -3 & 4 & 5 \\ 2 & -y/2 & y+3 \\ -6 & -2 & 4x-1 \end{bmatrix}$

   (a) -2 (b) -3 (c) 4
   (d) 5   (e) None of these

   1—Answer: a

4. Evaluate: $2\begin{bmatrix} 4 & 7 \\ 2 & 0 \end{bmatrix} - 3\begin{bmatrix} 1 & 2 \\ -1 & 5 \end{bmatrix}$

   (a) $\begin{bmatrix} 5 & 8 \\ 7 & -15 \end{bmatrix}$ (b) $\begin{bmatrix} 5 & 9 \\ 1 & -15 \end{bmatrix}$ (c) $-1\begin{bmatrix} 3 & 5 \\ 3 & 5 \end{bmatrix}$
   (d) $-1\begin{bmatrix} 5 & 9 \\ 1 & 5 \end{bmatrix}$ (e) None of these

   1—Answer: a

5. Evaluate: $3\begin{bmatrix} 2 & 7 \\ 9 & -1 \end{bmatrix} + 3\begin{bmatrix} 4 & 1 \\ 6 & 2 \end{bmatrix}$

   1—Answer: $\begin{bmatrix} 18 & 24 \\ 45 & 3 \end{bmatrix}$

6. Evaluate: $3\begin{bmatrix} 7 & 5 & 2 \\ -1 & 1 & 0 \\ 0 & 3 & 6 \end{bmatrix} + 2\begin{bmatrix} 1 & 0 & 0 \\ 0 & 1 & 0 \\ 0 & 0 & 1 \end{bmatrix} - 4\begin{bmatrix} 1 & -1 & 2 \\ 7 & 1 & -1 \\ 2 & 5 & 1 \end{bmatrix}$

1—Answer: $\begin{bmatrix} 19 & 19 & -2 \\ -31 & 1 & 4 \\ -8 & -11 & 16 \end{bmatrix}$

7. Given $A = \begin{bmatrix} 3 & 6 & -1 \\ 0 & 5 & 2 \end{bmatrix}$ and $B = \begin{bmatrix} 1 & 0 & 5 \\ -1 & 2 & 7 \end{bmatrix}$, find $3A - 2B$.

(a) $\begin{bmatrix} 7 & 18 & -13 \\ 2 & 11 & -8 \end{bmatrix}$

(b) $\begin{bmatrix} 7 & 18 & 2 \\ 0 & 11 & -8 \end{bmatrix}$

(c) $\begin{bmatrix} 11 & 18 & 7 \\ -2 & 19 & 20 \end{bmatrix}$

(d) $\begin{bmatrix} 7 & 18 & -13 \\ -2 & 9 & 20 \end{bmatrix}$

(e) None of these

1—Answer: a

8. Given $A = \begin{bmatrix} 1 & 2 & 3 \\ 4 & 7 & 1 \\ 0 & 3 & 2 \end{bmatrix}$ and $B = \begin{bmatrix} 0 & 0 & 1 \\ 1 & 4 & 0 \\ 2 & 3 & 7 \end{bmatrix}$, find $6A - 2B$.

(a) $\begin{bmatrix} 30 & 72 & 52 \\ 32 & 10 & 6 \\ 1 & 6 & -8 \end{bmatrix}$

(b) $\begin{bmatrix} 30 & 12 & 4 \\ 3 & 10 & 6 \\ 1 & 6 & -8 \end{bmatrix}$

(c) $\begin{bmatrix} 6 & 12 & 16 \\ 22 & 34 & 6 \\ -4 & 12 & -2 \end{bmatrix}$

(d) $\begin{bmatrix} 30 & 22 & 20 \\ 66 & 28 & 26 \\ 24 & 16 & 14 \end{bmatrix}$

(e) None of these

1—Answer: c

9. Given $A = \begin{bmatrix} 2 & 4 & -1 \\ 1 & 0 & 4 \\ 8 & 1 & 2 \end{bmatrix}$ and $B = \begin{bmatrix} 1 & 1 & 1 \\ -1 & 0 & 0 \\ 4 & 10 & -2 \end{bmatrix}$, find $2A - 2B$.

(a) $\begin{bmatrix} 2 & 6 & 4 \\ 4 & 0 & -8 \\ -8 & 18 & 1 \end{bmatrix}$

(b) $\begin{bmatrix} 2 & 6 & -4 \\ 4 & 0 & 8 \\ 8 & -18 & 8 \end{bmatrix}$

(c) $\begin{bmatrix} 0 & 6 & -4 \\ -4 & 0 & 4 \\ 4 & -9 & 0 \end{bmatrix}$

(d) $\begin{bmatrix} 1 & 3 & -2 \\ 0 & 0 & 4 \\ 4 & -9 & 4 \end{bmatrix}$

(e) None of these

1—Answer: b

10. Given $A = \begin{bmatrix} 3 & -2 & 4 \\ 0 & 0 & -1 \\ 3 & 2 & -1 \end{bmatrix}$ and $B = \begin{bmatrix} 3 & 1 & -1 \\ -1 & 0 & 0 \\ 2 & 4 & -2 \end{bmatrix}$, find $3A - 2B$.

(a) $\begin{bmatrix} 0 & -8 & 14 \\ 2 & 0 & -3 \\ 5 & -2 & 1 \end{bmatrix}$ (b) $\begin{bmatrix} 3 & -8 & 10 \\ 2 & 0 & -3 \\ 7 & -2 & -7 \end{bmatrix}$ (c) $\begin{bmatrix} 6 & -1 & 3 \\ -1 & 0 & -1 \\ 5 & 6 & -3 \end{bmatrix}$

(d) $\begin{bmatrix} 15 & -4 & 10 \\ -2 & 0 & -3 \\ 13 & 14 & 1 \end{bmatrix}$ (e) None of these

1—Answer: a

11. Given $A = \begin{bmatrix} 3 & 2 & 2 \\ -1 & 0 & -1 \\ 0 & 1 & 0 \end{bmatrix}$ and $B = \begin{bmatrix} 1 & 1 & 0 \\ 0 & 1 & 1 \\ 2 & -1 & 2 \end{bmatrix}$, find $2A - 3B$.

(a) $\begin{bmatrix} 8 & 7 & 4 \\ -2 & 3 & 1 \\ 6 & -1 & 6 \end{bmatrix}$ (b) $\begin{bmatrix} 8 & 7 & 4 \\ -2 & -3 & -5 \\ -6 & -1 & -6 \end{bmatrix}$ (c) $\begin{bmatrix} 3 & 1 & 4 \\ 2 & 3 & 5 \\ -6 & 5 & 6 \end{bmatrix}$

(d) $\begin{bmatrix} 3 & 1 & 4 \\ -2 & -3 & -5 \\ -6 & 5 & -6 \end{bmatrix}$ (e) None of these

1—Answer: d

12. Given $A = \begin{bmatrix} 1 & 0 & 3 \\ -1 & 2 & -2 \\ 1 & 1 & 2 \end{bmatrix}$ and $B = \begin{bmatrix} 1 & 1 & 0 \\ 3 & 1 & 2 \\ -1 & 1 & -1 \end{bmatrix}$, find $-2A + 5B$.

(a) $\begin{bmatrix} 3 & 5 & -6 \\ 17 & 1 & 14 \\ -7 & 3 & -9 \end{bmatrix}$ (b) $\begin{bmatrix} 3 & 5 & 6 \\ 13 & 9 & 6 \\ -3 & 8 & -1 \end{bmatrix}$ (c) $\begin{bmatrix} 7 & 5 & 6 \\ 13 & 9 & 6 \\ -3 & 8 & -1 \end{bmatrix}$

(d) $\begin{bmatrix} 3 & 5 & -6 \\ -17 & 0 & 14 \\ 7 & -3 & -9 \end{bmatrix}$ (e) None of these

1—Answer: a

13. If $A = \begin{bmatrix} 2 & -1 \\ 3 & 1 \end{bmatrix}$ and $B = \begin{bmatrix} 4 & 0 \\ -1 & -1 \end{bmatrix}$, find $A - 2B$.

(a) $\begin{bmatrix} -6 & -1 \\ 5 & -1 \end{bmatrix}$ (b) 13 (c) $\begin{bmatrix} -6 & -1 \\ 1 & 3 \end{bmatrix}$

(d) $\begin{bmatrix} -2 & -1 \\ 5 & 3 \end{bmatrix}$ (e) None of these

1—Answer: e

14. If $A = \begin{bmatrix} 2 & -1 \\ 3 & 1 \end{bmatrix}$ and $B = \begin{bmatrix} 4 & 0 \\ -1 & -1 \end{bmatrix}$, find $B - 2A$.

    (a) -14
    (b) $\begin{bmatrix} -6 & -1 \\ 5 & 3 \end{bmatrix}$
    (c) $\begin{bmatrix} 0 & -2 \\ -7 & -3 \end{bmatrix}$
    (d) $\begin{bmatrix} 0 & 2 \\ -7 & -3 \end{bmatrix}$
    (e) None of these

    1—Answer: d

15. If $A = \begin{bmatrix} 2 & -1 \\ -3 & 4 \end{bmatrix}$ and $B = \begin{bmatrix} -2 & 0 \\ -1 & 3 \end{bmatrix}$. Find C if $A + C = 2B$.

    1—Answer: $\begin{bmatrix} -6 & 1 \\ 1 & 2 \end{bmatrix}$

16. Use a graphing utility to find AB, given

    $A = \begin{bmatrix} 1 & 3 & 6 \\ 4 & 1 & 3 \end{bmatrix}$ and $B = \begin{bmatrix} 0 & 1 & 6 \\ 3 & -1 & 1 \\ 5 & 2 & 3 \end{bmatrix}$.

    1—T—Answer: $\begin{bmatrix} 39 & 10 & 27 \\ 28 & 9 & 34 \end{bmatrix}$

17. Use a graphing utility to find AB, given

    $A = \begin{bmatrix} 2 & 1 \\ 3 & -2 \end{bmatrix}$ and $B = \begin{bmatrix} -1 & 5 \\ 6 & 2 \end{bmatrix}$.

    (a) $\begin{bmatrix} 4 & 12 \\ -15 & 11 \end{bmatrix}$
    (b) $\begin{bmatrix} 13 & -11 \\ 18 & 2 \end{bmatrix}$
    (c) $\begin{bmatrix} 16 & 16 \\ -13 & 1 \end{bmatrix}$
    (d) $\begin{bmatrix} -2 & 5 \\ 18 & -4 \end{bmatrix}$
    (e) None of these

    1—T—Answer: a

18. Use a graphing utility to multiply: $\begin{bmatrix} 2 & 3 & 4 \\ -1 & 0 & 2 \end{bmatrix} \begin{bmatrix} -1 & 4 \\ 0 & 1 \\ 5 & 2 \end{bmatrix}$

    (a) $\begin{bmatrix} 18 & 19 \\ 11 & 0 \end{bmatrix}$
    (b) $\begin{bmatrix} -2 & 0 & 20 \\ -4 & 0 & 4 \end{bmatrix}$
    (c) $\begin{bmatrix} -6 & -3 & 4 \\ -2 & 0 & 6 \\ -1 & 2 & 4 \end{bmatrix}$
    (d) $\begin{bmatrix} -6 \\ 0 \\ 24 \end{bmatrix}$
    (e) None of these

    1—T—Answer: a

19. Use a graphing utility to find AB, given

$$A = \begin{bmatrix} 1 & -1 & 2 \\ 0 & 5 & 1 \\ -2 & 0 & -1 \end{bmatrix} \text{ and } B = \begin{bmatrix} -1 & 1 & 0 \\ 5 & -7 & 1 \\ 2 & 3 & -2 \end{bmatrix}.$$

(a) $\begin{bmatrix} -1 & -2 & 0 \\ 0 & -35 & 1 \\ -4 & 0 & 2 \end{bmatrix}$
(b) $\begin{bmatrix} -1 & 11 & 0 \\ 3 & -40 & 2 \\ 6 & 13 & 9 \end{bmatrix}$
(c) $\begin{bmatrix} 6 & 3 & -2 \\ 0 & -10 & -1 \\ -12 & 0 & 1 \end{bmatrix}$

(d) $\begin{bmatrix} -2 & 15 & -5 \\ 27 & -32 & 3 \\ 0 & -7 & 2 \end{bmatrix}$
(e) None of these

1—T—Answer: d

20. Use a graphing utility to find AB, given $A = \begin{bmatrix} 1 & 3 & 5 & 2 \\ -1 & 6 & 4 & 8 \end{bmatrix}$ and $B = \begin{bmatrix} 3 & 2 & 0 \\ 0 & 0 & 1 \\ 1 & 2 & -1 \\ 0 & 0 & 3 \end{bmatrix}.$

1—T—Answer: $AB = \begin{bmatrix} 8 & 12 & 4 \\ 1 & 6 & 26 \end{bmatrix}$

21. Given $A = \begin{bmatrix} 3 & -2 & 4 & 0 \\ 0 & 0 & -1 & 4 \\ 3 & 2 & -1 & -1 \end{bmatrix}$ and $B = \begin{bmatrix} 1 & 1 & 1 \\ -1 & 0 & 0 \\ 4 & 10 & -2 \end{bmatrix}$, find BA.

(a) $\begin{bmatrix} 21 & 14 & 10 & -4 \\ -3 & -2 & 1 & -1 \\ 6 & 4 & 11 & 7 \end{bmatrix}$
(b) $\begin{bmatrix} 21 & 43 & -5 \\ -4 & -10 & 2 \\ -3 & -7 & 5 \end{bmatrix}$
(c) $\begin{bmatrix} 6 & 0 & 2 & 3 \\ -3 & 2 & -4 & 0 \\ 6 & -12 & 8 & 42 \end{bmatrix}$

(d) Impossible
(e) None of these

1—Answer: c

22. Given $A = \begin{bmatrix} 3 & -2 & 4 & 0 \\ 0 & 0 & -1 & 4 \\ 3 & 2 & -1 & -1 \end{bmatrix}$ and $B = \begin{bmatrix} 1 & 1 & 1 \\ -1 & 0 & 0 \\ 4 & 10 & -2 \end{bmatrix}$, find AB.

(a) $\begin{bmatrix} 21 & 14 & 10 & -4 \\ -3 & -2 & 1 & -1 \\ 6 & 4 & 11 & 7 \end{bmatrix}$
(b) $\begin{bmatrix} 21 & 43 & -5 \\ -4 & -10 & 2 \\ -3 & -7 & 5 \\ 0 & 0 & 0 \end{bmatrix}$
(c) $\begin{bmatrix} 6 & 0 & 2 & 3 \\ -3 & 2 & -4 & 0 \\ 6 & -12 & 8 & 42 \end{bmatrix}$

(d) Impossible
(e) None of these

1—Answer: d

23. Given $A = \begin{bmatrix} 2 & 4 & -1 \\ 1 & 0 & 4 \\ 8 & 1 & 2 \end{bmatrix}$ and $B = \begin{bmatrix} 1 & 1 & 1 \\ -1 & 0 & 0 \\ 4 & 10 & -2 \end{bmatrix}$, find AB.

(a) $\begin{bmatrix} 11 & 5 & 5 \\ -2 & -4 & 1 \\ 2 & 14 & 32 \end{bmatrix}$
(b) $\begin{bmatrix} -6 & -8 & 4 \\ 17 & 41 & -7 \\ 15 & 28 & 4 \end{bmatrix}$
(c) $\begin{bmatrix} -5 & -8 & 4 \\ 17 & 41 & -7 \\ 15 & 28 & 4 \end{bmatrix}$

(d) Impossible
(e) None of these

1—Answer: b

24. Given $A = \begin{bmatrix} 2 & 4 & -1 & 3 \\ 1 & 0 & 4 & 0 \\ 8 & 1 & 2 & 1 \end{bmatrix}$ and $B = \begin{bmatrix} 3 & 1 & -1 \\ -1 & 0 & 0 \\ 2 & 4 & -2 \end{bmatrix}$, find BA.

(a) $\begin{bmatrix} -1 & 11 & -1 & 8 \\ -2 & -4 & 1 & -3 \\ -8 & 6 & 10 & 4 \end{bmatrix}$
(b) $\begin{bmatrix} 0 & -2 & 0 \\ 11 & 17 & -9 \\ 27 & 16 & -12 \end{bmatrix}$
(c) $\begin{bmatrix} -1 & -11 & 1 & 0 \\ -2 & 4 & -1 & -3 \\ 8 & -6 & 10 & -12 \end{bmatrix}$

(d) Impossible
(e) None of these

1—Answer: a

25. Given $A = \begin{bmatrix} 2 & 4 & -1 & 3 \\ 1 & 0 & 4 & 0 \\ 8 & 1 & 2 & 1 \end{bmatrix}$ and $B = \begin{bmatrix} 3 & 1 & -1 \\ -1 & 0 & 0 \\ 2 & 4 & -2 \end{bmatrix}$, find AB.

(a) $\begin{bmatrix} -1 & 11 & -1 & 8 \\ -2 & -4 & 1 & -3 \\ -8 & 6 & 10 & 4 \end{bmatrix}$
(b) $\begin{bmatrix} 0 & -2 & 0 \\ 11 & 17 & -9 \\ 27 & 16 & -12 \end{bmatrix}$
(c) $\begin{bmatrix} 6 & 4 & 1 & 3 \\ -1 & 0 & 0 & 0 \\ 16 & 4 & -4 & 0 \end{bmatrix}$

(d) Impossible
(e) None of these

1—Answer: d

26. Use a graphing utility to find AB, given $A = \begin{bmatrix} 3 & -2 & 4 \\ 0 & 0 & -1 \\ 3 & 2 & -1 \end{bmatrix}$ and $B = \begin{bmatrix} 3 & 1 & -1 \\ -1 & 0 & 0 \\ 2 & 4 & -2 \end{bmatrix}$.

(a) $\begin{bmatrix} 6 & -8 & 12 \\ -3 & 2 & -4 \\ 0 & -8 & 6 \end{bmatrix}$
(b) $\begin{bmatrix} 19 & 19 & -11 \\ -2 & -4 & 2 \\ 5 & -1 & -1 \end{bmatrix}$
(c) $\begin{bmatrix} 9 & -2 & -4 \\ 0 & 0 & 0 \\ 6 & 8 & 2 \end{bmatrix}$

(d) Impossible
(e) None of these

1—T—Answer: b

27. Use a graphing utility to find AB, given $A = \begin{bmatrix} 2 & 0 & 1 & 2 \\ 0 & 1 & 0 & 1 \\ -1 & -2 & 0 & 0 \end{bmatrix}$ and $B = \begin{bmatrix} 1 & 1 & 0 \\ 0 & 1 & 1 \\ 2 & -1 & 2 \end{bmatrix}$.

(a) $\begin{bmatrix} 2 & 0 & 1 & 2 \\ 0 & 1 & 0 & 0 \\ -2 & 2 & 0 & 0 \end{bmatrix}$
(b) $\begin{bmatrix} 4 & 2 & 3 \\ -1 & -2 & -1 \\ 6 & -5 & -1 \end{bmatrix}$
(c) $\begin{bmatrix} 2 & 1 & 1 & 3 \\ -1 & -1 & 0 & 1 \\ 2 & -5 & 2 & 3 \end{bmatrix}$

(d) Impossible
(e) None of these

1—T—Answer: d

28. Use a graphing utility to find AB, given $A = \begin{bmatrix} 2 & 0 & 1 & 2 \\ 0 & 1 & 0 & 1 \\ -1 & -2 & 0 & 0 \end{bmatrix}$ and $B = \begin{bmatrix} 1 & 1 & 0 \\ 0 & 1 & 1 \\ 2 & -1 & 2 \end{bmatrix}$.

(a) $\begin{bmatrix} 2 & 1 & 1 & 3 \\ -1 & -1 & 0 & 1 \\ 2 & -5 & 2 & 3 \end{bmatrix}$ (b) $\begin{bmatrix} 4 & 1 & 2 \\ 0 & 1 & 1 \\ -1 & -3 & -2 \end{bmatrix}$ (c) $\begin{bmatrix} 6 & -4 & 0 \\ 0 & 1 & 1 \\ -4 & -3 & -4 \end{bmatrix}$

(d) Impossible (e) None of these

1—T—Answer: a

29. Use a graphing utility to find BA, given $A = \begin{bmatrix} 3 & 2 & 2 \\ -1 & 0 & -1 \\ 0 & 1 & 0 \end{bmatrix}$ and $B = \begin{bmatrix} 1 & 1 & 0 \\ 0 & 1 & 1 \\ 2 & -1 & 2 \end{bmatrix}$.

(a) $\begin{bmatrix} 3 & 2 & 0 \\ 0 & 0 & -1 \\ 0 & -1 & 0 \end{bmatrix}$ (b) $\begin{bmatrix} 7 & 3 & 6 \\ -3 & 0 & -2 \\ 0 & 1 & 1 \end{bmatrix}$ (c) $\begin{bmatrix} 2 & 2 & 1 \\ -1 & 1 & 1 \\ 7 & 6 & 5 \end{bmatrix}$

(d) Impossible (e) None of these

1—T—Answer: c

30. Use a graphing utility to find AB, given $A = \begin{bmatrix} 3 & 1 & 2 & 1 \\ 0 & 1 & 0 & 0 \\ 2 & 1 & 3 & -1 \end{bmatrix}$ and $B = \begin{bmatrix} 1 & 1 & 0 \\ 3 & 1 & 2 \\ -1 & 1 & -1 \end{bmatrix}$.

(a) $\begin{bmatrix} 3 & 2 & 2 & 1 \\ 13 & 6 & 12 & 1 \\ -5 & -1 & -5 & 0 \end{bmatrix}$ (b) $\begin{bmatrix} 4 & 6 & 0 \\ 3 & 1 & 2 \\ 2 & 6 & -1 \end{bmatrix}$ (c) $\begin{bmatrix} 3 & 1 & 0 \\ 0 & 1 & 2 \\ -2 & 1 & -3 \end{bmatrix}$

(d) Impossible (e) None of these

1—T—Answer: d

31. Given $A = \begin{bmatrix} 3 & 2 & 2 & 1 \\ 13 & 6 & 12 & 1 \\ -5 & -1 & -5 & 0 \end{bmatrix}$ and $B = \begin{bmatrix} 1 & 1 & 0 \\ 3 & 1 & 2 \\ -1 & 1 & -1 \end{bmatrix}$, find BA.

1—Answer: $\begin{bmatrix} 3 & 2 & 2 & 1 \\ 13 & 6 & 12 & 1 \\ -5 & -1 & -5 & 0 \end{bmatrix}$

32. Given $A = \begin{bmatrix} 1 & 0 & 3 \\ -1 & 2 & -2 \\ 1 & 1 & 2 \end{bmatrix}$ and $B = \begin{bmatrix} 1 & 1 & 0 \\ 3 & 1 & 2 \\ -1 & 1 & -1 \end{bmatrix}$, find BA.

1—Answer: $\begin{bmatrix} 0 & 2 & 1 \\ 4 & 4 & 11 \\ -3 & 1 & -7 \end{bmatrix}$

33. Solve for X given $A = \begin{bmatrix} 3 & 1 \\ -2 & 5 \end{bmatrix}$ and $B = \begin{bmatrix} 1 & 1 \\ 1 & 9 \end{bmatrix}$: $2X - A = B$

(a) 17
(b) $\begin{bmatrix} 5 & 1 \\ -5 & 1 \end{bmatrix}$
(c) $\begin{bmatrix} 4 & 2 \\ -1 & 14 \end{bmatrix}$
(d) $\begin{bmatrix} 2 & 1 \\ -\frac{1}{2} & 7 \end{bmatrix}$
(e) None of these

1—Answer: d

34. Find AB if $A = \begin{bmatrix} 2 & -1 & 0 \\ 3 & 4 & 1 \end{bmatrix}$ and $B = \begin{bmatrix} 0 & 1 \\ 4 & 3 \\ 5 & -1 \end{bmatrix}$.

(a) $\begin{bmatrix} -4 & -1 \\ 21 & 14 \end{bmatrix}$
(b) $\begin{bmatrix} 3 & 4 & 1 \\ 17 & 8 & 3 \\ 7 & -9 & -1 \end{bmatrix}$
(c) $\begin{bmatrix} -2 & -1 \\ 21 & 14 \end{bmatrix}$
(d) Cannot be done
(e) None of these

1—Answer: a

35. Let $A = \begin{bmatrix} 2 & -1 & 0 \\ 0 & 5 & 3 \\ 1 & -2 & -1 \end{bmatrix}$ and $B = \begin{bmatrix} 1 & 0 & -1 \\ 2 & 3 & 0 \end{bmatrix}$. Find BA.

1—Answer: $\begin{bmatrix} 1 & 1 & 1 \\ 4 & 13 & 9 \end{bmatrix}$

36. Solve for X given $A = \begin{bmatrix} 2 & 3 \\ 5 & -2 \end{bmatrix}$ and $B = \begin{bmatrix} 2 & 5 \\ -1 & 6 \end{bmatrix}$: $2X + A = B$

(a) $\begin{bmatrix} 2 & 4 \\ 2 & 2 \end{bmatrix}$
(b) $\begin{bmatrix} 0 & 1 \\ -3 & 4 \end{bmatrix}$
(c) $\begin{bmatrix} 4 & 8 \\ 4 & 4 \end{bmatrix}$
(d) $\begin{bmatrix} 0 & 2 \\ -6 & 8 \end{bmatrix}$
(e) None of these

1—Answer: b

37. Solve for X given $A = \begin{bmatrix} -3 & 4 \\ 2 & 6 \end{bmatrix}$ and $B = \begin{bmatrix} 6 & 2 \\ 7 & 0 \end{bmatrix}$: $3X - A = B$

(a) $\frac{-44}{9}$
(b) $\begin{bmatrix} -3 & \frac{2}{3} \\ -\frac{5}{3} & 2 \end{bmatrix}$
(c) -4
(d) $\begin{bmatrix} 1 & 2 \\ 3 & 2 \end{bmatrix}$
(e) None of these

1—Answer: d

Chapter 7 Matrices and Determinants

38. Write the matrix equation for the system of linear equations:

$$3x + 2y = 5$$
$$7x - 6y = 1$$

(a) $[x \quad y] = \begin{bmatrix} 3 & 7 \\ 2 & -6 \end{bmatrix} \begin{bmatrix} 5 \\ 1 \end{bmatrix}$
(b) $\begin{bmatrix} x \\ y \end{bmatrix} = \begin{bmatrix} 3 & 2 \\ 7 & -6 \end{bmatrix} \begin{bmatrix} 5 \\ 1 \end{bmatrix}$
(c) $\begin{bmatrix} 3 & 2 \\ 7 & -6 \end{bmatrix} \begin{bmatrix} x \\ y \end{bmatrix} = \begin{bmatrix} 5 \\ 1 \end{bmatrix}$

(d) $[x \quad y] \begin{bmatrix} 5 \\ 1 \end{bmatrix} = \begin{bmatrix} 3 & 7 \\ 2 & -6 \end{bmatrix}$
(e) None of these

1—Answer: c

39. Write the matrix equation for the system of linear equations:

$$2x_1 + x_2 \qquad\qquad = 0$$
$$2x_1 + x_2 + x_3 \qquad = -1$$
$$\qquad 2x_2 - x_3 + 3x_4 = 1$$
$$\qquad\qquad 2x_3 - 3x_4 = 4$$

(a) $\begin{bmatrix} 2 & 1 & 0 & 0 \\ 3 & 1 & 1 & 0 \\ 0 & 2 & -1 & 3 \\ 0 & 0 & 2 & -3 \end{bmatrix} \begin{bmatrix} x_1 \\ x_2 \\ x_3 \\ x_4 \end{bmatrix} = \begin{bmatrix} 0 \\ -1 \\ 1 \\ 4 \end{bmatrix}$
(b) $\begin{bmatrix} 2 & 3 & 0 & 0 \\ 1 & 1 & 2 & 0 \\ 0 & 2 & -1 & 2 \\ 0 & 0 & 3 & -3 \end{bmatrix} \begin{bmatrix} x_1 \\ x_2 \\ x_3 \\ x_4 \end{bmatrix} = \begin{bmatrix} 0 \\ -1 \\ 1 \\ 4 \end{bmatrix}$

(c) $\begin{bmatrix} 2 & 1 & 0 & 0 \\ 3 & 1 & 1 & 0 \\ 0 & 2 & -1 & 3 \\ 0 & 0 & 2 & -3 \end{bmatrix} \begin{bmatrix} 0 \\ -1 \\ 1 \\ 4 \end{bmatrix} = \begin{bmatrix} x_1 \\ x_2 \\ x_3 \\ x_4 \end{bmatrix}$
(d) $\begin{bmatrix} 2 & 3 & 0 & 0 \\ 1 & 1 & 2 & 0 \\ 0 & 1 & -1 & 2 \\ 0 & 0 & 3 & -3 \end{bmatrix} \begin{bmatrix} 0 \\ -1 \\ 1 \\ 4 \end{bmatrix} = \begin{bmatrix} x_1 \\ x_2 \\ x_3 \\ x_4 \end{bmatrix}$

(e) None of these

1—Answer: a

40. Write the matrix equation for the system of linear equations:

$$2x - 4y = 12$$
$$x + 5y = 16$$

(a) $[x \quad y] = \begin{bmatrix} 2 & -4 \\ 1 & 5 \end{bmatrix} \begin{bmatrix} 12 \\ 16 \end{bmatrix}$
(b) $\begin{bmatrix} 2 & -4 \\ 1 & 5 \end{bmatrix} \begin{bmatrix} x \\ y \end{bmatrix} = \begin{bmatrix} 12 \\ 16 \end{bmatrix}$

(c) $\begin{bmatrix} x \\ y \end{bmatrix} = \begin{bmatrix} 2 & -4 \\ 1 & 5 \end{bmatrix} \begin{bmatrix} 12 \\ 16 \end{bmatrix}$
(d) $[x \quad y] \begin{bmatrix} 12 \\ 16 \end{bmatrix} = \begin{bmatrix} 3 & 7 \\ 2 & -6 \end{bmatrix}$

(e) None of these

1—Answer: b

41. A farmer raises two crops, corn and wheat, which are shipped to three processors daily. The number of bushels of crop i that are shipped to processor j is represented by a_{ij} in the matrix.

$$A = \begin{bmatrix} 125 & 75 & 100 \\ 100 & 150 & 125 \end{bmatrix}$$

The profit per bushel is represented by the matrix

$$B = [\$1.25 \ \ \$0.85].$$

Find the product BA and state what each entry of the product represents.

2—Answer: $BA = [241.25 \ \ 221.25 \ \ 231.25]$

The farmer's total profit for the crops shipped to the first processor is $241.25, to the second processor is $221.25, and to the third processor is $231.25.

42. A farmer raises two crops, corn and wheat, which are shipped to three processors daily. The number of bushels of crop i that are shipped to processor j is represented by a_{ij} in the matrix.

$$A = \begin{bmatrix} 125 & 125 & 100 \\ 100 & 175 & 225 \end{bmatrix}$$

The profit per bushel is represented by the matrix

$$B = [\$1.25 \ \ \$0.85].$$

Find the product BA and state what each entry of the product represents.

2—Answer: $BA = [387.50 \ \ 305.00 \ \ 316.25]$

The farmer's total profit for the crops shipped to the first processor is $387.50, to the second processor is $305.00, and to the third processor is $316.25.

❏ 7.3 The Inverse of a Square Matrix

1. Determine which of the following matrices have inverses.

(a) $\begin{bmatrix} 1 & -2 \\ -3 & 6 \end{bmatrix}$

(b) $\begin{bmatrix} 3 \\ 2 \end{bmatrix}$

(c) $\begin{bmatrix} 3 & 4 & -1 \\ 2 & 1 & 0 \end{bmatrix}$

(d) $\begin{bmatrix} 6 & 1 \\ -2 & 4 \end{bmatrix}$

(e) None of these

1—Answer: d

2. Determine which of the following matrices have inverses.

    (a) $\begin{bmatrix} 2 & -5 \\ -4 & 10 \end{bmatrix}$
    (b) $\begin{bmatrix} 2 & 6 \\ -4 & 12 \end{bmatrix}$
    (c) $\begin{bmatrix} 0 \\ 2 \\ 5 \end{bmatrix}$
    (d) $\begin{bmatrix} 4 & 2 \\ 1 & 7 \\ 0 & -1 \end{bmatrix}$
    (e) None of these

    1—Answer: b

3. Determine which of the following matrices have inverses.

    (a) $\begin{bmatrix} -1 \\ 5 \\ 2 \end{bmatrix}$
    (b) $\begin{bmatrix} 4 & -2 \\ -2 & 1 \end{bmatrix}$
    (c) $\begin{bmatrix} -3 & 5 \\ 6 & -10 \end{bmatrix}$
    (d) $\begin{bmatrix} -3 & 5 \\ -4 & 2 \end{bmatrix}$
    (e) None of these

    1—Answer: d

4. Determine which of the following matrices have inverses.

    (a) $\begin{bmatrix} 1 & -7 \\ 2 & 14 \end{bmatrix}$
    (b) $\begin{bmatrix} 2 & -7 \\ -4 & 14 \end{bmatrix}$
    (c) $\begin{bmatrix} -6 & 9 \\ 4 & -6 \end{bmatrix}$
    (d) $\begin{bmatrix} 4 \\ 1 \\ 3 \end{bmatrix}$
    (e) None of these

    1—Answer: a

5. Determine which of the following matrices have inverses.

    (a) $\begin{bmatrix} -4 & -3 \\ 8 & 6 \end{bmatrix}$
    (b) $\begin{bmatrix} 6 & 1 \\ -12 & -2 \end{bmatrix}$
    (c) $\begin{bmatrix} -2 & -1 \\ 3 & 2 \end{bmatrix}$
    (d) $\begin{bmatrix} 3 & 4 & 1 \end{bmatrix}$
    (e) None of these

    1—Answer: c

6. Determine the matrix that has an inverse.

    (a) $\begin{bmatrix} 3 & 2 & 1 \\ 2 & -1 & -1 \\ 1 & 4 & 0 \end{bmatrix}$
    (b) $\begin{bmatrix} 4 & 2 \\ 2 & 1 \end{bmatrix}$
    (c) $\begin{bmatrix} 2 & 3 \\ 5 & -1 \\ 1 & 0 \end{bmatrix}$
    (d) $\begin{bmatrix} 1 \\ 2 \\ 3 \end{bmatrix}$
    (e) None of these

    1—Answer: a

7. Given $A = \begin{bmatrix} 1 & 2 \\ -3 & 5 \end{bmatrix}$, find A^{-1}.

(a) $\begin{bmatrix} -1 & -2 \\ 3 & -5 \end{bmatrix}$ (b) $\begin{bmatrix} 1 & \frac{1}{2} \\ -\frac{1}{3} & \frac{1}{5} \end{bmatrix}$ (c) $\begin{bmatrix} -\frac{4}{15} & \frac{1}{3} \\ -\frac{2}{3} & \frac{11}{15} \end{bmatrix}$

(d) $\begin{bmatrix} \frac{5}{11} & -\frac{2}{11} \\ \frac{3}{11} & \frac{1}{11} \end{bmatrix}$ (e) None of these

1—Answer: d

8. Given $A = \begin{bmatrix} 5 & 1 \\ -2 & 3 \end{bmatrix}$, find A^{-1}.

(a) $\begin{bmatrix} \frac{5}{17} & \frac{1}{17} \\ -\frac{2}{17} & \frac{3}{17} \end{bmatrix}$ (b) $\begin{bmatrix} \frac{3}{17} & -\frac{1}{17} \\ \frac{2}{17} & \frac{5}{17} \end{bmatrix}$ (c) $\begin{bmatrix} 85 & 17 \\ 34 & 51 \end{bmatrix}$

(d) $\begin{bmatrix} -\frac{2}{5} & \frac{1}{3} \\ \frac{3}{2} & -\frac{1}{2} \end{bmatrix}$ (e) None of these

1—Answer: b

9. Given $A = \begin{bmatrix} 1 & 2 & 2 \\ 4 & 1 & 3 \\ -1 & 5 & 0 \end{bmatrix}$, find A^{-1}.

(a) $\frac{1}{21} \begin{bmatrix} -15 & 10 & 4 \\ -3 & 2 & 5 \\ 21 & -7 & -7 \end{bmatrix}$ (b) $\frac{1}{21} \begin{bmatrix} -15 & -22 & 28 \\ -3 & 27 & 25 \\ 21 & -7 & -7 \end{bmatrix}$ (c) $\begin{bmatrix} 1 & 0 & 0 \\ 0 & 1 & 0 \\ 0 & 0 & 1 \end{bmatrix}$

(d) $\begin{bmatrix} 1 & \frac{1}{2} & \frac{1}{2} \\ \frac{1}{4} & 1 & \frac{1}{3} \\ -1 & \frac{1}{5} & 0 \end{bmatrix}$ (e) None of these

1—Answer: a

10. Given $A = \begin{bmatrix} 1 & 3 & -1 \\ 0 & 2 & 1 \\ -1 & 1 & -2 \end{bmatrix}$, find A^{-1}.

1—Answer: $\begin{bmatrix} \frac{1}{2} & -\frac{1}{2} & -\frac{1}{2} \\ \frac{1}{10} & \frac{3}{10} & \frac{1}{10} \\ -\frac{1}{5} & \frac{2}{5} & -\frac{1}{5} \end{bmatrix}$

11. Given $A = \begin{bmatrix} 1 & 5 & -1 \\ 2 & 3 & -2 \\ -1 & -4 & 3 \end{bmatrix}$, find A^{-1}.

1—Answer: $\begin{bmatrix} -\frac{1}{14} & \frac{11}{14} & \frac{1}{2} \\ \frac{2}{7} & -\frac{1}{7} & 0 \\ \frac{5}{14} & \frac{1}{14} & \frac{1}{2} \end{bmatrix}$

12. Given $A = \begin{bmatrix} 1 & 1 & 1 \\ -1 & 0 & 0 \\ 4 & 10 & -2 \end{bmatrix}$, find A^{-1}.

   1—Answer: $\begin{bmatrix} 0 & -1 & 0 \\ \frac{1}{6} & \frac{1}{2} & \frac{1}{12} \\ \frac{5}{6} & \frac{1}{2} & -\frac{1}{12} \end{bmatrix}$

13. Given $C = \begin{bmatrix} 1 & 1 & 0 \\ 3 & 1 & 2 \\ -1 & 1 & -1 \end{bmatrix}$, find C^{-1}.

   (a) $\begin{bmatrix} 1 & 1 & 0 \\ \frac{1}{3} & 1 & \frac{1}{2} \\ -1 & 1 & -1 \end{bmatrix}$
   (b) $\begin{bmatrix} -1 & -1 & 0 \\ -3 & -1 & -2 \\ 1 & -1 & 1 \end{bmatrix}$
   (c) $\begin{bmatrix} \frac{3}{2} & -\frac{1}{2} & -1 \\ -\frac{1}{2} & \frac{1}{2} & 1 \\ -2 & 1 & 1 \end{bmatrix}$

   (d) $\begin{bmatrix} -\frac{3}{2} & 1 & -1 \\ \frac{1}{2} & 2 & 1 \\ -2 & 1 & 1 \end{bmatrix}$
   (e) None of these

   1—Answer: c

14. Given $A = \begin{bmatrix} 1 & 3 \\ 2 & 1 \end{bmatrix}$, find A^{-1}.

   (a) $\begin{bmatrix} 1 & \frac{1}{3} \\ \frac{1}{2} & 1 \end{bmatrix}$
   (b) $\begin{bmatrix} 1 & -3 \\ -2 & 1 \end{bmatrix}$
   (c) $\begin{bmatrix} \frac{1}{5} & \frac{3}{5} \\ \frac{2}{5} & \frac{1}{5} \end{bmatrix}$

   (d) $\begin{bmatrix} -\frac{1}{5} & \frac{3}{5} \\ \frac{2}{5} & -\frac{1}{5} \end{bmatrix}$
   (e) None of these

   1—Answer: d

15. Find the inverse of $A = \begin{bmatrix} 3 & 2 \\ 1 & 4 \end{bmatrix}$.

   (a) $\begin{bmatrix} \frac{2}{5} & -\frac{1}{5} \\ -\frac{1}{10} & \frac{3}{10} \end{bmatrix}$
   (b) $\begin{bmatrix} \frac{1}{3} & \frac{1}{2} \\ 1 & \frac{1}{4} \end{bmatrix}$
   (c) $\begin{bmatrix} 4 & -2 \\ -1 & 3 \end{bmatrix}$

   (d) $\begin{bmatrix} -3 & 1 \\ 2 & -4 \end{bmatrix}$
   (e) None of these

   1—Answer: a

Section 7.3 The Inverse of a Square Matrix 369

16. Find the inverse of $A = \begin{bmatrix} 2 & 3 \\ 1 & 2 \end{bmatrix}$.

(a) $\begin{bmatrix} \frac{1}{2} & \frac{1}{3} \\ 1 & \frac{1}{2} \end{bmatrix}$
(b) $\begin{bmatrix} 2 & -3 \\ -1 & 2 \end{bmatrix}$
(c) $\begin{bmatrix} -2 & 1 \\ 3 & -2 \end{bmatrix}$

(d) $\begin{bmatrix} \frac{2}{7} & -\frac{3}{7} \\ -\frac{1}{7} & \frac{2}{7} \end{bmatrix}$
(e) None of these

1—Answer: b

17. Find the inverse of $A = \begin{bmatrix} 2 & 3 \\ -3 & -3 \end{bmatrix}$.

(a) $\begin{bmatrix} -3 & -3 \\ 3 & 2 \end{bmatrix}$
(b) $\begin{bmatrix} \frac{1}{2} & \frac{1}{3} \\ -\frac{1}{3} & -\frac{1}{3} \end{bmatrix}$
(c) $\begin{bmatrix} \frac{1}{4} & \frac{1}{4} \\ -\frac{1}{4} & -\frac{1}{3} \end{bmatrix}$

(d) $\begin{bmatrix} -1 & -1 \\ 1 & \frac{2}{3} \end{bmatrix}$
(e) None of these

1—Answer: d

18. Find A^{-1} if $A = \begin{bmatrix} 2 & 0 & 3 \\ -1 & 0 & 2 \\ 0 & 1 & 1 \end{bmatrix}$.

(a) $\begin{bmatrix} 2 & -3 & 0 \\ -1 & -2 & 7 \\ 1 & 2 & 0 \end{bmatrix}$
(b) $\frac{1}{7}\begin{bmatrix} 2 & -3 & 0 \\ -1 & -2 & 7 \\ 1 & 2 & 0 \end{bmatrix}$
(c) $\begin{bmatrix} \frac{1}{2} & 0 & \frac{1}{3} \\ -1 & 0 & \frac{1}{2} \\ 0 & 1 & 1 \end{bmatrix}$

(d) $\begin{bmatrix} 2 & -1 & 0 \\ 0 & 0 & 1 \\ 3 & 2 & 1 \end{bmatrix}$
(e) None of these

1—Answer: b

19. Find A^{-1} if $A = \begin{bmatrix} 1 & -1 & -1 \\ 5 & 0 & 20 \\ 0 & 10 & -20 \end{bmatrix}$.

1—Answer: $\begin{bmatrix} \frac{4}{7} & \frac{3}{35} & \frac{2}{35} \\ -\frac{2}{7} & \frac{2}{35} & \frac{1}{14} \\ -\frac{1}{7} & \frac{1}{35} & -\frac{1}{70} \end{bmatrix}$

20. Use a graphing utility to find A^{-1} if it exists, given $A = \begin{bmatrix} 1 & 2 \\ 3 & 4 \end{bmatrix}$.

1—T—Answer: $\begin{bmatrix} -2 & 1 \\ 1.5 & -0.5 \end{bmatrix}$

21. Use a graphing utility to find B^{-1} if it exists, given

$$B = \begin{bmatrix} 2 & -1 & 1 \\ 1 & 2 & -1 \\ 3 & -4 & 2 \end{bmatrix}.$$

1—T—Answer: $\begin{bmatrix} 0 & 0.4 & 0.2 \\ 1 & -0.2 & -0.6 \\ 2 & -1 & -1 \end{bmatrix}$

22. Use a graphing utility to find C^{-1} if it exists, given

$$C = \begin{bmatrix} 1 & 0 & -2 & 3 \\ 1 & 1 & -3 & 2 \\ 2 & 1 & 0 & 3 \\ 0 & 1 & 0 & -3 \end{bmatrix}.$$

1—T—Answer: $\begin{bmatrix} 2.25 & -1.5 & 0.125 & 1.375 \\ -2.25 & 1.5 & 0.375 & -0.875 \\ -0.5 & 0 & 0.25 & -0.25 \\ -0.75 & 0.5 & 0.125 & -0.625 \end{bmatrix}$

23. Use a graphing utility to find A^{-1} if it exists, given

$$A = \begin{bmatrix} 1 & 1 & 1 \\ 0 & 1 & 0 \\ 2 & -1 & 3 \end{bmatrix}.$$

1—T—Answer: $\begin{bmatrix} 3 & -4 & -1 \\ 0 & 1 & 0 \\ -2 & 3 & 1 \end{bmatrix}$

24. Use a graphing utility to find A^{-1} if it exists, given

$$A = \begin{bmatrix} 1 & 4 & 3 \\ 2 & 6 & 1 \\ 1 & 0 & 3 \end{bmatrix}.$$

1—T—Answer: $\begin{bmatrix} -0.9 & 0.6 & 0.7 \\ 0.25 & 0 & -0.25 \\ 0.3 & -0.2 & 0.1 \end{bmatrix}$

25. Use a graphing utility to find B^{-1} if it exists, given

$$B = \begin{bmatrix} 1 & 0 & 0 \\ 0 & -1 & 0 \\ 1 & 0 & 3 \end{bmatrix}.$$

1—T—Answer: $\begin{bmatrix} 1 & 0 & 0 \\ 0 & -1 & 0 \\ -1 & 0 & 1 \end{bmatrix}$

26. Given the system of linear equations with coefficient matrix A, use A^{-1} to find (x, y, z, w).

$$\begin{aligned} x + y + z + w &= 0 \\ 2x - y &= -2 \\ 3z - 2w &= 0 \\ y - 3z &= 6 \end{aligned} \qquad A^{-1} = \begin{bmatrix} \frac{3}{14} & \frac{11}{28} & \frac{3}{28} & \frac{5}{28} \\ \frac{3}{7} & -\frac{3}{14} & \frac{3}{14} & \frac{5}{14} \\ \frac{1}{7} & -\frac{1}{14} & \frac{1}{14} & -\frac{3}{14} \\ \frac{3}{14} & -\frac{3}{28} & -\frac{11}{28} & -\frac{9}{28} \end{bmatrix}$$

(a) $(10, -15, -5, 0)$ (b) $\left(\frac{2}{7}, \frac{18}{7}, -\frac{8}{7}, -\frac{12}{7}\right)$ (c) $\left(\frac{3}{11}, \frac{18}{11}, -\frac{7}{11}, \frac{16}{11}\right)$

(d) $(0, 1, -1, 0)$ (e) None of these

1—**Answer: b**

27. Given the system of linear equations with coefficient matrix A, use A^{-1} to find (x, y, z, w).

$$\begin{aligned} x + 2y &= 1 \\ 3x + 8y + 5w &= 0 \\ x + 4y + 3z + 10w &= -1 \\ x - 3z &= -1 \end{aligned} \qquad A^{-1} = \begin{bmatrix} 3 & -1 & \frac{1}{2} & \frac{1}{2} \\ -1 & \frac{1}{2} & -\frac{1}{4} & -\frac{1}{4} \\ 1 & -\frac{1}{3} & \frac{1}{6} & -\frac{1}{6} \\ -\frac{1}{5} & 0 & \frac{1}{10} & \frac{1}{10} \end{bmatrix}$$

(a) $\left(2, -\frac{1}{2}, 1, -\frac{2}{5}\right)$ (b) $(3, -1, -1, 0)$ (c) $\left(\frac{2}{3}, \frac{1}{6}, -\frac{5}{9}, \frac{2}{3}\right)$

(d) $\left(1, 0, \frac{2}{3}, -\frac{5}{3}\right)$ (e) None of these

1—**Answer: a**

28. Given a system of linear equations with coefficient matrix A, use A^{-1} to find (x, y, z).

$$\begin{aligned} 3x + 2y + z &= 5 \\ x - 4y &= 6 \\ x - y + 3z &= 6 \end{aligned} \qquad A^{-1} = \frac{1}{39}\begin{bmatrix} 12 & 7 & -4 \\ 3 & -8 & -1 \\ -3 & -5 & 14 \end{bmatrix}$$

(a) $(-2, -1, 1)$ (b) $(-1, 6, -2)$ (c) $(2, -1, 1)$

(d) $(6, -1, 3)$ (e) None of these

1—**Answer: c**

29. Given a system of linear equations with coefficient matrix A, use A^{-1} to find (x, y, z).

$$\begin{aligned} 3x + y - z &= -11 \\ x - y - z &= -1 \\ x + 2y + 3z &= 3 \end{aligned} \qquad A^{-1} = \frac{1}{10}\begin{bmatrix} 1 & 5 & 2 \\ 4 & -10 & -2 \\ -3 & 5 & 4 \end{bmatrix}$$

(a) $(3, -2, 0)$ (b) $(2, -2, 3)$ (c) $(-1, 5, 1)$

(d) $(-3, 0, 2)$ (e) None of these

1—**Answer: d**

30. Given the system of linear equations with coefficient matrix A, use A^{-1} to find (x, y, z).

$$4x - y + z = 5$$
$$x - 4y + z = 8$$
$$2x + 2y - 3z = -12$$

$$A^{-1} = \frac{1}{45}\begin{bmatrix} 10 & -1 & 3 \\ 5 & -14 & -3 \\ 10 & -10 & -15 \end{bmatrix}$$

(a) $(0, -1, 4)$ (b) $(0, 4, -2)$ (c) $(1, -1, 2)$

(d) $(3, 0, 2)$ (e) None of these

1—Answer: e

31. Given $A = \begin{bmatrix} 1 & 2 \\ -3 & 5 \end{bmatrix}$, find A^{-1}.

(a) $\begin{bmatrix} -1 & -2 \\ 3 & -5 \end{bmatrix}$ (b) $\begin{bmatrix} 1 & \frac{1}{2} \\ -\frac{1}{3} & \frac{1}{5} \end{bmatrix}$ (c) $\begin{bmatrix} -\frac{4}{15} & \frac{1}{3} \\ -\frac{2}{3} & \frac{11}{15} \end{bmatrix}$

(d) $\begin{bmatrix} \frac{5}{11} & -\frac{2}{11} \\ \frac{3}{11} & \frac{1}{11} \end{bmatrix}$ (e) None of these

1—Answer: d

32. Use the inverse matrix to solve the system of linear equations.

$$2x + 2y = 12$$
$$x + 3y = 16$$

1—Answer: $(1, 5)$

33. Use an inverse matrix to solve the system of linear equations.

$$x + y + z = 5$$
$$2x - y + z = 1$$
$$-x + y + 3z = 1$$

(a) $(2, 3, 0)$ (b) $(-3, 0, 8)$ (c) $(2, 2, 1)$

(d) $(-3, 6, 2)$ (e) None of these

1—Answer: a

34. Use a graphing utility to solve (if possible) the system of linear equations.

$$x + y + z = 0$$
$$x + 2y + z = 1$$
$$2x + y + z = -1$$

(a) $(2, -3, 1)$ (b) $(-3, -5, 8)$ (c) $(-1, 1, 0)$

(d) $(-2, 1, 1)$ (e) None of these

1—T—Answer: c

35. Use an inverse matrix to solve the system of linear equations.

$$2x + 3y + 2z = 0$$
$$x - 6z = 4$$
$$x + y - 2z = 1$$

1—Answer: $\left(7, -5, \frac{1}{2}\right)$

36. Use a graphing utility to solve (if possible) the system of linear equations.

$$5x + y + 2z = 0$$
$$x - 3z = -2$$
$$2x + y + z = 6$$

1—T—Answer: $(-2, 10, 0)$

37. Use a graphing utility to solve the system of linear equations.

$$x - y + z = 6$$
$$2x + z = 4$$
$$-x - y - z = -4$$

(a) $(4, 2, 4)$ (b) $(3, -1, 2)$ (c) $(1, -7, -2)$

(d) $(2, -4, 0)$ (e) None of these

1—T—Answer: b

38. Use a graphing utility to solve (if possible) the system of linear equations.

$$x + y + z = 4$$
$$2x + y + z = 6$$
$$x + y + 2z = 9$$

(a) $(2, 2, 0)$ (b) $(-2, 3, 3)$ (c) $(2, -3, 5)$

(d) $(1, -1, 4)$ (e) None of these

1—T—Answer: c

39. Use an inverse matrix to solve the system of linear equations.

$$3x + 2y + z = 1$$
$$x - y = 10$$
$$-x + 2z = 5$$

2—Answer: $\left(\frac{37}{11}, -\frac{73}{11}, \frac{46}{11}\right)$

40. Use the inverse matrix method to solve the system of linear equations.

$6x + 6y - 5z = 11$

$3x + 6y - z = 6$

$9x - 3y + z = 0$

2—**Answer:** $\left(\frac{1}{3}, \frac{2}{3}, -1\right)$

41. A small business borrows $90,000; some at 7%, some at 9%, and the rest at 10% simple interest. The annual interest is $7110, and twice as much is borrowed at 7% as at 9%. Set up a system where x, y, and z represent the amounts borrowed at 7%, 9%, and 10%, respectively, then use the inverse of the coefficient matrix to find (x, y, z).

(a) ($46,000, $23,000, $21,000)   (b) ($54,000, $27,000, $9000)

(c) ($18,000, $36,000, $36,000)   (d) ($41,000, $20,500, $28,500)

(e) None of these

2—**Answer:** b

42. A small business borrows $550,000; some at 8%, and the rest at 11% simple interest. The annual interest is $53,600, and one-third of the amount borrowed at 11% is borrowed at 7%. Set up a system where x, y, and z represent the amounts borrowed at 7%, 8%, and 11%, respectively, then use the inverse of the coefficient matrix to find y.

(a) $70,000   (b) $90,000   (c) $120,000

(d) $130,000   (e) None of these

2—**Answer:** a

43. A small business borrows $175,000; some at 7%, some at 10%, and the rest at 11% simple interest. The annual interest is $15,800, and the amount borrowed at 7% exceeds the amount borrowed at 10% by $20,000. Set up a system where x, y, and z represent the amounts borrowed at 7%, 10%, and 11%, respectively, then use the inverse of the coefficient matrix to find x.

(a) $90,000   (b) $40,000   (c) $75,000

(d) $60,000   (e) None of these

2—**Answer:** c

7.4 The Determinant of a Square Matrix

1. Evaluate: $\begin{vmatrix} 7 & -1 \\ 6 & -2 \end{vmatrix}$

   (a) -20 (b) -8 (c) 8

   (d) 20 (e) None of these

   1—Answer: b

2. Evaluate: $\begin{vmatrix} 6 & 4 \\ 2 & -1 \end{vmatrix}$

   (a) 2 (b) -14 (c) -2

   (d) 14 (e) None of these

   1—Answer: b

3. Find the determinant of the matrix: $\begin{bmatrix} 3 & -4 \\ 2 & 6 \end{bmatrix}$

   (a) 10 (b) 26 (c) -26

   (d) -10 (e) None of these

   1—Answer: b

4. Find the determinant of the matrix: $\begin{bmatrix} 3 & -1 \\ 6 & 2 \end{bmatrix}$

   (a) 12 (b) -12 (c) 0

   (d) 9 (e) None of these

   1—Answer: a

5. Find the determinant of the matrix: $\begin{bmatrix} 6 & -4 \\ 2 & -1 \end{bmatrix}$

   (a) 2 (b) -2 (c) 14

   (d) -14 (e) None of these

   1—Answer: a

6. Evaluate: $\begin{vmatrix} 0 & 2 & 3 \\ 1 & -1 & 4 \\ 3 & 0 & 2 \end{vmatrix}$

   (a) 9   (b) 19   (c) 29

   (d) 0   (e) None of these

   1—Answer: c

7. Find the determinant of the matrix $\begin{bmatrix} 0 & -1 & 2 \\ 3 & 5 & 0 \\ 1 & -1 & 3 \end{bmatrix}$

   (a) 25   (b) −25   (c) 7

   (d) −7   (e) None of these

   1—Answer: d

8. Find the determinant of the matrix: $\begin{bmatrix} 2 & 3 & -1 \\ 0 & 5 & 0 \\ -1 & 1 & 2 \end{bmatrix}$

   1—Answer: 15

9. Find the determinant of the matrix: $\begin{bmatrix} 3 & 0 & 1 \\ -1 & 4 & -1 \\ 5 & -2 & 0 \end{bmatrix}$

   1—Answer: −24

10. Find the determinant of the matrix: $\begin{bmatrix} 3 & -1 & 6 \\ 2 & 0 & 4 \\ 1 & 6 & 2 \end{bmatrix}$

    1—Answer: 0

11. Find the minor M_{23} for the matrix: $\begin{bmatrix} 3 & 1 & -2 \\ 0 & 2 & 3 \\ 1 & -2 & -2 \end{bmatrix}$

    (a) 7   (b) 9   (c) −9

    (d) 13   (e) None of these

    1—Answer: e

12. Find the minor M_{23} for the matrix: $\begin{bmatrix} 1 & 4 & -2 \\ 3 & -1 & 1 \\ 5 & 2 & 7 \end{bmatrix}$

    (a) −95   (b) −7   (c) −18

    (d) 7   (e) None of these

    1—Answer: c

13. Find the minor M_{13} for the matrix: $\begin{bmatrix} 2 & 1 & -2 \\ -6 & -1 & 3 \\ 1 & 4 & 5 \end{bmatrix}$

(a) 16 (b) 1 (c) -23
(d) 5 (e) None of these

1—Answer: c

14. Find the minor M_{12} for the matrix: $\begin{bmatrix} 3 & 4 & 2 \\ -2 & -3 & 1 \\ 1 & 2 & -1 \end{bmatrix}$

(a) -8 (b) 0 (c) 3
(d) 1 (e) None of these

1—Answer: d

15. Find the minor M_{32} for the matrix: $\begin{bmatrix} 5 & 1 & 2 \\ -1 & -4 & 1 \\ 2 & 5 & -3 \end{bmatrix}$

(a) 7 (b) 3 (c) 23
(d) 27 (e) None of these

1—Answer: a

16. Find the cofactor C_{23} for the matrix: $\begin{bmatrix} 3 & 1 & -2 \\ 0 & 2 & 3 \\ 1 & -2 & -2 \end{bmatrix}$

(a) 7 (b) 9 (c) -9
(d) 13 (e) None of these

1—Answer: a

17. Find the cofactor C_{21} for the matrix: $\begin{bmatrix} 3 & 1 & -2 \\ 0 & 2 & 3 \\ 1 & -2 & -2 \end{bmatrix}$

(a) 2 (b) -6 (c) 6
(d) 0 (e) None of these

1—Answer: c

18. Find the cofactor C_{23} for the matrix: $\begin{bmatrix} 2 & 1 & 3 \\ -1 & 4 & 2 \\ 6 & -2 & -1 \end{bmatrix}$

(a) -10 (b) -7 (c) 2
(d) -1 (e) None of these

1—Answer: e

19. Find the cofactor C_{32} for the matrix: $\begin{bmatrix} 1 & 4 & 1 \\ -6 & 7 & 1 \\ 2 & -5 & 2 \end{bmatrix}$

   (a) -7 (b) 5 (c) -10
   (d) -6 (e) None of these

   1—Answer: a

20. Find the cofactor C_{32} for the matrix: $\begin{bmatrix} 7 & 5 & 4 \\ 6 & -2 & 1 \\ 3 & 4 & 2 \end{bmatrix}$

   (a) -43 (b) -13 (c) 17
   (d) 31 (e) None of these

   1—Answer: c

21. Find the cofactor C_{13} for the matrix: $\begin{bmatrix} -1 & 5 & 1 \\ 2 & -6 & -1 \\ 4 & 3 & 5 \end{bmatrix}$

   (a) 30 (b) 18 (c) 1
   (d) 11 (e) None of these

   1—Answer: a

22. Evaluate: $\begin{vmatrix} 1 & -1 & 0 & 2 \\ 0 & 5 & 7 & 3 \\ 0 & 0 & 4 & 1 \\ 0 & 0 & 0 & 1 \end{vmatrix}$

   1—Answer: 20

23. Find the determinant of the matrix by the method of expansion by cofactors.
   $\begin{bmatrix} 0 & 5 & 2 & 0 \\ 1 & 0 & -1 & 0 \\ 3 & -2 & 0 & -1 \\ -1 & 1 & 0 & 0 \end{bmatrix}$

   (a) 0 (b) 7 (c) -16
   (d) -12 (e) None of these

   1—Answer: b

24. Find the determinant of the matrix: $\begin{bmatrix} 3 & 0 & 5 & 1 \\ 0 & -1 & 0 & -1 \\ -1 & 1 & 0 & 3 \\ 0 & 3 & 0 & 3 \end{bmatrix}$

   1—Answer: 0

25. Find the determinant of the matrix. Use the method of expansion by cofactors.

$$\begin{bmatrix} 0 & 1 & -1 & 2 \\ 1 & -1 & 0 & 3 \\ 0 & 2 & 0 & 1 \\ -1 & 1 & 2 & -2 \end{bmatrix}$$

(a) -48 (b) 12 (c) -8

(d) 2 (e) None of these

1—Answer: c

26. Find the determinant of the matrix. Use the method of expansion by cofactors.

$$\begin{bmatrix} 0 & -2 & 0 & 4 \\ 1 & 1 & -2 & 3 \\ -1 & 3 & 5 & 1 \\ 3 & 0 & 1 & 0 \end{bmatrix}$$

(a) 276 (b) -258 (c) -102

(d) 1125 (e) None of these

1—Answer: b

27. Use the matrix capabilities of a graphing utility to find the determinant of the matrix:

$$\begin{bmatrix} 5 & -1 & 0 & 2 \\ 0 & 4 & 7 & 3 \\ 0 & 0 & 1 & 1 \\ 0 & 0 & 0 & 1 \end{bmatrix}$$

(a) 11 (b) -11 (c) -20

(d) 20 (e) None of these

1—T—Answer: d

28. Use the matrix capabilities of a graphing utility to find the determinant of the matrix:

$$\begin{bmatrix} 1 & 0 & 0 & 0 \\ 4 & 6 & 0 & 0 \\ -7 & 5 & -5 & 0 \\ 3 & 2 & 3 & -1 \end{bmatrix}$$

(a) 11 (b) -30 (c) 30

(d) -11 (e) None of these

1—T—Answer: c

29. Use the matrix capabilities of a graphing utility to find the determinant of the matrix:

$$\begin{bmatrix} 1 & 5 & 0 & 6 \\ 0 & -1 & 1 & 0 \\ -2 & 2 & 3 & 0 \\ 0 & 0 & 7 & 2 \end{bmatrix}$$

(a) -54 (b) 74 (c) -74

(d) 54 (e) None of these

1—T—Answer: d

Chapter 7 — Matrices and Determinants

30. Use the matrix capabilities of a utility to find the determinant of the matrix:
$$\begin{bmatrix} 4 & 0 & 0 & 6 \\ 0 & 2 & 1 & 1 \\ -1 & 5 & -3 & 1 \\ 0 & 7 & 2 & 4 \end{bmatrix}$$

(a) 250 (b) −22 (c) 118
(d) 133 (e) None of these

1—T—**Answer:** b

31. Solve for x: $\begin{vmatrix} 5+2x & 3 \\ x-1 & -6 \end{vmatrix} = 3$

(a) 34 (b) −2 (c) $-\frac{12}{5}$
(d) −2, 34 (e) None of these

2—**Answer:** b

32. Solve for x: $\begin{vmatrix} 7-2x & -1 \\ 5x+2 & 4 \end{vmatrix} = 3$

(a) 9 (b) −11 (c) $\frac{23}{13}$
(d) 9, −11 (e) None of these

2—**Answer:** a

33. Solve for x: $\begin{vmatrix} 5+2x & -2 \\ 7-x & 3 \end{vmatrix} = 5$

(a) 1 (b) $\frac{1}{2}$ (c) −6
(d) −3 (e) None of these

2—**Answer:** c

34. Solve for x: $\begin{vmatrix} 3x-1 & -1 \\ 4 & x+1 \end{vmatrix} = 4$

(a) 4 (b) 0 (c) $\frac{-1 \pm 2\sqrt{7}}{3}$
(d) $-1, \frac{1}{3}$ (e) None of these

2—**Answer:** d

35. Solve for x: $\begin{vmatrix} x-1 & 3 \\ 3 & x+1 \end{vmatrix} = 6$

(a) ±1 (b) ±4 (c) $\pm\sqrt{2}i$
(d) 5 (e) None of these

2—**Answer:** b

36. Solve for x: $\begin{vmatrix} x-1 & x \\ 2 & x+1 \end{vmatrix} = 7$

(a) $-4, 2$ (b) 0 (c) $5, 0$

(d) $1, -3$ (e) None of these

2—Answer: e

37. Evaluate the determinant: $\begin{vmatrix} 4u & 1 \\ 1 & 4v \end{vmatrix}$

(a) uv (b) $4u + 4v - 1$ (c) $16uv - 1$

(d) $4uv - 2$ (e) None of these

1—Answer: c

38. Evaluate the determinant: $\begin{vmatrix} e^{-x} & e^{2x} \\ -e^{-x} & 2e^{2x} \end{vmatrix}$

(a) $3e^{-2x^2}$ (b) $3e^x$ (c) $3e^{-2x}$

(d) $2e^{2x}$ (e) None of these

2—Answer: b

39. Evaluate the determinant: $\begin{vmatrix} x & \ln x \\ 1 & 1/x \end{vmatrix}$

(a) $\ln x - 1$ (b) $1 - \ln x$ (c) $-\ln x$ (d) $\ln x$ (e) None of these

1—Answer: b

40. Evaluate the determinant: $\begin{vmatrix} x \ln x & x \\ 1 + \ln x & 1 \end{vmatrix}$

1—Answer: $-x$

41. Evaluate the determinant: $\begin{vmatrix} e^{2x} & e^x \\ 2e^{2x} & e^x \end{vmatrix}$

1—Answer: $-e^{3x}$

42. Use the matrix capabilities of a graphing utility to evaluate:

$$\begin{vmatrix} 3 & -2 & 4 & 3 \\ 2 & -1 & 0 & 4 \\ -2 & 0 & 1 & 5 \\ -2 & -3 & 0 & 2 \end{vmatrix}$$

1—T—Answer: 270

43. Use the matrix capabilities of a graphing utility to evaluate:

$$\begin{vmatrix} 10 & -3 & 2 & 4 & 5 \\ -1 & 0 & -3 & 2 & 3 \\ 4 & -1 & -2 & 0 & 4 \\ 9 & 4 & -3 & 6 & 2 \\ -2 & 0 & -4 & 4 & 5 \end{vmatrix}$$

1—T—Answer: 672

44. Use the matrix capabilities of a graphing utility to evaluate:

$$\begin{vmatrix} 0 & 1 & 2 & 3 \\ 1 & 2 & 3 & 4 \\ 2 & 3 & 4 & 5 \\ 3 & 4 & 5 & 6 \end{vmatrix}$$

1—T—Answer: 0

45. Use the matrix capabilities of a graphing utility to evaluate:

$$\begin{vmatrix} 1 & 2 & 3 \\ 2 & 3 & 1 \\ 3 & 1 & 2 \end{vmatrix}$$

1—T—Answer: -18

7.5 Applications of Matrices and Determinants

1. Use Cramer's Rule to solve for y in the system of linear equations:

$$3x + 2y + 4z = 12$$
$$x - y + z = 3$$
$$2x + 7y - z = 9$$

(a) $y = \dfrac{\begin{vmatrix} 3 & 2 & 4 \\ 1 & -1 & 1 \\ 2 & 7 & -1 \end{vmatrix}}{\begin{vmatrix} 3 & 12 & 4 \\ 1 & 3 & 1 \\ 2 & 9 & -1 \end{vmatrix}}$

(b) $y = \dfrac{\begin{vmatrix} 3 & 12 & 4 \\ 1 & 3 & 1 \\ 2 & 9 & -1 \end{vmatrix}}{\begin{vmatrix} 3 & 2 & 4 \\ 1 & -1 & 1 \\ 2 & 7 & -1 \end{vmatrix}}$

(c) $y = \dfrac{\begin{vmatrix} 3 & 4 & 12 \\ 1 & -1 & 3 \\ 2 & 7 & 9 \end{vmatrix}}{\begin{vmatrix} 2 & 4 & 3 \\ -1 & 1 & 1 \\ 7 & -1 & 2 \end{vmatrix}}$

(d) $y = \begin{vmatrix} 3 & 2 & 4 \\ 1 & -1 & 1 \\ 2 & 7 & -1 \end{vmatrix} \begin{vmatrix} 2 & 4 & 12 \\ -1 & 1 & 3 \\ 7 & -1 & 9 \end{vmatrix}$

(e) None of these

1—Answer: b

2. Use Cramer's Rule to solve for x in the system of linear equations:

$$2x + 3y + 4z = 7$$
$$x - 2y + 5z = 3$$
$$4x + 5z = 9$$

(a) $x = \dfrac{\begin{vmatrix} 7 & 3 & 4 \\ 3 & -2 & 5 \\ 9 & 0 & 5 \end{vmatrix}}{\begin{vmatrix} 2 & 3 & 4 \\ 1 & -2 & 5 \\ 4 & 0 & 5 \end{vmatrix}}$

(b) $x = \dfrac{\begin{vmatrix} 2 & 3 & 4 \\ 1 & -2 & 5 \\ 4 & 0 & 5 \end{vmatrix}}{\begin{vmatrix} 7 & 3 & 4 \\ 3 & -2 & 5 \\ 9 & 0 & 5 \end{vmatrix}}$

(c) $x = \begin{vmatrix} 2 & 3 & 4 \\ 1 & -2 & 5 \\ 4 & 0 & 5 \end{vmatrix} \begin{vmatrix} 7 & 3 & 4 \\ 3 & -2 & 5 \\ 9 & 0 & 5 \end{vmatrix}$

(d) $x = \begin{vmatrix} 2 & 3 & 7 \\ 1 & -2 & 3 \\ 4 & 0 & 9 \end{vmatrix} \begin{vmatrix} 2 & 3 & 4 \\ 1 & -2 & 5 \\ 4 & 0 & 5 \end{vmatrix}$

(e) None of these

1—Answer: a

3. Use Cramer's Rule to solve for y in the system of linear equations:

   $3x + 2y - 10z = 5$

   $x - y + z = 10$

   $-7x + 2z = 1$

   (a) $y = \dfrac{\begin{vmatrix} 3 & 5 & -10 \\ 1 & 10 & 1 \\ -7 & 1 & 2 \end{vmatrix}}{\begin{vmatrix} 3 & 2 & -10 \\ 1 & -1 & 1 \\ -7 & 0 & 2 \end{vmatrix}}$

   (b) $y = \dfrac{\begin{vmatrix} 3 & 2 & -10 \\ 1 & -1 & 1 \\ -7 & 0 & 2 \end{vmatrix}}{\begin{vmatrix} 3 & 2 & -10 \\ 1 & 10 & 1 \\ -7 & 1 & 2 \end{vmatrix}}$

   (c) $y = \dfrac{\begin{vmatrix} 2 & 5 \\ -1 & 10 \\ 0 & 1 \end{vmatrix}}{\begin{vmatrix} 3 & -10 \\ 1 & 1 \\ -7 & 2 \end{vmatrix}}$

   (d) $y = 5\begin{vmatrix} 3 & 2 & -10 \\ 1 & -1 & 1 \\ -7 & 0 & 2 \end{vmatrix} + 10\begin{vmatrix} 3 & 2 & -10 \\ 1 & -1 & 1 \\ -7 & 0 & 2 \end{vmatrix} + 1\begin{vmatrix} 3 & 2 & -10 \\ 1 & -1 & 1 \\ -7 & 0 & 2 \end{vmatrix}$

   (e) None of these

   1—Answer: a

4. Use Cramer's Rule to solve the system of linear equations.

   $4x + 6y + 2z = 15$

   $x - y + 4z = -3$

   $3x + 2y + 2z = 6$

   2—Answer: $(1, 2, -\tfrac{1}{2})$

5. Use Cramer's Rule to solve the system of linear equations.

   $5x + 5y + 4z = 4$

   $10x - 5y + 2z = 11$

   $5x - 5y + 2z = 7$

   2—Answer: $(\tfrac{4}{5}, -\tfrac{2}{5}, \tfrac{1}{2})$

6. Use Cramer's Rule to solve for y:

$$3x - 2y + 2z = 3$$
$$x + 4y - z = 2$$
$$x + y + z = 6$$

(a) $y = \dfrac{\begin{vmatrix} 3 & -2 & 2 \\ 1 & 4 & -1 \\ 1 & 1 & 1 \end{vmatrix}}{\begin{vmatrix} 3 & 3 & 2 \\ 1 & 2 & -2 \\ 1 & 6 & 1 \end{vmatrix}}$

(b) $y = \dfrac{\begin{vmatrix} 3 & 3 & 2 \\ 1 & 2 & -1 \\ 1 & 6 & 1 \end{vmatrix}}{\begin{vmatrix} 3 & -2 & 2 \\ 1 & 4 & -1 \\ 1 & 1 & 1 \end{vmatrix}}$

(c) $y = \dfrac{\begin{vmatrix} 3 & 2 & 3 \\ 1 & -1 & 2 \\ 1 & 1 & 6 \end{vmatrix}}{\begin{vmatrix} 3 & -2 & 2 \\ 1 & 4 & -1 \\ 1 & 1 & 1 \end{vmatrix}}$

(d) $y = \dfrac{\begin{vmatrix} 3 & 3 & -2 \\ 1 & 2 & 4 \\ 1 & 6 & 1 \end{vmatrix}}{\begin{vmatrix} 3 & 1 & 1 \\ -2 & 4 & 1 \\ 2 & -1 & 1 \end{vmatrix}}$

(e) None of these

1—Answer: b

7. Use Cramer's Rule to solve for z:

$$2x - y + z = -3$$
$$x + y + z = 4$$
$$3x - 2y + 5z = 1$$

(a) $z = \dfrac{\begin{vmatrix} 2 & -1 & 1 \\ 1 & 1 & 1 \\ 3 & -2 & 5 \end{vmatrix}}{\begin{vmatrix} 2 & -1 & -3 \\ 1 & 1 & 4 \\ 3 & -2 & 1 \end{vmatrix}}$

(b) $y = \dfrac{\begin{vmatrix} 2 & -1 & 1 \\ 1 & 1 & 1 \\ 3 & -2 & 5 \end{vmatrix}}{\begin{vmatrix} -3 & 2 & -1 \\ 4 & 1 & 1 \\ 1 & 3 & -2 \end{vmatrix}}$

(c) $z = \dfrac{\begin{vmatrix} 2 & -1 & -3 \\ 1 & 1 & 4 \\ 3 & -2 & 1 \end{vmatrix}}{\begin{vmatrix} 2 & -1 & 1 \\ 1 & 1 & 1 \\ 3 & -2 & 5 \end{vmatrix}}$

(d) $z = \dfrac{\begin{vmatrix} -3 & -3 & 1 \\ 4 & 4 & 1 \\ 1 & 1 & 5 \end{vmatrix}}{\begin{vmatrix} 2 & -1 & -3 \\ 1 & 1 & 4 \\ 3 & -2 & 1 \end{vmatrix}}$

(e) None of these

1—Answer: c

8. Use a determinant to find the area of the triangle with vertices $(2, -1)$, $(3, 2)$, and $(5, 0)$.

(a) 4 (b) 8 (c) 6 (d) 12 (e) None of these

2—Answer: a

9. Use a determinant to find the area of the triangle with vertices $(-1, 4)$, $(2, 6)$, and $(1, 0)$.

    (a) 0   (b) 3   (c) 4   (d) 8   (e) None of these

    2—Answer: d

10. Use a determinant to find the area of the triangle with vertices $(-1, -3)$, $(5, 6)$, and $(0, 2)$.

    2—Answer: $\frac{21}{2}$

11. Use a determinant to calculate the area of the triangle with vertices $(1, 3)$, $(7, 2)$, and $(9, 5)$.

    (a) 30   (b) 25   (c) 20   (d) 10   (e) None of these

    1—Answer: d

12. Use a determinant to calculate the area of the triangle with vertices $(2, -4)$, $(4, -1)$, and $(7, -3)$.

    (a) $\frac{13}{2}$ (b) 13   (c) 26   (d) 17   (e) None of these

    1—Answer: a

13. Use a determinant to calculate the area of the triangle with vertices $(-5, 2)$, $(0, 7)$, and $(-4, 10)$.

    (a) $\frac{39}{2}$ (b) 70   (c) $\frac{35}{2}$ (d) 35   (e) None of these

    1—Answer: c

14. Use a determinant to calculate the area of the triangle with vertices $(-3, -10)$, $(-5, -2)$, and $(5, 1)$.

    (a) 86   (b) 43   (c) 172   (d) $\frac{43}{2}$ (e) None of these

    1—Answer: b

15. Use a determinant to ascertain whether the points are collinear: $(-7, 4)$, $(-1, 2)$, and $(2, 1)$.

    1—Answer: Yes

16. Use a determinant to ascertain whether the points are collinear: $(2, 2)$, $(0, 5)$, and $(4, -4)$.

    1—Answer: Yes

17. Use a determinant to ascertain whether the points are collinear: $(12, 6)$, $(-1, -3)$, and $(0, -1)$.

    1—Answer: No

18. Use a determinant to ascertain whether the points are collinear: $(-3, -8)$, $(6, 4)$, $(12, 12)$.

    1—Answer: Yes

19. Use a determinant to ascertain whether the points are collinear: $(-3, 6)$, $(1, 2)$, $(5, -2)$.

    1—Answer: Yes

20. Use a determinant to ascertain whether the points are collinear: $(-2, -3)$, $(1, -5)$, $(3, -1)$.

    1—Answer: No

21. Use a determinant to ascertain whether the points are collinear:
    $(-1, -1), (2, -3), (-2, 2)$

    1—Answer: No

22. Use a determinant to find an equation of the line passing through the points $(-1, 3)$ and $(0, -1)$.

    (a) $2x + y = 1$
    (b) $x - y = 0$
    (c) $x - 2y = 1$
    (d) $2x - y = 1$
    (e) None of these

    1—Answer: d

23. Use a determinant to find an equation of the line passing through the points $(4, -2)$ and $(-3, 1)$.

    (a) $3x - 7y + 2 = 0$
    (b) $7x + 3y + 2 = 0$
    (c) $3x + 7y + 2 = 0$
    (d) $3x + 7y - 2 = 0$
    (e) None of these

    1—Answer: c

24. Use a determinant to find an equation of the line passing through the points $(2, -5)$ and $(-1, -1)$.

    1—Answer: $4x + 3y = -7$

25. Use a determinant to find an equation of the line passing through the points $(1, 2)$ and $(-1, 6)$.

    (a) $x + 2y - 4 = 0$
    (b) $y = 2x$
    (c) $y = 3x^2 - 2x + 1$
    (d) $2x + y - 4 = 0$
    (e) None of these

    1—Answer: d

26. Use a determinant to find an equation of the line that passes through the center of the circle $(x - 4)^2 + (y + 1)^2 = 5$ and the point $(1, 3)$.

    (a) $2x - 5y + 13 = 0$
    (b) $3x + 4y - 13 = 0$
    (c) $4x + 3y - 13 = 0$
    (d) $5x - 2y + 13 = 0$
    (e) None of these

    2—Answer: c

27. Use a determinant to find an equation of the line that passes through the center of the circle $(x + 3)^2 + (y - 2)^2 = 7$ and the point $(1, 4)$.

    (a) $x - 2y + 7 = 0$
    (b) $2y - x + 7 = 0$
    (c) $x^2 + 3x - 2y + 4 = 0$
    (d) $3x + y - 7 = 0$
    (e) None of these

    2—Answer: a

28. Use a determinant to find an equation of the line that goes through the vertices of the parabolas $y = (x - 2)^2 + 3$ and $y = (x - 5)^2 - 4$.

    (a) $x + y - 1 = 0$
    (b) $7x + 3y - 23 = 0$
    (c) $3y + 7x + 23 = 0$
    (d) $x - y + 1 = 0$
    (e) None of these

    2—Answer: b

29. Use a matrix A to encode the message: H A P P Y   D A Y S

    $$A = \begin{bmatrix} 1 & -2 & 2 \\ -1 & 1 & 3 \\ 1 & -1 & -4 \end{bmatrix}$$

    2—Answer: 23 -31 -45 -9 -7 107 28 -32 -89 19 -38 38

30. Use the inverse of the matrix $A = \begin{bmatrix} 1 & -2 & 2 \\ -1 & 1 & 3 \\ 1 & -1 & -4 \end{bmatrix}$ to decode the cryptogram.

21 −40 14 2 −11 32 16 −16 −66 15 −20 −31

2—Answer: S P R I N G B R E A K

31. Find the uncoded row matrix of order 1×3 for the message SELL IT FAST, then encode the message using $A = \begin{bmatrix} 1 & -1 & 0 \\ 1 & 0 & 3 \\ -2 & 1 & -1 \end{bmatrix}$.

1—Answer: 0 −7 3 −6 −3 −9 8 −14 −6 −20 19 37

32. Find the uncoded row matrix of order 1×3 for the message I LOVE MATH, then encode the message using $A = \begin{bmatrix} 1 & -1 & 0 \\ 1 & 0 & 3 \\ -2 & 1 & -1 \end{bmatrix}$.

1—Answer: −15 3 −12 27 −10 61 11 1 38 28 −20 24

33. Find the uncoded row matrix of order 1×3 for the message CALL ME LATER, then encode the message using $A = \begin{bmatrix} 1 & -1 & 0 \\ 1 & 0 & 3 \\ -2 & 1 & -1 \end{bmatrix}$.

   (a) 16 −12 4 4 3 −13 9 −13 3 1 9 −9 −20 −18 5

   (b) −20 9 −9 −14 1 −13 −19 7 −12 11 4 55 18 −18 0

   (c) −5 12 19 −3 6 6 2 13 −6 −6 3 −1 0 2 2

   (d) 3 1 12 12 0 13 5 0 12 1 20 5 18 0 0

   (e) None of these

1—Answer: b

34. Find the uncoded row matrix of order 1×3 for the message BE BACK SOON, then encode the message using $A = \begin{bmatrix} 1 & -1 & 0 \\ 1 & 0 & 3 \\ -2 & 1 & -1 \end{bmatrix}$.

   (a) −2 15 1 0 −3 −27 −2 8 19 15 −3 1

   (b) 7 −2 15 −3 1 0 −27 8 −19 2 −1 31

   (c) 2 5 0 2 1 3 11 0 19 15 15 14

   (d) 0 2 5 −3 1 0 0 −27 16 2 1 1

   (e) None of these

1—Answer: b

35. Decode the cryptogram:
    129 −85 −38 −75 70 25 −9 18 3 188 −141 −58 using
    $$A = \begin{bmatrix} 13 & -10 & -4 \\ -6 & 5 & 2 \\ 3 & -2 & -1 \end{bmatrix}$$

    (a) WELCOME HOME (b) CLOSING OUT (c) OUT TO LUNCH
    (d) PLEASE HURRY (e) None of these

    1—Answer: c

36. Use a graphing utility and Cramer's Rule to solve (if possible) the system of equations.
    $$2x + y + z = -2$$
    $$x + 2y + 3z = -3$$
    $$x - 3y - z = -5$$

    2—T—Answer: $x = -1, y = 2, z = -2$

37. Use a graphing utility and Cramer's Rule to solve (if possible) the system of equations.
    $$x - y - 2z = 3$$
    $$y + 3z = -2$$
    $$3x + 4y - z = 11$$

    2—T—Answer: $x = 2, y = 1, z = -1$

38. Use a graphing utility and Cramer's Rule to solve (if possible) the system of equations.
    $$x + 3y + z = 4$$
    $$2x - y - 3z = 1$$
    $$4x + y + z = 5$$

    2—Answer: $x = 1, y = 1, z = 0$

39. Use a graphing utility and Cramer's Rule to solve (if possible) the system of equations.
    $$3x - y + 5z = 1$$
    $$2x + y + z = 0$$
    $$-x + y - 2z = 0$$

    2—T—Answer: $(-1, 1, 1)$

40. You are farming a triangular tract of land, as shown in the figure. From the northern most vertex A of the region, the distances to the other vertices are 2500 yards south and 1500 east to vertex B and 2000 yards south and 4000 yards east to vertex C. Use a graphing utility to approximate the number of square yards on the farm.

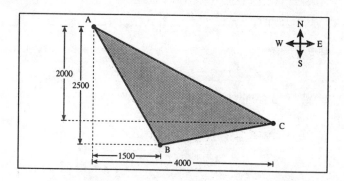

2—T—Answer: 1,750,000 square yards

41. You are bidding on a triangular plot of land, as shown in the figure. To estimate the number of square feet. You start at one vertex, walk 75 feet west and 45 feet north to the second vertex, and then walk 60 feet east and 15 feet north to the third vertex. Use a graphing utility, to determine how many square feet there are in the plot of land.

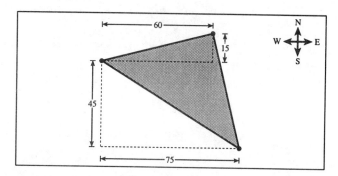

2—T—Answer: 1,912.5 square yards

CHAPTER EIGHT
Sequences and Probability

❑ 8.1 Sequences and Summation Notation

1. Find the first 5 terms of the sequence whose nth term is $a_n = (-1)^n(2n + 9)$. (Assume that n begins with 1.)

   (a) $-11, -13, -15, -17, -19 \ldots$
   (b) $-11, 13, -15, 17, -19, \ldots$
   (c) $-11, 2, -13, 4, -15, \ldots$
   (d) $-11, -24, -39, -56, -75, \ldots$
   (e) None of these

   1—Answer: b

2. Find the first 5 terms of the sequence whose nth terms is $a_n = n!$. (Assume that n begins with 0.)

   (a) 0, 1, 2, 6, 24
   (b) 0, 1, 2, 6, 12
   (c) 1, 1, 2, 6, 12
   (d) 1, 1, 2, 6, 24
   (e) None of these

   1—Answer: d

3. Find the first 5 terms of the sequence whose nth term is $a_n = 1 - \dfrac{1}{n}$. (Assume that n begins with 1.)

   (a) $\dfrac{1}{2}, \dfrac{1}{4}, \dfrac{1}{8}, \dfrac{1}{16}, \dfrac{1}{32}$
   (b) $0, \dfrac{1}{2}, \dfrac{2}{3}, \dfrac{3}{4}, \dfrac{4}{5}$
   (c) $0, \dfrac{1}{2}, \dfrac{1}{3}, \dfrac{1}{4}, \dfrac{1}{5}$
   (d) $1, \dfrac{1}{2}, \dfrac{2}{3}, \dfrac{3}{4}, \dfrac{4}{5}$
   (e) None of these

   1—Answer: b

4. Write the first 5 terms of the sequence whose nth term is $a_n = \dfrac{n!}{(n + 2)!}$. (Assume that n begins with 0.)

   1—Answer: $\dfrac{1}{2}, \dfrac{1}{6}, \dfrac{1}{12}, \dfrac{1}{20}, \dfrac{1}{30}$

5. Write the first five terms of the sequence whose nth term is $a_n = \dfrac{n}{n^2 + 1}$. (Assume that n begins with 1.)

   1—Answer: $\dfrac{1}{2}, \dfrac{2}{5}, \dfrac{3}{10}, \dfrac{4}{17}, \dfrac{5}{26}$

6. Write the first five terms of the sequence whose nth term is $a_n = \dfrac{n - 2}{n^2 + 1}$. (Assume that n begins with 1.)

   1—Answer: $-\dfrac{1}{2}, 0, \dfrac{1}{10}, \dfrac{2}{17}, \dfrac{3}{26}$

7. Write the first five terms of the sequence whose nth term is $a_n = \dfrac{(-1)^n}{n!}$. (Assume that n begins with 1.)

  1—Answer: $-1, \dfrac{1}{2}, -\dfrac{1}{6}, \dfrac{1}{24}, -\dfrac{1}{120}$

8. Write the first five terms of the sequence whose nth term is $a_n = \dfrac{n!}{(n+2)!}$. (Assume that n begins with 0.)

  1—Answer: $\dfrac{1}{2}, \dfrac{1}{6}, \dfrac{1}{12}, \dfrac{1}{20}, \dfrac{1}{30}$

9. Use a graphing utility to graph the first ten terms of the sequence. (Assume n begins with 1.)

  $a_n = 2 + \dfrac{1}{n}$

  1—T—Answer: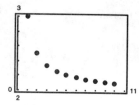

10. Use a graphing utility to graph the first ten terms of the sequence. (Assume n begins with 1.)

  $a_n = \dfrac{3n}{n+1}$

  1—T—Answer: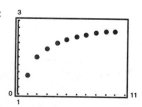

11. Use a graphing utility to graph the first ten terms of the sequence. (Assume n begins with 1.)

  $a_n = 4(1/2)^n$

  1—T—Answer: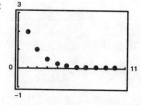

12. Simplify: $\dfrac{8!}{5!}$

    (a) $\dfrac{8}{5}$ (b) 336   (c) 56   (d) 48   (e) None of these

    1—Answer: b

13. Simplify: $\dfrac{6!}{4!2!}$

    (a) $\dfrac{3}{4}$ (b) $\dfrac{1}{8}$ (c) 15   (d) 30   (e) None of these

    1—Answer: c

14. Simplify: $\dfrac{(2n)!}{(2n-2)!}$

    2—0—Answer: $4n^2 + 2n$

15. Simplify: $\dfrac{3(4!)}{7!}$

    (a) 95,040   (b) $\dfrac{15}{14}$ (c) $\dfrac{1}{70}$ (d) $\dfrac{1}{7}$ (e) None of these

    1—Answer: c

16. Simplify: $\dfrac{(2n)!}{(2n-3)!}$

    (a) $\dfrac{2n}{2n-3}$ (b) $2n(2n-1)(2n-2)$ (c) $2n(n-1)(n-2)$
    (d) $\dfrac{2}{(n-1)(n-2)}$ (e) None of these

    1—Answer: b

17. Simplify: $\dfrac{(n+1)!}{(n-1)!}$

    (a) $n+1$ (b) $n-1$ (c) $\dfrac{n+1}{n-1}$
    (d) $n^2 + n$ (e) None of these

    1—Answer: d

18. Find a formula for the *n*th term of the sequence. (Assume that *n* begins with 1.)

    $\dfrac{3}{2}, \dfrac{6}{4}, \dfrac{9}{12}, \dfrac{12}{48}, \dfrac{15}{240}, \ldots$

    (a) $\dfrac{3n}{2n!}$ (b) $\dfrac{3n}{n(n+1)}$ (c) $\dfrac{3n}{n!(n+1)!}$
    (d) $\dfrac{3n}{2(n-1)!}$ (e) None of these

    1—Answer: a

19. Find a formula for the nth term of the sequence. (Assume that n begins with 1.)

$$\frac{1}{4}, \frac{2}{9}, \frac{3}{16}, \frac{4}{25}, \ldots$$

(a) $a_n = 1 - \frac{3n}{n^2}$

(b) $a_n = \frac{n}{(n+1)!}$

(c) $a_n = \frac{1}{2} + \frac{n}{(n+1)^3}$

(d) $a_n = \frac{n}{(n+1)^2}$

(e) None of these

1—Answer: d

20. Find a formula for the nth term of the sequence. (Assume that n begins with 1.)

$$\frac{2}{1}, \frac{4}{1}, \frac{6}{2}, \frac{8}{6}, \frac{10}{24}, \ldots$$

(a) $\frac{2^n}{(n+1)!}$

(b) $\frac{3 - 2^n}{n(2^n)}$

(c) $\frac{2n}{2n - 1}$

(d) $\frac{2n}{(n-1)!}$

(e) None of these

1—Answer: d

21. Find a formula for the nth term of the sequence. (Assume that n begins with 1.)

$$\frac{2}{1}, \frac{3}{2}, \frac{4}{3}, \frac{5}{4}, \frac{6}{5}, \ldots$$

(a) $\frac{n+1}{n-1}$

(b) $2 - \frac{1}{n}$

(c) $1 + \frac{1}{n}$

(d) $\frac{2n-1}{n}$

(e) None of these

1—Answer: c

22. Find a formula for the nth term of the sequence. (Assume that n begins with 1.)

$$1, 1, \frac{1}{2}, \frac{1}{6}, \frac{1}{24}, \ldots$$

(a) $\frac{n+1}{n!}$

(b) $\frac{n}{n!}$

(c) $\frac{n}{2n!}$

(d) $\frac{1}{2^n}$

(e) None of these

1—Answer: b

23. Find the sum: $\sum_{n=3}^{6} \frac{3}{n-2}$

(a) $\frac{12}{9}$

(b) $\frac{25}{4}$

(c) $\frac{3}{16}$

(d) $\frac{1}{2}$

(e) None of these

1—Answer: b

24. Use a calculator to find the sum.

$$\sum_{i=1}^{5} (10 - 2i)$$

1—T—Answer: 20

25. Use a calculator to find the sum.
$$\sum_{k=0}^{5} \frac{k!}{2}$$
1—T—Answer: 77

26. Use a calculator to find the sum.
$$\sum_{j=1}^{4} \frac{3}{j+2}$$
1—T—Answer: 2.85

27. Use a calculator to find the sum.
$$\sum_{i=0}^{4} \frac{(-1)^i}{i+2}$$
1—T—Answer: $0.38\overline{3}$

28. Find the sum: $\sum_{n=1}^{4} \frac{n+1}{n+2}$

(a) $\frac{61}{20}$ (b) $\frac{31}{20}$ (c) $\frac{143}{60}$

(d) $\frac{131}{60}$ (e) None of these

1—Answer: a

29. Find the sum: $\sum_{i=1}^{4} (1-i)$

(a) -3 (b) -6 (c) 6   (d) -5 (e) None of these

1—Answer: b

30. Find the sum: $\sum_{k=2}^{6} (-1)^k (2k)$

(a) 40   (b) -4 (c) 6   (d) 8   (e) None of these

1—Answer: d

31. Find the sum: $\sum_{i=0}^{3} i!$

(a) 9   (b) 6   (c) 10   (d) 7   (e) None of these

1—Answer: c

32. Use sigma notation to write the sum: $\frac{2}{1} + \frac{3}{2} + \frac{4}{3} + \cdots + \frac{7}{6}$

(a) $\sum_{n=1}^{7} \frac{n}{n-1}$ (b) $\sum_{n=1}^{6} \frac{n+1}{n}$ (c) $\sum_{n=1}^{7} \frac{n}{n+1}$

(d) $\sum_{n=2}^{n} \frac{n}{n-1}$ (e) None of these

1—Answer: b

33. Use sigma notation to write the sum: $\dfrac{3}{1} + \dfrac{3}{4} + \dfrac{3}{9} + \dfrac{3}{16} + \dfrac{3}{25}$

   (a) $\displaystyle\sum_{i=1}^{5} \dfrac{1}{i^2}$ 　　　　(b) $\displaystyle\sum_{i=1}^{5} \dfrac{3}{4i}$ 　　　　(c) $\displaystyle\sum_{i=1}^{6} \dfrac{3}{i^2}$

   (d) $\displaystyle\sum_{i=1}^{5} \dfrac{3}{i^2}$ 　　　　(e) None of these

   1—Answer: d

34. Use sigma notation to write the sum: $\dfrac{2}{3} + \dfrac{4}{4} + \dfrac{6}{5} + \dfrac{8}{6} + \cdots + \dfrac{14}{9}$

   (a) $\displaystyle\sum_{n=1}^{7} \dfrac{2n}{n+2}$ 　　　　(b) $\displaystyle\sum_{n=2}^{8} \dfrac{n+2}{n+1}$ 　　　　(c) $\displaystyle\sum_{n=0}^{6} \dfrac{n+2}{n+3}$

   (d) $\displaystyle\sum_{n=3}^{9} \dfrac{n-1}{n}$ 　　　　(e) None of these

   1—Answer: a

35. Use sigma notation to write the sum: $\dfrac{1}{2} + \dfrac{2}{6} + \dfrac{3}{24} + \dfrac{4}{120} + \dfrac{5}{720}$

   1—Answer: $\displaystyle\sum_{n=1}^{5} \dfrac{n}{(n+1)!}$

36. Use sigma notation to write the sum: $\dfrac{4}{2} + \dfrac{5}{4} + \dfrac{6}{6} + \dfrac{7}{8}$

   (a) $\displaystyle\sum_{n=1}^{4} \dfrac{n+3}{2n}$ 　　　　(b) $\displaystyle\sum_{n=1}^{4} \dfrac{4n-1}{2n}$ 　　　　(c) $\displaystyle\sum_{n=1}^{4} \dfrac{2n+2}{2n}$

   (d) $\displaystyle\sum_{n=1}^{4} \dfrac{2^{n+1}+1}{n+2}$ 　　　　(e) None of these

   1—Answer: a

37. Find the 10th term of the sequence: $\dfrac{3^1}{2^0}, \dfrac{3^2}{2^1}, \dfrac{3^3}{2^2}, \dfrac{3^4}{2^3}, \dfrac{3^5}{2^4}, \ldots$

   (a) $\dfrac{3^{10}}{2^{10}}$ 　　　　(b) $\dfrac{3^{11}}{2^{10}}$ 　　　　(c) $\dfrac{3^{10}}{2^9}$

   (d) $\dfrac{3^{11}}{2^{11}}$ 　　　　(e) None of these

   1—Answer: c

38. Find the 40th term of the sequence whose nth term is $a_n = 500\left(1 + \dfrac{0.095}{12}\right)^n$.

   (a) 363.83 　　　　(b) 752.19 　　　　(c) 690.84

   (d) 685.42 　　　　(e) None of these

   1—Answer: d

39. Find the fifth term of the sequence whose nth term is $a_n = 2(3^{n-1})$. (Assume that n begins with 1.)

   (a) 486 　　(b) −486 　　(c) $\dfrac{1}{162}$ 　　(d) 162 　　(e) None of these

   1—Answer: d

40. Find the third term of the sequence whose nth term is $a_n = \dfrac{(-1)^{n+1}}{n}$. (Assume that n begins with 1.)

(a) $\dfrac{1}{3}$ (b) $\dfrac{1}{81}$ (c) $-\dfrac{1}{3}$ (d) $\dfrac{1}{27}$ (e) None of these

1—Answer: a

41. Find the tenth term of the sequence whose nth term is $a_n = \dfrac{2n+1}{5+3(n-1)}$.

(a) $\dfrac{21}{32}$ (b) $\dfrac{21}{72}$ (c) $\dfrac{22}{32}$ (d) $\dfrac{19}{29}$ (e) None of these

1—Answer: a

42. Find the fifth term of the sequence whose nth term is $a_n = \dfrac{(-1)^{n+1}}{(n+1)^2}$. (Assume that n begins with 1.)

(a) $-\dfrac{1}{25}$ (b) $\dfrac{1}{5}$ (c) $\dfrac{1}{36}$ (d) $\dfrac{1}{6}$ (e) None of these

1—Answer: c

43. A deposit of $2000 is made in an account that earns 6% interest compounded monthly. The balance in the account after n months is given by

$$A_n = 2000\left(1 + \frac{0.06}{12}\right)^n, \; n = 1, 2, 3, \ldots$$

Find the balance in this account after 8 years ($n = 96$).

(a) $3552.17   (b) $2793.52   (c) $2163.73

(d) $3228.29   (e) None of these

1—Answer: d

44. A deposit of $4000 is made in an account that earns 7% interest compounded quarterly. The balance in the account after n months is given by

$$A_n = 4000\left(1 + \frac{0.07}{4}\right)^n, \; n = 1, 2, 3, \ldots$$

Find the balance in this account after 8 years ($n = 32$).

(a) $6968.85   (b) $4595.53   (c) $27,439.59

(d) $9874.31   (e) None of these

1—Answer: a

❏ 8.2 Arithmetic Sequences

1. Which of the following is an arithmetic sequence?

   (a) 1, 3, 9, 27, 81, . . .
   (b) 1, 16, 36, 64, 100, . . .
   (c) 2, 11, 20, 29, 38, . . .
   (d) 3, −5, 7, −9, 11, . . .
   (e) None of these

   1—Answer: c

2. Which of the following is an arithmetic sequence?

   (a) 2, 4, 8, 16, 32, . . .
   (b) −2, 4, −8, 16, −32, . . .
   (c) 3, 6, 9, 12, 15, . . .
   (d) All of these
   (e) None of these

   1—Answer: c

3. Which of the following is an arithmetic sequence?

   (a) 1, 3, 5, 7, 9, . . .
   (b) 4, 7, 10, 13, 16, . . .
   (c) −10, −6, −2, 2, 6, . . .
   (d) All of these
   (e) None of these

   1—Answer: d

4. Which of the following is an arithmetic sequence?

   (a) 2, 3, 5, 7, 11, . . .
   (b) $\frac{1}{2}, \frac{1}{3}, \frac{1}{4}, \frac{1}{5}, \frac{1}{6}, \ldots$
   (c) $\frac{1}{2}, \frac{1}{4}, \frac{1}{6}, \frac{1}{8}, \frac{1}{10}, \ldots$
   (d) All of these
   (e) None of these

   1—Answer: e

5. Which of the following is an arithmetic sequence?

   (a) $-\frac{1}{2}, -1, -\frac{3}{2}, -2, -\frac{5}{2}, \ldots$
   (b) $\frac{2}{5}, \frac{2}{25}, \frac{2}{125}, \frac{2}{625}, \frac{2}{3125}, \ldots$
   (c) −2, 2, −2, 2, −2, . . .
   (d) All of these
   (e) None of these

   1—Answer: a

6. Find the first 5 terms of the arithmetic sequence with $a_1 = 23$ and $d = -\frac{1}{2}$.

   (a) 23, $22\frac{1}{2}$, 22, $21\frac{1}{2}$, 21
   (b) 23, $23\frac{1}{2}$, 24, $24\frac{1}{2}$, 25
   (c) 23, $-\frac{23}{2}, \frac{23}{4}, -\frac{23}{8}, \frac{23}{16}$
   (d) 23, −46, 92, −184, 368
   (e) None of these

   1—Answer: a

7. Find the first 4 terms of the arithmetic sequence with $a_1 = 4$ and $d = -3$.

   (a) 4, $-\frac{4}{3}, \frac{4}{9}, -\frac{4}{27}, \ldots$
   (b) 4, 1, −2, −5, . . .
   (c) 4, 7, 10, 13, . . .
   (d) 4, −12, 36, −108, . . .
   (e) None of these

   1—Answer: b

8. Find the first 4 terms of the arithmetic sequence with $a_1 = 4$ and $d = \frac{1}{2}$.
   - (a) $4, 2, 1, \frac{1}{2}, \ldots$
   - (b) $4, 8, 16, 32, \ldots$
   - (c) $4, -2, 1, -\frac{1}{2}, \ldots$
   - (d) $4, 0, -4, -8$
   - (e) None of these

   1—Answer: e

9. Find the first 4 terms of the arithmetic sequence with $a_1 = 4$ and $d = -2$.
   - (a) $4, 2, 0, -2, \ldots$
   - (b) $4, 6, 8, 10, \ldots$
   - (c) $4, -2, 1, -\frac{1}{2}, \ldots$
   - (d) $4, -8, 16, -32, \ldots$
   - (e) None of these

   1—Answer: a

10. Find the first 4 terms of the arithmetic sequence with $a_1 = 3$ and $d = -3$.
    - (a) $3, 0, -3, -6, \ldots$
    - (b) $3, 6, 9, 12, \ldots$
    - (c) $3, -1, \frac{1}{3}, -\frac{1}{9}, \ldots$
    - (d) $3, -9, 27, -81, \ldots$
    - (e) None of these

    1—Answer: a

11. Find a_n for the arithmetic sequence with $a_1 = 5$, $d = -4$, and $n = 98$.
    - (a) -392
    - (b) -387
    - (c) -383
    - (d) 393
    - (e) None of these

    1—Answer: c

12. Find a_n for the arithmetic sequence with $a_1 = 12$, $d = \frac{1}{3}$, and $n = 52$.

    1—Answer: 29

13. Find the 99th term of the arithmetic sequence with $a_1 = 7$ and $d = -3$. (Assume that n begins with 1.)
    - (a) -287
    - (b) -290
    - (c) -293
    - (d) -297
    - (e) None of these

    1—Answer: a

14. Find the 30th term of the arithmetic sequence with $a_1 = -5$ and $d = \frac{1}{3}$. (Assume that n begins with 1.)

    1—Answer: $\frac{14}{3}$

15. Find the ninth term of the arithmetic sequence with $a_1 = 4$ and $d = 10$. (Assume that n begins with 1.)
    - (a) 94
    - (b) 84
    - (c) 46
    - (d) 49
    - (e) None of these

    1—Answer: b

16. Find the eighth term of the arithmetic sequence with $a_1 = 5$ and $d = 8$. (Assume that n begins with 1.)

    (a) 69   (b) 48   (c) 61   (d) 104   (e) None of these

    1—Answer: c

17. Find the seventeenth term of the arithmetic sequence with $a_1 = 2$ and $d = 7$. (Assume that n begins with 1.)

    1—Answer: 114

18. Find the sum of the first 50 terms of the arithmetic sequence:
    25, 35, 45, 55, 65, . . .

    (a) 27,000   (b) 13,750   (c) 12,875
    (d) 13,500   (e) None of these

    1—Answer: d

19. Find the sum of the first 50 positive integers that are multiples of 3.

    (a) 7500   (b) 3900   (c) 3825
    (d) 7650   (e) None of these

    1—Answer: c

20. Find the sum of the first 40 integers that are multiples of 4.

    1—Answer: 3280

21. Find the sum of the first 30 terms of the sequence:
    $\sqrt{2}, 2\sqrt{2}, 3\sqrt{2}, 4\sqrt{2}, 5\sqrt{2}, \ldots$

    2—Answer: $465\sqrt{2}$

22. Find the sum of the first n terms of the arithmetic sequence:
    $-4, 5, 14, 23, 32, \ldots$

    (a) $9n - 4$ (b) $\dfrac{n(9n - 17)}{2}$ (c) $\dfrac{1 - 9^n}{2}$
    (d) $\dfrac{n^2 - 9n}{2}$ (e) None of these

    1—Answer: b

23. Find the sum of the first 18 terms of the arithmetic sequence whose nth term is $a_n = 3n - 1$. (Assume that n begins with 1.)

    (a) 495   (b) 53   (c) 459   (d) 445   (e) None of these

    1—Answer: a

24. Find the sum of the first 19 terms of the arithmetic sequence whose nth term is $a_n = n + 1$. (Assume that n begins with 1.)

1—Answer: 209

25. Find a formula for a_n for the arithmetic sequence with $a_1 = 5$ and $d = -4$. (Assume that n begins with 1.)

(a) $a_n = -4n + 9$ (b) $a_n = -4_n + 5$ (c) $a_n = 5n - 4$

(d) $a_n = 9n - 4$ (e) None of these

1—Answer: a

26. Find a formula for a_n for the arithmetic sequence with $a_3 = 15$ and $d = -2$. (Assume that n begins with 1.)

(a) $a_n = -2n + 9$ (b) $a_n = -2n + 19$ (c) $a_n = -2n + 21$

(d) $a_n = -2n + 15$ (e) None of these

1—Answer: c

27. Find a formula for a_n for the arithmetic sequence with $a_2 = 12$ and $d = -3$. (Assume that n begins with 1.)

(a) $a_n = 12n - 3$ (b) $a_n = -3n + 15$ (c) $a_n = -3n + 12$

(d) $a_n = -3n + 18$ (e) None of these

1—Answer: d

28. Find a formula for a_n for the arithmetic sequence with $a_2 = 15$ and $d = \frac{3}{2}$. (Assume that n begins with 1.)

1—Answer: $a_n = -\frac{3}{2}n + 12$

29. Find the sum: $\sum_{n=1}^{500} (3n + 5)$

(a) 756,500   (b) 376,250   (c) 752,500

(d) 378,250   (e) None of these

1—Answer: d

30. Find the sum: $\sum_{i=1}^{7} 2(i + 1)$

1—Answer: 70

31. Determine the seating capacity of an auditorium with 25 rows of seats if there are 20 seats in the first row, 24 seats in the second row, 28 seats in the third row, and so on.

(a) 1200 (b) 1500 (c) 1700 (d) 1900 (e) None of these

2—**Answer:** c

32. Determine the seating capacity of an auditorium with 30 rows of seats if there are 25 seats in the first row, 28 seats in the second row, 31 seats in the third row, and so on.

(a) 1635 (b) 1792 (c) 2055 (d) 3125 (e) None of these

2—**Answer:** c

33. Determine the seating capacity of an auditorium with 28 rows of seats if there are 24 seats in the first row, 28 seats in the second row, 32 seats in the third row, and so on.

(a) 2128 (b) 2240 (c) 2296 (d) 2184 (e) None of these

2—**Answer:** d

34. A small business sells $8000 worth of products during its first year. The owner of the business has set a goal of increasing annual sales by $3500 each year for 11 years. Assuming that this goal is met, find the total sales during the first 12 years this business is in operation.

2—**Answer:** $327,000

35. A small business sells $6000 worth of products during its first year. The owner of the business has set a goal of increasing annual sales by $2500 each year for 7 years. Assuming that this goal is met, find the total sales during the first 8 years this business is in operation.

2—**Answer:** $118,000

8.3 Geometric Sequences

1. Which of the following is a geometric sequence?
   - (a) $1, -3, 5, -7, 9, \ldots$
   - (b) $6, 3, 0, -3, -6, \ldots$
   - (c) $2, 4, 8, 16, 32, \ldots$
   - (d) $-1, 0, -1, 0, -1, \ldots$
   - (e) None of these

   1—Answer: c

2. Determine whether the sequence $3, -2, \frac{4}{3}, -\frac{8}{9}, \frac{16}{27}, \ldots$ is geometric. If it is, find its common ratio.

   1—Answer: Yes, $r = -\dfrac{2}{3}$

3. Which of the following is a geometric sequence?
   - (a) $-2, 0, 2, 4, 6, \ldots$
   - (b) $2, 4, 8, 16, 32, \ldots$
   - (c) $2, 7, 3, 8, 4, \ldots$
   - (d) $2, \frac{1}{2}, \frac{1}{4}, \frac{1}{6}, \frac{1}{8}, \ldots$
   - (e) None of these

   1—Answer: b

4. Which of the following is a geometric sequence?
   - (a) $-1, -3, -5, -7, -9, \ldots$
   - (b) $2, 3, 5, 7, 11, \ldots$
   - (c) $1, 2, 4, 7, 11, 16, \ldots$
   - (d) $-2, 4, -8, 16, -32, \ldots$
   - (e) None of these

   1—Answer: d

5. Find the first five terms of the geometric sequence with $a_1 = 2$ and $r = \frac{2}{3}$.
   - (a) $2, \frac{4}{3}, \frac{8}{9}, \frac{16}{27}, \frac{32}{81}$
   - (b) $2, 3, \frac{9}{2}, \frac{27}{4}, \frac{81}{8}$
   - (c) $2, \frac{8}{3}, \frac{10}{3}, 4, \frac{14}{3}$
   - (d) $2, \frac{4}{3}, \frac{2}{3}, -\frac{2}{3}$
   - (e) None of these

   1—Answer: a

6. Find the first five terms of the geometric sequence with $a_1 = 3$ and $r = \frac{3}{2}$.
   - (a) $3, \frac{9}{2}, \frac{27}{4}, \frac{81}{8}, \frac{243}{16}$
   - (b) $3, 2, \frac{4}{3}, \frac{8}{9}, \frac{16}{27}$
   - (c) $3, \frac{9}{2}, 6, \frac{15}{2}, 9$
   - (d) $3, \frac{3}{2}, 0, -\frac{3}{2}, -3$
   - (e) None of these

   1—Answer: a

7. Write the first five terms of the geometric sequence with $a_1 = 3$ and $r = \frac{1}{2}$.
   - (a) $3, 3\frac{1}{2}, 4, 4\frac{1}{2}, 5$
   - (b) $3, 2\frac{1}{2}, 2, 1\frac{1}{2}, 1$
   - (c) $3, \frac{3}{2}, \frac{3}{4}, \frac{3}{8}, \frac{3}{16}$
   - (d) $3, \frac{3}{2}, \frac{3}{4}, \frac{3}{6}, \frac{3}{8}$
   - (e) None of these

   1—Answer: c

Chapter 8 Sequences and Probability

8. Write the first five terms of the geometric sequence with $a_1 = -3$ and $r = \frac{2}{3}$.

   (a) $-3, -2\frac{1}{3}, -1\frac{2}{3}, -1, -\frac{1}{3}$
   (b) $-3, -3\frac{2}{3}, -4\frac{1}{3}, -5, -5\frac{2}{3}$
   (c) $-3, -\frac{9}{2}, -\frac{27}{4}, -\frac{81}{8}, -\frac{243}{16}$
   (d) $-3, -2, -\frac{4}{3}, -\frac{8}{9}, -\frac{16}{27}$
   (e) None of these

   1—Answer: d

9. Find the 20th term of the geometric sequence with $a_1 = 5$ and $r = 1.1$.

   (a) 1.1665
   (b) 37.0012
   (c) 33.6375
   (d) 30.5795
   (e) None of these

   1—Answer: d

10. Find the 23rd term of the geometric sequence with $a_1 = -23$ and $r = \sqrt{2}$.

    (a) $-47104\sqrt{2}$
    (b) $-2048\sqrt{2}$
    (c) -2048
    (d) -47104
    (e) None of these

    1—Answer: d

11. Find the 28th term of the geometric sequence: 2, 2.4, 2.88, 3.456, 4.1472, . . .

    1—Answer: 274.7411

12. Find the 14th term of the geometric sequence with $a_1 = -11$ and $r = \sqrt{3}$.

    1—Answer: $-8019\sqrt{3}$

13. Find the sum: $\sum_{n=0}^{10} 2\left(\frac{3}{5}\right)^n$. Round you answer to four decimal places.

    (a) 4.9698
    (b) 5.0000
    (c) 4.9819
    (d) 55.0000
    (e) None of these

    1—Answer: c

14. Find the sum: $\sum_{j=0}^{40} 3(1.05)^j$. Round your answer to two decimal places.

    (a) 383.52
    (b) 362.40
    (c) 984.00
    (d) 22.18
    (e) None of these

    1—Answer: a

15. Find the sum: $\sum_{k=1}^{10} 4\left(\frac{3}{2}\right)^{k-1}$. Round your answer to three decimal places.

    1—Answer: 453.320

16. Find the sum: $\sum_{n=1}^{15} 3\left(\frac{5}{4}\right)^n$. Round your answer to two decimal places.

    (a) 329.06
    (b) 260.85
    (c) 322.61
    (d) 271.15
    (e) None of these

    1—Answer: a

17. Find the sum of the first 30 terms in the sequence. $2, \frac{5}{2}, \frac{25}{8}, \frac{125}{32}, \frac{625}{128}, \ldots$ Round your answer to two decimal places.

    (a) 791.25
    (b) 5161.88
    (c) 6454.35
    (d) 7116.42
    (e) None of these

    1—Answer: c

18. Find the sum of the first 30 terms in the sequence.
    $\sqrt{2}, 2\sqrt{2}, 3\sqrt{2}, 4\sqrt{2}, 5\sqrt{2}, \ldots$

    1—Answer: $465\sqrt{2}$

19. Find the sum of the first six terms of the geometric sequence with $a_1 = 2$ and $a_2 = -4$.

    (a) -42
    (b) $\frac{130}{3}$
    (c) 42
    (d) $-\frac{62}{3}$
    (e) None of these

    1—Answer: a

20. Find the sum of the first 10 terms of the geometric sequence with $a_1 = 3$ and $a_2 = \frac{3}{2}$. Round to three decimal places.

    (a) 5.994
    (b) 7.286
    (c) 6.984
    (d) 9.117
    (e) None of these

    2—Answer: a

21. Find a formula for the nth term of the geometric sequence with $a_1 = 2$ and $r = -\frac{1}{3}$. (Assume that n begins with 1.)

    (a) $a_n = \left(-\frac{2}{3}\right)^n$
    (b) $a_n = 2 - \frac{1}{3}n$
    (c) $a_n = 2\left(-\frac{1}{3}\right)^{n-1}$
    (d) $a_n = 2\left(-\frac{1}{3}\right)^n$
    (e) None of these

    1—Answer: c

22. Find a formula for the nth term of the geometric sequence with $a_1 = 4$ and $r = \frac{1}{3}$. (Assume that n begins with 1.)

   (a) $a_n = \left(\frac{1}{3}\right)^n$
   (b) $a_n = 4\left(\frac{1}{3}\right)^{n-1}$
   (c) $a_n = 4\left(\frac{1}{3}\right)^n$

   (d) $a_n = 4 + \left(\frac{1}{3}\right)^n$
   (e) None of these

   1—Answer: b

23. Find a formula for the nth term of the geometric series with $a_1 = \frac{1}{2}$ and $r = -\frac{1}{3}$. (Assume that n begins at 1.)

   (a) $a_n = \frac{1}{2}\left(-\frac{1}{3}\right)^{n-1}$
   (b) $a_n = \frac{1}{2}\left(-\frac{1}{3}\right)^n$
   (c) $a_n = \left(-\frac{1}{6}\right)^n$

   (d) $a_n = \frac{1}{2} - \frac{1}{3^n}$
   (e) None of these

   1—Answer: a

24. Find a formula for the nth term of the geometric series with $a_1 = \frac{2}{3}$ and $r = -\frac{1}{5}$. (Assume that n begins at 1.)

   (a) $a_n = \frac{2}{3}\left(-\frac{1}{5}\right)^n$
   (b) $a_n = \left(-\frac{2}{15}\right)^n$
   (c) $a_n = -\frac{1}{5}\left(\frac{2}{3}\right)^n$

   (d) $a_n = \frac{2}{3}\left(-\frac{1}{5}\right)^{n-1}$
   (e) None of these

   1—Answer: d

25. Find the common ratio of the geometric series: $-4, 3, -\frac{9}{4}, \frac{27}{16}, -\frac{81}{16}, \ldots$

   (a) $\frac{3}{4}$
   (b) $-\frac{3}{4}$
   (c) $\frac{4}{3}$
   (d) $-\frac{4}{3}$
   (e) None of these

   1—Answer: b

26. Find the sum of the infinite geometric sequence: $-7, -\frac{7}{3}, -\frac{7}{9}, -\frac{7}{27}, \ldots$

   (a) -5
   (b) $-\frac{21}{4}$
   (c) $-\frac{5}{2}$
   (d) $-\frac{21}{2}$
   (e) None of these

   1—Answer: d

27. Find the sum of the infinite geometric sequence: $1, 0.9, 0.81, 0.729, \ldots$

   (a) 23
   (b) 90
   (c) 10
   (d) 57
   (e) None of these

   1—Answer: c

28. Find the sum of the infinite geometric sequence: $1, \frac{1}{3}, \frac{1}{9}, \frac{1}{27}, \ldots$

   (a) $\frac{3}{2}$
   (b) 3
   (c) $\frac{5}{3}$
   (d) $\frac{5}{2}$
   (e) None of these

   1—Answer: a

29. Find the sum of the infinite geometric sequence with $a_1 = 9$ and $r = 0.7$.

   1—Answer: 30

30. Evaluate: $\sum_{n=0}^{\infty} 2\left(\dfrac{1}{2}\right)^n = 2 + 1 + \dfrac{1}{2} + \dfrac{1}{4} + \dfrac{1}{8} + \ldots$

   (a) 4   (b) 6   (c) 8   (d) 10   (e) None of these

   1—Answer: a

31. Evaluate: $\sum_{n=0}^{\infty} 4\left(\dfrac{2}{3}\right)^n = 4 + \dfrac{8}{3} + \dfrac{16}{9} + \dfrac{32}{27} + \ldots$

   (a) 8   (b) 10   (c) 12   (d) 14   (e) None of these

   1—Answer: c

32. Evaluate: $\sum_{n=0}^{\infty} 3\left(-\dfrac{1}{2}\right)^n$

   (a) 6   (b) 4   (c) 2   (d) 0   (e) None of these

   1—Answer: c

33. Evaluate: $\sum_{n=0}^{\infty} 5\left(-\dfrac{2}{3}\right)^n$

   (a) 6   (b) 3   (c) 1   (d) 0   (e) None of these

   1—M—Answer: b

34. An individual buys a $100,000 term life insurance policy. During the next five years the value of the policy will depreciate at the rate of 4% per year. (That is, at the end of year the depreciated value is 96% of the value at the beginning of the year.) Find the depreciated value of the policy at the end of five years.

   (a) $80,000   (b) $84,934.66   (c) $81,537.27   (d) $78,275.78   (e) None of these

   2—Answer: c

35. Suppose in 1985 you accepted a job at $17,000. If you receive a 5% raise in salary each year, what will be your salary in 1995?

   (a) $25,116.74   (b) $26,372.58   (c) $27,691.21   (d) $29,075.77   (e) None of these

   2—Answer: c

36. A city of 500,000 people is growing at a rate of 1% per year. Find a formula for the nth term of the geometric sequence that gives the population n years from now. Then estimate the population 20 years from now.

   2—Answer: $a_n = 500{,}000(1.01)^n$; 610,095

37. Suppose that you accept a job that pays a salary of $35,000 the first year. During the next 39 years, suppose you receive a 5% raise each year. Write a summation formula to represent your total salary after working n years. What would your total salary be over the 40-year period?

   2—**Answer:** $\sum_{i=1}^{n} 35{,}000(1.05)^{i-1}$; $S_{40} = \$4{,}227{,}992.10$

38. Suppose that you accept a job that pays a salary of $35,000 the first year. During the next 39 years, suppose you receive a 6% raise each year. Write a summation formula to represent your total salary after working n years. What would your total salary be over the 40-year period?

   2—**Answer:** $\sum_{i=1}^{n} 35{,}000(1.06)^{i-1}$; $S_{40} = \$5{,}416{,}668.80$

❏ 8.4 Mathematical Induction

1. Find the sum using the formulas for the sums of powers of integers: $\sum_{n=1}^{7} n^5$

   (a) 4219   (b) 4676   (c) 29008

   (d) 61776   (e) None of these

   1—**Answer:** c

2. Find the sum using the formulas for the sums of powers of integers: $\sum_{n=1}^{8} (n^2 - n^3)$

   (a) −994   (b) −1092   (c) −1296

   (d) −1538   (e) None of these

   1—**Answer:** b

3. Find the sum using the formulas for the sums of powers of integers: $\sum_{n=1}^{19} i^3$

   (a) 36,100   (b) 3581   (c) 44,100

   (d) 1,687,998   (e) None of these

   1—**Answer:** a

4. Find the sum using the formulas for the sums of powers of integers: $\sum_{n=1}^{15} i^4$

   (a) 2,299,200   (b) 1,337,340   (c) 445,780

   (d) 178,312   (e) None of these

   1—**Answer:** d

5. Find the sum using the formulas for the sums of powers of integers: $\sum_{n=1}^{50} (n^2 - n)$
   (a) 42,925
   (b) 41,650
   (c) 44,100
   (d) 43,150
   (e) None of these

   1—Answer: b

6. Find the sum using the formulas for the sums of powers of integers: $\sum_{n=1}^{15} 4n^2$
   (a) 4960
   (b) 1240
   (c) 73,810
   (d) 74,400
   (e) None of these

   1—Answer: a

7. Find the sum using the formulas for the sums of powers of integers: $\sum_{n=1}^{20} 3n^2$

   1—Answer: 8610

8. Find the sum using the formulas for the sums of powers of integers: $\sum_{n=1}^{19} 2n^2$

   1—Answer: 4940

9. Find the sum using the formulas for the sums of powers of integers: $\sum_{i=1}^{20} i^4$
   (a) 718,312
   (b) 4,933,320
   (c) 44,100
   (d) 722,666
   (e) None of these

   1—Answer: d

10. Identify S_{k+1} given $S_k = \dfrac{2k-1}{3k(k+1)}$.
    (a) $\dfrac{2k+1}{(3k+1)(k+1)}$
    (b) $\dfrac{2k+1}{3(k+1)(k+2)}$
    (c) $\dfrac{2k}{3(k+1)(k+2)}$
    (d) $\dfrac{2k}{(3k+1)(k+2)}$
    (e) None of these

    1—Answer: b

11. Identify S_{k+1} given $S_k = \dfrac{3}{k(k+2)}$.
    (a) $\dfrac{3}{(k+1)(k+3)}$
    (b) $\dfrac{k^2+2k+3}{k+2}$
    (c) $\dfrac{3}{(k+1)(k+2)}$
    (d) $\dfrac{k^2+5}{k(k+2)}$
    (e) None of these

    1—Answer: a

12. Identify S_{k+1} given $S_k = \dfrac{(k+1)^2}{k(k-1)}$.

    (a) $\dfrac{2k^2 + 2k}{k(k+1)}$ (b) $\dfrac{(k+2)^2}{k(k-1)}$ (c) $\dfrac{(k+1)^2 + k(k-1)}{k(k-1)}$

    (d) $\dfrac{(k+2)^2}{k(k+1)}$ (e) None of these

    1—Answer: d

13. Identify S_{k+1} given $S_k = \dfrac{k(2k-1)}{3}$.

    (a) $\dfrac{2k(k+1)}{3}$ (b) $\dfrac{2k^2 - k + 3}{3}$ (c) $\dfrac{(k+1)(2k+1)}{3}$

    (d) $\dfrac{2k^2 - k + 1}{3}$ (e) None of these

    1—Answer: c

14. Identify S_{k+1} given $S_k = k(3k - 1)$.

    (a) $(k+1)(3k+2)$ (b) $3k(k+1)$ (c) $k(3k-1) + 1$

    (d) $3k^2 + 1$ (e) None of these

    1—Answer: a

15. Identify S_{k+1} given $S_k = k^2(k+1)^2$.

    (a) $(k^2 + 1)(k+2)^2$ (b) $k^2(k+1)^2 + 1$ (c) $(k+1)^2(k-1)^2$

    (d) $(k+1)^2(k+2)^2$ (e) None of these

    1—Answer: d

16. Identify S_{k+1} given $S_k = \dfrac{k}{2}(3k + 2)$.

    (a) $\dfrac{3(k+1)^2}{2}$ (b) $\dfrac{k+1}{2}(3k+2)$ (c) $\dfrac{3k^2 + 2k + 2}{2}$

    (d) $\dfrac{k+1}{2}(3k+5)$ (e) None of these

    1—Answer: d

17. Identify S_{k+1} given $S_k = (k+1)(k-1)$.

    (a) $k(k+2)$ (b) k^2 (c) $k^2 - 2$

    (d) $(k+2)(k+1)$ (e) None of these

    1—Answer: a

18. Identify S_{k+1} given $S_k = \dfrac{k(k+1)}{2}$.

(a) $\dfrac{k^2 + k + 2}{2}$
(b) $\dfrac{k^2 + 2}{2}$
(c) $\dfrac{(k+1)(k+2)}{2}$
(d) $\dfrac{(k+1)(k+2)}{2k+1}$
(e) None of these

1—Answer: c

19. Prove by mathematical induction: $1 + 2 + 2^2 + 2^3 + \cdots + 2^{n-1} = 2^n - 1$

2—Answer: S_1: $2^1 - 1 = 2 - 1 = 1$

S_k: $1 + 2 + 2^2 + 2^3 + \cdots + 2^{k-1} = 2^k - 1$

S_{k+1}: $1 + 2 + 2^2 + 2^3 + \cdots + 2^k = 2^{k+1} - 1$

Assuming S_k, we have:

$$(1 + 2 + 2^2 + 2^3 + \cdots + 2^{k-1}) + 2^{(k+1)-1} = (1 + 2 + \cdots + 2^{k-1}) + 2^k$$
$$= (2^k - 1) + 2^k$$
$$= 2^{k+1} - 1$$

Hence, the formula is valid for all $n \geq 1$.

20. Prove by mathematical induction:
$1^2 + 2^2 + 3^2 + \cdots + n^2 = \dfrac{n(n+1)(2n+1)}{6}$

2—Answer: Let $n = 1$, then $\dfrac{n(n+1)(2n+1)}{6} = \dfrac{(1)(2)(3)}{6} = 1 = 1^2$.

For $n = k$, $S_k = 1^2 + 2^2 + 3^2 + \cdots + k^2 = \dfrac{k(k+1)(2k+1)}{6}$.

For $n = k + 1$, $S_{k+1} = \dfrac{k(k+1)(2k+1)}{6} + (k+1)^2$

$$= (k+1)\left[\dfrac{k(2k+1) + 6(k+1)}{6}\right]$$

$$= \dfrac{(k+1)(2k^2 + 7k + 6)}{6}$$

$$= \dfrac{(k+1)(k+2)(2k+3)}{6}$$

$$= \dfrac{(k+1)[(k+1)+1][2(k+1)+1]}{6}.$$

Thus, the formula is valid.

21. Use mathematical induction to prove

$$\frac{1}{1 \cdot 3} + \frac{1}{3 \cdot 5} + \frac{1}{5 \cdot 7} + \cdots + \frac{1}{(2n-1)(2n+1)} = \frac{n}{2n+1}.$$

2—Answer:

When $n = 1$, the formula is valid since $S_1 = \frac{1}{1 \cdot 3} = \frac{1}{3}$.

For $n = k$, assume S_k is true. $S_k = \frac{k}{2k+1}$

For $n = k + 1$, show $S_{k+1} = \frac{k+1}{2k+3}$.

$$S_{k+1} = \left[\frac{1}{1 \cdot 3} + \frac{1}{3 \cdot 5} + \frac{1}{5 \cdot 7} + \cdots + \frac{1}{(2k-1)(2k+1)}\right] + \frac{1}{(2k+1)(2k+3)}$$

$$= S_k + \frac{1}{(2k+1)(2k+3)}$$

$$= \frac{k}{2k+1} + \frac{1}{(2k+1)(2k+3)}$$

$$= \frac{k(2k+3) + 1}{(2k+1)(2k+3)}$$

$$= \frac{2k^2 + 3k + 1}{(2k+1)(2k+3)}$$

$$= \frac{(2k+1)(k+1)}{(2k+1)(2k+3)}$$

$$= \frac{k+1}{2k+3}$$

Thus, the formula is valid.

22. Use mathematical induction to prove
$$\frac{1}{1 \cdot 2} + \frac{1}{2 \cdot 3} + \frac{1}{3 \cdot 4} + \cdots + \frac{1}{n(n+1)} = \frac{n}{n+1}.$$

2—Answer:

Let $n = 1$, $S_1 = \frac{1}{1 \cdot 2} = \frac{1}{2}$.

For $n = k$, assume $S_k = \frac{k}{k+1}$ is true.

For $n = k + 1$ show $S_{k+1} = \frac{k+1}{k+2}$.

$$S_{k+1} = \left[\frac{1}{1 \cdot 2} + \frac{1}{2 \cdot 3} + \frac{1}{3 \cdot 4} + \cdots + \frac{1}{k(k+1)} \right] + \frac{1}{(k+1)(k+2)}$$

$$= S_k + \frac{1}{(k+1)(k+2)}$$

$$= \frac{k}{k+1} + \frac{1}{(k+1)(k+2)}$$

$$= \frac{k^2 + 2k + 1}{(k+1)(k+2)}$$

$$= \frac{(k+1)^2}{(k+1)(k+2)}$$

$$= \frac{k+1}{k+2}$$

Thus, the formula is valid.

23. Use mathematical induction to prove
$$6 + 11 + 16 + 21 + \cdots + (5n + 1) = \frac{n(7 + 5n)}{2}.$$

2—Answer: For $n = 1$, $S_1 = \frac{1(7 + 5)}{2} = 6$.

For $n = k$, assume $S_k = \frac{k(7 + 5k)}{2}$ is true.

For $n = k + 1$ show $S_{k+1} = \frac{(k+1)(5k+12)}{2}$.

$$S_{k+1} = [6 + 11 + 16 + \cdots + (5k + 1)] + [5(k+1) + 1]$$

$$= S_k + (5k + 6)$$

$$= \frac{k(7 + 5k)}{2} + (5k + 6)$$

$$= \frac{7k + 5k^2}{2} + \frac{10k + 12}{2}$$

$$= \frac{5k^2 + 17k + 12}{2}$$

$$= \frac{(k+1)(5k+12)}{2}$$

Thus, the formula is valid.

24. Use mathematical induction to prove

$$-3 + 1 + 5 + 9 + 13 + \ldots + (4n - 7) = n(2n - 5).$$

2—Answer: For $n = 1$, $S_1 = -3 = 1(2 - 5) = -3$.

For $n = k$, assume $S_k = k(2k - 5)$.

For $n = k + 1$ show $S_{k+1} = (k + 1)(2k - 3)$.

$$\begin{aligned}
S_{k+1} &= -3 + 1 + 5 + 9 + 13 + \cdots + (4k - 7) + (4k - 3) \\
&= S_k + (4k - 3) \\
&= k(2k - 5) + (4k - 3) \\
&= 2k^2 - k - 3 \\
&= (k + 1)(2k - 3)
\end{aligned}$$

Thus, the formula is valid.

25. Use mathematical induction to prove $n < 3^n$ for all positive integers n.

2—Answer: For $n = 1$, $1 < 3$.

For $n = k$, assume $k < 3^k$.

For $n = k + 1$, show $k + 1 < 3^{k+1}$.

$$k < 3^k$$

$$k + 1 < 3^k + 1 < 3^k + 3^k$$

$$k + 1 < 2(3^k) < 3(3^k)$$

$$k + 1 < 3^{k+1}$$

Hence, $n < 3^n$ for $n \geq 1$.

26. Use mathematical induction to prove
$$\frac{1}{2} + \frac{1}{4} + \frac{1}{8} + \cdots + \frac{1}{2^n} < 1 \text{ for } n \geq 1.$$

2—Answer: For $n = 1$, $\frac{1}{2} < 1$.

For $n = k$, assume $\frac{1}{2} + \frac{1}{4} + \frac{1}{8} + \cdots + \frac{1}{2^k} < 1$

For $n = k + 1$ show $\frac{1}{2} + \frac{1}{4} + \frac{1}{8} + \cdots + \frac{1}{2^k} + \frac{1}{2^{k+1}} < 1$ is true.

$$\frac{1}{2} + \frac{1}{4} + \frac{1}{8} + \cdots + \frac{1}{2^k} < 1$$
$$\frac{1}{2}\left(\frac{1}{2} + \frac{1}{4} + \frac{1}{8} + \cdots + \frac{1}{2^k}\right) < \frac{1}{2} \cdot 1$$
$$\frac{1}{4} + \frac{1}{8} + \frac{1}{16} + \cdots + \frac{1}{2^{k+1}} < \frac{1}{2}$$
$$\frac{1}{2} + \frac{1}{4} + \frac{1}{8} + \cdots + \frac{1}{2^{k+1}} < \frac{1}{2} + \frac{1}{2}$$
$$\frac{1}{2} + \frac{1}{2} + \frac{1}{8} + \cdots + \frac{1}{2^{k+1}} < 1$$

Hence, $\frac{1}{2} + \frac{1}{4} + \ldots + \frac{1}{2^n} < 1$ for $n \geq 1$.

27. Use mathematical induction to prove
$$\frac{1}{3} + \frac{1}{9} + \frac{1}{27} + \cdots + \frac{1}{3^n} < 1 \text{ for } n \geq 1.$$

2—Answer: For $n = 1$, $\frac{1}{3} < 1$.

For $n = k$, assume $\frac{1}{3} + \frac{1}{9} + \frac{1}{27} + \cdots + \frac{1}{3^k} < 1$ is true.

For $n = k + 1$ show $\frac{1}{3} + \frac{1}{9} + \frac{1}{27} + \cdots + \frac{1}{3^k} + \frac{1}{3^{k+1}} < 1$.

$$\frac{1}{3} + \frac{1}{9} + \frac{1}{27} + \cdots + \frac{1}{3^k} < 1$$
$$\frac{1}{3}\left[\frac{1}{3} + \frac{1}{9} + \frac{1}{27} + \cdots + \frac{1}{3^k}\right] < \frac{1}{3} \cdot 1$$
$$\frac{1}{9} + \frac{1}{27} + \frac{1}{81} + \cdots + \frac{1}{3^{k+1}} < \frac{1}{3}$$
$$\frac{1}{3} + \frac{1}{9} + \frac{1}{27} + \frac{1}{81} + \cdots + \frac{1}{3^{k+1}} < \frac{2}{3} < 1$$

Hence, $\frac{1}{3} + \frac{1}{9} + \frac{1}{27} + \cdots + \frac{1}{3^n} < 1$ for $n \geq 1$.

28. Find a quadratic model for the sequence with the indicated terms.

$a_0 = 1, a_1 = 5, a_2 = 13$

1—Answer: $a_n = 2n^2 + 2n + 1$

29. Find a quadratic model for the sequence with the indicated terms.

$a_0 = -1, a_1 = \frac{3}{2}, a_2 = 8$

1—Answer: $a_n = 2n^2 + \frac{1}{2}n - 1$

30. Find a quadratic model for the sequence with the indicated terms.

$a_0 = -2, a_1 = 5, a_2 = 22$

1—Answer: $a_n = 5n^2 + 2n - 2$

31. Determine whether the sequence has a linear model, quadratic model or neither.

$a_1 = 2, a_n = a_{n-1} + n$

(a) Linear   (b) Quadratic   (c) Neither

1—Answer: b

32. Determine whether the sequence has a linear model, quadratic model or neither.

$a_1 = 4, a_n = 2a_{n-1} + 3$

(a) Linear   (b) Quadratic   (c) Neither

1—Answer: a

33. Determine whether the sequence has a linear model, quadratic model or neither.

$a_1 = 2, a_n = (a_{n-1})^3$

(a) Linear   (b) Quadratic   (c) Neither

1—Answer: c

34. Determine whether the sequence has a linear model, quadratic model or neither.

$a_1 = 1, a_n = 2a_{n-1} + n$

(a) Linear   (b) Quadratic   (c) Neither

1—Answer: c

35. Determine whether the sequence has a linear model, quadratic model or neither.

$a_1 = 4, a_n = a_{n-1} - n$

(a) Linear   (b) Quadratic   (c) Neither

1—Answer: b

8.5 The Binomial Theorem

1. Evaluate: $_{12}C_{10}$

    (a) $\frac{1}{66}$ (b) 66 (c) 132

    (d) $\frac{1}{120}$ (e) None of these

    1—Answer: b

2. Evaluate: $_{10}C_3$

    (a) 1000 (b) 604,800 (c) 720

    (d) 120 (e) None of these

    1—Answer: d

3. Evaluate: $_{17}C_{14}$

    (a) 5.9×10^{13} (b) 842,771 (c) 4080

    (d) 680 (e) None of these

    1—Answer: d

4. Evaluate: $_{12}C_9$

    1—Answer: 220

5. Evaluate: $_{45}C_2$

    1—Answer: 990

6. Evaluate: $_6C_2$

    (a) 15 (b) 30 (c) 360

    (d) 12 (e) None of these

    1—Answer: a

7. Evaluate: $_9C_7$

    (a) 181,440 (b) 63 (c) 72

    (d) 36 (e) None of these

    1—Answer: d

8. Evaluate: $_8C_5$

    (a) 40 (b) 336 (c) 56

    (d) 6720 (e) None of these

    1—Answer: c

9. Evaluate: $_9C_5$

    (a) 15,120   (b) 126   (c) 3024

    (d) 45   (e) None of these

    1—**Answer:** b

10. Evaluate: $_6C_4$

    (a) 24   (b) 15   (c) 30

    (d) 360   (e) None of these

    1—**Answer:** b

11. Evaluate: $_7C_4$

    (a) 35   (b) 28   (c) 210

    (d) 840   (e) None of these

    1—**Answer:** a

12. Use the Binomial Theorem to expand then simplify: $(x - 3)^5$

    (a) $x^5 - 15x^4 + 30x^3 - 30x^2 + 15x - 243$ (b) $x^5 - 15x^4 + 900x^3 - 27{,}000x^2 + 50{,}625x - 243$

    (c) $x^5 - 15x^4 + 90x^3 - 270x^2 + 405x - 243$ (d) $x^5 - 3x^4 + 9x^3 - 27x^2 + 81x - 243$

    (e) None of these

    1—**Answer:** c

13. Use the Binomial Theorem to expand, then simplify: $(2x - 3)^3$

    (a) $8x^3 - 324x^2 + 324x - 27$ (b) $8x^3 - 36x^2 + 54x - 27$

    (c) $2x^3 - 18x^2 + 54x - 27$ (d) $8x^3 - 12x^2 + 27x - 27$

    (e) None of these

    1—**Answer:** b

14. Expand: $(3 - 2x)^3$

    (a) $27 - 3x + 3x^2 - 8x^3$ (b) $27 - 9x + 9x^2 - 8x^3$ (c) $27 - 27x + 6x^2 - 8x^3$

    (d) $27 - 54x + 36x^2 - 8x^3$ (e) None of these

    1—**Answer:** d

15. Use Pascal's Triangle to evaluate the complex number $(2 - i)^4$.

    (a) 17   (b) $-7 - 24i$ (c) $13 + 6i$

    (d) 15   (e) None of these

    1—**Answer:** b

16. Use the Binomial Theorem to expand, then simplify: $(i - 1)^4$

(a) $4 - 8i$ (b) $-4 - 8i$ (c) -4

(d) 6 (e) None of these

1—Answer: c

17. Use the Binomial Theorem to expand, then simplify: $(i - 1)^4$

(a) $-24 + 9i$ (b) $7 + 16i$ (c) $31 - 40i$

(d) $-7 - 24i$ (e) None of these

1—Answer: d

18. Use Pascal's Triangle to expand $(x - 2y)^4$.

1—Answer: $x^4 - 8x^3y + 24x^2y^2 - 32xy^3 + 16y^4$

19. Use Pascal's Triangle to expand $(2x - y)^3$.

(a) $2x^3 - 6x^2y + 6xy^2 - y^3$ (b) $8x^3 - 4x^2y + 2xy^2 - y^3$

(c) $2x^3 + 3x^2y + 3xy^2 + y^3$ (d) $8x^3 - 12x^2y + 6xy^2 - y^3$

(e) None of these

1—Answer: d

20. Use Pascal's Triangle to expand $(x + 3y)^3$.

(a) $x^3 + 6x^2y + 6xy^2 + 9y^3$ (b) $x^3 - 3x^2y + 6xy^2 - 27y^3$

(c) $x^3 + 9x^2y + 27xy^2 + 27y^3$ (d) $x^3 + 9x^2y + 9xy^2 + 3y^3$

(e) None of these

1—Answer: c

21. Use Pascal's Triangle to expand $(3x + y)^4$.

(a) $81x^4 + 108x^3y + 54x^2y^2 + 12xy^3 + y^4$ (b) $3x^4 + 12x^3y + 18x^2y^2 + 12xy^3 + y^4$

(c) $81x^4 + 1728x^3y + 324x^2y^2 + 12xy^3 + y^4$ (d) $3x^4 - 12x^3y + 324x^2y^2 + 12xy^3 - y^4$

(e) None of these

1—Answer: a

22. Use Pascal's Triangle to expand $(5x + 2y)^3$.

(a) $125x^3 + 450x^2y + 60xy^2 + 8y^3$ (b) $125x^3 + 150x^2y + 60xy^2 + 8y^3$

(c) $125x^3 + 50x^2y + 20xy^2 + 8y^3$ (d) $5x^3 + 60x^2y + 30xy^2 + 2y^2$

(e) None of these

1—Answer: b

23. Use Pascal's Triangle to expand $(2x + y)^5$.

    (a) $32x^5 + 16x^4y + 8x^3y^2 + 4x^2y^3 + 2xy^4 + y^5$ (b) $32x^5 + 80x^4y + 80x^3y^2 + 40x^2y^3 + 10xy^4 + y^5$

    (c) $2x^5 + 10x^4y + 20x^3y^2 + 20x^2y^3 + 10xy^4 + y^5$

    (d) $32x^5 + 10,000x^4y + 8,000x^3y^2 + 400x^2y^3 + 10xy^4 + y^5$

    (e) None of these

    1—Answer: b

24. Use Pascal's Triangle to expand: $(2a - b)^3$.

    (a) $8a^3 - 4a^2b + 2ab^2 - b^3$ (b) $8a^3 + 12a^2b + 6ab^2 + b^3$

    (c) $8a^3 - 12a^2b + 6ab^2 - b^3$ (d) $8a^3 - b^3$

    (e) None of these

    1—Answer: c

25. Use Pascal's Triangle to expand: $(3x + y)^3$.

    (a) $27x^3 + 27x^2y + 9xy^2 + y^3$ (b) $27x^3 + 27x^2y + 9xy^2 - y^3$ (c) $27x^3 + 9x^2y + 3xy^2 + y^3$

    (d) $27x^3 + y^3$ (e) None of these

    1—Answer: a

26. Find the coefficient of x^4y^3 in the expansion of $(x + 2y)^7$.

    (a) 35   (b) 8   (c) 1,680   (d) 280   (e) None of these

    1—Answer: d

27. Find the coefficient of x^4y^3 in the expansion of $(2x + y)^7$.

    1—Answer: 560

28. Determine the coefficient of x^5y^7 in the expansion of $(5x + 2y)^{12}$.

    (a) 316,800,000   (b) 400,000   (c) 792

    (d) 7920   (e) None of these

    1—Answer: a

29. Determine the coefficient of x^3y^5 in the expansion of $(3x + 2y)^8$.

    (a) 336   (b) 48,384   (c) 864

    (d) 52,488   (e) None of these

    1—Answer: b

30. Determine the coefficient of x^2y^7 in the expansion of $(3x - 2y)^9$.

(a) −1152 (b) 1152 (c) 41,472

(d) −41,472 (e) None of these

1—Answer: d

31. Determine the coefficient of the x^2y^7 in the expansion of $(7x - 2y)^9$.

1—Answer: −225,792

32. Find the 4th term in the expansion: $\left(\dfrac{1}{4} + \dfrac{3}{4}\right)^5$

1—Answer: $\dfrac{270}{4^5}$

33. Find the 5th term in the expansion: $\left(\dfrac{1}{3} + \dfrac{2}{3}\right)^5$

(a) $\dfrac{120}{243}$ (b) $\dfrac{16}{243}$ (c) $\dfrac{80}{243}$

(d) $\dfrac{32}{243}$ (e) None of these

1—Answer: a

34. Find the 4th term in the expansion: $(0.2 + 0.8)^5$

(a) 0.0064 (b) 0.4096 (c) 0.0512

(d) 0.2048 (e) None of these

1—Answer: d

35. Find the sum of the first 3 terms in the expansion of $(1 + 0.03)^7$.

(a) 1.2289 (b) 1.2299 (c) 1.1935

(d) 1.2415 (e) None of these

1—Answer: a

36. Find the sum of the first 3 terms in the expansion of $(1 + 0.02)^7$.

(a) 1.1260 (b) 1.1487 (c) 1.1484

(d) 1.1540 (e) None of these

1—Answer: c

37. Find the sum of the first 3 terms in the expansion of $(3 + 0.02)^7$.

(a) 2291.1240 (b) 2291.1012 (c) 2275.9380

(d) 2291.1141 (e) None of these

1—Answer: b

38. Use a graphing utility to graph f and g on the same viewing rectangle. What is the relationship between the two graphs? Use the Binomial Theorem to write g in standard form.

$$f(x) = x^3 - 2x \qquad g(x) = f(x + 3)$$

2—T—Answer:

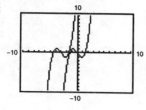

g is a horizontal translation of f 2 units to the left;

$$g(x) = x^3 + 9x^2 + 27x + 27 - 2x - 6$$
$$= x^3 + 9x^2 + 25x + 21$$

39. Use a graphing utility to graph f and g on the same viewing rectangle. What is the relationship between the two graphs? Use the Binomial Theorem to write g in standard form.

$$f(x) = x^3 - 2x^2 \qquad g(x) = f(x - 1)$$

2—T—Answer:

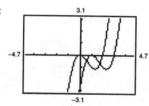

g is a horizontal translation of f 1 unit to the right;

$$g(x) = x^3 - 3x^2 + 3x - 1 - 2(x^2 - 2x + 1)$$
$$= x^3 - 5x^2 + 7x - 3$$

40. Use the Binomial Theorem to expand and simplify: $\left(\sqrt{x} + 2\right)^3$

1—Answer: $x^{3/2} + 6x + 12x^{1/2} + 8 = x\sqrt{x} + 6x + 12\sqrt{x} + 8$

41. Use the Binomial Theorem to expand and simplify: $\left(\sqrt[3]{x} - 2\right)^3$

1—Answer: $x - 6x^{2/3} + 12x^{1/3} - 8 = x - 6\sqrt[3]{x^2} + 12\sqrt[3]{x} - 8$

42. Use the Binomial Theorem to expand and simplify: $(2\sqrt{y} - 3)^4$

   1—Answer: $16y^2 + 96y^{3/2} + 196y + 216y^{1/2} + 81$
   $= 16y^2 + 96y\sqrt{y} + 196y + 216\sqrt{y} + 81$

43. Use the Binomial Theorem to expand and simplify: $(2\sqrt{t} - \sqrt{x})^3$

   1—Answer: $8t^{3/2} - 12tx^{1/2} + 6t^{1/2} - x^{3/2} = 8t\sqrt{t} - 12t\sqrt{x} + 6x\sqrt{t} - x\sqrt{x}$

❑ 8.6 Counting Principles

1. Evaluate: $_{10}P_6$
   - (a) 5040
   - (b) 151,200
   - (c) 210
   - (d) 60
   - (e) None of these

   1—Answer: b

2. Evaluate: $_{14}P_4$
   - (a) 24,024
   - (b) 8008
   - (c) 5040
   - (d) 720
   - (e) None of these

   1—Answer: a

3. Evaluate: $_7P_4$
   - (a) 840
   - (b) 35
   - (c) 10,920
   - (d) 210
   - (e) None of these

   1—Answer: a

4. Evaluate: $_{20}P_3$
   - (a) 1140
   - (b) 116,280
   - (c) 6840
   - (d) 4.05×10^{17}
   - (e) None of these

   1—Answer: c

5. Evaluate: $_7P_2$
   - (a) 2520
   - (b) 210
   - (c) 21
   - (d) 42
   - (e) None of these

   1—Answer: d

6. Evaluate: $_{14}P_3$

   (a) 1.45×10^{10} (b) 24,024 (c) 2184
   (d) 364 (e) None of these

   1—Answer: c

7. Find the number of distinguishable permutations using the letters LETTERFILE.

   (a) 3,628,800 (b) 1024 (c) 151,200
   (d) 5040 (e) None of these

   1—Answer: c

8. In how many distinguishable ways can the letters MISSISSIPPI be arranged?

   1—Answer: 34,650

9. Find the number of distinguishable ways the letters OKEECHOBEE can be arranged.

   (a) 75,600 (b) 3,628,800 (c) 151,200
   (d) 1,814,400 (e) None of these

   1—Answer: a

10. Find the number of distinguishable permutations using the letters in the word ARKANSAS.

    (a) 40,320 (b) 13,440 (c) 6720 (d) 3360 (e) None of these

    1—Answer: d

11. Find the number of distinguishable permutations using the letters in the word CALCULATOR.

    (a) 3,628,800 (b) 43,360 (c) 435,600 (d) 907,200 (e) None of these

    1—Answer: c

12. Find the number of distinguishable permutations using the letters in the word MATHEMATICS.

    1—Answer: 4,989,600

13. Find the number of distinguishable permutations with the following letters:

    {A, A, A, B, B, C, C, C, C,}

    (a) 2520 (b) 1260 (c) 362,880 (d) 288 (e) None of these

    1—Answer: b

14. How many different ways (subject orders) can three algebra books, two trigonometry books and two arithmetic books be arranged on a shelf?

    (a) 5040   (b) 210   (c) 128   (d) 823,543   (e) None of these

1—Answer: b

15. How many different ways can three chocolate, four strawberry, and two butterscotch sundaes be served to nine people?

1—Answer: 1260 ways

16. An organization consisting of 54 members is going to elect four officers. No person may hold more than one office. How many different outcomes are possible?

    (a) 354,294   (b) 8,503,056   (c) 316,251

    (d) 7,590,024   (e) None of these

2—Answer: d

17. A class of nine students line up single file for lunch. How many different ways can this occur if the six boys in the class must line up first?

    (a) 18   (b) 60,480   (c) 4320   (d) 504   (e) None of these

1—Answer: c

18. How many ways can a ten-question multiple choice test be answered if each question has five possible answers?

    (a) 50   (b) 120   (c) 3,628,800   (d) 9,765,625   (e) None of these

1—Answer: d

19. An organization consisting of 36 members is going to elect three officers. No person may hold more than one office. How many different outcomes are possible?

    (a) 7140   (b) 42,840   (c) 6.2×10^{40}

    (d) 3.7×10^{41} (e) None of these

1—Answer: b

20. How many ways can an eight-question multiple choice test be answered if each question has five possible answers?

    (a) 390,625   (b) 4,838,400   (c) 40,320   (d) 120   (e) None of these

1—Answer: a

21. There are 20 girls in a beauty pageant. A queen, a first runner-up and a second runner-up are to be chosen. How many different outcomes are possible?

    (a) 1140   (b) 6840   (c) 2.4×10^{18}

    (d) 4.1×10^{17} (e) None of these

    1—Answer: b

22. A group of six students are seated in a single row at a football game. In how many different orders can they be seated?

    1—Answer: 720

23. The flags of seven different countries are to be displayed in a row. In how many different orders can they be flown?

    (a) 5040   (b) 1258   (c) 128   (d) 49   (e) None of these

    1—Answer: a

24. Seven members of a family line up to have their picture taken. In how many different ways can they be arranged?

    (a) 49   (b) 128   (c) 5040   (d) 1258   (e) None of these

    1—Answer: c

25. If there are ten questions on a test, how many different versions of the same test can be made by rearranging the questions?

    (a) 1024   (b) 100   (c) 30,240   (d) 3,628,800   (e) None of these

    1—Answer: d

26. Eight sailboats are entered in a race. In how many ways can they finish?

    (a) 6720   (b) 256   (c) 40,320   (d) 1680   (e) None of these

    1—Answer: c

27. Six girls are chosen as cheerleaders. In how many different orders could they have been chosen?

    (a) 720   (b) 64   (c) 46,656   (d) 17,280   (e) None of these

    1—Answer: a

28. There are seven possible digits in a phone number. How many different phone numbers are possible if the first digit cannot be 0 and no digit can be used more than once?

    (a) 128   (b) 181,440   (c) 544,320   (d) 5040   (e) None of these

    1—Answer: c

29. How many different ways (subject order) can seven algebra books, five trigonometry books, and four calculus books be arranged on a shelf?

(a) 140 (b) 1,441,440 (c) 65,536

(d) 2.1×10^{13} (e) None of these

1—Answer: b

30. A license plate number consists of three letters followed by three digit. How many distinct license plate numbers can be formed?

(a) 17,576,000 (b) 30,844,800 (c) 11,232,000

(d) 12,812,904 (e) None of these

1—Answer: a

31. A group of nine students line up single file for lunch. How many different ways can this occur if the five boys must line up first?

(a) 362,880 (b) 2880 (c) 15,120 (d) 126 (e) None of these

1—Answer: b

32. A random number generator selects an integer from 1 to 50. Find the number of ways a square number can occur.

(a) 99,884,400 (b) 5040 (c) 128 (d) 7 (e) None of these

1—Answer: d

33. Determine the number of ways the last four digits of a telephone number can be arranged if the first four digits cannot be 0.

(a) 10,000 (b) 5040 (c) 9000 (d) 4536 (e) None of these

1—Answer: c

34. If a license plate number consists of two letters followed by two digits, how many different license plate numbers are possible?

(a) 58,500 (b) 67,600 (c) 256 (d) 24 (e) None of these

1—Answer: b

35. A scrabble tray contains the tiles FERSXAI. How many different four-letter arrangements ("words") can be made?

1—Answer: 840

36. A ship has six flags available for signaling. If a signal consists of hoisting three of the flags, how many different signals are possible?

1—Answer: 120

37. A random number generator selects an integer from 1 to 20. Find the number of way in which a number that is a multiple of three can be selected.

    (a) 6 (b) 720 (c) 5 (d) 120 (e) None of these

    1—Answer: a

38. A random number generator selects two integers from 1 to 20. Find the number of ways that the sum of these two integers is 8.

    (a) 4 (b) 7 (c) 9 (d) 6 (e) None of these

    1—Answer: b

39. An auto license plate is made using two letters followed by three digits. How many license plates are possible?

    (a) 676,000 (b) 468,000 (c) 82 (d) 1,757,600 (e) None of these

    1—Answer: a

40. Determine the number of possible 5 digit ZIP codes.

    (a) 120 (b) 90,000 (c) 3,628,800 (d) 100,000 (e) None of these

    1—Answer: d

41. Determine the number of seven digit telephone numbers that can be formed under the condition that each of the first three digits cannot be 0.

    1—Answer: 7,290,000

42. How many different three-letter arrangements ("words") can be made from the letters ABCDEFG?

    (a) 24 (b) 35 (c) 840 (d) 210 (e) None of these

    1—Answer: d

43. A ship has eight flags available for signaling. If a signal consists of hoisting three of the flags, how many different signals are possible?

    (a) 1680 (b) 336 (c) 56 (d) 40,320 (e) None of these

    1—Answer: b

44. If a special at a diner offers a choice of one each of two appetizers, four entrees and five desserts, how many distinct meals are possible under the special?

    (a) 20 (b) 40 (c) 60 (d) 80 (e) None of these

    1—Answer: b

45. The chief designer for a large auto company is considering four different radiator grilles, two different headlight styles and five different front fender designs. How many front-end designs can be made using these three characteristics?

    (a) 80   (b) 60   (c) 40   (d) 20   (e) None of these

    1—Answer: c

46. If a menu at a diner offers a choice of three appetizers, six entrees and four desserts, how many distinct meals are possible?

    (a) 72   (b) 216   (c) 24   (d) 103,680   (e) None of these

    1—Answer: a

47. If a woman's wardrobe consists of two jackets, three skirts, and five blouses, how many different outfits consisting of a jacket, skirt, and blouse can be made?

    (a) 1440   (b) 10   (c) 90   (d) 30   (e) None of these

    1—Answer: d

48. In how many ways can a subcommittee of five people be selected from a committee of ten people?

    (a) 252   (b) 30,240   (c) 6048   (d) 1260   (e) None of these

    1—Answer: a

49. A record club offers new customers six free selections from a list of 130 different recordings. How many different introductory offers are possible?

    1—Answer: $\dfrac{130!}{6!124!} = 5,963,412,000$

50. In how many ways can a subcommittee of six people be selected for a committee of 12 people?

    (a) 665,280   (b) 924   (c) 720   (d) 520   (e) None of these

    1—Answer: b

51. In how many ways can a committee of nine people be selected from a group of 12 people?

    (a) 1320   (b) 79,833,600   (c) 362,880   (d) 220   (e) None of these

    1—Answer: d

52. How many ways can four girls be picked from a group of 30 girls?

    1—Answer: 27,405

53. A committee composed of three math majors and four science majors is to be selected from a group of 20 math majors and 16 science majors. How many different committees can be formed?

    (a) 2,074,800
    (b) 6840
    (c) 4.05×10^{17}
    (d) 320
    (e) None of these

    2—Answer: a

54. At a Boy Scout jamboree there are 12 senior patrol leaders, 10 assistant senior patrol leaders, one assistant patrol leader, three patrol leaders, and four regular scouts can be formed?

    (a) 3.2×10^{11}
    (b) 1.7×10^{12}
    (c) 2.4×10^{9}
    (d) 1.4×10^{9}
    (e) None of these

    2—Answer: b

55. A band director is taking auditions for a special pep band which requires three trumpets, one trombone, one saxophone, and two clarinets. There were five trumpet players, four trombone players, three saxophone players and 10 clarinet players who came to audition. How many possible combinations does the band director have?

    (a) 100
    (b) 600
    (c) 5400
    (d) 32,400
    (e) None of these

    2—Answer: c

56. In how many ways can a committee consisting of two deacons and four regular church members be formed in a church that has five deacons and 120 regular members?

    1—Answer: 82,145,700

57. A small college needs four additional faculty members: a mathematician, two chemists, and an engineer. In how many ways can these positions be filled if there are two applicants for mathematics, six applicants for chemistry and three applicants for engineering?

    (a) 180
    (b) 330
    (c) 90
    (d) 36
    (e) None of these

    2—Answer: c

58. In how many ways can a committee of two boys and three girls be formed from a group of 10 boys and 12 girls?

    (a) 9900
    (b) 1320
    (c) 118,800
    (d) 265
    (e) None of these

    1—Answer: a

59. In how many ways can a committee of three boys and three girls be formed from a group of 10 boys and 12 girls?

(a) 340 (b) 2040 (c) 26,400 (d) 960,400 (e) None of these

1—Answer: c

60. In how many ways can a committee of two boys and four girls be formed from a group of six boys and nine girls?

(a) 141 (b) 90,720 (c) 3054 (d) 1890 (e) None of these

1—Answer: d

61. Find the number of diagonals in a heptagon (7-sided polygon). A line segment connecting any two nonadjacent vertices is called a diagonal of the polygon.

(a) 21 (b) 14 (c) 35 (d) 28 (e) None of these

2—Answer: b

62. Find the number of diagonals in a nonagon (9-sided polygon). A line segment connecting any two nonadjacent vertices is called a diagonal of the polygon.

(a) 56 (b) 36 (c) 27 (d) 18 (e) None of these

2—Answer: c

❑ 8.7 Probability

1. Describe the sample space: "A number is chosen at random from the numbers one to five inclusive."

(a) {1, 2, 3, 4} (b) 1 (c) $\frac{1}{5}$

(d) {1, 2, 3, 4, 5} (e) None of these

1—Answer: d

2. Describe the sample space: "A student must select the correct answer to a multiple choice test question given five possible answers."

(a) The correct answer (b) The five possible answers (c) $\frac{1}{5}$

(d) The four wrong answers (e) None of these

1—Answer: b

3. Describe the sample space: "A letter is selected from the word MATHEMATICS."

(a) One of the letters (b) $\frac{1}{11}$

(c) (M, A, T, H, E, M, A, T, I, C, S) (d) {M, A, T, H, E, I, C, S}

(e) None of these

1—Answer: d

4. Describe the sample space: "A letter is selected from the word FINITE."
   - (a) {F, I, N, T, E}
   - (b) One of the letters
   - (c) (F, I, N, I, T, E)
   - (d) $\frac{1}{6}$
   - (e) None of these

   1—**Answer:** a

5. Describe the sample space: "A ball is selected from a bag containing seven balls, numbered from 1 to 7."
   - (a) The ball numbered 3
   - (b) $\frac{1}{7}$
   - (c) The set of 7 balls
   - (d) The balls numbered 3, 4, and 7
   - (e) None of these

   1—**Answer:** c

6. A card is drawn at random from a standard deck of 52 playing cards. Find the probability that the card is a spade.
   - (a) $\frac{1}{13}$
   - (b) $\frac{1}{4}$
   - (c) $\frac{12}{13}$
   - (d) $\frac{3}{4}$
   - (e) None of these

   1—**Answer:** b

7. A card is drawn at random from a standard deck of 52 playing cards. Find the probability that the card is a 10 or an ace.
   - (a) $\frac{2}{13}$
   - (b) $\frac{1}{169}$
   - (c) $\frac{4}{13}$
   - (d) $\frac{1}{4}$
   - (e) None of these

   1—**Answer:** a

8. A card is drawn at random from a standard deck of 52 playing cards. Find the probability that the card is an ace or spade.
   - (a) $\frac{17}{52}$
   - (b) $\frac{4}{13}$
   - (c) $\frac{1}{52}$
   - (d) $\frac{2}{13}$
   - (e) None of these

   1—**Answer:** b

9. Two cards are randomly selected from a standard deck of 52 playing cards. Find the probability that one card will be an ace and the other will be a 10.
   - (a) $\frac{1}{52}$
   - (b) $\frac{8}{663}$
   - (c) $\frac{1}{169}$
   - (d) $\frac{2}{13}$
   - (e) None of these

   1—**Answer:** c

10. Find the probability of choosing an E when selecting a letter at random from those in the word COLLEGE.
    - (a) $\frac{2}{7}$
    - (b) $\frac{1}{5}$
    - (c) $\frac{2}{5}$
    - (d) $\frac{1}{7}$
    - (e) None of these

    1—**Answer:** a

11. Find the probability of choosing an A, B, or N when selecting a letter at random from those in the word BANANA.
    - (a) $\frac{1}{26}$
    - (b) 0
    - (c) 1
    - (d) $\frac{1}{2}$
    - (e) None of these

    1—**Answer:** c

12. In a group of 10 children, 3 have blond hair and 7 have brown hair. If a child is chosen at random, what is the probability that the child will have brown hair?

    1—**Answer:** $\frac{7}{10}$

13. A bag contains four red balls and seven white balls. If a ball is drawn at random, what is the probability that it is a red ball?

    (a) $\frac{1}{4}$ (b) $\frac{4}{7}$ (c) $\frac{1}{11}$ (d) $\frac{4}{11}$ (e) None of these

    1—Answer: d

14. A bag contains nine red balls and six white balls. If one ball is drawn at random from the bag, what is the probability that is a red ball?

    (a) $\frac{1}{9}$ (b) $\frac{3}{5}$ (c) $\frac{2}{5}$

    (d) $\frac{3}{2}$ (e) None of these

    1—Answer: b

15. A bag contains nine red balls numbered 1-9 and six white balls numbered 10-15. If one ball is drawn at random, what is the probability that the number on it is even?

    (a) $\frac{3}{5}$ (b) $\frac{7}{15}$ (c) $\frac{4}{15}$

    (d) $\frac{1}{5}$ (e) None of these

    1—Answer: b

16. A bag contains nine red balls numbered 1-9 and six white balls numbered 10-15. If one ball is drawn at random, what is the probability that the number on it is divisible by three?

    (a) $\frac{1}{3}$ (b) $\frac{1}{2}$ (c) $\frac{2}{3}$

    (d) $\frac{2}{5}$ (e) None of these

    1—Answer: a

17. A bag contains nine red balls numbered 1-9 and six white balls numbered 10-15. If one ball is drawn at random, what is the probability that the number on it is divisible by five?

    (a) 3   (b) $\frac{2}{3}$ (c) $\frac{3}{5}$

    (d) $\frac{1}{5}$ (e) None of these

    1—Answer: d

18. A die is tossed three times. What is the probability that a two will come up all three times?

    (a) $\frac{1}{172}$ (b) $\frac{1}{18}$ (c) $\frac{1}{120}$

    (d) $\frac{1}{216}$ (e) None of these

    1—Answer: d

19. What is the probability of drawing a white marble from a box containing six white, three red and five black marbles?

    (a) $\frac{1}{6}$ (b) $\frac{1}{14}$ (c) $\frac{3}{7}$

    (d) $\frac{1}{15}$ (e) None of these

    1—Answer: c

20. A fair coin is tossed four times. What is the probability of getting heads on all four tosses?

   1—Answer: $\frac{1}{16}$

21. A fair coin is tossed four times. What is the probability of getting a head on the first toss and tails on the other three tosses?

   (a) $\frac{1}{8}$ (b) $\frac{1}{16}$ (c) $\frac{1}{12}$
   (d) $\frac{3}{16}$ (e) None of these

   1—Answer: b

22. A fair coin is tossed four times. What is the probability of getting exactly one head?

   (a) $\frac{1}{2}$ (b) $\frac{1}{4}$ (c) $\frac{1}{8}$
   (d) $\frac{1}{16}$ (e) None of these

   1—Answer: b

23. Two six-sided dice are tossed. What is the probability that the total is 11?

   (a) $\frac{1}{18}$ (b) $\frac{1}{6}$ (c) $\frac{1}{8}$
   (d) $\frac{2}{15}$ (e) None of these

   1—Answer: a

24. Two integers from 0 to 9 inclusive are chosen by a random number generator. What is the probability of choosing the number 2 both times?

   (a) $\frac{1}{10}$ (b) $\frac{1}{100}$ (c) $\frac{1}{50}$
   (d) $\frac{4}{5}$ (e) None of these

   1—Answer: b

25. Two integers (between 1 and 40 inclusive) are chosen by a random number generator. What is the probability that both numbers chosen are divisible by 4?

   1—Answer: $\frac{1}{16}$

26. What is the probability that in a group of 6 people at least 2 will have their birthdays within the same week?

   (a) 0.74   (b) 0.50   (c) 0.26
   (d) 0.47   (e) None of these

   2—Answer: c

27. What is the probability that 2 people chosen at random from a group of 8 married couples are married to each other?

   1—Answer: $\frac{1}{15}$

28. There are 5 red and 4 black balls in a box. If 3 balls are picked without replacement, what is the probability that at least one of them is red?

   1—Answer: $\frac{20}{21}$

29. A box holds 12 white, 5 red, and 6 black marbles. If 2 marbles are picked at random, without replacement, what is the probability that they will both be black?

   (a) $\frac{36}{529}$ (b) $\frac{247}{506}$ (c) $\frac{15}{253}$

   (d) $\frac{6}{23}$ (e) None of these

   1—Answer: c

30. Five cards are drawn from an ordinary deck of 52 playing cards. What is the probability of getting exactly one ace? Round your answer to three decimal places.

   (a) 0.060 (b) 0.299 (c) 0.064

   (d) 0.341 (e) None of these

   1—Answer: b

31. Two six-sided dice are tossed. What is the probability that the total is seven?

   (a) $\frac{1}{2}$ (b) $\frac{7}{12}$ (c) $\frac{1}{6}$

   (d) $\frac{1}{3}$ (e) None of these

   1—Answer: c

32. Two six-sided dice are tossed. What is the probability that the total is nine?

   (a) $\frac{1}{6}$ (b) $\frac{1}{9}$ (c) $\frac{1}{8}$

   (d) $\frac{1}{3}$ (e) None of these

   1—Answer: b

33. Two six-sided dice are tossed. What is the probability that the total is 12?

   (a) $\frac{1}{12}$ (b) $\frac{1}{18}$ (c) $\frac{1}{36}$

   (d) $\frac{1}{30}$ (e) None of these

   1—Answer: c

34. Two six-sided dice are tossed. What is the probability that the total is ten?

   (a) $\frac{1}{12}$ (b) $\frac{1}{4}$ (c) $\frac{1}{18}$

   (d) $\frac{1}{8}$ (e) None of these

   1—Answer: a

35. Two cards are drawn with replacement from a box containing six blue cards numbered 1-6 and eleven white cards numbered 7-17. What is the probability that both cards are even-numbered?

   (a) $\frac{64}{289}$ (b) $\frac{4}{17}$ (c) $\frac{1}{34}$

   (d) $\frac{7}{34}$ (e) None of these

   1—Answer: a

36. Two cards are drawn with replacement from a box containing six blue cards numbered 1-6 and eleven white cards numbered 7-17. What is the probability that the first card is odd-numbered and the second card is white?

   (a) $\frac{99}{136}$ (b) $\frac{99}{272}$ (c) $\frac{99}{289}$
   (d) $\frac{20}{289}$ (e) None of these

   1—Answer: c

37. Drawing from a standard deck of 52 cards, what is the probability that the card is an eight or a face card?

   (a) $\frac{3}{169}$ (b) $\frac{4}{13}$ (c) $\frac{2}{13}$
   (d) $\frac{1}{48}$ (e) None of these

   1—Answer: b

38. Drawing from a standard deck of 52 cards, what is the probability that the card is an ace, king or queen?

   (a) $\frac{3}{13}$ (b) $\frac{1}{13}$ (c) $\frac{1}{2197}$
   (d) $\frac{1}{64}$ (e) None of these

   1—Answer: a

39. Drawing from a standard deck of 52 cards, what is the probability that the card is a five or a red jack?

   (a) $\frac{1}{13}$ (b) $\frac{3}{26}$ (c) $\frac{3}{13}$
   (d) $\frac{1}{238}$ (e) None of these

   1—Answer: b

40. A small business college has 800 seniors, 700 juniors, 900 sophomores and 1,200 freshmen. If a student is randomly selected, what is the probability that the student is a junior or senior?

   (a) $\frac{13}{36}$ (b) $\frac{1}{4}$ (c) $\frac{5}{12}$
   (d) $\frac{5}{162}$ (e) None of these

   1—Answer: c

41. A small business college has 800 seniors, 700 juniors, 900 sophomores and 1,200 freshmen. If a student is randomly selected, what is the probability that the student is a freshman or senior?

   (a) $\frac{1}{4}$ (b) $\frac{1}{54}$ (c) $\frac{5}{12}$
   (d) $\frac{5}{9}$ (e) None of these

   1—Answer: d

42. A small business college has 400 seniors, 300 juniors, 500 sophomores and 600 freshmen. If a student is randomly selected, what is the probability that the student is a junior or a senior?

   (a) $\frac{1}{4}$ (b) $\frac{2}{9}$ (c) $\frac{7}{18}$
   (d) $\frac{1}{27}$ (e) None of these

   1—Answer: c

43. A small business college has 400 seniors, 300 juniors, 500 sophomores and 600 freshmen. If a student is randomly selected, what is the probability that the student is a freshmen or a senior?

(a) $\frac{2}{27}$ (b) $\frac{5}{9}$ (c) $\frac{1}{4}$

(d) $\frac{5}{18}$ (e) None of these

1—Answer: b

44. In drawing one card from a standard deck of 52 playing cards, what is the probability of obtaining a king or a club?

(a) $\frac{1}{52}$ (b) $\frac{17}{52}$ (c) $\frac{4}{13}$

(d) $\frac{2}{13}$ (e) None of these

1—Answer: c

45. In drawing one card from a standard deck of 52 playing cards, what is the probability of obtaining an ace or a heart?

(a) $\frac{4}{13}$ (b) $\frac{1}{52}$ (c) $\frac{17}{52}$

(d) $\frac{2}{13}$ (e) None of these

1—Answer: a

46. In drawing one card from a standard deck of 52 playing cards, what is the probability of obtaining a queen or a diamond?

(a) $\frac{1}{52}$ (b) $\frac{2}{13}$ (c) $\frac{4}{13}$

(d) $\frac{17}{52}$ (e) None of these

1—Answer: c

47. In drawing one card from a standard deck of 52 playing cards, what is the probability that it is a jack or a spade?

(a) $\frac{1}{52}$ (b) $\frac{17}{52}$ (c) $\frac{2}{13}$

(d) $\frac{4}{13}$ (e) None of these

1—Answer: d

48. In drawing one card from a standard deck of 52 playing cards, what is the probability that it is a diamond or a face card?

(a) $\frac{11}{26}$ (b) $\frac{17}{52}$ (c) $\frac{3}{13}$

(d) $\frac{1}{13}$ (e) None of these

1—Answer: a

49. Before an election, a sample of 120,000 people throughout the county showed that 79,386 people would vote for Candidate A. If a person from the sample is chosen at random, what is the probability that the person is one of the people who said they would not vote for Candidate A?

(a) 0.66 (b) 0.34 (c) 0.47

(d) 0.53 (e) None of these

1—Answer: b

50. If the probability of getting a rotten apple in a basket of apples is 12%, what is the probability of getting 3 good apples choosing one from each of the three different baskets?

(a) 0.9983
(b) 0.0017
(c) 0.8800
(d) 0.6815
(e) None of these

1—Answer: d

51. In an experiment in which 2 six-sided dice are tossed, what is the probability of *not* getting a sum of 10?

1—Answer: $\frac{11}{12}$

52. If $P(A) = \frac{6}{11}$, find $P(A')$.

(a) 0
(b) 1
(c) $\frac{5}{11}$
(d) $\frac{5}{6}$
(e) None of these

1—Answer: c

53. A "doctored" die is tossed 100,000 times and comes up six on 35,861 rolls. Find the probability of rolling a number other than 6 with this die.

(a) 0.36
(b) 0.64
(c) 0.17
(d) 0.83
(e) None of these

1—Answer: b

54. A sample of nursing homes in a state reveals that 112,000 of 218,000 residents are female. If a nursing home resident is chosen at random from this state, what is the probability that the resident is male?

1—Answer: $\frac{53}{109}$

55. There are 120 40-watt bulbs, 200 60-watt bulbs and 80 100-watt bulbs in an art gallery. All of the 60-watt bulbs and half of the 100-watt bulbs are transparent and the rest are not. While installing the bulbs, a worker dropped one of them. What is the probability that the broken bulb was not a transparent 100-watt bulb?

(a) $\frac{1}{10}$
(b) $\frac{1}{5}$
(c) $\frac{9}{10}$
(d) $\frac{2}{5}$
(e) None of these

1—Answer: c

Test A
Chapter P

Name _____ Date _____

Class _____ Section _____

1. Determine how many natural numbers there are in the set:
   $\{-3, -\frac{1}{2}, 2, 0.3535\ldots\}$.
   - (a) 1
   - (b) 2
   - (c) 3
   - (d) 4
   - (e) None of the numbers are natural numbers.

2. Use inequality notation to describe the set of real numbers that are less than 4 and at least -2.
   - (a) $-2 < x < 4$
   - (b) $-2 < x \leq 4$
   - (c) $-2 \leq x < 4$
   - (d) $-2 \leq x \leq 4$
   - (e) None of these

3. Evaluate: $(2^3 \cdot 3^2)^{-1}$.
   - (a) -72
   - (b) $\frac{1}{46,656}$
   - (c) $-\frac{1}{36}$
   - (d) $\frac{1}{72}$
   - (e) None of these

4. Simplify: $\left(\dfrac{x^{-3}y^2}{z}\right)^{-4}$.
   - (a) $\dfrac{z^4}{x^7 y^6}$
   - (b) $\dfrac{y^2 z^4}{x^7}$
   - (c) $\dfrac{x^{12} z^4}{y^8}$
   - (d) $\dfrac{z^4}{x^{12} y^8}$
   - (e) None of these

5. Simplify: $\sqrt[3]{-625 x^7 y^5}$.
   - (a) $5xy\sqrt[3]{-5x^4 y^2}$
   - (b) $-5xy\sqrt[3]{5x^4 y^2}$
   - (c) $-125 x^2 y \sqrt[3]{5xy^2}$
   - (d) $-5x^2 y \sqrt[3]{5xy^2}$
   - (e) Does not simplify

6. Simplify: $(-4x^2 + 2x) - (5x^3 + 2x^2 - 1) + (x^2 + 1)$.
   - (a) $-9x^3 + 5x^2$
   - (b) $5x^3 - x^2 + 2x + 1$
   - (c) $6x^6$
   - (d) $-5x^3 - 5x^2 + 2x + 2$
   - (e) None of these

7. Represent the area of the region as a polynomial in standard form.

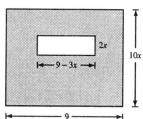

   - (a) $6x^2 - 72x$
   - (b) $48x - 18$
   - (c) $90x - 6x^2$
   - (d) $90x$
   - (e) None of these

8. Factor: $9x^2 - 42x + 49$.

   (a) $(3x - 7)^2$ (b) $(9x - 7)^2$ (c) $(3x + 7)^2$

   (d) $x(9x - 42 + 49)$ (e) None of these

9. Reduce: $\dfrac{x^2 - 8x + 12}{5x - 30}$.

   (a) $\dfrac{x - 2}{5}$ (b) $\dfrac{2 + x}{-5}$ (c) $\dfrac{x + 2}{5}$

   (d) $\dfrac{-x - 2}{5}$ (e) None of these

10. Multiply, then simplify: $\dfrac{2 - x}{x^2 + 4} \cdot \dfrac{x + 2}{x^2 + 5x - 14}$.

    (a) $-\dfrac{x + 2}{(x^2 + 4)(x + 7)}$ (b) $\dfrac{1}{(x + 2)(x + 7)}$ (c) $\dfrac{x + 2}{(x^2 + 4)(x + 7)}$

    (d) $\dfrac{-1}{(x + 2)(x + 7)}$ (e) None of these

11. Subtract, then simplify: $\dfrac{3}{x^2 + 2x + 1} - \dfrac{1}{x + 1}$.

    (a) $\dfrac{4 - x}{x^2 + 2x + 1}$ (b) $\dfrac{-x^2 + 5x + 2}{(x + 1)(x^2 + 2x + 1)}$ (c) $\dfrac{-x^2 + x + 2}{(x + 2x + 1)(x + 1)}$

    (d) $\dfrac{2 - x}{x^2 + 2x + 1}$ (e) None of these

12. Write as a sum of terms: $\dfrac{x - 2x^2 + x^3}{\sqrt{x}}$.

    (a) $x - 2x^2 + x^3 - x^{1/2}$ (b) $x^{-1/2} + 2x^{-2} - x^{-3} + x^{1/2}$ (c) $x^{1/2} - 2x^{1/2} + x^{3/2}$

    (d) $x^{1/2} - 2x^{3/2} + x^{5/2}$ (e) None of these

13. The triangle shown in the figure has vertices at the points $(-1, -1)$, $(-1, 2)$, and $(1, 1)$. Shift the triangle 3 units to the right and 2 units down and find the vertices of the shifted triangle.

    (a) $(2, 1), (2, 4), (4, 3)$

    (b) $(-4, 1), (-4, 4), (-2, 3)$

    (c) $(-4, -3), (-4, 0), (-2, -1)$

    (d) $(2, -3), (2, 0), (4, -1)$

    (e) None of these

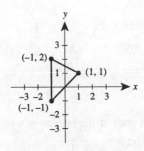

14. Find the distance between the origin and the midpoint of the two points (3, 3) and (3, 5).

(a) $3\sqrt{2}$ (b) 7   (c) $\sqrt{34}$

(d) 5   (e) None of these

15. In a football game, the quarterback throws a pass from the 8-yard line, 15 yards from the sideline. The pass is caught on the 43-yard line, 3 yards from the same sideline. How long was the pass? (Assume the pass and the reception are on the same side of midfield.)

(a) 40 yards   (b) 36.1 yards   (c) 39.4 yards

(d) 37 yards   (e) None of these

442   Chapter P   Prerequisites

Test B   Name _____ Date _____
Chapter P   Class _____ Section _____

1. Determine how many integers there are in the set:
   $\{-3, -\frac{1}{2}, 2, 0.3535\ldots\}$.
   (a) 1
   (b) 2
   (c) 3
   (d) 4
   (e) None of the numbers are integers.

2. Use inequality notation to describe the set of real numbers that are more than 5 and at most 10.
   (a) $5 \leq x \leq 10$
   (b) $5 \leq x < 10$
   (c) $5 < x < 10$
   (d) $5 < x \leq 10$
   (e) None of these

3. Evaluate: $(6^{-2})(3^0)(2^3)$.
   (a) $\frac{2}{9}$
   (b) -288
   (c) -216
   (d) $\frac{1}{6}$
   (e) None of these

4. Simplify: $\left(\dfrac{3x^2 y^3}{xw^{-2}}\right)^3$.
   (a) $9w^6 x^3 y^9$
   (b) $9w^{-8} x^8 y^{27}$
   (c) $27w^6 x^3 y^9$
   (d) $3w^6 x^3 y^9$
   (e) None of these

5. Simplify: $\sqrt{75x^2 y^{-4}}$.
   (a) $\dfrac{5\sqrt{3}x}{y^2}$
   (b) $\dfrac{3\sqrt{5}|x|}{y^2}$
   (c) $5\sqrt{3}|x|y^2$
   (d) $\dfrac{5\sqrt{3}|x|}{y^2}$
   (e) None of these

6. Simplify: $(2x^2 + 3x - 1) + (x^3 + x^2 + 5) - (2x^2 - 5x + 7)$.
   (a) $x^3 - 2x + x^2 + 4$
   (b) $x^3 + x^2 + 8x - 3$
   (c) $8x^5 + 4$
   (d) $63x^3 + 5x^6$
   (e) None of these

7. Represent the area of the region as a polynomial in standard form.
   (a) $140x$
   (b) $144x - 4x^2$
   (c) $55x$
   (d) $120x^2$
   (e) None of these

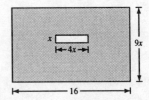

8. Factor: $3x^2 - 13x - 16$.

    (a) $(3x - 16)(x - 1)$
    (b) $(3x + 16)(x - 1)$
    (c) $(x + 16)(3x - 1)$
    (d) $(x - 16)(3x + 1)$
    (e) None of these

9. Reduce: $\dfrac{x^2 - 7x + 10}{x^2 - 8x + 15}$.

    (a) $\dfrac{x + 2}{x + 3}$
    (b) $\dfrac{x - 2}{x - 3}$
    (c) $\dfrac{x + 3}{x + 2}$
    (d) $\dfrac{x - 2}{x + 3}$
    (e) None of these

10. Divide: $\dfrac{x + y}{x^3 - x^2} \div \dfrac{x^2 + y^2}{x^2 - x}$.

    (a) $\dfrac{1}{x(x + y)}$
    (b) $\dfrac{x + y}{x(x^2 + y^2)}$
    (c) $\dfrac{x(x^2 + y^2)}{x + y}$
    (d) $-x$
    (e) None of these

11. Add, then simplify: $\dfrac{3}{x^2 + x - 2} + \dfrac{x}{x^2 - x - 6}$.

    (a) $\dfrac{x^2 + 2x - 9}{(x - 3)(x + 2)(x - 1)}$
    (b) $\dfrac{x^3 + 3x^2 - 5x - 6}{(x^2 + x - 2)(x^2 - x - 6)}$
    (c) $\dfrac{4x - 10}{(x - 3)(x + 2)(x - 1)}$
    (d) $\dfrac{3 + x}{2x^2 - 8}$
    (e) None of these

12. Write as a sum of terms: $\dfrac{4x^3 - 3x^2 + 1}{x^{3/2}}$.

    (a) $4x^{9/2} - 3x^3 + x^{3/2}$
    (b) $4x^3 - 3x^2 + 1 - x^{3/2}$
    (c) $4x^{1/2} - 3x^{-1/2} + x^{-3/2}$
    (d) $4x^{-3/2} - 3x^{-1/2} + x^{-3/2}$
    (e) None of these

13. The triangle shown in the figure has vertices at the points $(-1, 2)$, $(1, 2)$ and $(0, 0)$. Shift the triangle 2 units up and find the vertices of the shifted triangle.

    (a) $(1, 2), (3, 2), (2, 0)$
    (b) $(-1, 4), (1, 4), (0, 2)$
    (c) $(-1, 0), (1, 0), (0, -2)$
    (d) $(-3, 2), (-1, 2), (-2, 2)$
    (e) None of these

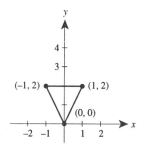

14. Find the distance between the origin and the midpoint of the two points (2, 7) and (6, 5).

    (a) 10
    (b) $2\sqrt{13}$
    (c) $4\sqrt{13}$
    (d) $2\sqrt{5}$
    (e) None of these

15. In a football game, the quarterback throws a pass from the 3-yard line, 10 yards from the sideline. The pass is caught on the 43-yard line, 40 yards from the same sideline. How long was the pass? (Assume the pass and the reception are on the same side of midfield.)

    (a) 60 yards
    (b) 55 yards
    (c) 50 yards
    (d) 45 yards
    (e) None of these

Test C Name _____ Date _____
Chapter P Class _____ Section _____

1. Determine how many integers there are in the set:
   $\{5, -16, \frac{2}{3}, 0\}$.
   (a) 4
   (b) 3
   (c) 2
   (d) 1
   (e) None of the numbers are integers.

2. Use inequality notation to describe the set of real numbers that are at least -1 and at most 3.
   (a) $-1 \le x \le 3$
   (b) $-1 < x < 3$
   (c) $-1 \le x < 3$
   (d) $-1 < x \le 3$
   (e) None of these

3. Evaluate: $(4)^{-2}(3)^0(-1)^2$.
   (a) 0
   (b) -8
   (c) -16
   (d) $\frac{1}{16}$
   (e) None of these

4. Simplify: $\left(\dfrac{x^{-5}y^2}{z^2}\right)^{-3}$.
   (a) $\dfrac{x^{-8}y^{-1}}{z^{-1}}$
   (b) $\dfrac{x^{15}z^6}{y^6}$
   (c) $\dfrac{z^5}{x^{-2}y^{-1}}$
   (d) $\dfrac{x^{-15}y^6}{z^6}$
   (e) None of these

5. Simplify: $\sqrt[3]{24x^4y^5}$.
   (a) $3x^2y^2\sqrt[3]{6x^2y^3}$
   (b) $8xy\sqrt[3]{3xy}$
   (c) $2xy\sqrt[3]{6xy^2}$
   (d) $2xy\sqrt[3]{3xy^2}$
   (e) None of these

6. Simplify: $(3x^2 - 2x) + (7x^3 - 2x^2 + 1) - (16x^2 - 7)$.
   (a) $7x^3 - 15x^2 - 2x + 8$
   (b) $7x^3 + 15x^2 - 6$
   (c) $7x^3 - 15x^2 - 2x - 6$
   (d) $15x^5 + 7x^3 - 2x - 6$
   (e) None of these

7. Represent the area of the region as a polynomial in standard form.
   (a) $-2x + 18$
   (b) $12x + 14$
   (c) 14
   (d) $11x + 14$
   (e) None of these

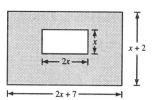

8. Factor: $3x^2 - 19x - 14$.
   (a) $(3x + 2)(x - 7)$
   (b) $(3x - 7)(x + 2)$
   (c) $(3x - 2)(x + 7)$
   (d) $(3x + 7)(x - 2)$
   (e) None of these

9. Reduce: $\dfrac{x^2 + 3x - 10}{x^2 + 2x - 15}$.
   (a) $\dfrac{x - 2}{x - 3}$
   (b) $\dfrac{x - 3}{x - 2}$
   (c) $\dfrac{x + 2}{x - 3}$
   (d) $\dfrac{x + 2}{x + 3}$
   (e) None of these

10. Divide: $\dfrac{4x - 16}{5x + 15} \div \dfrac{4 - x}{2x + 6}$.
    (a) 0
    (b) $-\dfrac{4(x - 4)}{5(x + 3)^2}$
    (c) $-\dfrac{8}{5}$
    (d) $-\dfrac{3}{10}$
    (e) None of these

11. Add, then simplify: $\dfrac{x + 1}{x^2 + x - 2} + \dfrac{x + 3}{x^2 - 4x + 3}$.
    (a) $\dfrac{2x^2 + 3}{(x - 1)(x - 3)(x + 2)}$
    (b) $\dfrac{2x^3 + x^2 - 3}{(x^2 + x - 2)(x^2 - 4x + 3)}$
    (c) $\dfrac{2x + 4}{-3x + 1}$
    (d) $\dfrac{2x^2 + 3x + 3}{(x - 1)(x - 3)(x + 2)}$
    (e) None of these

12. Write as a sum of terms: $\dfrac{3x^4 - 2x^2 + 1}{\sqrt[3]{x}}$.
    (a) $3x^4 - 2x^2 + x^{-3}$
    (b) $3x^4 - 2x^2 + 1 - x^{1/3}$
    (c) $3x^{11/3} - 2x^{5/3} + x^{-1/3}$
    (d) $3x - 2^{-1} + x^{-1/3}$
    (e) None of these

13. The triangle shown in the figure has vertices at the points $(-1, 2)$, $(1, 2)$, and $(0, 0)$. Shift the triangle 3 units to the left and find the vertices of the shifted triangle.
    (a) $(-1, -1), (1, -1), (0, -3)$
    (b) $(-4, 2), (-2, 2), (-3, 0)$
    (c) $(-1, -1), (-1, 1), (-3, 0)$
    (d) $(2, 2), (-2, 2), (-3, 0)$
    (e) None of these

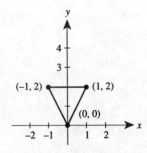

14. Find the distance between the origin and the midpoint of the two points $(5, 7)$ and $(-3, 1)$.

  (a) $\sqrt{17}$ (b) 5   (c) $\sqrt{10}$

  (d) 4   (e) None of these

15. A homeowner needs to determine the distance y from the peak to the lower edge of the roof on his garage. He knows the distance from the ground to the peak is 13 feet and the distance from the lower edge of the roof to the ground is 10 feet. Find y if the garage is 18 feet wide. (Round to 1 decimal place.)

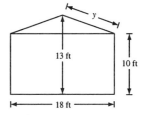

  (a) 10.1 feet   (b) 9.8 feet

  (c) 9.5 feet   (d) 9.3 feet

  (e) None of these

Test D Name_____ Date_____
Chapter P Class_____ Section_____

1. Use a calculator to find the decimal form of the rational number: $\frac{10}{11}$.

2. Use a calculator to order the numbers from smallest to largest: $\left\{\frac{45}{99}, \frac{2}{5}, \frac{7}{5}, \frac{152}{333}, \frac{23}{5}\right\}$.

3. Write the expression as a repeated multiplication: $(2x)^3$.

4. Simplify: $(3x^2y^3z)^{-2}(xy^4)$.

5. Perform the operation and simplify: $\dfrac{x^{4/3}y^{1/3}}{(xy)^{2/3}}$.

6. Write in standard form: $(3x^2 + 2x) + x(1 - 7x) + (2x + 5)$.

7. Find the area of the shaded region.

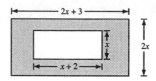

8. Factor: $14x^2 - 19x - 3$.

9. Reduce to lowest terms: $\dfrac{2x^2 + 5x - 3}{6x - 3}$.

10. Multiply, then simplify: $\dfrac{x^2 - 5x + 4}{x^2 + 4} \cdot \dfrac{x + 2}{x^2 + 3x - 4}$.

11. A marble is tossed in to a box whose base is shown. Find the probability that the marble will come to rest in the shaded portion of the box.

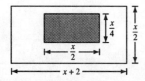

12. Simplify: $\frac{1}{8}(3x + 1)^{3/2} + \frac{1}{4}(3x + 1)^{1/2}$.

13. Find the distance between the points $(3, -1)$ and $(7, 2)$.

14. Find the midpoint of the line segment joining $(6, 9)$ and $(-3, 1)$.

15. Find the length of the hypotenuse of the right triangle determined by the points $(1, 1)$, $(-2, 1)$, and $(-2, 4)$.

Test E Name _____ Date _____

Chapter P Class _____ Section _____

1. Use a calculator to find the decimal form of the rational number: $\frac{173}{330}$.

2. Use a calculator to order the numbers from smallest to largest: $\left\{\frac{13}{2}, \frac{28}{5}, \frac{650}{99}, 6.56, 6.065\right\}$.

3. Write the expression as a repeated multiplication: $(3y)^4$.

4. Simplify: $(-2x^2)^5(5x^3)^{-2}$.

5. Perform the operation and simplify: $\dfrac{x^{-1/2} \cdot x^{1/3}}{x^2 \cdot x^{-3}}$.

6. Write in standard form: $3x^2 - 2x(1 + 3x - x^2)$.

7. Find the area of the shaded region.

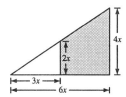

8. Factor: $35x^2 + 9x - 2$.

9. Reduce to lowest terms: $\dfrac{4x - 2x^2}{x^2 + x - 6}$.

10. Divide, then simplify: $\dfrac{x + 1}{x^2 - 1} \div \dfrac{x^2 + 1}{x - 1}$.

11. A marble is tossed in to a triangle whose base is shown. Find the probability that the marble will come to rest in the shaded portion of the triangle.

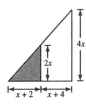

12. Simplify: $\dfrac{3x\left(\dfrac{5}{2}\right)(2x - 1)^{3/2} - (2x - 1)^{5/2}(3)}{(3x)^2}$.

13. Find the distance between the points $(3, 5)$ and $(-2, -1)$.

14. Find the midpoint of the line segment joining $(-6, -2)$ and $(5, -1)$.

15. Find the length of the hypotenuse of the right triangle determined by the points $(-1, 1)$, $(3, 1)$, and $(3, -3)$.

Test A
Chapter 1

Name _____ Date _____
Class _____ Section _____

1. Identify the graph of the equation: $y = |x + 7|$.

   (a)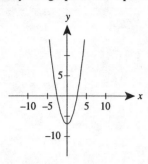

   (b)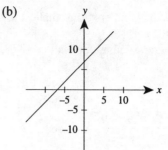

   (c)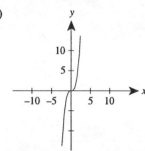

   (d)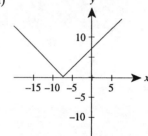

   (e) None of these

2. Use a graphing utility to graph $y = 6 + 3x - 2x^2$. Use the standard viewing rectangle.

   (a)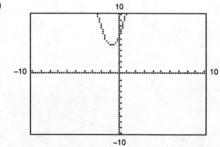

   (b)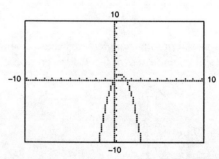

   (c)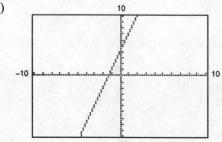

   (d)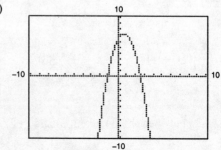

   (e) None of these

3. Determine the standard equation of the circle with radius 3 and center $(3, -2)$.
   (a) $(x - 3)^2 + (y - 2)^2 = 3$
   (b) $(x - 3)(y - 2) = 9$
   (c) $(x - 3)^2 + (y + 2)^2 = \sqrt{3}$
   (d) $(x - 3)^2 + (y + 2)^2 = 9$
   (e) None of these

4. Solve for x: $13x - 9 = 3x + 10$.
   (a) $\frac{19}{10}$
   (b) $\frac{10}{19}$
   (c) $\frac{1}{10}$
   (d) $\frac{19}{16}$
   (e) None of these

5. Solve for x: $3 - \dfrac{4x + 5}{x - 2} = \dfrac{7x - 9}{x - 2}$.
   (a) $-\dfrac{1}{4}$
   (b) -1
   (c) 1
   (d) No solution
   (e) None of these

   2—Answer: a

6. The length of a rectangular room is 2 feet longer than the width. Write an algebraic expression for the perimeter of a room with width x feet.
   (a) $x(x + 2)$
   (b) $4x + 2$
   (c) $2x + 2(x + 2)$
   (d) $x + (x + 2)$
   (e) None of these

7. Ann invested $8000 in a fund that pays $2\frac{1}{2}\%$ more simple interest per year than a similar fund in which her husband had invested $10,000. At the end of a year their interest totaled $1690.00. What rate of interest did Ann receive?

8. Solve by completing the square: $x^2 + 4x - 2 = 0$.
   (a) $2 \pm \sqrt{6}$
   (b) $2 \pm \sqrt{2}$
   (c) $-2 \pm \sqrt{2}$
   (d) $-2 \pm \sqrt{6}$
   (e) None of these

9. Solve: $3x^2 - 6x + 2 = 0$.
   (a) $\dfrac{3 \pm \sqrt{3}}{3}$
   (b) $1 \pm \sqrt{3}$
   (c) $\dfrac{3 \pm \sqrt{15}}{3}$
   (d) $\dfrac{1}{3}, 2$
   (e) None of these

10. Multiply: $(3 + 7i)(6 - 2i)$.
    (a) $18 + 22i$
    (b) $4 + 48i$
    (c) $4 + 36i$
    (d) $32 + 36i$
    (e) None of these

11. Solve for x: $3x - 2\sqrt{x} - 5 = 0$.
    - (a) $\frac{5}{3}$
    - (b) $-1, \frac{5}{3}$
    - (c) $1, \frac{25}{9}$
    - (d) $\frac{25}{9}$
    - (e) None of these

12. Solve: $5x + 6 > 7x + 9$.
    - (a) $\left(-\frac{3}{2}, \infty\right)$
    - (b) $\left(\frac{6}{5}, \frac{9}{7}\right)$
    - (c) $\left(-\infty, -\frac{3}{2}\right)$
    - (d) $\left(\frac{3}{2}, \infty\right)$
    - (e) None of these

13. Solve the inequality: $(x + 3)^2 \geq 4$.
    - (a) $[1, 5]$
    - (b) $(-\infty, 5]$
    - (c) $(-\infty, -5] \cup [-1, \infty)$
    - (d) $[-5, -1]$
    - (e) None of these

14. Use a graphing utility to solve: $x^2 + 4x + 2 \leq 0$.
    - (a) $[-3.41, \infty)$
    - (b) $[-3.41, -0.59]$
    - (c) $(-\infty, -3.41] \cup [-0.59, \infty)$
    - (d) $[-2, 2]$
    - (e) None of these

15. Solve the inequality: $\dfrac{3}{x - 2} \leq \dfrac{5}{x + 2}$.
    - (a) $(-\infty, -2) \cup (2, 6)$
    - (b) $(-2, 2) \cup [6, \infty)$
    - (c) $[8, \infty)$
    - (d) $(-2, 2) \cup [8, \infty)$
    - (e) None of these

Test B

Chapter 1

1. Identify the graph of the equation: $y = \sqrt{2 - x}$.

   (a)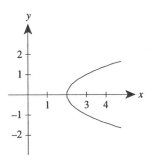

   (b)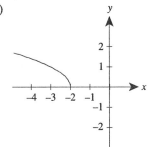

   (c)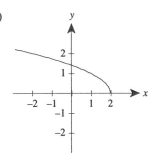

   (d)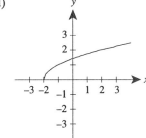

   (e) None of these

2. Use a graphing utility to graph $y = x\sqrt{5 - x}$. Use the standard viewing rectangle.

   (a)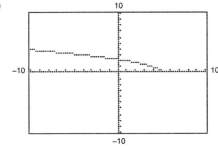

   (b)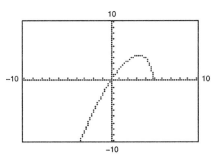

   (c)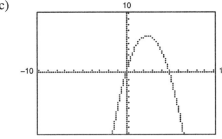

   (d)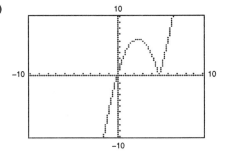

   (e) None of these

3. Determine the standard equation of the circle with radius 5 and center $(-4, -3)$.
   (a) $(x - 4)^2 + (y - 3)^2 = 25$
   (b) $(x - 4)(x - 3) = 5$
   (c) $(x + 4)^2 + (y + 3)^2 = 5$
   (d) $(x + 4)^2 + (y - 3)^2 = 25$
   (e) None of these

4. Solve for x: $7 - 3x + 2 = 4x - 1$.
   (a) 5
   (b) 10
   (c) $\frac{7}{10}$
   (d) $\frac{10}{7}$
   (e) None of these

5. Solve for x: $\frac{7x}{x - 2} + \frac{2x}{x + 2} = 9$.
   (a) $-\frac{18}{5}$
   (b) $\frac{2}{3}$
   (c) $-\frac{2}{5}$
   (d) $\frac{5}{18}$
   (e) None of these

6. A jacket is discounted by 20%. Write an algebraic expression for the sale price of a jacket that originally sells for x dollars.
   (a) $x - 0.20x$
   (b) $x - 0.20$
   (c) $x + 0.20x$
   (d) $0.20x - x$
   (e) None of these

7. Two trains traveling the same speed leave the city. The southbound train reaches its destination in 45 minutes. The eastbound train reaches its destination in 1 hour. How fast were the trains traveling if their destinations are 88 miles apart?
   (a) 70.4 mph
   (b) 1.2 mph
   (c) 2 mph
   (d) 49.6 mph
   (e) None of these

8. Solve by completing the square: $x^2 - 6x + 1 = 0$.
   (a) $3 \pm \sqrt{26}$
   (b) $3 \pm \sqrt{10}$
   (c) $3 \pm \sqrt{17}$
   (d) $3 \pm 2\sqrt{2}$
   (e) None of these

9. Solve: $(x - 1)^2 = 3x + 5$.
   (a) 1, 4
   (b) $\frac{5 \pm \sqrt{39}}{2}$
   (c) $\frac{5 \pm \sqrt{41}}{2}$
   (d) $-1, 6$
   (e) None of these

10. Multiply: $(3 - \sqrt{-4})(7 + \sqrt{-9})$.
    (a) $15 + 23i$
    (b) $27 - 5i$
    (c) $27 + 5i$
    (d) $15 + 5i$
    (e) None of these

11. Solve for x: $\sqrt{2 - 5x} = 5x$.
    - (a) $\frac{1}{5}$
    - (b) $-\frac{2}{5}$
    - (c) $\frac{1}{5}, -\frac{2}{5}$
    - (d) $\frac{1}{10}$
    - (e) None of these

12. Solve the inequality: $4 - 3x \geq 5x + 12$.
    - (a) $(-\infty, -1]$
    - (b) $(-\infty, 8]$
    - (c) $[-1, \infty)$
    - (d) $(-\infty, -2)$
    - (e) None of these

13. Solve the inequality: $(x - 1)^2 \leq 25$.
    - (a) $[-4, 6]$
    - (b) $(-\infty, -4] \cup [6, \infty)$
    - (c) $(-\infty, -6] \cup [4, \infty)$
    - (d) $[-6, 4]$
    - (e) None of these

14. Use a graphing utility to solve $x^2 + x - 3 < 0$.
    - (a) $(-\infty, -3) \cup (1, \infty)$
    - (b) $(-\infty, -2.30) \cup (1.30, \infty)$
    - (c) $(-2.30, 1.30)$
    - (d) $(-3, 1)$
    - (e) None of these

15. Solve the inequality: $\dfrac{2}{x - 1} \leq \dfrac{3}{x + 1}$.
    - (a) $(-1, 1) \cup [5, \infty)$
    - (b) $(-\infty, -1) \cup (1, 5]$
    - (c) $[5, \infty)$
    - (d) Empty set
    - (e) None of these

Test C
Chapter 1

1. Identify the graph of the equation: $y = \sqrt{x + 3}$.

   (a)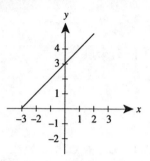

   (b)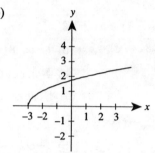

   (c)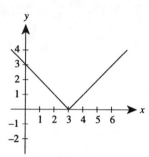

   (d)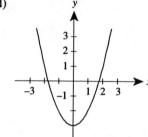

   (e) None of these

2. Use a graphing utility to graph $y = x^3 + 4x$. Use the standard viewing rectangle.

   (a)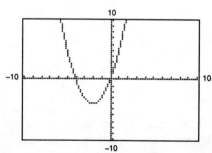

   (b)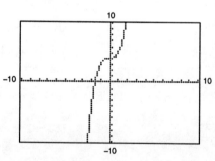

   (c)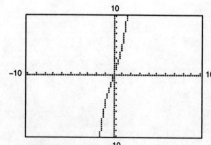

   (d)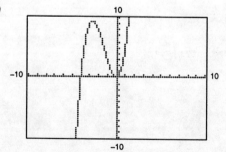

   (e) None of these

3. Determine the standard equation of the circle with radius 2 and center $(-1, 3)$.

   (a) $(x - 1)^2 + (y - 3)^2 = 2$
   (b) $(x - 1)^2 + (y - 3)^2 = 4$
   (c) $(x + 1)^2 + (y - 3)^2 = 2$
   (d) $(x + 1)^2 + (y - 3)^2 = 4$
   (e) None of these

4. Solve for x: $4 + 7x - 3x + 2 = 8x + 6$.

   (a) No solution
   (b) 0
   (c) 1
   (d) 2
   (e) None of these

5. Solve for x: $\dfrac{1}{x - 2} + \dfrac{3}{x + 3} = \dfrac{4}{x^2 + x - 6}$.

   (a) $\dfrac{4}{7}$
   (b) 3
   (c) $\dfrac{7}{4}$
   (d) 1
   (e) None of these

6. A stereo is discounted 40%. Write an algebraic expression for the sale price of a stereo that originally sells for x dollars.

   (a) $x - 0.40$
   (b) $x - 0.40$
   (c) $x + 0.40$
   (d) $0.40x$
   (e) None of these

7. Two brothers, Bob and Bill, live 450 miles apart. Starting at the same time they plan to drive until they meet. Bill averages 10 miles per hour faster than Bob who averages 50 mph. How long will it take them to meet?

   (a) $3\frac{2}{15}$ hours
   (b) $3\frac{9}{11}$ hours
   (c) $4\frac{1}{11}$ hours
   (d) $4\frac{1}{2}$ hours
   (e) None of these

8. Solve by completing the square: $1 - x = x(x + 3)$.

   (a) $x = -\dfrac{3}{2} \pm \dfrac{\sqrt{13}}{2}$
   (b) $-2 \pm \sqrt{5}$
   (c) $-1 \pm \sqrt{2}$
   (d) $\dfrac{3}{2} \pm \dfrac{\sqrt{11}}{2}$
   (e) None of these

9. Solve: $4x^2 + 12x = 135$.

   (a) $-\dfrac{9}{2}, \dfrac{15}{2}$
   (b) $-\dfrac{5}{2}, \dfrac{3}{2}$
   (c) $-\dfrac{15}{2}, \dfrac{9}{2}$
   (d) $\dfrac{-3 \pm \sqrt{6}}{2}$
   (e) None of these

10. Multiply: $(4 - \sqrt{-9})^2$.

    (a) 7
    (b) $7 - 24i$
    (c) $25 - 24i$
    (d) $7 - 12i$
    (e) None of these

11. Solve for x: $\sqrt{15x + 4} = 4 - \sqrt{2x + 3}$.
    (a) 3
    (b) $\frac{11}{169}$
    (c) $3, \frac{11}{169}$
    (d) $-3, -\frac{11}{169}$
    (e) None of these

12. Solve the inequality: $6 - 5x \leq x - 6$.
    (a) $[0, \infty)$
    (b) $[2, \infty)$
    (c) $(-\infty, 2]$
    (d) $(-\infty, 0)$
    (e) None of these

13. Solve the inequality: $3x^3 - 6x^2 > 0$.
    (a) $(-\infty, 0) \cup (2, \infty)$
    (b) $(0, 2)$
    (c) $(-\infty, 0)$
    (d) $(2, \infty)$
    (e) None of these

14. Use a graphing utility to solve $x^2 - 4x + 2 < 0$.
    (a) $(-3.41, -0.59)$
    (b) $(0.59, 3.41)$
    (c) $(-\infty, -3.41) \cup (-0.59, \infty)$
    (d) $(-\infty, 0.59) \cup (3.41, \infty)$
    (e) None of these

15. Solve the inequality: $\dfrac{2}{x + 2} \geq \dfrac{3}{x - 1}$.
    (a) $[-8, \infty)$
    (b) $[-8, -2) \cup (1, \infty)$
    (c) $(-\infty, -8] \cup (-2, 1)$
    (d) $(-\infty, -8]$
    (e) None of these

Test D Name _____ Date _____

Chapter 1 Class _____ Section _____

1. Use a graphing utility to graph $y = x^2 - 5x + 6$. Use a standard setting. Approximate any intercepts.

2. Determine the standard equation of the circle with radius 7 and center $(-2, -4)$.

3. Solve the equation $3(x + 3) = 2 - (1 - 2x)$.

4. Solve $0.55 + 13.9(2.1 - x) = 14$ for x. Round your results to two decimal places.

5. Write an algebraic expression for the time to travel 525 miles at an average speed of r miles per hour.

6. A truck driver averaged 60 mph on a 600 mile trip and averaged 40 mph on the return trip. What was the average speed for the round trip?

    (a) 55 mph   (b) 45 mph   (c) 48 mph

    (d) 50 mph   (e) None of these

7. Solve for x: $4x^2 + 12x + 9 = 0$.

8. Solve for x: $x^2 - 3x + \dfrac{3}{2} = 0$.

9. Divide, then write the result in standard form: $\dfrac{3 + 7i}{3 - 7i}$.

10. Solve for x: $3x + 5 = \sqrt{2 - 2x}$.

11. Use a graphing utility to graph $y = x^3 - 7x^2 + 12x$. Approximate any x-intercepts. Set $y = 0$ and solve the equation.

12. Graph the solution: $|3x - 1| > 9$.

13. Solve the inequality: $(x - 2)^2 \leq 9$.

14. Find the domain of $\sqrt{36 - x^2}$.

15. A projectile is fired straight upward from ground level with an initial velocity of 64 feet per second. When will the height be less than 48 feet?

Test E Name _____ Date _____
Chapter 1 Class _____ Section _____

1. Use a graphing utility to graph $y = x^2 - 2x - 3$. Use a standard setting. Approximate any intercepts.

2. Determine the standard equation of the circle with radius 16 and center $\left(0, \frac{1}{2}\right)$.

3. Solve the equation $3(x - 6) = 2 + 2(x - 5)$.

4. Solve $0.68 + 12.5(3.2 - x) = 16$ for x. Round your results to two decimal places.

5. Write an expression for the area of the region.

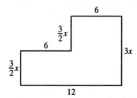

6. Two cars, starting together, travel in opposite directions on a highway, one at 55 mph and the other at 45 mph. How far apart are they after $2\frac{1}{2}$ hours?

7. Solve for x: $3x^2 + 19x - 14 = 0$.

8. Solve for x: $-3x^2 + 4x + 6 = 0$.

9. Divide, then write the result in standard form: $\dfrac{3 - 7i}{3 + 7i}$.

10. Solve for x: $\sqrt{x + 1} = 9 - \sqrt{x}$.

11. Use a graphing utility to graph $y = |x^2 - 1| - 5$. Approximate any x-intercepts. Set $y = 0$ and solve the equation.

12. Solve the inequality algebraically: $|x + 5| \leq 2$.

13. Solve the inequality: $x^2 - x > 6$.

14. Find the domain of $\sqrt{16 - 4x^2}$.

15. A projectile is fired straight upward from ground level with an initial velocity of 64 feet per second. When will the height be more than 48 feet?

Test A

Chapter 2

Name _____ Date _____

Class _____ Section _____

1. Find the slope of the line passing through (6, 10) and (−1, 4).

   (a) $\frac{7}{6}$ (b) $-\frac{7}{6}$ (c) $\frac{6}{7}$

   (d) $-\frac{6}{7}$ (e) None of these

2. What is the slope of a line that is perpendicular to the line given by $2x + 3y + 9 = 0$?

   (a) $\frac{2}{3}$ (b) $-\frac{2}{3}$ (c) $\frac{3}{2}$

   (d) $-\frac{3}{2}$ (e) None of these

3. Find the equation of the line that passes through (1, 3) and is perpendicular to the line $2x + 3y + 5 = 0$.

   (a) $3x - 2y + 3 = 0$ (b) $2x + 3y - 11 = 0$ (c) $2x + 3y - 9 = 0$

   (d) $3x - 2y - 7 = 0$ (e) None of these

4. In which of the following equations is y a function of x?

   (a) $3y + 2x - 9 = 17$ (b) $2x^2 + x = 4y$

   (c) Both a and b (d) Neither a nor b

5. Given $f(x) = \begin{cases} 7x - 10, & x \leq 2 \\ x^2 + 6, & x > 2 \end{cases}$, find $f(0)$.

   (a) −10 (b) 0 (c) −4

   (d) 6 (e) None of these

6. Which of the functions fits the data?

x	−2	0	1	3	5	10
y	−6	0	3	9	15	30

   (a) $f(x) = x^3$ (b) $f(x) = \sqrt[3]{x}$ (c) $f(x) = |x|^3$

   (d) $f(x) = 3x$ (e) None of these

7. Find the domain of the function $g(x) = \dfrac{5x}{x^2 - 7x + 12}$.

8. Use the vertical line test to determine in which case y is a function of x.

   (a)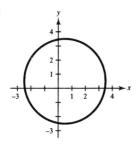

   (b)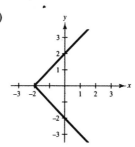

   (c)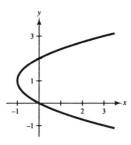

   (d)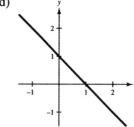

   (e) None of these

9. Describe the transformation of the graph of $f(x) = x^2$ for the graph of $g(x) = (x + 9)^2$.

   (a) Vertical shift 9 units up
   (b) Vertical shift 9 units down
   (c) Horizontal shift 9 units to the right
   (d) Horizontal shift 9 units to the left
   (e) None of these

10. Given $f(x) = x - 2$ and $g(x) = 6 - 2x$, find $(f + g)(-2)$.

    (a) 6
    (b) 2
    (c) -2
    (d) -14
    (e) None of these

11. Given $f(x) = x^2 - 2x$ and $g(x) = 3x + 2x$, find $(f \circ g)(x)$.

    (a) $4x^2 + 8x + 3$
    (b) $2x^2 - 4x + 3$
    (c) $2x^3 - x^2 - 6x$
    (d) $3x^2 + x$
    (e) None of these

12. Graphically, determine which sets of functions are not inverses of each other.

    (a) $f(x) = 9 + x$

    $g(x) = 9 - x$

    (b) $f(x) = x^2$

    $g(x) = -x^2$

    (c) $f(x) = \dfrac{x+3}{3}$

    $g(x) = \dfrac{3}{x+3}$

    (d) All of these are inverses of each other.

    (e) None of these are inverses of each other.

13. Find the inverse of the function: $f(x) = \dfrac{x+3}{2}$.

    (a) $2x - \dfrac{2}{3}$

    (b) $\dfrac{2}{x+3}$

    (c) $2x - 6$

    (d) $2x - 3$

    (e) None of these

Test B Name _____ Date _____
Chapter 2 Class _____ Section _____

1. Find the slope of the line passing through $(-1, 16)$ and $(4, 2)$.
   - (a) $-\frac{5}{14}$
   - (b) $-\frac{14}{5}$
   - (c) $\frac{5}{14}$
   - (d) $\frac{14}{5}$
   - (e) None of these

2. What is the slope of the line parallel to the line $4x - 2y = 9$.
   - (a) $\frac{9}{2}$
   - (b) $\frac{9}{4}$
   - (c) $-\frac{2}{4}$
   - (d) 2
   - (e) None of these

3. Find the equation of the line that passes through $(2, -1)$ and is parallel to the line $2x + 7y = 5$.
   - (a) $2x - 7y - 11 = 0$
   - (b) $2x + 7y + 3 = 0$
   - (c) $2x + 7y - 12 = 0$
   - (d) $7x - 2y - 16 = 0$
   - (e) None of these

4. In which of the following equations is y a function of x?
   - (a) $2x + 3y - 1 = 0$
   - (b) $x^2 + 3y^2 = 7$
   - (c) $2x^2y = 7$
   - (d) Both a and b
   - (e) Both a and c

5. Given $f(x) = \begin{cases} 3x + 4, & x \leq 2 \\ x^2 + 1, & x > 2 \end{cases}$, find $f(3)$.
   - (a) 13
   - (b) 10
   - (c) 5
   - (d) 3
   - (e) None of these

6. For what values of x does $f(x) = g(x)$? $f(x) = 3x + 1$ $g(x) = x^2 - 3$
   - (a) 0
   - (b) 4, 1
   - (c) $-4, -1$
   - (d) $4, -1$
   - (e) None of these

7. Find the domain of the function $f(x) = \dfrac{2x - 1}{2x + 1}$.

8. Use the vertical line test to determine in which case y is a function of x.

   (a)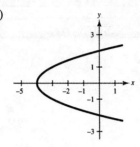

   (b)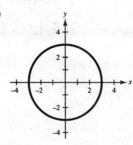

   (c)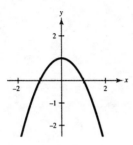

   (d)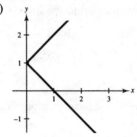

   (e) None of these

9. Describe the transformation of the graph of $f(x) = |x|$ for the graph of $g(x) = |x| - 20$.

   (a) Vertical shift 20 units up
   (b) Vertical shift 20 units down
   (c) Horizontal shift 20 units to the right
   (d) Horizontal shift 20 units to the left
   (e) None of these

10. Given $f(x) = x$ and $g(x) = x^2 - 7$, find $(fg)(3)$.

    (a) -13
    (b) 29
    (c) 5
    (d) 6
    (e) None of these

11. Given $f(x) = 4 - 2x^2$ and $g(x) = 2 - x$, find $(f \circ g)(x)$.

    (a) $4x^2 - 16x + 20$
    (b) $2x^2 - 4$
    (c) $2x^2 - 2$
    (d) $-2x^3 - 4x^2 - 4x + 8$
    (e) None of these

12. Graphically determine which sets of functions are not inverses of each other.

   (a) $f(x) = x + 5$
   $g(x) = x - 5$

   (b) $f(x) = x^3$
   $g(x) = \sqrt[3]{x}$

   (c) $f(x) = \dfrac{x+2}{4}$
   $g(x) = 4x - 2$

   (d) All of these are inverses of each other.

   (e) None of these are inverses of each other.

13. Find the inverse of the function: $f(x) = \dfrac{4 + 5x}{7}$.

   (a) $\dfrac{7}{5}(x - 4)$

   (b) $\dfrac{1}{5}(7x - 4)$

   (c) $-\dfrac{7}{4} - \dfrac{7}{5x}$

   (d) $\dfrac{7}{4 + 5x}$

   (e) None of these

Test C

Chapter 2

Name _____ Date _____

Class _____ Section _____

1. Find the slope of the line passing through $(3, -2)$ and $(5, 7)$.
   - (a) $-\frac{9}{7}$
   - (b) $\frac{9}{2}$
   - (c) $\frac{5}{2}$
   - (d) $\frac{2}{9}$
   - (e) None of these

2. Find the slope of the line perpendicular to the line $3x - 4y = 12$.
   - (a) Undefined
   - (b) 0
   - (c) $\frac{4}{3}$
   - (d) $-\frac{3}{4}$
   - (e) None of these

3. Find an equation of the line that passes through $(6, 2)$ and is perpendicular to the line $3x + 2y = 2$.
   - (a) $y = -\frac{3}{2}x + 11$
   - (b) $y = -\frac{2}{3}x + 6$
   - (c) $y = \frac{3}{2}x - 7$
   - (d) $y = \frac{2}{3}x - 2$
   - (e) None of these

4. In which of the following equations is y a function of x?
   - (a) $3y + 2x - 7 = 0$
   - (b) $5x^2y = 9 - 2x$
   - (c) $3x^2 - 4y^2 = 9$
   - (d) $x = 3y^2 - 1$
   - (e) Both a and b

5. Given $f(x) = \begin{cases} 2x - 1, & x \leq -2 \\ x + 6, & x > -2 \end{cases}$, find $f(-6)$.
   - (a) -11
   - (b) -13
   - (c) 0
   - (d) 11
   - (e) None of these

6. Find all real values of x for which $f(x) = g(x)$: $f(x) = x^4 + 3x^2$ $g(x) = 7x^2$
   - (a) $-4, 0, 4$
   - (b) $0, 2$
   - (c) 2
   - (d) $0, 2, -2$
   - (e) None of these

7. Find the domain of the function: $f(x) = \dfrac{1}{x + 2}$.

8. Use the vertical line test to determine in which case y is a function of x.

(a)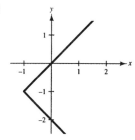

(b)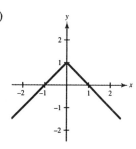

(c)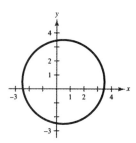

(d)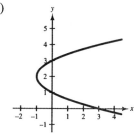

(e) None of these

9. Describe the transformation of the graph of $f(x) = \sqrt{x}$ for the graph of $g(x) = \sqrt{x - 5}$.

   (a) Vertical shift 5 units up
   (b) Vertical shift 5 units down
   (c) Horizontal shift 5 units to the right
   (d) Horizontal shift 5 units to the left
   (e) None of these

10. Given $f(x) = 9x + 1$ and $g(x) = 4 - x$, find $(f - x)(5)$.

    (a) 37
    (b) 47
    (c) 55
    (d) −46
    (e) None of these

11. Given $f(x) = \dfrac{1}{x^2}$ and $g(x) = \sqrt{x^2 + 4}$, find $(f \circ g)(x)$.

    (a) $\dfrac{1}{x^2 + 4}$
    (b) $\dfrac{1}{\sqrt{x^2 + 4}}$
    (c) $x^2 + 4$
    (d) $\dfrac{1}{x^2\sqrt{x^2 + 4}}$
    (e) None of these

12. Graphically determine which sets of functions are not inverses of each other.

(a) $f(x) = \dfrac{1}{2}x - 1$

$g(x) = 2x + 1$

(b) $f(x) = \sqrt[5]{x}$

$g(x) = \dfrac{1}{\sqrt[5]{x}}$

(c) $f(x) = \dfrac{x-1}{5}$

$g(x) = x + \dfrac{1}{5}$

(d) All of these are inverses of each other.

(e) None of these are inverses of each other.

13. Find the inverse of the function: $f(x) = \dfrac{2}{3x+1}$.

(a) $\dfrac{3x-1}{2}$
(b) $\dfrac{2-x}{3x}$
(c) $\dfrac{3x+1}{2}$
(d) $\dfrac{1-x}{2}$
(e) None of these

Test D Name _____ Date _____
Chapter 2 Class _____ Section _____

1. Find the slope of the line passing through $(5, 9)$ and $(-1, -3)$.

2. Find an equation of the line that passes through $(8, 17)$ and is perpendicular to the line $x + 2y = 2$.

3. Morgan Sporting Goods had net sales of $150,000 in January of this past year. In March, their net sales were $300,000. Assuming that their sales are increasing linearly, write an equation of net sales, S, in terms of the month using $t = 1$ for January.

4. Select which of the two equations represents y as a function of x and specify why the other does not.
    (a) $x^2 + 5y - x = 7$
    (b) $y^2 - 5x = 7$

5. $F(x) = -1 + 2x - x^2$. Find $F(k + 1)$ and simplify.

6. Find the domain of the function: $g(x) = \dfrac{x}{x^2 + 1}$.
    (a) All real $x \neq 1$
    (b) All real $x \neq -1$
    (c) All real $x \neq 0$
    (d) All real x
    (e) None of these

7. Determine the domain and range of the function $f(x) = 3 - |x|$.

8. Determine the interval(s) over which the function is increasing: $y = \frac{2}{3}x^3 - x^2$.

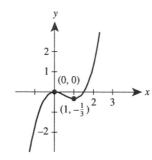

9. Given the graph of $y = x^4$ sketch the graph of $y = (x-2)^4 + 6$.

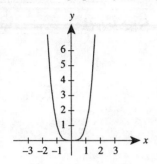

10. Use a graphing utility to graph $(f/g)(x)$ if $f(x) = 2x^2 - x$ and $g(x) = x$.

11. Given $f(x) = \dfrac{1}{x}$ and $g(x) = \dfrac{1}{x}$, find $(f \circ g)(9)$.

12. Graphically, determine whether the functions $f(x) = (x+1)^3$ and $g(x) = \sqrt[3]{x} - 1$ are inverses of each other.

13. Given $f(x) = 2x^2 + 1$ for $x \geq 0$, find $f^{-1}(x)$.

Test E Name _____ Date _____
Chapter 2 Class _____ Section _____

1. Find the slope of the line passing through $(3, 7)$ and $(-1, -2)$.

2. Find an equation of the line that passes through $(3, 5)$ and is perpendicular to the line $x + 3y = 6$.

3. Curtis Area Schools had an enrollment of 2800 students in 1980 and 12,600 in 1988. Assuming the growth is linear, write an equation of the enrollment, E, in terms of the year using $t = 0$ for 1980.

4. Select which of the two equations represents y as a function of x and specify why the other does not.

    (a) $|x| + y = 4$ \qquad (b) $x + |y| = 4$

5. $F(x) = 5 + 2x - x^2$. Find $F(k + 1) - F(k)$ and simplify.

6. Find the domain of the function $h(x) = \dfrac{x + 4}{x(x - 5)}$.

7. Determine the domain and range of the function $f(x) = 3 - x^2$.

8. Determine the interval(s) over which the function is increasing: $y = \dfrac{1}{x^2 - 1}$.

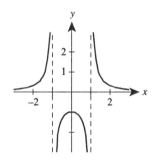

9. Given the graph of $y = x^2$ sketch the graph of $y = (x + 3)^2 - 1$.

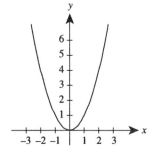

10. Use a graphing utility to graph $(fg)(x)$ if $f(x) = \dfrac{1}{x}$ and $g(x) = x + 2$.

11. Given $f(x) = x^3 + 4$ and $g(x) = \sqrt[3]{x}$, find $(f \circ g)(-3)$.

12. Graphically, determine whether the functions $f(x) = \sqrt{x^2 - 5}$ and $g(x) = x^2 + 5$ are inverses of each other.

13. Given $f(x) = \dfrac{2x + 1}{3}$, find $f^{-1}(x)$.

Test A
Chapter 3

Name _____ Date _____

Class _____ Section _____

1. Write in the form $y = a(x - h)^2 + k$: $y = 2x^2 + 16x + 9$
   (a) $y = 2(x + 4)^2 - 7$
   (b) $y = 2(x + 2)^2 + 5$
   (c) $y = 2(x + 4)^2 - 23$
   (d) $y = 2(x + 8)^2 + 73$
   (e) None of these

2. Find the quadratic function that has a maximum point at $(-1, 17)$ and passes through $(7, 1)$.
   (a) $y = \frac{1}{4}(-x^2 - 2x + 16)$
   (b) $y = -\frac{1}{4}(x + 1)^2 + 17$
   (c) $y = (x - 7)^2 + 1$
   (d) $y = (x - 1)^2 + 17$
   (e) None of these

3. Determine the left and right behavior of the graph: $y = 4x^2 - 2x + 1$.
   (a) Up to the left, down to the right
   (b) Down to the left, up to the right
   (c) Up to the left, up to the right
   (d) Down to the left, down to the right
   (e) None of these

4. Find a polynomial function with the given zeros: $-2, -2, 1, 3$
   (a) $f(x) = (x + 1)(x + 3)(x - 2)$
   (b) $f(x) = (x - 2)^2(x - 1)(x - 3)$
   (c) $f(x) = (x + 2)(x - 1)(x - 3)$
   (d) $f(x) = (x + 2)^2(x - 1)(x - 3)$
   (e) None of these

5. Divide $(9x^3 - 6x^2 - 8x - 3) \div (3x + 2)$
   (a) $3x^2 - \frac{8}{3}x - \frac{7/3}{3x + 2}$
   (b) $3x^2 - 4x - 2 + \frac{7}{3x + 2}$
   (c) $3x^2 - 4x - \frac{3}{3x + 2}$
   (d) $3x^2 - 4x - \frac{16}{3} + \frac{23/3}{3x + 2}$
   (e) None of these

6. Use synthetic division to determine which of the following is a solution of the equation: $3x^4 - 2x^3 + 26x^2 - 18x - 9 = 0$
   (a) 3
   (b) 1
   (c) -3
   (d) $\frac{1}{3}$
   (e) None of these

7. Use Descarte's Rule of Signs to determine the possible number of positive and negative zeros: $f(x) = 5x^4 - 3x^3 - 4x + 2$
   (a) 2 positive, 2 negative
   (b) 2 or 0 positive, 0 negative
   (c) 4 positive, 0 negative
   (d) 0 positive, 4 negative
   (e) None of these

8. Find all of the real zeros of the function: $f(x) = 4x^3 - 3x - 1$
   (a) $1, -\frac{1}{2}$
   (b) $1, \frac{1}{2}, -\frac{1}{2}$
   (c) $\frac{1}{2}, 1$
   (d) 1
   (e) None of these

9. Write as a product of linear factors: $f(x) = x^4 - 5x^3 + 8x^2 - 20x + 16$
   (a) $(x + 2)(x - 2)(x - 4)(x - 1)$
   (b) $(x + 4)(x + 1)(x - 2i)(x + 2i)$
   (c) $(x - 4)(x - 1)(x + 2i)(x - 2i)$
   (d) $(x + 4)(x + 1)(x + 2i)(x + 2i)$
   (e) None of these

10. Find all of the zeros of the function: $f(x) = x^3 - \frac{9}{2}x^2 + \frac{11}{2}x - \frac{3}{2}$
    (a) $\frac{3}{2}, \frac{3 \pm \sqrt{5}}{2}$
    (b) $1, -2, 3$
    (c) $\frac{3}{2}, \frac{3 \pm \sqrt{5}i}{2}$
    (d) $\frac{3}{2}, 1 \pm i$
    (e) None of these

11. Which of the following sentences describes the formula to find the volume of a right circular cone, $V = \frac{\pi r^2 h}{3}$?
    (a) The volume of a right circular cone is directly proportional to the height and jointly proportional to the square of the radius.
    (b) The volume of a right circular cone is indirectly proportional to the height and radius.
    (c) The volume of a right circular cone varies inversely with the height and the square of the radius.
    (d) The volume of a right circular cone is jointly proportional to the height and the square of the radius.
    (e) None of these

12. V varies jointly with P and the square of Q and inversely with the cube of S. $V = 8$ when $P = 12$, $Q = 4$ and $S = 2$. Find V when $P = 6$, $Q = 1$ and $S = 1$.
    (a) 16   (b) 8   (c) 4   (d) 2   (e) None of these

Test B

Chapter 3

Name _____ Date _____

Class _____ Section _____

1. Write in the form $y = a(x - h)^2 + k$: $y = -2x^2 - 4x - 5$
   (a) $y = -2(x - 1)^2 - 2$
   (b) $y = (2x - 2)^2 - 1$
   (c) $y = -2(x + 2)^2 - 1$
   (d) $y = -2(x + 1)^2 - 3$
   (e) None of these

2. Find the quadratic function that has a minimum at $(1, -2)$ and passes through $(0, 0)$.
   (a) $y = 2(x - 1)^2 - 2$
   (b) $y = 2(x + 1)^2 - 2$
   (c) $y = -2(x - 1)^2 + 2$
   (d) $y = -2(x + 1)^2 + 2$
   (e) None of these

3. Determine the left and right behavior of the graph: $f(x) = -x^5 + 2x^2 - 1$.
   (a) Up to the left, down to the right
   (b) Down to the left, up to the right
   (c) Up to the left, up to the right
   (d) Down to the left, down to the right
   (e) None of these

4. Find a polynomial function with zeros: $1, 0, -3$
   (a) $f(x) = x(x - 3)^3(x + 1)^2$
   (b) $f(x) = x^2(x - 1)(x + 3)$
   (c) $f(x) = x(x - 3)(x - 1)$
   (d) $f(x) = (x - 1)(x + 3)^2$

5. Divide: $(6x^3 + 7x^2 - 15x + 6) \div (2x - 1)$
   (a) $3x^2 + 2x - \dfrac{17}{2} - \dfrac{5}{2(2x - 1)}$
   (b) $3x^2 + 5x - 5 + \dfrac{1}{2x - 1}$
   (c) $3x^2 + 5x + 5 + \dfrac{11}{2x - 1}$
   (d) $3x^2 + 4x - 17 + \dfrac{29/2}{2x - 1}$
   (e) None of these

6. Use synthetic division to determine which of the following are solutions of the equation: $3x^3 - 11x^2 - 6x + 8 = 0$
   (a) $\tfrac{2}{3}$
   (b) -1
   (c) 4
   (d) All of these
   (e) None of these

7. Use Descarte's Rule of Signs to determine the possible number of positive and negative zeros: $f(x) = x^3 + 2x - 1$
   (a) 1 positive, 0 negative
   (b) 0 positive, 1 negative
   (c) 3 or 1 positive, 0 or 2 negative
   (d) 1 positive, 2 negative
   (e) None of these

8. Find all of the real roots: $x^3 - 7x + 6 = 0$

   (a) $-3, 1, 2$ (b) $-2, -1, 3$ (c) $-6, -1, 1$

   (d) $-1, 1, 6$ (e) None of these

9. Write as a product of linear factors: $f(x) = x^4 - 6x^3 - 4x^2 + 40x + 32$

   (a) $(x - 4)(x + 2)(x + 2 + \sqrt{8})(x + 2 - \sqrt{8})$ (b) $(x + 4)(x - 2)(x - 2 + \sqrt{8})(x - 2 - \sqrt{8})$

   (c) $(x - 4)(x - 2)(x - 2 + \sqrt{8})(x - 2 - \sqrt{8})$ (d) $(x + 4)(x + 2)(x + 2 + \sqrt{8})(x + 2 - \sqrt{8})$

   (e) None of these

10. Find all of the zeros of the function: $f(x) = x^3 + 6x^2 + 12x + 7$

    (a) $-1, 7$ (b) $-1, \dfrac{-5 \pm \sqrt{3}}{2}$ (c) -1

    (d) $-1, \dfrac{-5\sqrt{3}i}{2}$ (e) None of these

11. Which of the following sentences describes the formula to find the lateral surface area of a right circular cylinder, $S = 2\pi rh$?

    (a) The lateral surface area of a right circular cylinder is inversely proportional to the radius and the height.

    (b) The lateral surface area of a right circular cylinder is jointly proportional to the radius and the height.

    (c) The lateral surface area of a right circular cylinder is directly proportional to the radius and the height.

    (d) The lateral surface area of a right circular cylinder varies indirectly with the radius and height.

    (e) None of these

12. x varies jointly with y and the cube of z and inversely with w. $x = 64$ when $y = 2, z = 4$ and $w = 6$. Find x when $y = 1, z = 3$ and $w = 15$.

    (a) $\dfrac{84}{165}$ (b) $\dfrac{3}{7}$ (c) 16 (d) $\dfrac{27}{5}$ (e) None of these

Test C

Chapter 3

Name _____ Date _____

Class _____ Section _____

1. Write in the form $y = a(x - h)^2 + k$: $y = 3x^2 + 12x + 17$
   (a) $y = (x + 2)^2 + \frac{13}{3}$
   (b) $y = 3(x + 2)^2 + 21$
   (c) $y = 3(x + 2)^2 + 5$
   (d) $y = (x + 2)^2 + \frac{29}{3}$
   (e) None of these

2. Find the quadratic function whose graph opens upward and has x-intercepts at $(-4, 0)$ and $(1, 0)$.
   (a) $y = x^2 + 3x - 4$
   (b) $y = 4 - 3x - x^2$
   (c) $y = x^2 + 5x + 4$
   (d) $y = x^2 - 3x - 4$
   (e) None of these

3. Determine the left and right behavior of the graph: $f(x) = 3x^5 - 7x^2 + 2$.
   (a) Down to the left, up to the right
   (b) Down to the left, up to the right
   (c) Up to the left, up to the right
   (d) Down to the left, down to the right
   (e) None of these

4. Find a polynomial function with zeros: $0, 1, -2$
   (a) $f(x) = x(x - 1)(x - 2)$
   (b) $f(x) = x(x + 1)(x + 2)$
   (c) $f(x) = (x - 1)(x + 2)$
   (d) $f(x) = x^2(x - 1)(x + 2)$
   (e) None of these

5. Divide: $(3x^4 + 2x^3 - 3x + 1) \div (x^2 + 1)$
   (a) $3x^2 + 2x + 3 - \dfrac{5x + 2}{x^2 + 1}$
   (b) $3x^2 + 2x - 3 + \dfrac{-5x + 4}{x^2 + 1}$
   (c) $3x^2 - x^2 - 4 + \dfrac{5}{x^2 + 1}$
   (d) $3x^2 - x + 1 + \dfrac{-4x + 5}{x^2 + 1}$
   (e) None of these

6. Use synthetic division to determine which of the following is a solution of the equation: $6x^4 - 11x^3 - 10x^2 + 19x - 6 = 0$
   (a) 2
   (b) 3
   (c) -2
   (d) -3
   (e) None of these

7. Use Descarte's Rule of Signs to determine the possible number of positive and negative zeros: $f(x) = 6x^5 - 6x^3 + 10x + 5$
   (a) 4 or 2 or 0 positive, 1 negative
   (b) 3 or 1 positive, 2 or 4 negative
   (c) 2 or 0 positive, 3 or 1 negative
   (d) 1 positive, 0 negative
   (e) None of these

8. Find all of the real roots: $2x^3 + 5x^2 - x - 6 = 0$

    (a) $-3, -1, 1$ (b) $-1, \frac{3}{2}, 2$ (c) $-2, -\frac{3}{2}, 1$

    (d) $-6, 2, 5$ (e) None of these

9. Write as a product of linear factors: $f(x) = x^4 + 2x^3 - 5x^2 - 18x - 36$

    (a) $(x + 3)(x + 3)(x + 1 + \sqrt{3})(x + 1 - \sqrt{3})$ (b) $(x - 3)(x - 3)(x - 1 + \sqrt{6})(x - 1 - \sqrt{6})$

    (c) $(x - 3)(x + 3)(x + 1 + \sqrt{3}i)(x + 1 - \sqrt{3}i)$ (d) $(x - 3)(x + 3)(x - 1 + \sqrt{3}i)(x - 1 - \sqrt{3}i)$

    (e) None of these

10. Find all of the zeros of the function: $f(x) = x^4 - 5x^3 + 8x^2 - 20x + 16$

    (a) $1, 4, \pm 2$ (b) $-4, -1, \pm 2i$ (c) $1, 4, \pm 2i$

    (d) $-4, -1, -2i, -2i$ (e) None of these

11. Write a sentence using variation terminology to describe the formula to find the volume of a right circular cylinder, $V = \pi r^2 h$.

    (a) The volume is directly proportional to radius and height.

    (b) The volume is inversely proportional to the height and the square of the radius.

    (c) The volume is jointly proportional to the height and the square of the radius.

    (d) The volume varies inversely with the height and the square of the radius.

    (e) None of these

12. z varies directly as the square of x and inversely as y. $z = \frac{3}{2}$ when $x = 3$ and $y = 4$. Find z when $x = 12$ and $y = 6$.

    (a) $\frac{207}{5}$ (b) 16 (c) 3 (d) $\frac{21}{2}$ (e) None of these

Test D

Chapter 3

Name _____ Date _____

Class _____ Section _____

1. Write the standard form of the equation of the parabola.

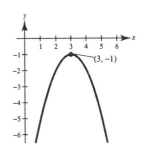

2. Find the quadratic function that has a maximum at $(-1, 2)$ and passes through $(0, 1)$.

3. Determine the left and right behavior of the graph: $f(x) = -4x^3 + 3x^2 - 1$.

4. An open box is to be made from a 16-inch square piece of material by cutting equal squares from each corner and turning up the sides. Verify the volume of the box is $V(x) = 4x(8 - x)^2$. Sketch the graph of the function using a graphing utility and use the graph to estimate the volume of x for which $V(x)$ is maximum.

5. Divide: $(6x^4 - 4x^3 + x^2 + 10x - 1) \div (3x + 1)$

6. A rectangular room has a volume of $4x^3 - 7x^2 - 16x + 3$ cubic feet. The height of the room is $x - 3$. Find the algebraic expression for the number of square feet of floor space in the room.

7. List the possible rational zeros of the function: $f(x) = 2x^4 - 3x^2 + 15$

8. List the possible rational zeros of the function. Then use a graphing utility to graph the function to eliminate some of the possible zeros. Finally determine all real zeros of the function: $f(x) = 3x^3 - x^2 - 12x + 4$

9. Write as a product of linear factors: $x^2 - 16$

10. Use the fact that $1 - 2i$ is zero of f to find the remaining zeros:
$f(x) = x^3 - 3x^2 + 7x - 5$

11. x varies inversely with the square of y. If $x = 1$ when $y = 5$, find the constant of proportionality.

12. The table shows the per capita consumption of broccoli, y (in pounds), for the years 1980 through 1989.

Year, x	1980	1981	1982	1983	1984	1985	1986	1987	1988	1989
Pounds, y	1.6	1.8	2.2	2.3	2.7	2.9	3.5	3.6	4.2	4.5

Construct a scatter plot for the data, let $t = 0$ represent 1980. Find the least squares regression line that fits this data, and graph the linear model on the same set of axes as the scatter plot.

Test E

Chapter 3

Name _____ Date _____

Class _____ Section _____

1. Write the form $y = a(x - h)^2 + k$: $y = -x^2 + 3x - 2$.

2. Find the quadratic function whose graph opens upward and has x-intercepts at $(0, 0)$ and $(6, 0)$.

3. Determine the left and right behavior of the graph: $f(x) = 3x^4 + 2x^3 + 7x^2 + x - 1$.

4. An open box is to be made from a 20-inch square piece of material by cutting equal squares from each corner and turning up the sides. Verify the volume of the box is $V(x) = 4x(10 - x)^2$. Sketch the graph of the function using a graphing utility and use the graph to estimate the volume of x for which $V(x)$ is maximum.

5. Divide: $(2x^4 + 7x - 2) \div (x^2 + 3)$.

6. A rectangular room has a volume of $2x^3 - 17x + 3$ cubic feet. The height of the room is $x + 3$. Find the algebraic expression for the number of square feet of floor space in the room.

7. Given $f(x) = 3x^3 + 4x - 1$, determine whether $x = -2$ is an upper bound for the zeros of f, a lower bound for the zeros of f, or neither.

8. List the possible rational zeros of the function. Then use a graphing utility to graph the function to eliminate some of the possible zeros. Finally determine all real zeros of the function: $f(x) = 2x^3 - x^2 - 18x + 9$.

9. Write as a product of linear factors: $f(x) = x^4 - 100$.

10. Use the fact that $2 \pm \sqrt{3}i$ is a zero of f to find the remaining zeros: $f(x) = x^4 - 6x^3 + 12x^2 - 2x - 21$.

11. x varies jointly with y and the square of z and inversely with w. If $x = 4/3$ when $y = 1$, $z = -2$, and $w = 7$, find x when $y = 2$, $z = -1$, and $w = 6$.

12. The table shows the total amount, y in millions of dollars, spent by the federal government on mathematical research from 1980 through 1990.

Year, x	1980	1981	1982	1983	1984	1985	1986	1987	1988	1989	1990
Amount, y	91	118	128	134	151	184	185	205	212	230	245

Construct a scatter plot for the data, let $t = 0$ represent 1980. Find the least squares regression line that fits the data and graph the model on the same set of axes as the scatter plot.

Test A

Chapter 4

Name _____ Date _____
Class _____ Section _____

1. Find the domain: $f(x) = \dfrac{x+2}{x^2 - 3x + 2}$

   (a) All reals except $x = -2, 1, 2$
   (b) All reals except $x = -2$
   (c) All reals except $x = 1, 2$
   (d) All reals
   (e) None of these

2. Find the vertical asymptote(s): $f(x) = \dfrac{1}{(x+2)(x-5)}$

   (a) $x = -2, x = 5$
   (b) $y = 1$
   (c) $y = 0$
   (d) $y = 1, y = 0$
   (e) None of these

3. Find the horizontal asymptote(s): $f(x) = \dfrac{x^2 - 1}{x^2 + 9}$

   (a) $y = 1$
   (b) $y = 0$
   (c) $x = 1$
   (d) $x = \pm 1$
   (e) None of these

4. Match the graph with the correct function.

   (a) $f(x) = \dfrac{1}{2x+1}$
   (b) $f(x) = \dfrac{x-1}{2x+1}$
   (c) $f(x) = \dfrac{x^2 + 2x + 2}{2x - 1}$
   (d) $f(x) = \dfrac{x^3 + 2x^2 + x - 2}{2x + 1}$
   (e) None of these

   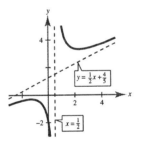

5. Find the slant asymptote: $f(x) = \dfrac{3x^2 + 2x - 1}{x - 1}$

   (a) $y = -3x + 5$
   (b) $y = 3x + 5$
   (c) $y = 3x - 5$
   (d) $y = -3x - 5$
   (e) None of these

6. Find the partial fraction decomposition: $\dfrac{4x + 23}{x^2 - x - 6}$

   (a) $\dfrac{4x}{x+2} + \dfrac{23}{x-3}$
   (b) $\dfrac{2}{x+2} - \dfrac{13}{x-3}$
   (c) $\dfrac{5}{x-3} - \dfrac{2}{x+2}$
   (d) $\dfrac{7}{x-3} - \dfrac{3}{x+2}$
   (e) None of these

7. Match the graph with the correct equation.

   (a) $\dfrac{x^2}{1} + \dfrac{y^2}{3} = 1$

   (b) $\dfrac{x^2}{3} + \dfrac{y^2}{1} = 1$

   (c) $\dfrac{x^2}{9} + \dfrac{y^2}{1} = 1$

   (d) $\dfrac{x^2}{2} + \dfrac{y^2}{9} = 1$

   (e) None of these

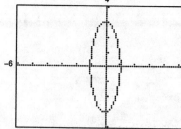

8. Find an equation of the hyperbola with center at (0, 0), vertices at (±3, 0), and foci at (±3√5, 0).

   (a) $\dfrac{x^2}{9} - \dfrac{y^2}{45} = 1$

   (b) $\dfrac{y^2}{9} - \dfrac{x^2}{45} = 1$

   (c) $\dfrac{x^2}{9} - \dfrac{y^2}{36} = 1$

   (d) $\dfrac{x^2}{9} - \dfrac{y^2}{54} = 1$

   (e) None of these

9. Write in standard form: $4x^2 + 9y^2 - 8x + 72y + 4 = 0$

   (a) $\dfrac{(x-1)^2}{36} + \dfrac{(y+4)^2}{16} = 1$

   (b) $\dfrac{(x-1)^2}{144} + \dfrac{(y+8)^2}{64} = 1$

   (c) $\dfrac{(x-1)^2}{13/4} + \dfrac{(y+4)^2}{13/9} = 1$

   (d) $\dfrac{(x-4)^2}{327} + \dfrac{(y+36)^2}{436/3} = 1$

   (e) None of these

10. Classify the graph of $3x^2 + 6x - 4y + 12 = 0$.

    (a) Circle

    (b) Hyperbola

    (c) Ellipse

    (d) Parabola

    (e) None of these

Test B

Chapter 4

Name _____ Date _____

Class _____ Section _____

1. Find the domain: $f(x) = \dfrac{x^2}{x+1}$

   (a) All reals (b) All reals except $x = -1$ (c) All reals except $x = 0$

   (d) All reals except $x = -1, 0$ (e) None of these

2. Find the vertical asymptote(s): $f(x) = \dfrac{x+3}{(x-2)(x+5)}$

   (a) $y = 2, y = -5, y = -3$ (b) $x = 2, x = -5, x = -3, x = 1$

   (c) $x = 1$ (d) $x = 2, x = -5$ (e) None of these

3. Find the horizontal asymptote(s): $f(x) = \dfrac{3x-2}{x+2}$

   (a) $y = 0$ (b) $x = -2$ (c) $x = \frac{1}{3}$

   (d) $y = 3$ (e) None of these

4. Match the graph with the correct function.

   (a) $f(x) = \dfrac{x+3}{x-1}$ (b) $f(x) = x + 3$

   (c) $f(x) = \dfrac{x-1}{x^2+2x-3}$ (d) $f(x) = \dfrac{x^2+2x-3}{x-1}$

   (e) None of these

   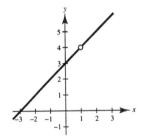

5. Find the slant asymptote: $f(x) = \dfrac{x^3 + 7x^2 - 1}{x^2 + 1}$

   (a) $y = 1$ (b) $y = x + 7$ (c) $y = x - 8$

   (d) $y = x + 1$ (e) None of these

6. Find the partial fraction decomposition: $\dfrac{7}{3x^2 + 5x - 2}$

   (a) $\dfrac{6}{(3x-2)} + \dfrac{1}{x-1}$ (b) $\dfrac{3}{3x-1} - \dfrac{1}{x+2}$ (c) $\dfrac{6}{x+2} - \dfrac{18}{3x-1}$

   (d) $\dfrac{4}{3x-1} + \dfrac{2}{x+2}$ (e) None of these

7. Match the graph with the correct equation.

   (a) $\dfrac{x^2}{16} - \dfrac{y^2}{4} = 1$ (b) $\dfrac{x^2}{4} - \dfrac{y^2}{16} = 1$

   (c) $\dfrac{y^2}{16} - \dfrac{x^2}{4} = 1$ (d) $\dfrac{y^2}{4} - \dfrac{x^2}{16} = 1$

   (e) None of these

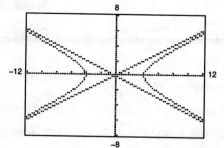

8. Find an equation of the hyperbola with center at $(0, 0)$, vertices at $(0, \pm 9)$, and asymptotes $y = \pm \tfrac{9}{2} x$.

   (a) $\dfrac{x^2}{2} - \dfrac{y^2}{9} = 1$ (b) $\dfrac{x^2}{81} - \dfrac{y^2}{4} = 1$ (c) $\dfrac{y^2}{81} - \dfrac{x^2}{4} = 1$

   (d) $\dfrac{x^2}{77} - \dfrac{y^2}{4} = 1$ (e) None of these

9. Write in standard form: $9x^2 - 4y^2 - 54x + 8y + 41 = 0$

   (a) $\dfrac{(y+1)^2}{41/4} - \dfrac{(x-3)^2}{41/9} = 1$ (b) $\dfrac{(x-3)^2}{31/9} - \dfrac{(y-1)^2}{31/4} = 1$

   (c) $\dfrac{(x-3)^2}{4} - \dfrac{(y-1)^2}{9} = 1$ (d) $\dfrac{(y+1)^2}{14} + \dfrac{(x-3)^2}{63/2} = 1$

   (e) None of these

10. Classify the graph of $2x^2 - 5y^2 + 4x - 6 = 0$.

    (a) Circle   (b) Hyperbola   (c) Ellipse

    (d) Parabola   (e) None of these

Test C

Chapter 4

Name _____ Date _____
Class _____ Section _____

1. Find the domain: $f(x) = \dfrac{x^3 - 1}{x^2 - 4}$

   (a) All reals (b) All reals except $x = 2$ (c) All reals except $x = 1$
   (d) All reals except $x = 1, 2$ (e) None of these

2. Find the vertical asymptote(s): $f(x) = \dfrac{x + 2}{x^2 - 9}$

   (a) $x = 3$ (b) $x = -2, x = -3, x = 3$ (c) $y = 0, x = -2$
   (d) $x = -3, x = 3$ (e) None of these

3. Find the horizontal asymptote(s): $f(x) = \dfrac{x^2 - 4}{x^2 - 9}$

   (a) $x = \pm 3$ (b) $x = \pm 3$ (c) $y = 1$
   (d) $y = 0$ (e) None of these

4. Match the graph with the correct function.

   (a) $f(x) = \dfrac{x - 5}{x + 3}$ (b) $f(x) = \dfrac{5 - x}{x + 3}$

   (c) $f(x) = \dfrac{x + 5}{x + 3}$ (d) $f(x) = \dfrac{x + 5}{x + 3}$

   (e) None of these

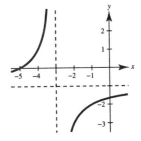

5. Find the slant asymptote: $f(x) = \dfrac{x^2 + 2x - 1}{x - 1}$

   (a) $y = 1$ (b) $y = x - 1$ (c) $y = x + 1$
   (d) $y = x + 3$ (e) None of these

6. Find the partial fraction decomposition: $\dfrac{5x + 3}{x^2 - 3x - 10}$

   (a) $\dfrac{2}{x + 5} - \dfrac{7}{x - 2}$ (b) $\dfrac{7}{x - 5} - \dfrac{2}{x + 2}$ (c) $\dfrac{2}{x - 5} + \dfrac{3}{x + 2}$

   (d) $\dfrac{4}{x - 5} + \dfrac{1}{x + 2}$ (e) None of these

7. Match the graph with the correct equation.

   (a) $\dfrac{x^2}{4} + \dfrac{y^2}{2} = 1$

   (b) $\dfrac{y^2}{4} + \dfrac{x^2}{2} = 1$

   (c) $\dfrac{x^2}{16} + \dfrac{y^2}{4} = 1$

   (d) $\dfrac{y^2}{16} - \dfrac{x^2}{4} = 1$

   (e) None of these

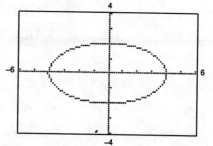

8. Find an equation of the hyperbola with vertices $(\pm 12, 0)$ and foci $(\pm 13, 0)$.

   (a) $\dfrac{x^2}{144} - \dfrac{y^2}{169} = 1$

   (b) $\dfrac{x^2}{169} - \dfrac{y^2}{144} = 1$

   (c) $\dfrac{y^2}{169} - \dfrac{x^2}{25} = 1$

   (d) $\dfrac{x^2}{144} - \dfrac{y^2}{25} = 1$

   (e) None of these

9. Write in standard form: $4x^2 - 5y^2 - 16x - 30y - 9 = 0$

   (a) $\dfrac{(x-4)^2}{11} - \dfrac{(y-3)^2}{4} = 1$

   (b) $\dfrac{(y+3)^2}{4} - \dfrac{(x-2)^2}{5} = 1$

   (c) $\dfrac{(y-3)^2}{6} - \dfrac{(x+2)^2}{9} = 1$

   (d) $\dfrac{(x+2)^2}{4} - \dfrac{(y+3)^2}{6} = 1$

10. Classify the graph of $3x^2 + 3y^2 - 4x + 5y - 16 = 0$.

    (a) Circle

    (b) Parabola

    (c) Ellipse

    (d) Hyperbola

    (e) None of these

Test D

Chapter 4

Name _____ Date _____

Class _____ Section _____

1. Find the vertical asymptote(s): $f(x) = \dfrac{x^2 - 9}{x^2 - 6x + 8}$

2. Find the domain: $f(x) = \dfrac{x}{x^2 + 3x - 4}$

3. Find the horizontal asymptote(s): $f(x) = \dfrac{3x^2 + 2x - 16}{x^2 - 7}$

4. Use a graphing utility to graph $f(x) = \dfrac{2}{x - 1}$.

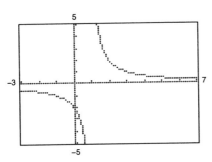

5. The population of a bacteria culture is given by

    $P = \dfrac{20t + 2}{5 + 0.2t}, \quad t \geq 0$

    where t is time in hours. Use a graphing utility using the indicated range setting, graph the function and determine the value which the population approaches.

    Xmin = 0
    Xmax = 200
    Xscl = 25
    Ymin = 0
    Ymax = 125
    Yscl = 25

6. Find the partial fraction decomposition: $\dfrac{3x^2 - 7x + 1}{(x - 1)^3}$

7. Use a graphing utility to graph $x^2 = 24(y - 2)$.

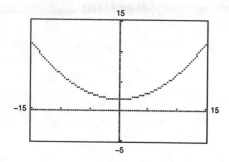

8. Find the standard equation of the ellipse with center at (0, 0), one focus at (3, 0), and a major axis of length 12.

9. Find an equation of the ellipse with foci at (0, 2), and vertices at (0, 0), and (0, 10).

10. Given the equation $2x^2 + 2xy + y^2 - 1 = 0$, use the quadratic formula to solve for y, and use a graphing utility to graph the resulting equations. Identify the conic.

Test E

Chapter 4

Name _____ Date _____

Class _____ Section _____

1. Find the vertical asymptote(s): $f(x) = \dfrac{x^2 - 4}{x^2 - 6x + 5}$

2. Find the domain: $f(x) = \dfrac{4 + x}{x^2 - 10}$

3. Find the horizontal asymptote(s): $f(x) = \dfrac{x^2 - 3}{(x - 2)(x + 1)}$

4. Use a graphing utility to graph $f(x) = \dfrac{x}{x^2 - 1}$.

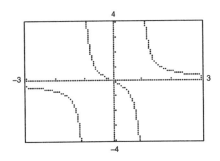

5. The population of a bacteria culture is given by

   $$P = \dfrac{20t + 2}{10 + 0.02t}, \quad t \geq 0$$

   where t is time in hours. Use a graphing utility using the indicated range setting, graph the function and determine the value which the population approaches.

   | Xmin = 0 |
   | Xmax = 15000 |
   | Xscl = 1000 |
   | Ymin = 0 |
   | Ymax = 1250 |
   | Yscl = 250 |

6. Find the partial fraction decomposition: $\dfrac{5x^2 + 12x + 10}{(x + 1)^3}$

7. Use a graphing utility to graph $x^2 + 5y^2 = 5$

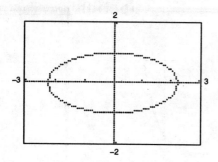

8. Find the standard equation of the ellipse that passes through the point $(2, 6\sqrt{2})$ and has end points of $(\pm 6, 0)$ on the minor axis.

9. Find the standard equation of the hyperbola with center at $(2, 5)$, one focus at $(2, 15)$, and transverse axis of length 12.

10. Given the equation $5x^2 - 4xy + y^2 - 1 = 0$, use the quadratic formula to solve for y, and use a graphing utility to graph the resulting equations. Identify the conic.

Test A

Chapter 5

Name _____ Date _____

Class _____ Section _____

1. Match the graph with the correction function.

    (a) $f(x) = 4^x - 5$ (b) $f(x) = 4^x + 5$

    (c) $f(x) = 4^{-x} + 5$ (d) $f(x) = 4^{-x} - 5$

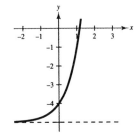

2. A certain population increases according to the model $P(t) = 250e^{0.47t}$. Use the model to determine the population when $t = 5$. Round your answer to the nearest integer.

    (a) 40 (b) 1597 (c) 1998

    (d) 2621 (e) None of these

3. Write the exponential form: $\log_b 37 = 2$

    (a) $37^2 = b$ (b) $2^b = 37$ (c) $b = 10$

    (d) $b^2 = 37$ (e) None of these

4. Match the graph with the correct function.

    (a) $f(x) = -3 + \ln x$ (b) $f(x) = 3 + \ln x$

    (c) $f(x) = \ln(x - 3)$ (d) $f(x) = \ln(x + 3)$

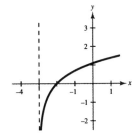

5. Write as the logarithm of a single quantity: $\frac{1}{4} \log_b 16 - 2 \log_b 5 + \log_b 7$

    (a) $\frac{14}{25}$ (b) $\log_b \frac{2}{175}$ (c) 1

    (d) $\log_b \frac{14}{25}$ (e) None of these

6. Evaluate $\log_a 16$, given that $\log_a 2 = 0.4307$.

    (a) 0.0344 (b) 1.7228 (c) 4.4307

    (d) 1.8168 (e) None of these

7. Solve for x: $\log(3x + 7) + \log(x - 2) = 1$
   - (a) $\frac{8}{3}$
   - (b) $3, -\frac{8}{3}$
   - (c) 2
   - (d) $2, -\frac{5}{3}$
   - (e) None of these

8. Use a graphing utility to graph $f(x) = 5e^{x+1} - 10$ and approximate its zero accurate to three decimal places.
   - (a) 1.693
   - (b) -0.307
   - (c) 0.588
   - (d) -1.693
   - (e) None of these

9. The ice trays in a freezer are filled with water at 68° F. The freezer maintains a temperature of 20° F. According to Newton's Law of Cooling, the water temperature T is related to the time t (in hours) by the equation

   $$kt = \ln \frac{T - 20}{68 - 20}.$$

   After 1 hour, the water temperature in the ice trays is 49° F. Use the fact that $T = 49$ when $t = 1$ to find how long it takes the water to freeze (water freezes at 32° F).
   - (a) 3.27 hours
   - (b) 2.75 hours
   - (c) 5.10 hours
   - (d) 1.17 hours
   - (e) None of these

10. The spread of a flu virus through a certain population is modeled by

    $$y = \frac{1000}{1 + 990e^{-0.7t}},$$

    where y is the total number infected after t days. In how many days will 820 people be infected with the virus?
    - (a) 10 days
    - (b) 11 days
    - (c) 12 days
    - (d) 13 days
    - (e) None of these

Test B

Chapter 5

Name _____ Date _____

Class _____ Section _____

1. Match the graph with the correct function.

    (a) $y = 3^{x-1}$
    (b) $y = 3^x - 1$
    (c) $y = 3^{1-x}$
    (d) $y = 3^{-x} - 1$

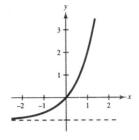

2. A certain population increases according to the model $P(t) = 250e^{0.47t}$. Use the model to determine the population when $t = 10$. Round your answer to the nearest integer.

    (a) 400
    (b) 4091
    (c) 27,487
    (d) 23,716
    (e) None of these

3. Write the exponential form: $\log_b 7 = 13$

    (a) $7^{13} = b$
    (b) $b^{13} = 7$
    (c) $b^7 = 13$
    (d) $7^b = 13$
    (e) None of these

4. Match the graph with the correct function.

    (a) $f(x) = 3 + \log x$
    (b) $f(x) = \log(x + 3)$
    (c) $f(x) = \frac{1}{3} \log x$
    (d) $f(x) = 3 \log x$

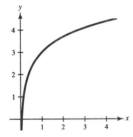

5. Write as the logarithm of a single quantity: $\frac{1}{2}[\ln(x + 1) + 2\ln(x - 1)] + \frac{1}{3} \ln x$

    (a) $\ln \sqrt[3]{x} \sqrt{(x+1)(x^2-1)}$
    (b) $\ln \sqrt[3]{x} \sqrt{x^2 - 1}$
    (c) $\ln \sqrt{x(x^2 - 1)}$
    (d) $\ln \sqrt[3]{x(x+1)(x-1)^2}$
    (e) None of these

6. Evaluate $\log_a 18$, given that $\log_a 2 = 0.2789$, $\log_a 3 = 0.4421$.

    (a) 1.1631
    (b) 0.2466
    (c) 0.0349
    (d) 1.4420
    (e) None of these

7. Solve for x: $\ln(7 - x) + \ln(3x + 5) = \ln(24x)$

    (a) $\frac{6}{11}$
    (b) $\frac{7}{3}$
    (c) $\frac{7}{3}, -5$
    (d) $\frac{6}{11}, 5$
    (e) None of these

8. Use a graphing utility to graph $f(x) = 2e^{x/2} - 14$ and approximate its zero accurate to three decimal places.

   (a) 2.639   (b) 0.973   (c) 1.946

   (d) 3.892   (e) None of these

9. The ice trays in a freezer are filled with water at 60° F. The freezer maintains a temperature of 20° F. According to Newton's Law of Cooling, the water temperature T is related to the time t (in hours) by the equation

$$kt = \ln \frac{T - 20}{60 - 20}.$$

   After 1 hour, the water temperature in the ice trays is 44° F. Use the fact that $T = 44$ when $t = 1$ to find how long it takes the water to freeze (water freezes at 32° F).

   (a) 2.4 hours   (b) 3.2 hours   (c) 1.7 hours

   (d) 5.1 hours   (e) None of these

10. The spread of a flu virus through a certain population is modeled by

$$y = \frac{1000}{1 + 990e^{-0.7t}},$$

   where y is the total number infected after t days. In how many days will 690 people be infected with the virus?

   (a) 10 days   (b) 11 days   (c) 12 days

   (d) 13 days   (e) None of these

Test C

Chapter 5

Name _____ Date _____

Class _____ Section _____

1. Match the graph with the correction function.

   (a) $f(x) = \left(\frac{1}{2}\right)^x - 1$ (b) $f(x) = 3^{-x^2} - 1$

   (c) $f(x) = 3^{x+1}$ (d) $f(x) = 4^{-x}$

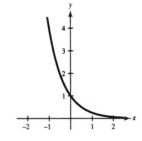

2. A certain population increases according to the model $P(t) = 250e^{0.47t}$. Use the model to determine the population when $t = 8$. Round your answer to the nearest integer.

   (a) 400 (b) 2621 (c) 10,737

   (d) 27,487 (e) None of these

3. Write the exponential form: $\log_7 b = 12$

   (a) $7^{12} = b$ (b) $b^7 = 12$ (c) $7^b = 12$

   (d) $b^{12} = 7$ (e) None of these

4. Match the graph with the correct function.

   (a) $f(x) = e^x$ (b) $f(x) = e^{x-1}$

   (c) $f(x) = \ln x$ (d) $f(x) = \ln(x - 1)$

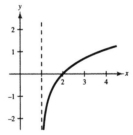

5. Write as the logarithm of a single quantity: $\log_2(x - 2) + \log_2(x + 2)$

   (a) $-2 + 2\log_2 x$ (b) $\log_2(x^2 - 4)$ (c) $2 \log_2 x$

   (d) $\log_2 2x$ (e) None of these

6. Evaluate $\log_a \frac{9}{2}$, given that $\log_a 2 = 0.2789$, $\log_a 3 = 0.4421$.

   (a) -0.0834 (b) 1.1631 (c) -0.3264

   (d) 0.6053 (e) None of these

7. Solve for x: $\log(7 - x) - \log(3x + 2) = 1$

   (a) $\frac{19}{31}$ (b) $-\frac{13}{31}$ (c) $-\frac{27}{29}$

   (d) $\frac{9}{4}$ (e) None of these

8. Use a graphing utility to graph $f(x) = 2e^{x-5} - 10$ and approximate its zero accurate to three decimal places.

   (a) 0.322   (b) −3.391   (c) 6.609

   (d) 8.047   (e) None of these

9. The ice trays in a freezer are filled with water at 50° F. The freezer maintains a temperature of 0° F. According to Newton's Law of Cooling, the water temperature T is related to the time t (in hours) by the equation

$$kt = \ln \frac{T}{50}.$$

   After 1 hour, the water temperature in the ice trays is 43° F. Use the fact that $T = 43$ when $t = 1$ to find how long it takes the water to freeze (water freezes at 32° F).

   (a) 2.4 hours   (b) 3.0 hours   (c) 3.6 hours

   (d) 2.1 hours   (e) None of these

10. The spread of a flu virus through a certain population is modeled by

$$y = \frac{1000}{1 + 990e^{-0.7t}},$$

   where y is the total number infected after t days. In how many days will 900 people be infected with the virus?

   (a) 11 days   (b) 13 days   (c) 15 days

   (d) 17 days   (e) None of these

Test D

Chapter 5

Name _____ Date _____

Class _____ Section _____

1. Without using a graphing utility sketch the graph of $f(x) = 3^x - 5$.

2. A certain population decreases according to the equation $y = 300 - 5e^{0.2t}$. Find the initial population and the population (to the nearest integer) when $t = 10$.

3. Write the logarithmic form: $5^2 = 25$

4. Sketch the graph: $f(x) = 1 + \log_5 x$

5. Write as a sum, difference, or multiple of logarithms: $\ln \dfrac{5x}{\sqrt[3]{x^2 + 1}}$

6. Evaluate $\log_b \left(\dfrac{14}{3b}\right)$, given that $\log_b 2 = 0.2789$, $\log_b 3 = 0.4421$, and $\log_b 7 = 0.7831$.

7. Solve for x: $\log x + \log(x + 3) = 1$

8. Use a graphing utility to graph $f(x) = 10e^{2x+1} - 5$ and approximate its zero accurate to three decimal places.

9. Find the constant k so that the exponential function $y = 3e^{kt}$ passes through the points $(0, 3)$ and $(3, 5)$.

10. The demand equation for a certain product is given by $p = 450 - 0.4e^{0.007x}$. Find the demand x if the price charged is $300.

Test E

Chapter 5

1. Without using a graphing utility sketch the graph of $f(x) = 3^x - 2$.

2. A certain population grows according to the equation $y = 40e^{0.025t}$. Find the initial population and the population (to the nearest integer) when $t = 50$.

3. Write the logarithmic form: $3^5 = 243$

4. Sketch the graph: $y = \ln(1 - x)$

5. Write as the logarithm of a single quantity: $\dfrac{1}{5}[3 \log(x + 1) + 2 \log(x - 1) - \log 7]$

6. Evaluate $\log_b \sqrt{10b}$, given that $\log_b 2 = 0.3562$ and $\log_b 5 = 0.8271$.

7. Solve for x: $x^2 - 4x = \log_2 32$

8. Use a graphing utility to graph $f(x) = 4e^{x-5/2} - 2$ and approximate its zero accurate to three decimal places.

9. Find the constant k so that the exponential function $y = 2e^{kt}$ passes through the points $(0, 2)$ and $(2, 5)$.

10. The demand equation for a certain product is given by $p = 450 - 0.4e^{0.007x}$. Find the demand x if the price charged is $250.

Test A
Chapter 6

Name _____ Date _____
Class _____ Section _____

1. Solve the system by the method of substitution:

    $x + y = 1$
    $x^2 + 3y^2 = 21$

    (a) $\left(\frac{3}{2}, -3\right)$ (b) $\left(3, -\frac{3}{2}\right)$ (c) $\left(-\frac{3}{2}, \frac{5}{2}\right), (3, -2)$

    (d) $\left(\frac{3}{2}, -\frac{1}{2}\right), (-3, 4)$ (e) No solution

2. Find the number of points of intersection of the graphs:

    $x^2 + y = 3$
    $x^2 + y^2 = 1$

    (a) 4   (b) 3   (c) 2

    (d) 1   (e) 0

3. Solve the linear system by the method of elimination:

    $7x - 3y = 26$
    $2x + 5y = 25$

    (a) $\left(-5, -\frac{61}{3}\right)$ (b) $(5, 3)$ (c) Infinitely many solutions

    (d) No solution   (e) None of these

4. Solve the following system of equations for x:

    $\dfrac{3}{x} - \dfrac{2}{y} = 5$

    $\dfrac{1}{x} + \dfrac{4}{y} = 4$

    (a) $\dfrac{1}{2}$ (b) 2   (c) 5

    (d) $\dfrac{1}{5}$ (e) None of these

5. Use Gaussian elimination to solve the system of equations:

    $x - 6y + z = 1$
    $-x + 2y - 4z = 3$
    $7x - 10y + 3z = -25$

    (a) $(5, 1, 2)$ (b) $(-5, -1, 0)$ (c) $(-1, 3, 1)$

    (d) No solution   (e) None of these

6. Find an equation of the parabola, $y = ax^2 + bx + c$, that passes through $(0, 5)$, $(2, -5)$, and $(-3, -40)$.

   (a) $y = 3x^2 - 2x - 7$ (b) $y = -4x^2 + 3x + 5$ (c) $y = 4x^2 + 3x + 5$

   (d) $y = 9x^2 - 121$ (e) None of these

7. Match the graph with the correct inequality.

   (a) $y < x^2 + 3x - 1$ (b) $y > x^2 + 3x - 1$

   (c) $y \leq x^2 + 3x - 1$ (d) $y \geq x^2 + 3x - 1$

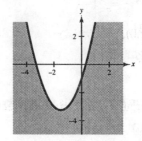

8. Match the graph with the correct system of inequalities.

   (a) $x + 2y \leq 6$ (b) $x + 2y \geq 6$
       $x - y \leq 2$ $x - y \geq 2$
       $y \geq 0$ $x \geq 0$

   (c) $x + 2y \leq 6$ (d) $x + 2y \geq 6$
       $x - y \geq 2$ $x - y \geq 2$
       $x \geq 0$ $y \geq 0$

   (e) None of these

9. Find the maximum value of the objective function $z = 5x + 6y$ subject to the constraints:

   $$x \geq 0$$
   $$y \geq 0$$
   $$x + 2y \leq 8$$
   $$3x + 3y \leq 15$$

   (a) 30   (b) 28

   (c) 25   (d) 24

   (e) None of these

10. A company produces two models of calculators at two different plants. In one day Plant A can produce 140 of Model I and 35 of Model II. In one day Plant B can produce 60 of Model I and 90 of Model II. The company needs to produce at least 460 Model I and 340 of Model II. Find the minimum cost. Assume it costs $1200 per day to operate Plant A and $900 per day for Plant B.

    (a) $C = \$11{,}640$ (b) $C = \$8730$ (c) $C = \$5100$

    (d) $C = \$3948$ (e) None of these

Test B
Chapter 6

Name _____ Date _____

Class _____ Section _____

1. Solve the system by the method of substitution:

    $2x^2 + 2y^2 = 7$

    $x + y^2 = 7$

    (a) $(2.8, 2.0), (-0.5, 7.3)$ (b) $(4.6, 1.5), (-2.6, 3.1)$ (c) $(2.8, -0.5)$

    (d) $(4.6), (-2.6)$ (e) No solution

2. Find the number of points of intersection of the graphs:

    $x^2 + y^2 = 2$

    $2x + y = 10$

    (a) 4   (b) 3   (c) 2

    (d) 1   (e) 0

3. Solve the linear system by the method of elimination:

    $2x + 4y = 7$

    $3x + 6y = 5$

    (a) $\left(1, \frac{5}{4}\right)$ (b) $(0, 0)$ (c) Infinitely many solutions

    (d) No solution   (e) None of these

4. Solve the following system of equations for y:

    $\dfrac{2}{x} + \dfrac{3}{y} = 7$

    $\dfrac{3}{x} - \dfrac{1}{y} = 16$

    (a) 5   (b) -1 (c) $\dfrac{1}{5}$

    (d) 2   (e) None of these

5. Solve the system of linear equations:

    $x - y + z = 5$

    $3x + 2y - z = -2$

    $2x + y + 3z = 10$

    (a) $(1, -1, 3)$ (b) $(2, -5, -2)$ (c) $(-1, 7, 13)$

    (d) $(3, -9, -7)$ (e) No solution

6. Find the value of c in the quadratic equation, $y = ax^2 + bx + c$, if its graph passes through the points $(1, 0)$, $(-1, -6)$, and $(2, 9)$.

   (a) -5 (b) -4 (c) 3

   (d) 11 (e) None of these

7. Match the graph with the correct inequality.

   (a) $y > -2$ (b) $y < -2$

   (c) $x > -2$ (d) $x \geq -2$

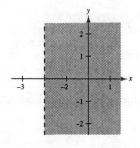

8. Match the graph with the correct system of inequalities.

   (a) $x + 2y \leq 4$ (b) $x + 2y \geq 4$
   $x \leq y$ $x \leq y$
   $x \geq 0$ $y \geq 0$

   (c) $x + 2y \leq 4$ (d) $x + 2y \leq 4$
   $x \leq y$ $y \leq x$
   $y \geq 0$ $y \geq 0$

   (e) None of these

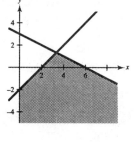

9. Find the maximum value of the objective function $z = 10x + 8y$ subject to the constraints:

   $x \geq 0$
   $y \geq 0$
   $x + y \leq 5$
   $3x + y \leq 12$
   $-2x + y \leq 2$

   (a) 40 (b) 50

   (c) 42 (d) 47

   (e) None of these

10. A company produces two models of calculators at two different plants. In one day Plant A can produce 60 of Model I and 70 of Model II. In one day Plant B can produce 80 of Model I and 40 of Model II. The company needs to produce at least 460 Model I and 340 of Model II. Find the minimum cost. Assume it costs $1200 per day to operate Plant A and $900 per day for Plant B.

    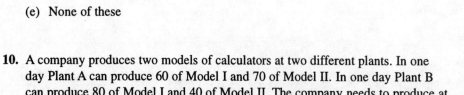

Test C
Chapter 6

1. Solve the system by the method of substitution:

   $x^2 + 2y = 6$

   $2x + y = 3$

   (a) $(4, -5)$ (b) $(2, 1)$ (c) $(0, 3)$

   (d) $(0, 3)$ and $(4, -5)$ (e) None of these

2. Find the number of points of intersection:

   $x^2 + y^2 = 5$

   $x + 2y - 5 = 0$

   (a) 4 (b) 3 (c) 2

   (d) 1 (e) 0

3. Solve the linear system by the method of elimination:

   $6x - 5y = 4$

   $3x + 2y = 1$

   (a) $\left(\frac{13}{27}, -\frac{2}{9}\right)$ (b) $\left(-\frac{2}{9}, -\frac{8}{5}\right)$ (c) $\left(-\frac{8}{5}, -\frac{68}{25}\right)$

   (d) $\left(2, \frac{8}{5}\right)$ (e) None of these

4. Solve the following system of equations for x:

   $\dfrac{5}{x} - \dfrac{3}{y} = 2$

   $\dfrac{2}{x} + \dfrac{5}{y} = -24$

   (a) $-\dfrac{1}{2}$ (b) $-\dfrac{1}{4}$ (c) 5

   (d) $-\dfrac{1}{3}$ (e) None of these

5. Solve the system of linear equations:

   $6x - 9y + 4z = -7$

   $2x + 6y - z = 6$

   $4x - 3y + 2z = -2$

   (a) $\left(\dfrac{1}{2}, \dfrac{2}{3}, -1\right)$ (b) $\left(\dfrac{11}{21}, 1, -\dfrac{2}{7}\right)$ (c) $\left(a, \dfrac{31a}{15}, \dfrac{44a}{5}\right)$

   (d) No solution (e) None of these

6. Find the value of b in the quadratic equation, $y = ax^2 + bx + c$, if its graph passes through the points $(-1, 4)$, $(1, -2)$, and $(2, -2)$.

   (a) -3 (b) 2 (c) -2
   (d) -1 (e) None of these

7. Match the graph with the correct inequality.

   (a) $3x - 4y < 12$ (b) $3x - 4y \le 12$
   (c) $3x - 4y > 12$ (d) $3x - 4y \ge 12$

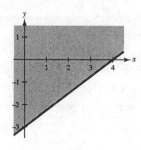

8. Match the graph with the correct system of inequalities.

   (a) $x + 2y \le 4$ (b) $x + 2y \ge 4$
   $x \le y$ $y \le x$
   $x \ge 0$ $y \ge 0$

   (c) $x + 2y \ge 4$ (d) $x + 2y \ge 4$
   $y \le x$ $x \le y$
   $y \ge 0$ $y \ge 0$

   (e) None of these

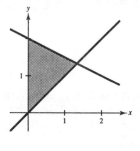

9. Find the maximum value of the objective function $z = 3x + 2y$ subject to the constraints:

   $x \ge 0$
   $y \ge 0$
   $x + y \le 4$
   $x + 3y \le 6$

   (a) 12 (b) 11
   (c) 10 (d) 9
   (e) None of these

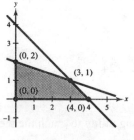

10. A company produces two models of calculators at two different plants. In one day Plant A can produce 70 of Model I and 40 of Model II. In one day Plant B can produce 80 of Model I and 90 of Model II. The company needs to produce at least 1370 Model I and 1270 Model II. Find the minimum cost. Assume it costs $900 per day to operate Plant A and $1200 per day for Plant B.

    (a) $C = \$32{,}570$ (b) $C = \$28{,}575$ (c) $C = \$20{,}550$
    (d) $C = \$19{,}500$ (e) None of these

Test D
Chapter 6

Name _____ Date _____

Class _____ Section _____

1. Solve the system by the method of substitution:

   $y = \dfrac{1}{x}$

   $x + 5y = 6$

2. Use a graphing utility to find all points of intersection of the graphs:

   $(x - 3)^2 + y^2 = 4$

   $-2x + y^2 = 0$

3. Solve this system by method of elimination and verify the solution with a graphing utility:

   $2x - 5y = -4$

   $4x + 3y = 5$

4. How many liters of a 40% solution of acid must be combined with a 15% solution to obtain 30 liters of a 20% solution?

5. Use Gaussian elimination to solve the system of equations:

   $x + 2y + z = 6$

   $2x - y + 3z = -2$

   $x + y - 2z = 0$

6. Find an equation of the parabola, $y = ax^2 + bx + c$, that passes through $(0, -5)$, $(2, 1)$, and $(-1, -14)$. Verify your result with a graphing utility.

7. Use a graphing utility to graph the inequality

   $x^2 + (y - 1)^2 \leq 25$.

8. Sketch the graph of the system of inequalities.

   $2x + 3y \leq 6$

   $x - 2y \geq -2$

9. Find the maximum value of the objective function $C = 3x + 2y$ subject to the constraints:

$$x \geq 0$$
$$y \geq 0$$
$$3x + 4y \leq 25$$
$$3x - y \leq 5$$

10. A merchant plans to sell two models of an item at costs of $350 and $400. The $350 model yields a profit of $85 and the $400 model yields a profit of $90. The total demand per month for the two models will not exceed 150. Find the number of units of each model that should be stocked each month in order to maximize the profit. Assume the merchant can invest no more than $56,000 for inventory of these items.

Test E

Chapter 6

Name _____ Date _____

Class _____ Section _____

1. Solve the system by the method of substitution:

    $2x^2 - y = -2$

    $x - y = -2$

2. Use a graphing utility to find all points of intersection of the graphs:

    $x^2 - 4x + y = 0$

    $x - y = 0$

3. Solve this system by method of elimination and verify the solution with a graphing utility:

    $6x + y = -2$

    $4x - 3y = 17$

4. The perimeter of a rectangle is 91 feet and the length is 8 feet more than twice the width. Find the dimensions of the rectangle.

5. Solve the system of linear equations:

    $x + 3y + z = 0$

    $5x - y + z + w = 0$

    $2x + 2z + w = 2$

    $3x + 2z - w = 10$

6. Find an equation of the parabola, $y = ax^2 + bx + c$, that passes through $(1, 1), (-1, 11)$, and $(3, 23)$. Verify your result with a graphing utility.

7. Use a graphing utility to graph the inequality

    $3x^2 + y \geq 6$.

8. Sketch the graph of the system of inequalities.

    $2y - 3x \leq 10$

    $2y \geq x^2$

9. Find the maximum value of the objective function $z = 3x + 2y$ subject to the constraints:

$$x \geq 0$$
$$y \geq 0$$
$$3x + 4y \leq 25$$
$$3x - y \leq 5$$

10. A merchant plans to sell two models of an item at costs of $350 and $500. The $350 model yields a profit of $45 and the $500 model yields a profit of $60. The total demand per month for the two models will not exceed 145. Find the number of units of each model that should be stocked each month in order to maximize the profit. Assume the merchant can invest no more than $56,000 for inventory of these items.

Test A

Chapter 7

Name _____ Date _____

Class _____ Section _____

1. Write the matrix in reduced row-echelon form: $\begin{bmatrix} 3 & 1 & 1 & 7 \\ 1 & -2 & 0 & 5 \\ 1 & 1 & 2 & 6 \end{bmatrix}$

   (a) $\begin{bmatrix} 1 & 1 & 2 & 6 \\ 0 & 3 & 2 & 1 \\ 0 & 0 & 11 & 31 \end{bmatrix}$

   (b) $\begin{bmatrix} 1 & 0 & 0 & \frac{21}{11} \\ 0 & 1 & 0 & -\frac{17}{11} \\ 0 & 0 & 1 & \frac{31}{11} \end{bmatrix}$

   (c) $\begin{bmatrix} 1 & -2 & 0 & 5 \\ 0 & 1 & \frac{1}{7} & -\frac{8}{7} \\ 0 & 0 & 11 & 31 \end{bmatrix}$

   (d) $\begin{bmatrix} 1 & 0 & 0 & \frac{3}{11} \\ 0 & 1 & 0 & \frac{7}{11} \\ 0 & 0 & 1 & -\frac{14}{11} \end{bmatrix}$

   (e) None of these

2. Use Gaussian elimination with back-substitution or Gauss-Jordan elimination to solve the following system of linear equations.

   $3x + 2y + z = 7$

   $x - y + z = 6$

   $x + z = 5$

   (a) $(2, -1, 3)$ (b) $\left(1, -\frac{1}{2}, 5\right)$ (c) $(-1, 1, 2)$

   (d) $(0, 4, -1)$ (e) None of these

3. Given $A = \begin{bmatrix} 3 & 6 & -1 \\ 0 & 5 & 2 \end{bmatrix}$ and $B = \begin{bmatrix} 1 & 0 & 5 \\ -1 & 2 & 7 \end{bmatrix}$, find $3A - 2B$.

   (a) $\begin{bmatrix} 7 & 18 & -13 \\ 2 & 11 & -8 \end{bmatrix}$

   (b) $\begin{bmatrix} 7 & 18 & 2 \\ 0 & 11 & -8 \end{bmatrix}$

   (c) $\begin{bmatrix} 11 & 18 & 7 \\ -2 & 19 & 20 \end{bmatrix}$

   (d) $\begin{bmatrix} 7 & 18 & -13 \\ -2 & 9 & 20 \end{bmatrix}$

   (e) None of these

4. Use a graphing utility to find AB, given $A = \begin{bmatrix} 3 & -2 & 4 \\ 0 & 0 & -1 \\ 3 & 2 & -1 \end{bmatrix}$ and $B = \begin{bmatrix} 3 & 1 & -1 \\ -1 & 0 & 0 \\ 2 & 4 & -2 \end{bmatrix}$.

   (a) $\begin{bmatrix} 6 & -8 & 12 \\ -3 & 2 & -4 \\ 0 & -8 & 6 \end{bmatrix}$

   (b) $\begin{bmatrix} 19 & 19 & -11 \\ -2 & -4 & 2 \\ 5 & -1 & -1 \end{bmatrix}$

   (c) $\begin{bmatrix} 9 & -2 & -4 \\ 0 & 0 & 0 \\ 6 & 8 & 2 \end{bmatrix}$

   (d) Impossible   (e) None of these

5. Find the inverse of $A = \begin{bmatrix} 3 & 2 \\ 1 & 4 \end{bmatrix}$.

(a) $\begin{bmatrix} \frac{2}{5} & -\frac{1}{5} \\ -\frac{1}{10} & \frac{3}{10} \end{bmatrix}$
(b) $\begin{bmatrix} \frac{1}{3} & \frac{1}{2} \\ 1 & \frac{1}{4} \end{bmatrix}$
(c) $\begin{bmatrix} 4 & -2 \\ -1 & 3 \end{bmatrix}$

(d) $\begin{bmatrix} -3 & 1 \\ 2 & -4 \end{bmatrix}$
(e) None of these

6. Given a system of linear equations with coefficient matrix A, use A^{-1} to find (x, y, z).

$$3x + 2y + z = 5$$
$$x - 4y = 6$$
$$x - y + 3z = 6$$

$A^{-1} = \frac{1}{39} \begin{bmatrix} 12 & 7 & -4 \\ 3 & -8 & -1 \\ -3 & -5 & 14 \end{bmatrix}$

(a) $(-2, -1, 1)$
(b) $(-1, 6, -2)$
(c) $(2, -1, 1)$
(d) $(6, -1, 3)$
(e) None of these

7. Find the determinant of the matrix: $\begin{bmatrix} 3 & -4 \\ 2 & 6 \end{bmatrix}$

(a) 10
(b) 26
(c) -26
(d) -10
(e) None of these

8. Use the matrix capabilities of a graphing utility to find the determinant of the matrix:

$\begin{bmatrix} 5 & -1 & 0 & 2 \\ 0 & 4 & 7 & 3 \\ 0 & 0 & 1 & 1 \\ 0 & 0 & 0 & 1 \end{bmatrix}$

(a) 11
(b) -11
(c) -20
(d) 20
(e) None of these

9. Use Cramer's Rule to solve for y in the system of linear equations:

$$3x + 2y + 4z = 12$$
$$x - y + z = 3$$
$$2x + 7y - z = 9$$

(a) $y = \dfrac{\begin{vmatrix} 3 & 2 & 4 \\ 1 & -1 & 1 \\ 2 & 7 & -1 \end{vmatrix}}{\begin{vmatrix} 3 & 12 & 4 \\ 1 & 3 & 1 \\ 2 & 9 & -1 \end{vmatrix}}$

(b) $y = \dfrac{\begin{vmatrix} 3 & 12 & 4 \\ 1 & 3 & 1 \\ 2 & 9 & -1 \end{vmatrix}}{\begin{vmatrix} 3 & 2 & 4 \\ 1 & -1 & 1 \\ 2 & 7 & -1 \end{vmatrix}}$

(c) $y = \dfrac{\begin{vmatrix} 3 & 4 & 12 \\ 1 & -1 & 3 \\ 2 & 7 & 9 \end{vmatrix}}{\begin{vmatrix} 2 & 4 & 3 \\ -1 & 1 & 1 \\ 7 & -1 & 2 \end{vmatrix}}$

(d) $y = \begin{vmatrix} 3 & 2 & 4 \\ 1 & -1 & 1 \\ 2 & 7 & -1 \end{vmatrix} \begin{vmatrix} 2 & 4 & 12 \\ -1 & 1 & 3 \\ 7 & -1 & 9 \end{vmatrix}$

(e) None of these

10. Find the uncoded row matrix of order 1×3 for the message CALL ME LATER, then encode the message using $A = \begin{bmatrix} 1 & -1 & 0 \\ 1 & 0 & 3 \\ -2 & 1 & -1 \end{bmatrix}$.

(a) 16 −12 4 4 3 −13 9 −13 3 1 9 −9 −20 −18 5

(b) −20 9 −9 −14 1 −13 −19 7 −12 11 4 55 18 −18 0

(c) −5 12 19 −3 6 6 2 13 −6 −6 3 −1 0 2 2

(d) 3 1 12 12 0 13 5 0 12 1 20 5 18 0 0

(e) None of these

Test B

Chapter 7

1. Write the matrix in reduced row-echelon form: $\begin{bmatrix} 3 & 6 & -2 & 28 \\ -2 & -4 & 5 & -37 \\ 1 & 2 & 9 & -39 \end{bmatrix}$

   (a) $\begin{bmatrix} 1 & 2 & 1 & 1 \\ 0 & 0 & 1 & -5 \\ 0 & 0 & 0 & 0 \end{bmatrix}$
   (b) $\begin{bmatrix} 0 & 0 & 0 & 0 \\ 1 & 2 & 0 & 6 \\ 0 & 0 & 1 & -5 \end{bmatrix}$
   (c) $\begin{bmatrix} 1 & 2 & 0 & 6 \\ 0 & 0 & 1 & -5 \\ 0 & 0 & 0 & 0 \end{bmatrix}$

   (d) $\begin{bmatrix} 1 & 2 & 1 & 1 \\ 0 & 0 & 1 & -5 \\ 0 & 0 & 0 & 3 \end{bmatrix}$
   (e) None of these

2. Use Gaussian elimination with back-substitution or Gauss-Jordan elimination to solve the following system of linear equations.

   $2x + y - z = -3$
   $4x - y + z = 6$
   $2x + 3y + 2z = -9$

   (a) $(1, -1, 4)$
   (b) $(\frac{1}{2}, 0, 4)$
   (c) $(\frac{1}{2}, 2, 0)$
   (d) $(\frac{3}{2}, -9, 0)$
   (e) None of these

3. Given $A = \begin{bmatrix} 1 & 2 & 3 \\ 4 & 7 & 1 \\ 0 & 3 & 2 \end{bmatrix}$ and $B = \begin{bmatrix} 0 & 0 & 1 \\ 1 & 4 & 0 \\ 2 & 3 & 7 \end{bmatrix}$, find $6A - 2B$.

   (a) $\begin{bmatrix} 30 & 72 & 52 \\ 32 & 10 & 6 \\ 1 & 6 & -8 \end{bmatrix}$
   (b) $\begin{bmatrix} 30 & 12 & 4 \\ 3 & 10 & 6 \\ 1 & 6 & -8 \end{bmatrix}$
   (c) $\begin{bmatrix} 6 & 12 & 16 \\ 22 & 34 & 6 \\ -4 & 12 & -2 \end{bmatrix}$

   (d) $\begin{bmatrix} 30 & 22 & 20 \\ 66 & 28 & 26 \\ 24 & 16 & 14 \end{bmatrix}$
   (e) None of these

4. Use a graphing utility to find AB, given $A = \begin{bmatrix} 2 & 0 & 1 & 2 \\ 0 & 1 & 0 & 1 \\ -1 & -2 & 0 & 0 \end{bmatrix}$ and $B = \begin{bmatrix} 1 & 1 & 0 \\ 0 & 1 & 1 \\ 2 & -1 & 2 \end{bmatrix}$.

   (a) $\begin{bmatrix} 2 & 0 & 1 & 2 \\ 0 & 1 & 0 & 0 \\ -2 & 2 & 0 & 0 \end{bmatrix}$
   (b) $\begin{bmatrix} 4 & 2 & 3 \\ -1 & -2 & -1 \\ 6 & -5 & -1 \end{bmatrix}$
   (c) $\begin{bmatrix} 2 & 1 & 1 & 3 \\ -1 & -1 & 0 & 1 \\ 2 & -5 & 2 & 3 \end{bmatrix}$

   (d) Impossible
   (e) None of these

5. Find the inverse of $A = \begin{bmatrix} 2 & 3 \\ 1 & 2 \end{bmatrix}$.

(a) $\begin{bmatrix} \frac{1}{2} & \frac{1}{3} \\ 1 & \frac{1}{2} \end{bmatrix}$

(b) $\begin{bmatrix} 2 & -3 \\ -1 & 2 \end{bmatrix}$

(c) $\begin{bmatrix} -2 & 1 \\ 3 & -2 \end{bmatrix}$

(d) $\begin{bmatrix} \frac{2}{7} & -\frac{3}{7} \\ -\frac{1}{7} & \frac{2}{7} \end{bmatrix}$

(e) None of these

6. Given a system of linear equations with coefficient matrix A, use A^{-1} to find (x, y, z).

$$3x + y - z = -11$$
$$x - y - z = -1$$
$$x + 2y + 3z = 3$$

$$A^{-1} = \frac{1}{10} \begin{bmatrix} 1 & 5 & 2 \\ 4 & -10 & -2 \\ -3 & 5 & 4 \end{bmatrix}$$

(a) $(3, -2, 0)$

(b) $(2, -2, 3)$

(c) $(-1, 5, 1)$

(d) $(-3, 0, 2)$

(e) None of these

7. Find the determinant of the matrix: $\begin{bmatrix} 3 & -1 \\ 6 & 2 \end{bmatrix}$

(a) 12

(b) -12

(c) 0

(d) 9

(e) None of these

8. Use the matrix capabilities of a graphing utility to find the determinant of the matrix:

$$\begin{bmatrix} 1 & 0 & 0 & 0 \\ 4 & 6 & 0 & 0 \\ -7 & 5 & -5 & 0 \\ 3 & 2 & 3 & -1 \end{bmatrix}$$

(a) 11

(b) -30

(c) 30

(d) -11

(e) None of these

9. Use Cramer's Rule to solve for y:

$$3x - 2y + 2z = 3$$
$$x + 4y - z = 2$$
$$x + y + z = 6$$

(a) $y = \dfrac{\begin{vmatrix} 3 & -2 & 2 \\ 1 & 4 & -1 \\ 1 & 1 & 1 \end{vmatrix}}{\begin{vmatrix} 3 & 3 & 2 \\ 1 & 2 & -2 \\ 1 & 6 & 1 \end{vmatrix}}$

(b) $y = \dfrac{\begin{vmatrix} 3 & 3 & 2 \\ 1 & 2 & -1 \\ 1 & 6 & 1 \end{vmatrix}}{\begin{vmatrix} 3 & -2 & 2 \\ 1 & 4 & -1 \\ 1 & 1 & 1 \end{vmatrix}}$

(c) $y = \dfrac{\begin{vmatrix} 3 & 2 & 3 \\ 1 & -1 & 2 \\ 1 & 1 & 6 \end{vmatrix}}{\begin{vmatrix} 3 & -2 & 2 \\ 1 & 4 & -1 \\ 1 & 1 & 1 \end{vmatrix}}$

(d) $y = \dfrac{\begin{vmatrix} 3 & 3 & -2 \\ 1 & 2 & 4 \\ 1 & 6 & 1 \end{vmatrix}}{\begin{vmatrix} 3 & 1 & 1 \\ -2 & 4 & 1 \\ 2 & -1 & 1 \end{vmatrix}}$

(e) None of these

10. Find the uncoded row matrix of order 1×3 for the message BE BACK SOON, then encode the message using $A = \begin{bmatrix} 1 & -1 & 0 \\ 1 & 0 & 3 \\ -2 & 1 & -1 \end{bmatrix}$.

(a) $-2\ 15\ 1\ 0\ -3\ -27\ -2\ 8\ 19\ 15\ -3\ 1$

(b) $7\ -2\ 15\ -3\ 1\ 0\ -27\ 8\ -19\ 2\ -1\ 31$

(c) $2\ 5\ 0\ 2\ 1\ 3\ 11\ 0\ 19\ 15\ 15\ 14$

(d) $0\ 2\ 5\ -3\ 1\ 0\ 0\ -27\ 16\ 2\ 1\ 1$

(e) None of these

Test C

Chapter 7

1. Write the matrix in reduced row-echelon form: $\begin{bmatrix} 1 & 3 & -8 & 13 \\ 2 & -1 & 6 & -19 \\ -5 & 1 & 2 & 44 \end{bmatrix}$

   (a) $\begin{bmatrix} 1 & 0 & 0 & -7 \\ 0 & 1 & 0 & 8 \\ 0 & 0 & 1 & \frac{1}{2} \end{bmatrix}$
   (b) $\begin{bmatrix} 1 & 0 & 6 & -4 \\ 0 & 1 & 2 & 9 \\ 0 & 0 & 2 & 1 \end{bmatrix}$
   (c) $\begin{bmatrix} 1 & 0 & 6 & -4 \\ 0 & 1 & -4 & 6 \\ 0 & 0 & 0 & 0 \end{bmatrix}$

   (d) $\begin{bmatrix} 1 & 1 & 8 & 5 \\ 0 & 1 & 2 & 9 \\ 0 & 0 & 2 & 1 \end{bmatrix}$
   (e) None of these

2. Use Gaussian elimination with back-substitution or Gauss-Jordan elimination to solve the following system of linear equations.

   $-2x + 3y - 4z = 4$
   $x - y - 5z = 0$
   $-2x + 4y + 5z = 9$

   (a) $(-3, 2, -1)$
   (b) $(3, -2, -1)$
   (c) $(2, 1, 1)$

   (d) $(0, 4, 2)$
   (e) None of these

3. Given $A = \begin{bmatrix} 2 & 4 & -1 \\ 1 & 0 & 4 \\ 8 & 1 & 2 \end{bmatrix}$ and $B = \begin{bmatrix} 1 & 1 & 1 \\ -1 & 0 & 0 \\ 4 & 10 & -2 \end{bmatrix}$, find $2A - 2B$.

   (a) $\begin{bmatrix} 2 & 6 & 4 \\ 4 & 0 & -8 \\ -8 & 18 & 1 \end{bmatrix}$
   (b) $\begin{bmatrix} 2 & 6 & -4 \\ 4 & 0 & 8 \\ 8 & -18 & 8 \end{bmatrix}$
   (c) $\begin{bmatrix} 0 & 6 & -4 \\ -4 & 0 & 4 \\ 4 & -9 & 0 \end{bmatrix}$

   (d) $\begin{bmatrix} 1 & 3 & -2 \\ 0 & 0 & 4 \\ 4 & -9 & 4 \end{bmatrix}$
   (e) None of these

4. Use a graphing utility to find AB, given $A = \begin{bmatrix} 2 & 0 & 1 & 2 \\ 0 & 1 & 0 & 1 \\ -1 & -2 & 0 & 0 \end{bmatrix}$ and $B = \begin{bmatrix} 1 & 1 & 0 \\ 0 & 1 & 1 \\ 2 & -1 & 2 \end{bmatrix}$.

   (a) $\begin{bmatrix} 2 & 1 & 1 & 3 \\ -1 & -1 & 0 & 1 \\ 2 & -5 & 2 & 3 \end{bmatrix}$
   (b) $\begin{bmatrix} 4 & 1 & 2 \\ 0 & 1 & 1 \\ -1 & -3 & -2 \end{bmatrix}$
   (c) $\begin{bmatrix} 6 & -4 & 0 \\ 0 & 1 & 1 \\ -4 & -3 & -4 \end{bmatrix}$

   (d) Impossible
   (e) None of these

5. Find the inverse of $A = \begin{bmatrix} 2 & 3 \\ -3 & -3 \end{bmatrix}$.

   (a) $\begin{bmatrix} -3 & -3 \\ 3 & 2 \end{bmatrix}$
   (b) $\begin{bmatrix} \frac{1}{2} & \frac{1}{3} \\ -\frac{1}{3} & -\frac{1}{3} \end{bmatrix}$
   (c) $\begin{bmatrix} \frac{1}{4} & \frac{1}{4} \\ -\frac{1}{4} & -\frac{1}{3} \end{bmatrix}$
   (d) $\begin{bmatrix} -1 & -1 \\ 1 & \frac{2}{3} \end{bmatrix}$
   (e) None of these

6. Given the system of linear equations with coefficient matrix A, use A^{-1} to find (x, y, z).

   $$\begin{aligned} 4x - y + z &= 5 \\ x - 4y + z &= 8 \\ 2x + 2y - 3z &= -12 \end{aligned} \qquad A^{-1} = \frac{1}{45}\begin{bmatrix} 10 & -1 & 3 \\ 5 & -14 & -3 \\ 10 & -10 & -15 \end{bmatrix}$$

   (a) $(0, -1, 4)$
   (b) $(0, 4, -2)$
   (c) $(1, -1, 2)$
   (d) $(3, 0, 2)$
   (e) None of these

7. Find the determinant of the matrix: $\begin{bmatrix} 6 & -4 \\ 2 & -1 \end{bmatrix}$

   (a) 2
   (b) -2
   (c) 14
   (d) -14
   (e) None of these

8. Use the matrix capabilities of a graphing utility to find the determinant of the matrix:
   $$\begin{bmatrix} 1 & 5 & 0 & 6 \\ 0 & -1 & 1 & 0 \\ -2 & 2 & 3 & 0 \\ 0 & 0 & 7 & 2 \end{bmatrix}$$

   (a) -54
   (b) 74
   (c) -74
   (d) 54
   (e) None of these

9. Find the determinant of the matrix $\begin{bmatrix} 0 & -1 & 2 \\ 3 & 5 & 0 \\ 1 & -1 & 3 \end{bmatrix}$

    (a) 25   (b) −25   (c) 7

    (d) −7   (e) None of these

10. Decode the cryptogram:
129 −85 −38 −75 70 25 −9 18 3 188 −141 −58 using
$A = \begin{bmatrix} 13 & -10 & -4 \\ -6 & 5 & 2 \\ 3 & -2 & -1 \end{bmatrix}$

    (a) WELCOME HOME   (b) CLOSING OUT   (c) OUT TO LUNCH

    (d) PLEASE HURRY   (e) None of these

Test D
Chapter 7

1. Write the matrix in reduced row-echelon form: $\begin{bmatrix} 1 & 2 & -1 & 3 \\ 7 & -1 & 0 & 2 \\ 3 & 2 & 1 & -1 \end{bmatrix}$

2. Find the equation of the parabola that passes through the given points. Use a graphing utility to verify your result.

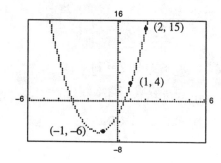

3. If $A = \begin{bmatrix} 2 & -1 \\ -3 & 4 \end{bmatrix}$ and $B = \begin{bmatrix} -2 & 0 \\ -1 & 3 \end{bmatrix}$. Find C if $A + C = 2B$.

4. Given $A = \begin{bmatrix} 3 & 2 & 2 & 1 \\ 13 & 6 & 12 & 1 \\ -5 & -1 & -5 & 0 \end{bmatrix}$ and $B = \begin{bmatrix} 1 & 1 & 0 \\ 3 & 1 & 2 \\ -1 & 1 & -1 \end{bmatrix}$, find BA.

5. Given $A = \begin{bmatrix} 1 & 3 & -1 \\ 0 & 2 & 1 \\ -1 & 1 & -2 \end{bmatrix}$, find A^{-1}.

6. Use an inverse matrix to solve the system of linear equations.

   $3x + 2y + z = 1$
   $x - y = 10$
   $-x + 2z = 5$

7. Evaluate the determinant: $\begin{vmatrix} x \ln x & x \\ 1 + \ln x & 1 \end{vmatrix}$

8. Use the matrix capabilities of a graphing utility to evaluate:

   $\begin{vmatrix} 3 & -2 & 4 & 3 \\ 2 & -1 & 0 & 4 \\ -2 & 0 & 1 & 5 \\ -2 & -3 & 0 & 2 \end{vmatrix}$

9. Use Cramer's Rule to solve the system of linear equations.

   $4x + 6y + 2z = 15$

   $x - y + 4z = -3$

   $3x + 2y + 2z = 6$

10. Use a graphing utility and Cramer's Rule to solve (if possible) the system of equations.

    $x - y - 2z = 3$

    $y + 3z = -2$

    $3x + 4y - z = 11$

Test E

Chapter 7

1. Write the matrix in reduced row-echelon form: $\begin{bmatrix} 21 & 14 & -7 & 10 \\ 7 & 7 & 7 & -1 \\ 3 & -14 & 28 & 23 \end{bmatrix}$

2. Find the equation of the parabola that passes through the given points. Use a graphing utility to verify your result.

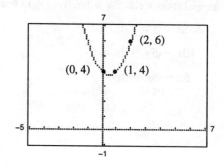

3. Use a graphing utility to find AB, given

$$A = \begin{bmatrix} 1 & 3 & 6 \\ 4 & 1 & 3 \end{bmatrix} \text{ and } B = \begin{bmatrix} 0 & 1 & 6 \\ 3 & -1 & 1 \\ 5 & 2 & 3 \end{bmatrix}.$$

4. Given $A = \begin{bmatrix} 1 & 0 & 3 \\ -1 & 2 & -2 \\ 1 & 1 & 2 \end{bmatrix}$ and $B = \begin{bmatrix} 1 & 1 & 0 \\ 3 & 1 & 2 \\ -1 & 1 & -1 \end{bmatrix}$, find BA.

5. Given $A = \begin{bmatrix} 1 & 5 & -1 \\ 2 & 3 & -2 \\ -1 & -4 & 3 \end{bmatrix}$, find A^{-1}.

6. Use the inverse matrix method to solve the system of linear equations.

$$6x + 6y - 5z = 11$$
$$3x + 6y - z = 6$$
$$9x - 3y + z = 0$$

7. Evaluate the determinant: $\begin{vmatrix} e^{2x} & e^x \\ 2e^{2x} & e^x \end{vmatrix}$

8. Use the matrix capabilities of a graphing utility to evaluate:

$$\begin{vmatrix} 10 & -3 & 2 & 4 & 5 \\ -1 & 0 & -3 & 2 & 3 \\ 4 & -1 & -2 & 0 & 4 \\ 9 & 4 & -3 & 6 & 2 \\ -2 & 0 & -4 & 4 & 5 \end{vmatrix}$$

9. Use Cramer's Rule to solve the system of linear equations.

$$5x + 5y + 4z = 4$$
$$10x - 5y + 2z = 11$$
$$5x - 5y + 2z = 7$$

10. Use a graphing utility and Cramer's Rule to solve (if possible) the system of equations.

$$x + 3y + z = 4$$
$$2x - y - 3z = 1$$
$$4x + y + z = 5$$

Test A
Chapter 8

Name _____ Date _____
Class _____ Section _____

1. Find the first 5 terms of the sequence whose nth term is $a_n = (-1)^n(2n + 9)$. (Assume that n begins with 1.)

    (a) $-11, -13, -15, -17, -19\ldots$
    (b) $-11, 13, -15, 17, -19, \ldots$
    (c) $-11, 2, -13, 4, -15, \ldots$
    (d) $-11, -24, -39, -56, -75, \ldots$
    (e) None of these

2. Find the sum: $\sum_{n=3}^{6} \frac{3}{n-2}$

    (a) $\frac{12}{9}$
    (b) $\frac{25}{4}$
    (c) $\frac{3}{16}$
    (d) $\frac{1}{2}$
    (e) None of these

3. Find a_n for the arithmetic sequence with $a_1 = 5$, $d = -4$, and $n = 98$.

    (a) -392
    (b) -387
    (c) -383
    (d) 393
    (e) None of these

4. Find a formula for a_n for the arithmetic sequence with $a_1 = 5$ and $d = -4$. (Assume that n begins with 1.)

    (a) $a_n = -4n + 9$
    (b) $a_n = -4_n + 5$
    (c) $a_n = 5n - 4$
    (d) $a_n = 9n - 4$
    (e) None of these

5. Write the first five terms of the geometric sequence with $a_1 = -3$ and $r = \frac{2}{3}$.

    (a) $-3, -2\frac{1}{3}, -1\frac{2}{3}, -1, -\frac{1}{3}$
    (b) $-3, -3\frac{2}{3}, -4\frac{1}{3}, -5, -5\frac{2}{3}$
    (c) $-3, -\frac{9}{2}, -\frac{27}{4}, -\frac{81}{8}, -\frac{243}{16}$
    (d) $-3, -2, -\frac{4}{3}, -\frac{8}{9}, -\frac{16}{27}$
    (e) None of these

6. Find the sum of the infinite geometric sequence: $-7, -\frac{7}{3}, -\frac{7}{9}, -\frac{7}{27}, \ldots$

    (a) -5
    (b) $-\frac{21}{4}$
    (c) $-\frac{5}{2}$
    (d) $-\frac{21}{2}$
    (e) None of these

7. Find the sum using the formulas for the sums of powers of integers: $\sum_{n=1}^{8} (n^2 - n^3)$

    (a) -994
    (b) -1092
    (c) -1296
    (d) -1538
    (e) None of these

8. Identify S_{k+1} given $S_k = \frac{k(2k-1)}{3}$.

(a) $\dfrac{2k(k+1)}{3}$ (b) $\dfrac{2k^2 - k + 3}{3}$ (c) $\dfrac{(k+1)(2k+1)}{3}$

(d) $\dfrac{2k^2 - k + 1}{3}$ (e) None of these

9. Use the Binomial Theorem to expand then simplify: $(x - 3)^5$

   (a) $x^5 - 15x^4 + 30x^3 - 30x^2 + 15x - 243$

   (b) $x^5 - 15x^4 + 900x^3 - 27{,}000x^2 + 50{,}625x - 243$

   (c) $x^5 - 15x^4 + 90x^3 - 270x^2 + 405x - 243$

   (d) $x^5 - 3x^4 + 9x^3 - 27x^2 + 81x - 243$

   (e) None of these

10. Determine the coefficient of $x^5 y^7$ in the expansion of $(5x + 2y)^{12}$.

    (a) 316,800,000   (b) 400,000   (c) 792

    (d) 7920   (e) None of these

11. Evaluate: $_{10}P_6$

    (a) 5040   (b) 151,200   (c) 210

    (d) 60   (e) None of these

12. Determine the number of ways the last four digits of a telephone number can be arranged if the first four digits cannot be 0.

    (a) 10,000   (b) 5040   (c) 9000   (d) 4536   (e) None of these

13. A card is drawn at random from a standard deck of 52 playing cards. Find the probability that the card is a 10 or an ace.

    (a) $\frac{2}{13}$ (b) $\frac{1}{169}$ (c) $\frac{4}{13}$ (d) $\frac{1}{4}$ (e) None of these

14. A small business college has 800 seniors, 700 juniors, 900 sophomores and 1200 freshmen. If a student is randomly selected, what is the probability that the student is a freshman or senior?

    (a) $\frac{1}{4}$ (b) $\frac{1}{54}$ (c) $\frac{5}{12}$

    (d) $\frac{5}{9}$ (e) None of these

Test B

Chapter 8

Name _____ Date _____

Class _____ Section _____

1. Find the first 5 terms of the sequence whose nth terms is $a_n = n!$. (Assume that n begins with 0.)

   (a) 0, 1, 2, 6, 24   (b) 0, 1, 2, 6, 12   (c) 1, 1, 2, 6, 12

   (d) 1, 1, 2, 6, 24   (e) None of these

2. Find the sum: $\sum_{n=1}^{4} \dfrac{n+1}{n+2}$

   (a) $\dfrac{61}{20}$ (b) $\dfrac{31}{20}$ (c) $\dfrac{143}{60}$

   (d) $\dfrac{131}{60}$ (e) None of these

3. Find the 99th term of the arithmetic sequence with $a_1 = 7$ and $d = -3$. (Assume that n begins with 1.)

   (a) -287 (b) -290 (c) -293

   (d) -297 (e) None of these

4. Find a formula for a_n for the arithmetic sequence with $a_3 = 15$ and $d = -2$. (Assume that n begins with 1.)

   (a) $a_n = -2n + 9$ (b) $a_n = -2n + 19$ (c) $a_n = -2n + 21$

   (d) $a_n = -2n + 15$ (e) None of these

5. Find the 20th term of the geometric sequence with $a_1 = 5$ and $r = 1.1$.

   (a) 1.1665   (b) 37.0012   (c) 33.6375

   (d) 30.5795   (e) None of these

6. Find the sum of the infinite geometric sequence: 1, 0.9, 0.81, 0.729, ...

   (a) 23   (b) 90   (c) 10   (d) 57   (e) None of these

7. Find the sum using the formulas for the sums of powers of integers: $\sum_{n=1}^{50} (n^2 - n)$

   (a) 42,925   (b) 41,650   (c) 44,100

   (d) 43,150   (e) None of these

8. Identify S_{k+1} given $S_k = k(3k - 1)$.
   (a) $(k + 1)(3k + 2)$
   (b) $3k(k + 1)$
   (c) $k(3k - 1) + 1$
   (d) $3k^2 + 1$
   (e) None of these

9. Use the Binomial Theorem to expand, then simplify: $(2x - 3)^3$
   (a) $8x^3 - 324x^2 + 324x - 27$
   (b) $8x^3 - 36x^2 + 54x - 27$
   (c) $2x^3 - 18x^2 + 54x - 27$
   (d) $8x^3 - 12x^2 + 27x - 27$
   (e) None of these

10. Determine the coefficient of x^3y^5 in the expansion of $(3x + 2y)^8$.
    (a) 336
    (b) 48,384
    (c) 864
    (d) 52,488
    (e) None of these

11. Evaluate: $_{14}P_4$
    (a) 24,024
    (b) 8008
    (c) 5040
    (d) 720
    (e) None of these

12. If a license plate number consists of two letters followed by two digits, how many different license plate numbers are possible?
    (a) 58,500
    (b) 67,600
    (c) 256
    (d) 24
    (e) None of these

13. A card is drawn at random from a standard deck of 52 playing cards. Find the probability that the card is an ace or spade.
    (a) $\frac{17}{52}$
    (b) $\frac{4}{13}$
    (c) $\frac{1}{52}$
    (d) $\frac{2}{13}$
    (e) None of these

14. A small business college has 400 seniors, 300 juniors, 500 sophomores and 600 freshmen. If a student is randomly selected, what is the probability that the student is a junior or a senior?
    (a) $\frac{1}{4}$
    (b) $\frac{2}{9}$
    (c) $\frac{7}{18}$
    (d) $\frac{1}{27}$
    (e) None of these

Test C

Chapter 8

1. Find the first 5 terms of the sequence whose nth term is $a_n = 1 - \dfrac{1}{n}$. (Assume that n begins with 1.)

   (a) $\dfrac{1}{2}, \dfrac{1}{4}, \dfrac{1}{8}, \dfrac{1}{16}, \dfrac{1}{32}$ (b) $0, \dfrac{1}{2}, \dfrac{2}{3}, \dfrac{3}{4}, \dfrac{4}{5}$ (c) $0, \dfrac{1}{2}, \dfrac{1}{3}, \dfrac{1}{4}, \dfrac{1}{5}$

   (d) $1, \dfrac{1}{2}, \dfrac{2}{3}, \dfrac{3}{4}, \dfrac{4}{5}$ (e) None of these

2. Find the sum: $\displaystyle\sum_{i=1}^{4} (1 - i)$

   (a) -3 (b) -6 (c) 6 (d) -5 (e) None of these

3. Find the ninth term of the arithmetic sequence with $a_1 = 4$ and $d = 10$. (Assume that n begins with 1.)

   (a) 94   (b) 84   (c) 46   (d) 49   (e) None of these

4. Find a formula for a_n for the arithmetic sequence with $a_2 = 12$ and $d = -3$. (Assume that n begins with 1.)

   (a) $a_n = 12n - 3$ (b) $a_n = -3n + 15$ (c) $a_n = -3n + 12$

   (d) $a_n = -3n + 18$ (e) None of these

5. Find the 23rd term of the geometric sequence with $a_1 = -23$ and $r = \sqrt{2}$.

   (a) $-47104\sqrt{2}$ (b) $-2048\sqrt{2}$ (c) -2048

   (d) -47104 (e) None of these

6. Find the sum of the infinite geometric sequence: $1, \dfrac{1}{3}, \dfrac{1}{9}, \dfrac{1}{27}, \ldots$

   (a) $\dfrac{3}{2}$ (b) 3 (c) $\dfrac{5}{3}$ (d) $\dfrac{5}{2}$ (e) None of these

7. Find the sum using the formulas for the sums of powers of integers: $\displaystyle\sum_{n=1}^{15} 4n^2$

   (a) 4960   (b) 1240   (c) 73,810

   (d) 74,400   (e) None of these

8. Identify S_{k+1} given $S_k = k^2(k + 1)^2$.

   (a) $(k^2 + 1)(k + 2)^2$ (b) $k^2(k + 1)^2 + 1$ (c) $(k + 1)^2(k - 1)^2$

   (d) $(k + 1)^2(k + 2)^2$ (e) None of these

9. Expand: $(3 - 2x)^3$

   (a) $27 - 3x + 3x^2 - 8x^3$ (b) $27 - 9x + 9x^2 - 8x^3$ (c) $27 - 27x + 6x^2 - 8x^3$

   (d) $27 - 54x + 36x^2 - 8x^3$ (e) None of these

10. Determine the coefficient of x^2y^7 in the expansion of $(3x - 2y)^9$.

    (a) -1152 (b) 1152 (c) $41,472$

    (d) $-41,472$ (e) None of these

11. Evaluate: $_7P_4$

    (a) 840 (b) 35 (c) $10,920$

    (d) 210 (e) None of these

12. An auto license plate is made using two letters followed by three digits. How many license plates are possible?

    (a) 676,000 (b) 468,000 (c) 82 (d) 1,757,600 (e) None of these

13. Find the probability of choosing an E when selecting a letter at random from those in the word COLLEGE.

    (a) $\frac{2}{7}$ (b) $\frac{1}{5}$ (c) $\frac{2}{5}$ (d) $\frac{1}{7}$ (e) None of these

14. A small business college has 400 seniors, 300 juniors, 500 sophomores and 600 freshmen. If a student is randomly selected, what is the probability that the student is a freshmen or a senior?

    (a) $\frac{2}{27}$ (b) $\frac{5}{9}$ (c) $\frac{1}{4}$

    (d) $\frac{5}{18}$ (e) None of these

Test D

Chapter 8

Name _____ Date _____

Class _____ Section _____

1. Write the first five terms of the sequence whose nth term is $a_n = \dfrac{n-2}{n^2+1}$. (Assume that n begins with 1.)

2. Use a calculator to find the sum.

    $$\sum_{i=1}^{5} (10 - 2i)$$

3. Find a_n for the arithmetic sequence with $a_1 = 12$, $d = \tfrac{1}{3}$, and $n = 52$.

4. Find the sum of the first 19 terms of the arithmetic sequence whose nth term is $a_n = n + 1$. (Assume that n begins with 1.)

5. Find the 28th term of the geometric sequence: 2, 2.4, 2.88, 3.456, 4.1472, ...

6. Find the sum of the infinite geometric sequence with $a_1 = 9$ and $r = 0.7$.

7. Find the sum using the formulas for the sums of powers of integers: $\sum_{n=1}^{20} 3n^2$

8. Prove by mathematical induction: $1 + 2 + 2^2 + 2^3 + \cdots + 2^{n-1} = 2^n - 1$

9. Use the Binomial Theorem to expand and simplify: $\left(\sqrt{x} + 2\right)^3$

10. Determine the coefficient of the $x^2 y^7$ in the expansion of $(7x - 2y)^9$.

11. In how many distinguishable ways can the letters MISSISSIPPI be arranged?

12. How many different ways can three chocolate, four strawberry, and two butterscotch sundaes be served to nine people?

13. In a group of 10 children, 3 have blond hair and 7 have brown hair. If a child is chosen at random, what is the probability that the child will have brown hair?

14. There are 5 red and 4 black balls in a box. If 3 balls are picked without replacement, what is the probability that at least one of them is red?

Test E

Chapter 8

Name _____ Date _____

Class _____ Section _____

1. Write the first five terms of the sequence whose nth term is $a_n = \dfrac{n!}{(n+2)!}$. (Assume that n begins with 0.)

2. Use a calculator to find the sum.

   $$\sum_{k=0}^{5} \dfrac{k!}{2}$$

3. Find the 30th term of the arithmetic sequence with $a_1 = -5$ and $d = \tfrac{1}{3}$. (Assume that n begins with 1.)

4. Find a formula for a_n for the arithmetic sequence with $a_2 = 15$ and $d = \tfrac{3}{2}$. (Assume that n begins with 1.)

5. Find the 14th term of the geometric sequence with $a_1 = -11$ and $r = \sqrt{3}$.

6. Find the sum of the first 30 terms in the sequence.
   $\sqrt{2}, 2\sqrt{2}, 3\sqrt{2}, 4\sqrt{2}, 5\sqrt{2}, \ldots$

7. Find the sum using the formulas for the sums of powers of integers: $\sum_{n=1}^{19} 2n^2$

8. Use mathematical induction to prove $n < 3^n$ for all positive integers n.

9. Use the Binomial Theorem to expand and simplify: $(2\sqrt{y} - 3)^4$

10. Find the coefficient of x^4y^3 in the expansion of $(2x + y)^7$.

11. Find the number of distinguishable permutations using the letters in the word MATHEMATICS.

12. A group of six students are seated in a single row at a football game. In how many different orders can they be seated?

13. A fair coin is tossed four times. What is the probability of getting heads on all four tosses?

14. A sample of nursing homes in a state reveals that 112,000 of 218,000 residents are female. If a nursing home resident is chosen at random from this state, what is the probability that the resident is male?

College Algebra
Multiple Choice

Name _____ Date _____
Class _____ Section _____

1. Simplify: $2x^2y\sqrt[3]{2x} + 7x^2\sqrt[3]{2xy^3} - 4\sqrt[3]{16x^7y^3}$.

   (a) $x^6y^3\sqrt[3]{2x}$
   (b) $x^2y\sqrt[3]{2x}$
   (c) $9x^2y\sqrt[3]{2x} - 8y\sqrt[3]{2x^7y}$
   (d) $2x^3y$
   (e) None of these

2. Simplify: $3x(5x + 2) - 14(2x^2 - x + 1)$.

   (a) $28x^2 - 3x - 14$
   (b) $13x^2 - 8x + 14$
   (c) $-13x^2 + 20x - 14$
   (d) $13x^2 - 20x + 14$
   (e) None of these

3. Represent the area of the region as a polynomial in standard form.

   (a) $84x$
   (b) $-10x^2 + 18x + 90$
   (c) $-6x^2 + 90x$
   (d) $10x + 36$
   (e) None of these

   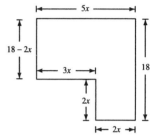

4. Factor completely: $3x^4 - 48$.

   (a) $3(x - 2)^2(x + 2)^2$
   (b) $3(x - 2)^4$
   (c) $3x^2(x - 4)^2$
   (d) $3(x^2 + 4)(x + 2)(x - 2)$
   (e) None of these

5. Add, then simplify: $\dfrac{2}{x^2 - 9} + \dfrac{5}{x^2 - x - 12}$.

   (a) $\dfrac{7}{(x^2 - 9)(x^2 - x - 12)}$
   (b) $\dfrac{7x^2 - x - 21}{(x^2 - 9)(x^2 - x - 12)}$
   (c) $\dfrac{7x - 7}{(x - 3)(x - 4)(x + 3)}$
   (d) $\dfrac{7x - 23}{(x - 3)(x + 3)(x - 4)}$
   (e) None of these

6. Match the equation with the graph.

   (a) $y = \sqrt{9 - x^2}$
   (b) $y = |x^2 - 9|$
   (c) $y = \sqrt{x^2 - 9}$
   (d) $y = (9 - x)^2$
   (e) None of these

   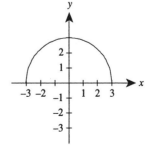

7. Solve for x: $\dfrac{1}{x-2} + \dfrac{3}{x+3} = \dfrac{4}{x^2+x-6}$.

   (a) $\dfrac{4}{7}$ (b) 3 (c) $\dfrac{7}{4}$

   (d) 1 (e) None of these

8. Two brothers, Bob and Bill, live 450 miles apart. Starting at the same time they plan to drive until they meet. Bill averages 10 miles per hour faster than Bob who averages 50 mph. How long will it take them to meet?

   (a) $3\frac{2}{15}$ hours (b) $3\frac{9}{11}$ hours (c) $4\frac{1}{11}$ hours

   (d) $4\frac{1}{2}$ hours (e) None of these

9. Solve for x: $2x^2 - 5x = x^2 + 1$.

   (a) 0, 5 (b) $\dfrac{4}{11}$ (c) $\dfrac{5 \pm \sqrt{29}}{2}$

   (d) $\dfrac{-5 \pm \sqrt{21}}{2}$ (e) None of these

10. Solve: $x^3 - 5x - 2x^2 + 10 = 0$.

    (a) $-2, \pm\sqrt{5}$ (b) $\pm\sqrt{5}$ (c) $2, \sqrt{5}$

    (d) $2, \pm\sqrt{5}$ (e) None of these

11. Graph the solution: $-6 < 7x + 2 \le 5$.

    (a)  (b)

    (c) (d)

    (e) None of these

12. Find the equation of the line that is perpendicular to $2x + 3y = 12$ but has the same y-intercept.

    (a) $2x + 3y = 8$ (b) $2x - 3y = 12$ (c) $2x + 3y = -12$

    (d) $3x - 2y = -8$ (e) None of these

13. Given $f(x) = x^2 - 3x + 4$, find $f(x+2) - f(2)$.

    (a) $x^2 - 3x + 4$ (b) $x^2 + x$ (c) $x^2 + x - 8$

    (d) $x^2 - 3x - 4$ (e) None of these

14. Find the range of the function: $y = \sqrt{9 - x^2}$.

(a) $(-\infty, -3], [3, \infty)$
(b) $[-3, 3]$
(c) $[0, 3]$
(d) $[3, \infty)$
(e) None of these

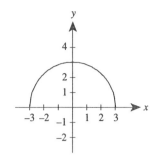

15. Given $f(x) = x^2 - 2x$ and $g(x) = 3x + 2x$, find $(f \circ g)(x)$.

(a) $4x^2 + 8x + 3$
(b) $2x^2 - 4x + 3$
(c) $2x^3 - x^2 - 6x$
(d) $3x^2 + x$
(e) None of these

16. Given $f(x) = 3x^3 - 1$, find $f^{-1}(x)$.

(a) $\dfrac{1}{3x^3 - 1}$
(b) $3x^{-1} - 1$
(c) $3(x + 1)$
(d) $\sqrt[3]{\dfrac{x + 1}{3}}$
(e) None of these

17. Write in the form $y = a(x - h)^2 + k$: $y = 2x^2 + 16x + 9$

(a) $y = 2(x + 4)^2 - 7$
(b) $y = 2(x + 2)^2 + 5$
(c) $y = 2(x + 4)^2 - 23$
(d) $y = 2(x + 8)^2 + 73$
(e) None of these

18. Find a polynomial function with the given zeros: $-2, -2, 1, 3$

(a) $f(x) = (x + 1)(x + 3)(x - 2)$
(b) $f(x) = (x - 2)^2(x - 1)(x - 3)$
(c) $f(x) = (x + 2)(x - 1)(x - 3)$
(d) $f(x) = (x + 2)^2(x - 1)(x - 3)$
(e) None of these

19. Find all of the real roots: $2x^3 + 5x^2 - x - 6 = 0$

(a) $-3, -1, 1$
(b) $-1, \frac{3}{2}, 2$
(c) $-2, -\frac{3}{2}, 1$
(d) $-6, 2, 5$
(e) None of these

20. Write as a product of linear factors: $f(x) = x^4 - 6x^3 - 4x^2 + 40x + 32$

(a) $(x - 4)(x + 2)(x + 2 + \sqrt{8})(x + 2 - \sqrt{8})$
(b) $(x + 4)(x - 2)(x - 2 + \sqrt{8})(x - 2 - \sqrt{8})$
(c) $(x - 4)(x - 2)(x - 2 + \sqrt{8})(x - 2 - \sqrt{8})$
(d) $(x + 4)(x + 2)(x + 2 + \sqrt{8})(x + 2 - \sqrt{8})$
(e) None of these

21. Find the vertical asymptote(s): $f(x) = \dfrac{x+3}{(x-2)(x+5)}$
    (a) $y = 2, y = -5, y = -3$
    (b) $x = 2, x = -5, x = -3, x = 1$
    (c) $x = 1$
    (d) $x = 2, x = -5$
    (e) None of these

22. The concentration of a mixture is given by
    $C = \dfrac{3x+8}{4(x+8)}.$
    Use a graphing utility using the indicated range setting to determine what the concentration approaches.
    (a) 75%
    (b) 25%
    (c) 50%
    (d) 33%
    (e) None of these

    Xmin = 0
    Xmax = 200
    Xscl = 50
    Ymin = 0
    Ymax = 1
    Yscl = .1

23. Find the partial fraction decomposition: $\dfrac{2x^2 + 6x - 11}{(x-3)(x+2)^2}$
    (a) $\dfrac{6}{x+2} + \dfrac{1}{(x+2)^2} - \dfrac{1}{x-3}$
    (b) $\dfrac{-3}{x+2} + \dfrac{5}{(x+2)^2} - \dfrac{2}{x-3}$
    (c) $\dfrac{2}{x+2} + \dfrac{7}{(x+2)^2} - \dfrac{11}{x-3}$
    (d) $\dfrac{1}{x+2} + \dfrac{3}{(x+2)^2} + \dfrac{1}{x-3}$
    (e) None of these

24. Match the graph with the correct equation.
    (a) $\dfrac{x^2}{4} + \dfrac{y^2}{2} = 1$
    (b) $\dfrac{y^2}{4} + \dfrac{x^2}{2} = 1$
    (c) $\dfrac{x^2}{16} + \dfrac{y^2}{4} = 1$
    (d) $\dfrac{y^2}{16} - \dfrac{x^2}{4} = 1$
    (e) None of these

    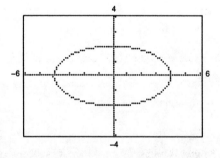

25. Find the center of the hyperbola: $3x^2 - 4y^2 - 6x - 16y + 7 = 0$
    (a) $(1, -2)$
    (b) $(4, 3)$
    (c) $(1, -8)$
    (d) $(3, -8)$
    (e) None of these

26. A certain population increases according to the model $P(t) = 250e^{0.47t}$. Use the model to determine the population when $t = 10$. Round your answer to the nearest integer.
    (a) 400
    (b) 4091
    (c) 27,487
    (d) 23,716
    (e) None of these

27. Find the domain of the function: $f(x) = 3 + \ln(x - 1)$

(a) $(-\infty, \infty)$ (b) $(0, \infty)$ (c) $(1, \infty)$

(d) $(3, \infty)$ (e) None of these

28. Write as a sum, difference, or multiple of logarithms: $\log_b \left(\dfrac{x^3 y^2}{\sqrt{w}} \right)$

(a) $x^3 + y^3 - \sqrt{w}$

(b) $\dfrac{1}{3} \log_b x + \dfrac{1}{2} \log_b y - 2 \log_b w$

(c) $3 \log_b x + 2 \log_b y - \dfrac{1}{2} \log_b w$

(d) $\dfrac{3 \log x + 2 \log y}{(1/2) \log w}$

(e) None of these

29. Solve for x: $\ln(7 - x) + \ln(3x + 5) = \ln(24x)$

(a) $\tfrac{6}{11}$ (b) $\tfrac{7}{3}$ (c) $\tfrac{7}{3}, -5$

(d) $\tfrac{6}{11}, 5$ (e) None of these

30. The spread of a flu virus through a certain population is modeled by

$$y = \frac{1000}{1 + 990 e^{-0.7t}},$$

where y is the total number infected after t days. In how many days will 530 people be infected with the virus?

(a) 13 days   (b) 12 days   (c) 11 days

(d) 10 days   (e) None of these

31. Solve the system by the method of substitution:

$$x^2 + 2y = 6$$
$$2x + y = 3$$

(a) $(4, -5)$ (b) $(2, 1)$ (c) $(0, 3)$

(d) $(0, 3)$ and $(4, -5)$ (e) None of these

32. Solve the system of linear equations:

$$6x - 9y + 4z = -7$$
$$2x + 6y - z = 6$$
$$4x - 3y + 2z = -2$$

(a) $\left(\dfrac{1}{2}, \dfrac{2}{3}, -1 \right)$ (b) $\left(\dfrac{11}{21}, 1, -\dfrac{2}{7} \right)$ (c) $\left(a, \dfrac{31a}{15}, \dfrac{44a}{5} \right)$

(d) No solution   (e) None of these

33. Find the maximum value of the objective function $z = 6x - 2y$ subject to the constraints:

$$x \geq 0$$
$$y \geq 0$$
$$x + y \leq 10$$
$$4x + y \geq 12$$

(a) 60   (b) 12   (c) 18

(d) 44   (e) None of these

34. Use a graphing utility to find AB, given $A = \begin{bmatrix} 3 & -2 & 4 \\ 0 & 0 & -1 \\ 3 & 2 & -1 \end{bmatrix}$ and $B = \begin{bmatrix} 3 & 1 & -1 \\ -1 & 0 & 0 \\ 2 & 4 & -2 \end{bmatrix}$.

(a) $\begin{bmatrix} 6 & -8 & 12 \\ -3 & 2 & -4 \\ 0 & -8 & 6 \end{bmatrix}$ (b) $\begin{bmatrix} 19 & 19 & -11 \\ -2 & -4 & 2 \\ 5 & -1 & -1 \end{bmatrix}$ (c) $\begin{bmatrix} 9 & -2 & -4 \\ 0 & 0 & 0 \\ 6 & 8 & 2 \end{bmatrix}$

(d) Impossible   (e) None of these

35. Given a system of linear equations with coefficient matrix A, use A^{-1} to find (x, y, z).

$$3x + y - z = -11$$
$$x - y - z = -1$$
$$x + 2y + 3z = 3$$

$$A^{-1} = \tfrac{1}{10}\begin{bmatrix} 1 & 5 & 2 \\ 4 & -10 & -2 \\ -3 & 5 & 4 \end{bmatrix}$$

(a) $(3, -2, 0)$ (b) $(2, -2, 3)$ (c) $(-1, 5, 1)$

(d) $(-3, 0, 2)$ (e) None of these

36. Use Cramer's Rule to solve for y in the system of linear equations:

$$3x + 2y + 4z = 12$$
$$x - y + z = 3$$
$$2x + 7y - z = 9$$

(a) $y = \dfrac{\begin{vmatrix} 3 & 2 & 4 \\ 1 & -1 & 1 \\ 2 & 7 & -1 \end{vmatrix}}{\begin{vmatrix} 3 & 12 & 4 \\ 1 & 3 & 1 \\ 2 & 9 & -1 \end{vmatrix}}$ (b) $y = \dfrac{\begin{vmatrix} 3 & 12 & 4 \\ 1 & 3 & 1 \\ 2 & 9 & -1 \end{vmatrix}}{\begin{vmatrix} 3 & 2 & 4 \\ 1 & -1 & 1 \\ 2 & 7 & -1 \end{vmatrix}}$ (c) $y = \dfrac{\begin{vmatrix} 3 & 4 & 12 \\ 1 & -1 & 3 \\ 2 & 7 & 9 \end{vmatrix}}{\begin{vmatrix} 2 & 4 & 3 \\ -1 & 1 & 1 \\ 7 & -1 & 2 \end{vmatrix}}$

(d) $y = \begin{vmatrix} 3 & 2 & 4 \\ 1 & -1 & 1 \\ 2 & 7 & -1 \end{vmatrix} \begin{vmatrix} 2 & 4 & 12 \\ -1 & 1 & 3 \\ 7 & -1 & 9 \end{vmatrix}$

(e) None of these

37. Use sigma notation to write the sum: $\dfrac{2}{3} + \dfrac{4}{4} + \dfrac{6}{5} + \dfrac{8}{6} + \ldots + \dfrac{14}{9}$

   (a) $\displaystyle\sum_{n=1}^{7} \dfrac{2n}{n+2}$
   (b) $\displaystyle\sum_{n=2}^{8} \dfrac{n+2}{n+1}$
   (c) $\displaystyle\sum_{n=0}^{6} \dfrac{n+2}{n+3}$
   (d) $\displaystyle\sum_{n=3}^{9} \dfrac{n-1}{n}$
   (e) None of these

38. Evaluate: $\displaystyle\sum_{n=0}^{\infty} 2\left(\dfrac{1}{2}\right)^n = 2 + 1 + \dfrac{1}{2} + \dfrac{1}{4} + \dfrac{1}{8} + \ldots$

   (a) 4   (b) 6   (c) 8   (d) 10   (e) None of these

39. The flags of seven different countries are to be displayed in a row. In how many different orders can they be flown?

   (a) 5040   (b) 1258   (c) 128   (d) 49   (e) None of these

40. A box holds 12 white, 5 red, and 6 black marbles. If 2 marbles are picked at random, without replacement, what is the probability that they will both be black?

   (a) $\dfrac{36}{529}$
   (b) $\dfrac{247}{506}$
   (c) $\dfrac{15}{253}$
   (d) $\dfrac{6}{23}$
   (e) None of these

College Algebra
Open Ended

Name _____ Date _____
Class _____ Section _____

1. Simplify: $(-2x^2)^5(5x^3)^{-2}$.

2. Write in standard form: $3x^2 - 2x(1 + 3x - x^2)$.

3. Expand: $[(x - 1) + y]^2$.

4. Factor: $3x - 24x^4$.

5. Subtract, then simplify: $\dfrac{3}{x} - \dfrac{9}{x + 1}$.

6. Identify the type(s) of symmetry: $x^2 + xy + y^2 = 0$.

7. Solve for x: $\dfrac{2x - 5}{x - 3} = \dfrac{4x + 1}{2x}$.

8. Two cars, starting together, travel in opposite directions on a highway, one at 55 mph and the other at 45 mph. How far apart are they after three hours and 12 minutes?

9. Solve for x: $\dfrac{1}{x - 1} + \dfrac{x}{x + 2} = 2$.

10. Solve for x: $2x^4 - 7x^2 + 5 = 0$.

11. Graph the solution: $-16 \leq 7 - 2x < 5$.

12. Find an equation of the line that passes through $(8, 17)$ and is perpendicular to the line $x + 2y = 2$.

13. Given $f(x) = 3x - 7$, find $f(x + 1) + f(2)$.

14. Find the domain and range of the function: $f(x) = |3 + x|$.

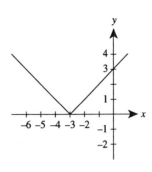

15. What sequence of transformations will yield the graph of $g(x) = \sqrt[4]{x - 3} + 2$ from the graph of $f(x) = \sqrt[4]{x}$?

16. Given $f(x) = 2x^2 + 1$ for $x \geq 0$, find $f^{-1}(x)$.

17. Write the form $y = a(x - h)^2 + k$: $y = -x^2 + 3x - 2$

18. List the possible rational zeros of the function. Then use a graphing utility to graph the function to eliminate some of the possible zeros. Finally determine all real zeros of the function: $f(x) = 2x^3 - 7x^2 + x + 10$

19. Write as a product of linear factors: $x^4 - 16$

20. Find a fourth degree polynomial function that has zeros: $1, -1, 0,$ and 2

21. Find the domain: $f(x) = \dfrac{x}{x^2 + 3x - 4}$

22. Find the vertical, horizontal, or slant asymptotes: $f(x) = \dfrac{x - 2}{x^2 - 2x - 3}$

23. Find the partial fraction decomposition: $\dfrac{12x^2 - 13x - 3}{(x - 1)^2(x + 3)}$

24. Use a graphing utility to graph: $\dfrac{x^2}{9} - \dfrac{y^2}{4} = 1$

25. Find the center of the ellipse: $5x^2 + 2y^2 - 20x + 24y + 82 = 0$

26. A certain population grows according to the equation $y = 40e^{0.025t}$. Find the initial population and the population (to the nearest integer) when $t = 50$.

27. Find the domain of the function: $f(x) = 3 - \log(x^2 - 1)$

28. Write as a sum, difference, or multiple of logarithms: $\ln \dfrac{5x}{\sqrt[3]{x^2 + 1}}$

29. Solve for x: $\log_3(x^2 + 5) = \log_3(4x^2 - 2x)$

30. The number N of bacteria in a culture is given by

    $N = 200e^{kt}$.

    If $N = 300$ when $t = 4$ hours, find k (to the nearest tenth) and then determine approximately how long it will take for the number of bacteria to triple in size.

31. Solve the system by the method of substitution:

    $2x^2 - y = -2$
    $x - y = -2$

32. Solve the system of linear equations:

    $x + y - z = -1$
    $2x + 3y - z = -2$
    $-3x - 2y + 2z = -3$

33. Find the minimum and maximum values of the objective function $z = 4x + 16y$ subject to the constraints:

    $x \geq 0$
    $y \geq 0$
    $3x + y \leq 23$
    $x + 2y \leq 16$
    $3x \geq 2y$

34. Given $A = \begin{bmatrix} 1 & 0 & 3 \\ -1 & 2 & -2 \\ 1 & 1 & 2 \end{bmatrix}$ and $B = \begin{bmatrix} 1 & 1 & 0 \\ 3 & 1 & 2 \\ -1 & 1 & -1 \end{bmatrix}$, find BA.

35. Use the inverse matrix to solve the system of linear equations.

    $2x + 2y = 12$
    $x + 3y = 16$

36. Use Cramer's Rule to solve the system of linear equations.

    $4x + 6y + 2z = 15$
    $x - y + 4z = -3$
    $3x + 2y + 2z = 6$

37. Use sigma notation to write the sum: $\dfrac{1}{2} + \dfrac{2}{6} + \dfrac{3}{24} + \dfrac{4}{120} + \dfrac{5}{720}$

38. Find the sum using the formulas for the sums of powers of integers: $\sum_{n=1}^{19} 2n^2$

39. A group of six students are seated in a single row at a football game. In how many different orders can they be seated?

40. There are 5 red and 4 black balls in a box. If 3 balls are picked without replacement, what is the probability that at least one of them is red?

Answers to CHAPTER P Tests

Test A

1. a
2. c
3. d
4. c
5. d
6. d
7. a
8. a
9. a
10. a
11. d
12. d
13. d
14. d
15. d

Test B

1. b
2. d
3. a
4. c
5. d
6. b
7. b
8. e
9. b
10. b
11. a
12. e
13. b
14. b
15. c

Test C

1. b
2. a
3. d
4. b
5. d
6. a
7. d
8. a
9. a
10. c
11. d
12. c
13. b
14. a
15. c

Test D

1. $0.\overline{90}$
2. $\left\{\dfrac{2}{5}, \dfrac{45}{99}, \dfrac{152}{333}, \dfrac{7}{5}, \dfrac{23}{5}\right\}$
3. $(2x)(2x)(2x)$
4. $\dfrac{1}{9x^3y^2z^2}$
5. $\dfrac{x^{2/3}}{y^{1/3}}$
6. $-4x^2 + 5x + 5$
7. $3x^2 + 4x$ square units
8. $(2x - 3)(7x + 1)$
9. $\dfrac{x + 3}{3}$
10. $\dfrac{x^2 - 2x - 8}{(x^2 + 4)(x + 4)}$
11. $\dfrac{x}{x + 2}$
12. $\dfrac{3}{8}(3x + 1)^{1/2}(x + 1)$
13. 5
14. $\left(\dfrac{3}{2}, 5\right)$
15. $3\sqrt{2}$

Test E

1. $0.5\overline{24}$
2. $\left\{\dfrac{28}{5}, 6.065, \dfrac{13}{2}, 6.56, \dfrac{650}{99}\right\}$
3. $(3y)(3y)(3y)(3y)$
4. $-\dfrac{32x^4}{25}$
5. $x^{5/6}$
6. $2x^3 - 3x^2 - 2x$
7. $9x^2$ square units
8. $(5x + 2)(7x - 1)$
9. $\dfrac{-2x}{x + 3}$
10. $\dfrac{1}{x^2 + 1}$
11. $\dfrac{x + 2}{4(x + 3)}$
12. $-\dfrac{(x + 2)(2x - 1)^{3/2}}{6x^2}$
13. $\sqrt{61}$
14. $\left(-\dfrac{1}{2}, -\dfrac{3}{2}\right)$
15. $4\sqrt{2}$

Answers to CHAPTER 1 Tests

Test A

1. d
2. d
3. d
4. a
5. a
6. c
7. a
8. d
9. a
10. d
11. d
12. c
13. c
14. b
15. d

Test B

1. c
2. b
3. e
4. d
5. a
6. a
7. c
8. d
9. c
10. b
11. a
12. a
13. a
14. c
15. a

Test C

1. b
2. c
3. d
4. b
5. c
6. b
7. b
8. b
9. c
10. b
11. b
12. b
13. d
14. b
15. c

Test D

1. 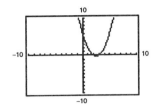   (2, 0); (3, 0); (0, 6)   2. $(x + 2)^2 + (y + 4)^2 = 49$

3. $x = -8$ 4. 1.13   5. $A = \left(\frac{3}{2}x\right)(6) + \left(\frac{3}{2}x\right)(12) = 27x$

6. 250 miles   7. $-\frac{3}{2}$ 8. $\frac{3 \pm \sqrt{3}}{2}$ 9. $-\frac{20}{29} + \frac{21}{29}i$

10. -1 11. 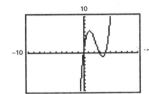   (0, 0); (3, 0); (4, 0); $x = 0, 3, 4$

12. [number line with $-\frac{8}{3}$ and $\frac{10}{3}$]   13. $[-1, 5]$ 14. $[-6, 6]$

15. $(0, 1) \cup (3, 4)$

Test E

1. 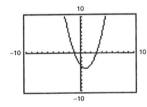 $(-1, 0); (3, 0); (0, -3)$ 2. $x^2 + \left(y - \frac{1}{2}\right)^2 = 256$

3. $x = 10$ 4. 1.97   5. $A = \frac{1}{2}a\left(\frac{3}{4}a + 1\right) = \frac{3}{8}a^2 + \frac{1}{2}a$

6. 175 miles   7. $-7, \frac{2}{3}$ 8. $\frac{2 + \sqrt{22}}{3}$ 9. $-\frac{20}{29} + \frac{21}{29}i$

10. $\frac{1600}{81}$ 11. 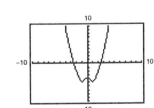 $\left(\frac{1}{2}, 0\right); x = \frac{1}{2}$

12. $-7 \leq x \leq -3$ 13. $(-\infty, -2) \cup (3, \infty)$ 14. $[-2, 2]$ 15. $(1, 3)$

Answers to CHAPTER 2 Tests

Test A

1. c
2. c
3. a
4. c
5. a
6. d
7. b
8. d
9. d
10. a
11. a
12. e
13. d

Test B

1. b
2. d
3. b
4. e
5. b
6. c
7. c
8. c
9. b
10. d
11. e
12. d
13. b

Test C

1. b
2. e
3. d
4. e
5. b
6. d
7. b
8. b
9. c
10. b
11. a
12. e
13. b

Test D

1. 2
2. $y = 2x + 1$
3. $S = 75,000t + 75,000$
4. a; in b some values of x yield more than one y.
5. $-k^2$
6. All real $x \neq 3, x \neq 4$
7. Domain: all real x; Range: $f(x) \leq 3$
8. $(-\infty, 0), (1, \infty)$

9.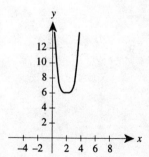

10.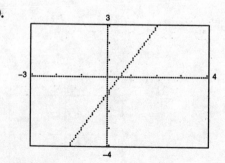

11. 9 **12.** Yes, they are inverses of each other. **13.** $f^{-1}(x) = \sqrt{\dfrac{x-1}{2}}$

Test E

1. $\dfrac{9}{4}$ **2.** $y = 3x - 4$ **3.** $E = 1225t + 2800$

4. a; in b some values of x yield more than one y. **5.** $1 - 2k$ **6.** All real $x \neq -\dfrac{1}{2}$

7. Domain: all real x; Range: $f(x) \leq 3$ **8.** $(-\infty, -1), (-1, 0)$

9. 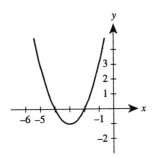   **10.**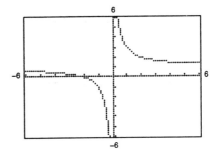

11. 1 **12.** No, they are not inverses of each other. **13.** $f^{-1}(x) = \dfrac{3x - 1}{2}$

Answers to CHAPTER 3 Tests

Test A

1. c
2. b
3. c
4. d
5. c
6. b
7. b
8. a
9. c
10. a
11. d
12. d

Test B

1. d
2. a
3. a
4. b
5. b
6. d
7. a
8. a
9. a
10. d
11. b
12. d

Test C

1. c
2. a
3. a
4. d
5. b
6. a
7. c
8. c
9. c
10. c
11. c
12. b

Test D

1. $f(x) = -(x - 3)^2 - 1$

2. $f(x) = -(x + 1)^2 + 2$

3. Up to the left, down to the right.

4. $v(x) = 4x(8 - x)^2$
   $x \approx 2.67$ inches
   $v(2.67) \approx 303.4$ in^3

5. $2x^3 - 2x^2 + x + 3 - \dfrac{4}{3x + 1}$

6. $4x^2 + 5x - 1$

7. $\pm\frac{1}{2}, \pm 1, \pm\frac{3}{2}, \pm\frac{5}{2}, \pm 3, \pm 5, \pm\frac{15}{2}, \pm 15$

8. $\pm\frac{1}{3}, \pm\frac{2}{3}, \pm 1, \pm\frac{4}{3}, \pm 2, \pm 4$;

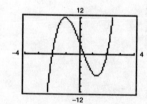

9. $(x + 2)(x - 2)(x + 2i)(x - 2i)$ **10.** $1, 1 \pm 2i$ **11.** 25

12. 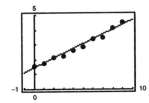   ; $y = 0.325x + 1.465$

Test E

1. $y = -\left(x - \frac{3}{2}\right)^2 + \frac{1}{4}$ **2.** $f(x) = x^2 - 6x$

3. Up to the left and right.   **4.** $v(x) = 4x(10 - x)^2$

$x \approx 3.33$ inches

$v(3.33) \approx 592.6$ in^3

5. $2x^2 - 6 + \dfrac{7x + 16}{x^2 + 3}$ **6.** $2x^2 - 6x + 1$

7. Lower bound   **8.** $\pm\frac{1}{2}, \pm 1, \pm\frac{3}{2}, \pm 3, \pm\frac{9}{2}, \pm 9$;

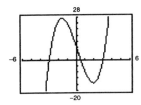

9. $(x + \sqrt{10})(x - \sqrt{10})(x + \sqrt{10}i)(x - \sqrt{10}i)$ **10.** $-1, 3, 2 \pm \sqrt{3}i$ **11.** $\dfrac{7}{9}$

12. 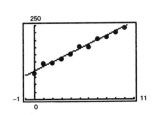   ; $y = 14.963x + 96.363$

Answers to CHAPTER 4 Tests

Test A

1. c
2. a
3. a
4. c
5. b
6. d
7. e
8. c
9. a
10. d

Test B

1. b
2. d
3. d
4. d
5. b
6. b
7. a
8. c
9. c
10. d

Test C

1. e
2. d
3. c
4. c
5. d
6. d
7. c
8. d
9. b
10. a

Test D

1. $x = 2, x = 4$
2. $x \neq -4, 1$
3. $y = 3$

4.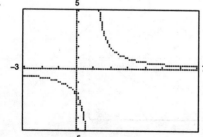
5. 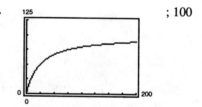 ; 100

6. $\dfrac{3}{x-1} - \dfrac{1}{(x-1)^2} - \dfrac{3}{(x-1)^3}$

7.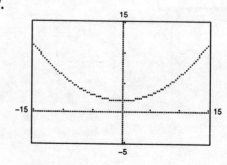

Chapter 4   Answers Keys to Chapter Tests   555

8. $\dfrac{x^2}{36} + \dfrac{y^2}{27} = 1$ 9. $25x^2 + 16y^2 - 160y = 0$

10. $y = -x \pm \sqrt{1 - x^2}$; 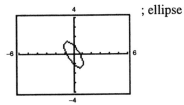 ; ellipse

Test E

1. $x = 1, x = 5$ 2. $x \ne \pm\sqrt{10}$ 3. $y = 1$

4. 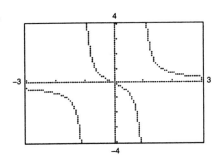   5. 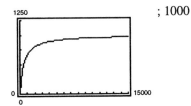 ; 1000

6. $\dfrac{5}{x+1} + \dfrac{2}{(x+1)^2} + \dfrac{3}{(x+1)^3}$ 7.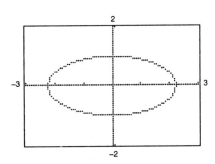

8. $\dfrac{x^2}{36} + \dfrac{y^2}{81} = 1$ 9. $\dfrac{(y-5)^2}{36} - \dfrac{(x-2)^2}{64} = 1$

10. $y = 2x \pm \sqrt{1 - x^2}$; 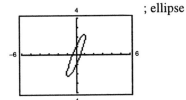 ; ellipse

Answers to CHAPTER 5 Tests

Test A

1. a 2. d 3. d 4. d
5. d 6. b 7. a 8. b
9. b 10. c

Test B

1. b 2. c 3. b 4. a
5. e 6. a 7. b 8. d
9. a 10. b

Test C

1. d 2. c 3. a 4. d
5. b 6. d 7. b 8. c
9. b 10. b

Test D

1.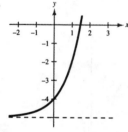

2. 295,263

3. $\log_3 25 = 2$

4.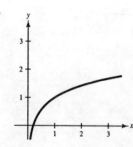

5. $\ln 5 + \ln x - \frac{1}{3}\ln(x^2 + 1)$

6. -0.3801

7. 2

8. -0.847

9. $k = \frac{1}{3} \ln \frac{5}{3}$

10. 847

Test E

1.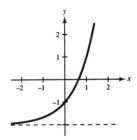

2. 40,140

3. $\log_3 243 = 5$

4.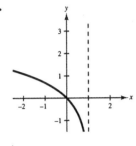

5. $\log \sqrt[5]{\dfrac{(x+1)^3(x-1)^2}{7}}$

6. 1.09165

7. $-1, 5$

8. 3.614

9. $k = \tfrac{1}{2} \ln \tfrac{5}{2}$

10. 888

Answers to CHAPTER 6 Tests

Test A

1. c   2. e   3. b   4. a
5. b   6. b   7. c   8. c
9. b   10. c

Test B

1. e   2. e   3. d   4. b
5. a   6. a   7. c   8. a
9. d   10. a

Test C

1. d   2. d   3. a   4. a
5. a   6. a   7. b   8. b
9. a   10. d

Test D

1. $(1, 1); \left(5, \frac{1}{5}\right)$

2. 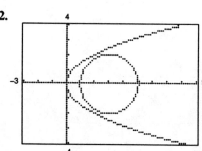 ; No points of intersection

3. $\left(\frac{1}{2}, 1\right)$;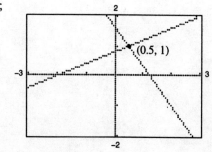

4. 6

5. $(-1, 3, 1)$

6. $y = -2x^2 + 7x - 5$;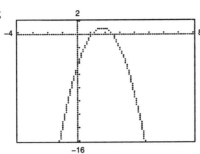

7.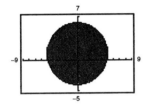

8.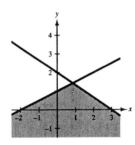

9. 17

10. 80 of the $350 model; 70 of the $400 model

Test E

1. $(0, 2), \left(\frac{1}{2}, \frac{5}{2}\right)$

2. 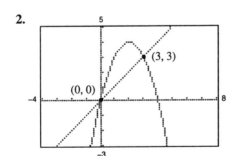 ; $(0, 0), (3, 3)$

3. $\left(\frac{1}{2}, -5\right)$;

4. $L = 33$ feet, $W = 12.5$ feet

5. $x = 0, y = -1, z = 3, w = -4$

6. $y = 4x^2 - 5x + 2$;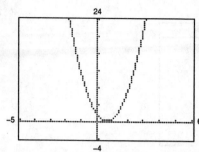

7.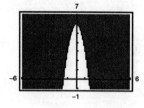

8.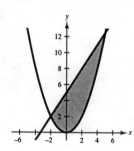

9. 17

10. 110 of the $350 model; 35 of the $500 model

Answers to CHAPTER 7 Tests

Test A

1. b   2. a   3. a   4. b
5. a   6. c   7. b   8. d
9. b   10. b

Test B

1. c   2. b   3. c   4. d
5. b   6. d   7. a   8. c
9. b   10. b

Test C

1. a   2. a   3. b   4. a
5. d   6. e   7. a   8. d
9. c   10. c

Test D

1. $\begin{bmatrix} 1 & 0 & 0 & \frac{5}{16} \\ 0 & 1 & 0 & \frac{3}{16} \\ 0 & 0 & 1 & -\frac{37}{16} \end{bmatrix}$ 2. $y = 2x^2 + 5x - 3$

3. $\begin{bmatrix} -6 & 1 \\ 1 & 2 \end{bmatrix}$ 4. $\begin{bmatrix} 3 & 2 & 2 & 1 \\ 13 & 6 & 12 & 1 \\ -5 & -1 & -5 & 0 \end{bmatrix}$

5. $\begin{bmatrix} \frac{1}{2} & -\frac{1}{2} & -\frac{1}{2} \\ \frac{1}{10} & \frac{3}{10} & \frac{1}{10} \\ -\frac{1}{5} & \frac{2}{5} & -\frac{1}{5} \end{bmatrix}$ 6. $\left(\frac{37}{11}, -\frac{73}{11}, \frac{46}{11}\right)$ 7. $-x$ 8. 270

9. $\left(1, 2, -\frac{1}{2}\right)$ 10. $x = 2, y = 1, z = -1$

Test E

1. $\begin{bmatrix} 1 & 0 & 0 & \frac{9}{7} \\ 0 & 1 & 0 & -\frac{9}{7} \\ 0 & 0 & 1 & -\frac{1}{7} \end{bmatrix}$

2. $y = x^2 - x + 4$

3. $\begin{bmatrix} 39 & 10 & 27 \\ 18 & 9 & 34 \end{bmatrix}$

4. $\begin{bmatrix} 0 & 2 & 1 \\ 4 & 4 & 11 \\ -3 & 1 & -7 \end{bmatrix}$

5. $\begin{bmatrix} -\frac{1}{14} & \frac{11}{14} & \frac{1}{2} \\ \frac{2}{7} & -\frac{1}{7} & 0 \\ \frac{5}{14} & \frac{1}{14} & \frac{1}{2} \end{bmatrix}$

6. $\left(\frac{1}{3}, \frac{2}{3}, -1\right)$

7. $-e^{3x}$

8. 672

9. $\left(\frac{4}{5}, -\frac{2}{5}, \frac{1}{2}\right)$

10. $x = 1, y = 1, z = 0$

Answers to CHAPTER 8 Tests

Test A

1. b 2. b 3. c 4. a
5. d 6. d 7. b 8. c
9. c 10. a 11. b 12. c
13. a 14. d

Test B

1. d 2. a 3. a 4. c
5. d 6. c 7. b 8. a
9. b 10. b 11. a 12. b
13. b 14. c

Test C

1. b 2. b 3. b 4. d
5. d 6. a 7. a 8. d
9. d 10. d 11. a 12. a
13. c 14. b

Test D

1. $-\frac{1}{2}, 0, \frac{1}{10}, \frac{2}{17}, \frac{3}{26}$ 2. 20   3. 29   4. 209

5. 274.7411   6. 30   7. 8610

8. S_1: $2^1 - 1 = 2 - 1 = 1$

   S_k: $1 + 2 + 2^2 + 2^3 + \ldots + 2^{k-1} = 2^k - 1$

   S_{k+1}: $1 + 2 + 2^2 + 2^3 + \ldots + 2^k = 2^{k+1} - 1$

   Assuming S_k, we have:

   $$(1 + 2 + 2^2 + 2^3 + \ldots + 2^{k-1}) + 2^{(k+1)-1} = (1 + 2 + \ldots + 2^{k-1}) + 2^k$$
   $$= (2^k - 1) + 2^k$$
   $$= 2^{k+1} - 1$$

   Hence, the formula is valid for all $n \geq 1$.

9. $x\sqrt{x} + 6x + 12\sqrt{x} + 8$ **10.** $-225{,}792$ **11.** $34{,}650$

12. 1260 **13.** $\frac{7}{10}$ **14.** $\frac{20}{21}$

Test E

1. $\frac{1}{2}, \frac{1}{6}, \frac{1}{12}, \frac{1}{20}, \frac{1}{30}$ **2.** 77 **3.** $\frac{14}{3}$ **4.** $a_n = -\frac{3}{2}n + 12$

5. $-8019\sqrt{3}$ **6.** $465\sqrt{2}$ **7.** 4940

8. For $n = 1$, $1 < 3$

For $n = k$, assume $k < 3^k$

For $n = k + 1$, show $k + 1 < 3^{k+1}$

$$k < 3^k$$

$$k + 1 < 3^k + 1 < 3^k + 3^k$$

$$k + 1 < 2(3^k) < 3(3^k)$$

$$k + 1 < 3^{k+1}$$

Hence $n < 3^n$ for $n \geq 1$.

9. $16y^2 + 96y\sqrt{y} + 196y + 216\sqrt{y} + 81$ **10.** 560 **11.** $4{,}989{,}600$

12. 720 **13.** $\frac{1}{16}$ **14.** $\frac{53}{109}$

College Algebra FINAL EXAM Answers

Multiple Choice Test

1. b	2. c	3. c	4. d
5. d	6. a	7. c	8. c
9. c	10. d	11. b	12. d
13. b	14. c	15. a	16. d
17. c	18. d	19. c	20. a
21. d	22. a	23. d	24. c
25. a	26. c	27. c	28. c
29. b	30. d	31. d	32. a
33. a	34. b	35. d	36. b
37. a	38. a	39. a	40. c

Open Ended Test

1. $-\dfrac{32x^4}{25}$

2. $2x^3 - 3x^2 - 2x$

3. $x^2 - 2x + 1 + 2xy - 2y + y^2$

4. $3x(1 - 2x)(1 + 2x + 4x^2)$

5. $\dfrac{3(1 - 2x)}{x(x + 1)}$

6. Symmetric to the origin

7. -3

8. 320 miles

9. $-1 \pm \sqrt{7}$

10. $\pm 1, \pm \dfrac{\sqrt{10}}{2}$

11.

12. $y = 2x + 1$

13. $3x - 5$

14. Domain: $(-\infty, \infty)$
    Range: $[0, \infty)$

15. Horizontal shift right 3 units
    Vertical shift up 2 units

16. $f^{-1}(x) = \sqrt{\dfrac{x - 1}{2}}$

17. $y = -\left(x - \dfrac{3}{2}\right)^2 + \dfrac{1}{4}$

18. $\pm \dfrac{1}{2}, \pm 1, \pm \dfrac{5}{2}, \pm 5, \pm 10$; $-1, 2, \dfrac{5}{2}$

19. $(x + 2)(x - 2)(x + 2i)(x - 2i)$

20. $f(x) = x^4 - 2x^3 - x^2 + 2x$

21. All real $x \neq \pm\sqrt{10}$

22. $x = -1, x = 3, y = 0$

23. $\dfrac{3}{x - 1} - \dfrac{1}{(x - 1)^2} + \dfrac{9}{x + 3}$

24.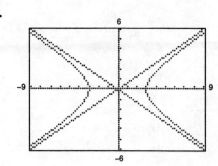

25. $(2, -6)$

26. $40; 140$

27. $(-\infty, -1), (1, \infty)$

28. $\ln 5 + \ln x - \frac{1}{3} \ln (x^2 + 1)$

29. $-1, \frac{5}{3}$

30. $k = 0.1, t \approx 11$ hours

31. $(0, 2), \left(\frac{1}{2}, \frac{5}{2}\right)$

32. $(5, -3, 3)$

33. $z = 0$, minimum; $z = 112$, maximum

34. $\begin{bmatrix} 0 & 2 & 1 \\ 4 & 4 & 11 \\ -3 & 1 & -7 \end{bmatrix}$

35. $(1, 5)$

36. $\left(1, 2, -\frac{1}{2}\right)$

37. $\sum_{n=1}^{5} \frac{n}{(n+1)!}$

38. 4940

39. 720

40. $\frac{20}{21}$